Birkhäuser Advances in Infectious Diseases

For further volumes:
http://www.springer.com/series/5444

Alan C. Jackson
Editor

Viral Infections of the Human Nervous System

Springer

Editor
Alan C. Jackson
Health Sciences Centre
University of Manitoba
Winnipeg
MB
Canada

ISBN 978-3-0348-0424-0 ISBN 978-3-0348-0425-7 (eBook)
DOI 10.1007/978-3-0348-0425-7
Springer Basel Heidelberg New York Dordrecht London

Library of Congress Control Number: 2012944984

Printed on acid-free paper

Springer is part of Springer Science+Business Media (www.springer.com)

Foreword

As a young Army Medical Corps officer, I was assigned to work in the Department of Virus Diseases of Walter Reed Army Institute of Research during the Asian influenza epidemic of 1957. At that time, we knew nothing of the genomic structure of influenza viruses and had no idea that we were working with a recombinant of a human and a duck virus. In the spring, the influenza epidemic waned. The focus of the diagnostic laboratory was shifted to the three clinical syndromes putatively caused by viral infections of the nervous system—aseptic meningitis, encephalitis, and paralytic poliomyelitis. In those days, rabies with its long incubation period, unique clinical features, and uniformly fatal course was regarded as a strange outlier.

Amazing how the landscape has changed over the past 50 years and how two very divergent paths evolved in clinical virology. The latency and reactivation of herpesviruses, the chronic infection with measles virus in the form of subacute sclerosing panencephalitis, the prominent fetal damage caused by rubella virus, the role of viruses in demyelinating diseases (postinfectious encephalomyelitis and progressive multifocal leukoencephalopathy), and the role of infectious prions in chronic degenerative diseases led to an expanding interest in viral infections of the human nervous system.

Conversely in the middle of the twentieth century, the interest in infectious diseases faded. The discovery of antibiotics and antiviral drugs, the eradication of smallpox, and the control of measles and poliomyelitis with vaccines all led to death knells for the specialty of infectious diseases. Infectious disease services were minimized. Prominent infectious disease physicians moved into "healthcare delivery" careers; several published obituaries for the specialty. Infectious diseases were disappearing as a specialty despite the foreboding of *new* diseases such as Legionnaire's disease, a paralytic form of enterovirus 71, and the evolution of an encephalitic strain of California virus in the Midwestern USA. Then in 1981, the surprising and frightening onslaught of acquired immune deficiency disease dramatically changed all of medicine and society.

Why are we now seeing new diseases every year? Greater surveillance and reporting is one explanation, but some new diseases are caused by mutations of familiar viruses, some result from transportation of exotic viruses to new sites, and some result from animal viruses that have been introduced into human populations. All these factors are propelled by the burgeoning global human population and its mobility and speed of global movement. Today, a new exotic virus transmitted to a human in Asia or Africa can be in your local airport or indeed at your church social within one incubation period or even a single day.

This book addresses many of the factors that have made the study of viral infections of the nervous system so compelling and raises intriguing questions that must be addressed over the next decades.

Baltimore, MD
March 2012

Richard T. Johnson

Preface

Viral infections of the nervous system are a challenging group of diseases for clinicians and for researchers. The pathogenetic mechanisms involved in this group of diseases are very diverse. Although some, like enteroviral meningitis, are common. However, many are rare and have limited and unpredictable distributions, both geographically and in time (e.g., Nipah virus infection). Specialized diagnostic investigations are often necessary for definitive diagnosis, although a presumptive diagnosis should often be suspected on the basis of the clinical features. Many of these infections are serious diseases with high morbidity or mortality or with fatal outcomes (e.g., Creutzfeldt–Jakob disease and rabies). A majority of the authors are neurologists and most have either a background or a distinguished career in basic neurovirology research, which gives them unique insights in writing about these diseases. Only further research will give us a better understanding of the basic mechanisms involved in all aspects of these infections, which will, hopefully, lead to future advances in their therapy.

My interest in the field of neurovirology became solidified when 30 years ago I first read Dr. Richard T. Johnson's book entitled *Viral Infections of the Nervous System* (Raven Press, 1982). Two years later, I became a postdoctoral fellow in Dr. Johnson's research laboratory at The Johns Hopkins University in Baltimore. I hope this volume will also stimulate the interest of young people in this intriguing field. I would like to thank Dr. Beatrice Menz at Springer Basel for giving me the opportunity of putting together a volume on these infections and to all of the expert contributors for their hardwork in preparing up-to-date chapters and sharing their expertise and insights on this diverse group of diseases. They have all done a superb job.

Winnipeg, MB, Canada
February 2012

Alan C. Jackson

Contents

Part IV Viral Zoonoses

Part V Unconventional Agents

List of Contributors

Allen J. Aksamit Department of Neurology, Mayo Clinic, Rochester, MN, USA

Banu Anlar Department of Pediatric Neurology, Hacettepe University, Ankara, Turkey

Li-Yen Chang Department of Medical Microbiology, Faculty of Medicine, Tropical Infectious Diseases Research and Education Centre, University of Malaya, Kuala Lumpur, Malaysia

Randall J. Cohrs Department of Neurology, University of Colorado School of Medicine, Aurora, CO, USA

Marc Fischer Arboviral Diseases Branch, Centers for Disease Control and Prevention, Fort Collins, CO, USA

Pr. Philippe Gasque Groupe de Recherche Infection et Immunopathologie, EA4517, INSERMU945, Université de la Réunion, St. Denis, Ile de la Réunion/ Reunion Island, France

Donald Gilden Department of Neurology, B182, University of Colorado School of Medicine, Aurora, CO, USA

Taylor Harrison Department of Neurology, Emory University, Grady Memorial Hospital, Atlanta, GA, USA

Alan C. Jackson Departments of Internal Medicine (Neurology) and of Medical Microbiology, University of Manitoba, Health Sciences Centre, Winnipeg, MB, Canada

Steven Jacobson Viral Immunology Section, NINDS/NIH, Bethesda, MD, USA

Richard T. Johnson Department of Neurology, The Johns Hopkins University School of Medicine and Department of Molecular Microbiology and Immunology, Johns Hopkins Bloomberg School of Public Health, Baltimore, MD, USA

Burk Jubelt Department of Neurology, SUNY Upstate Medical University, Syracuse, NY, USA

Raya Massoud Viral Immunology Section, NINDS/NIH, Bethesda, MD, USA

Maria A. Nagel Department of Neurology, University of Colorado School of Medicine, Aurora, CO, USA

Guey Chuen Perng Department of Pathology and Laboratory Medicine, Emory Vaccine Center, Emory University School of Medicine, Atlanta, GA, USA

Jeffrey Rumbaugh Department of Neurology, Emory University, Neurology (GEC), Atlanta Veterans Administration Medical Center, Decatur, GA, USA

James J. Sejvar Division of High-Consequence Pathogens and Pathology, Division of Vectorborne Infectious Diseases, National Center for Emerging and Zoonotic Infectious Diseases, Atlanta, GA, USA

Valerie L. Sim Centre for Prions and Protein Folding Diseases, University of Alberta, Edmonton, AB, Canada

Marylou V. Solbrig Departments of Internal Medicine (Neurology) and of Medical Microbiology, University of Manitoba, Health Sciences Centre, Winnipeg, MB, Canada

Tom Solomon Walton Centre NHS Foundation Trust and Institute of Infection and Global Health, University of Liverpool, Liverpool, UK

Israel Steiner Department of Neurology, Rabin Medical Center, Petach Tikva, Israel

Chong-Tin Tan Neurology Division, Faculty of Medicine, University of Malaya, Kuala Lumpur, Malaysia

Alex Tselis Department of Neurology, Wayne State University, Detroit, MI, USA

Lance Turtle Institute of Infection and Global Health, University of Liverpool, Liverpool, UK

William R. Tyor Department of Neurology, Emory University, Neurology (GEC), Atlanta Veterans Administration Medical Center, Decatur, GA, USA

Kalbiye Yalaz Department of Pediatric Neurology, Faculty of Medicine, Hacettepe University, Ankara, Turkey

Part I
Encephalitis

Measles Virus Infection and Subacute Sclerosing Panencephalitis

Banu Anlar and Kalbiye Yalaz

Abstract Measles virus can cause two acute neurological disorders: acute infectious encephalitis and postinfectious autoimmune encephalitis, each with a risk of about 1 in 1,000 measles cases. Two other rare neurological problems manifest after a latent period: subacute measles encephalitis occurring in immunocompromised individuals, and subacute sclerosing panencephalitis (SSPE) in immunocompetent hosts. SSPE develops 1–10 years after measles infection; it is usually progressive and fatal. Mental and behavioral changes, myoclonia, and ataxia are typical initial manifestations. The diagnosis is based on the demonstration of intrathecal anti-measles virus immunoglobulin G synthesis. Pathological examination of brain biopsy or autopsy material demonstrates inflammation, neuronal loss, gliosis, demyelination, and typically, inclusion bodies containing measles virus antigens or RNA. Treatment with inosiplex and interferons may induce temporary stabilization or remission in about 30–35 % of the cases. Immunization against measles virus and maintenance of immunization rates above 90 % in the population are of extreme importance for the prevention of these debilitating or fatal disorders.

Keywords Demyelinating • Immunoglobulin • Magnetic resonance imaging • Measles • Subacute sclerosing panencephalitis

Abbreviations

ADEM	Acute disseminated encephalomyelitis
CSF	Cerebrospinal fluid
EEG	Electroencephalography

B. Anlar (✉) • K. Yalaz
Department of Pediatric Neurology, Hacettepe University Faculty of Medicine,
Ankara 06100, Turkey
e-mail: banlar@hacettepe.edu.tr

A.C. Jackson (ed.), *Viral Infections of the Human Nervous System*,
Birkhäuser Advances in Infectious Diseases,
DOI 10.1007/978-3-0348-0425-7_1, © Springer Basel 2013

IL	Interleukin
MRI	Magnetic resonance imaging
MV	Measles virus
SSPE	Subacute sclerosing panencephalitis

1 Introduction

1.1 Measles Virus

Measles virus (MV) is a single-stranded negative-sense RNA virus in the Paramyxoviridae family. Its genome encodes eight proteins among which the haemagglutinin protein induces a strong neutralizing antibody response with life-long immunity (Moss and Griffin 2011).

MV infects humans only. Its cellular receptors are CD150 (signaling lymphocyte activation molecule, SLAM) on lymphocytes, and as shown recently, CD147 (extracellular matrix metalloproteinase inducer, EMMPRIN) on epithelial cells. CD46, a complement regulatory protein, is primarily a receptor for the vaccine virus; CD147, for wild-type MV, and CD150, for both (Naniche et al. 1993; Tatsuo et al. 2000; Watanabe et al. 2010).

1.2 Acute Measles Infection

After MV transmitted via respiratory droplets enters respiratory epithelial cells, acute infection starts in the respiratory tract, then spreads to organs including lymphoid tissue, skin, lungs, and liver. The clinical picture begins with cough, rhinitis, and conjunctivitis. Although this may appear as a common upper respiratory infection, the child with measles is sicker; the cough is more prominent with a "metallic" sound, and conjunctivitis is more remarkable. However in older children and adults the initial symptoms may be indistinguishable from common upper respiratory infections. Koplik's spots can be seen in the buccal mucosa for 24–48 h before rash erupts. After 3–4 days of prodromal period, an erythematous rash begins from the neck and face, and fever subsides. The rash proceeds caudally and fades with a brownish pink color within a week. Pediatricians currently practicing in North America may not be familiar with the clinical picture of the prodromal period. Of interest, measles can occur without rash in about 20 % of the cases and up to 60 % in subjects with preexisting low-titer antibodies (Cherian et al. 1984; Lisse et al. 1998; Prasad et al. 1995).

Widespread application of vaccination programs eliminated measles in many areas. However, epidemics have recently been observed both in developing and developed countries due to failure to maintain the 95 % immunization rate needed for eradication. Two doses of vaccine are recommended, although countries using one or two doses eliminated measles in similar periods (Sever et al. 2011). The first dose is given between 12 and 15 months (or 9 months during epidemics) and the second, between 4 and 6 years of age in non-endemic countries (Committee on Infectious Diseases and American Academy of Pediatrics 2011). During epidemics when risk of exposure exists, the second dose can be administered from 1 month after the first dose. In addition, mass campaigns every 3–4 years are required even in vaccinated populations in order to immunize the accumulating susceptible population of children against any imported virus. It is important to note that the risk of neurological adverse effects seemingly associated with measles vaccine is approximately one per million doses while the risk of neurological complications of measles is more than two per thousand, as described below (Shu et al. 2011). Measles is a highly contagious disease and continuous surveillance is mandatory.

Neurologic complications of measles include the following clinical pictures:

1. Acute measles encephalitis during active viral infection.
2. Acute post-measles encephalomyelitis that follows the infection.
3. Subacute measles encephalitis (SME) (measles inclusion body encephalitis) encountered in immunocompromised individuals.
4. Subacute sclerosing panencephalitis (SSPE) developing after a latent period of several years.

1.3 Acute Measles Encephalitis

This is the most common acute neurological complication of measles in children: the risk of encephalitis is 22/28,000 or 0.1–0.3 % of acute measles cases, and has not changed in the last 50 years (Laboccetta and Tornay 1964; European Centre for Disease Prevention and Control ECDC 2011; Stanescu et al. 2011; Filia et al. 2011).

During acute infection MV can infect cerebral endothelial cells and invade the central nervous system (Dittmar et al. 2008). The resulting clinical picture is encephalitis or encephalomyelitis indistinguishable from other viral encephalitides except for the presence of rash. Certain cases may not show any exanthema and can be suspected only by the presence of a concurrent epidemic, or diagnosed retrospectively. Neurological symptoms appear during or, rarely, before the rash. Fever, which typically subsides after the appearance of rash in measles, recurs along with irritability, lethargy, headache, confusion, convulsions, and loss of consciousness. Neurological examination may show signs of meningeal irritation. When there is accompanying myelitis and cerebellitis, bladder or bowel dysfunction and ataxia can be observed. Interestingly, febrile seizures, i.e., seizures triggered by fever only with no clinical or laboratory evidence of encephalitis, are not common in young children with measles. Therefore, convulsions associated with measles infection should suggest encephalitis.

The cerebrospinal fluid (CSF) reveals increased opening pressure, predominantly lymphocytic pleocytosis, and an increased protein level. Electroencephalogram (EEG) may show high voltage focal or diffuse slowing that usually persists for several months. Up to 50 % of the patients with acute measles have slowing of EEG activity without any neurological symptoms, a finding that can be considered as subclinical encephalitis (Gibbs et al. 1964). MR imaging may be normal, or show white and gray matter lesions involving the cerebral cortex or basal ganglia. Deep gray matter involvement has been suggested as a more severe variant with higher rate of complications and protracted course. Perivascular inflammation, microglial nodules, and neuronal degeneration are observed on pathological examination. MV RNA and antigens can be demonstrated in the brain tissue, as well as tumor necrosis factor-alpha mRNA (Plaza and Nuovo 2005). On the other hand, cytoplasmic and nuclear inclusion bodies and multinucleated giant cells, the typical findings of measles encephalitis, may be absent in patients with rapid, fulminant course.

As with other viruses, clinical overlap exists between acute infectious encephalitis and the postinfectious encephalitis described below. The nature and the timing of the symptoms are not always distinctive: in a series of 12 patients with encephalitis occurring 2–7 days after measles rash, four had gray matter involvement suggesting acute MV encephalitis whilst the others were more compatible with the postinfectious form (Kim et al. 2003).

Treatment is symptomatic and consists in preventing or treating hyperthermia, convulsions, secondary infections, and fluid and electrolyte disturbances. The use of corticosteroids in combination with intravenous immunoglobulin may improve the chances of recovery (Nakajima et al. 2008).

The course varies from a mild confusion recovering completely within several days to fulminant progression to coma and death within 24 h. In older literature, about 60 % of children recovered completely, 15 % died, and 25 % were left with sequelae such as mental retardation, epilepsy, behavioral problems, hearing loss, or motor deficits (Meyer and Byers 1952). Sequelae may be underestimated because certain behavioral and attention problems may go unnoticed until school age. On the other hand, a higher rate of recovery without sequelae is reported in more recent series (Kim et al. 2003).

1.4 Acute Post-measles Encephalomyelitis

This form of encephalopathy is more commonly known as acute disseminated encephalomyelitis (ADEM), and attributed to postinfectious autoimmune mechanisms. Neurological signs follow an interval of 3–10 (2–30) days after an acute viral infection. Measles was one of the frequent antecedent infections in ADEM series before the implementation of widespread immunization programs: 1/1,000 measles cases were reported to develop this clinical picture. The interval period tends to be shorter (2–4 days) in measles and other exanthematous diseases. Characteristically, the resurgence of fever in a child whose rash is fading should suggest this entity, especially if accompanied by headache and altered consciousness. Meningeal irritation, seizures, and less frequently, focal motor deficits, optic neuropathy, or myelopathy can be associated. Children <2 years old tend to have a rare but more fulminant variant: acute hemorrhagic leukoencephalitis, where morbidity and mortality are considerably higher.

The diagnosis is based on clinical signs and the history of measles, and supported by magnetic resonance imaging (MRI) showing bilateral, asymmetrical, patchy, frequently edematous white matter changes. Bilateral striatal lesions can also be observed (Lee et al. 2003). Cerebrospinal fluid may be normal or contain lymphocytes and neutrophils, or elevated protein levels. Pathologically, perivenular inflammation and myelin disruption are observed in rare cases where a brain biopsy or autopsy is performed.

As stated above, clinical features may not be distinguishable from acute encephalitis, especially in the presence of fever and the absence of a clear period of improvement after acute measles infection. The predominantly white matter involvement on MR images and the absence of significant pleocytosis may be taken as supportive of ADEM compared to infectious MV encephalitis.

1.5 Other Rare Acute Complications

Other complications of measles infection that are rarely observed or mentioned in the literature include optic neuropathy in the anterior or retrobulbar forms, manifesting as acute loss of vision during or even preceding measles (Srivastava and Nema 1963; Hirayama et al. 2010), acute transient cerebellar ataxia (Tyler 1957), isolated myelitis (Gunaratne et al. 2001), or peripheral neuropathy of the Guillain Barré type (Tomiyasu et al. 2009). Increased intracranial pressure, coma, decerebrate rigidity, herniation, and death were described in a case with no evidence of encephalitis on pathological examination, possibly due to inflammatory toxic molecules in a host with a specific, yet undefined immune deficit (Tyler 1957). On the other end of the spectrum, aseptic meningitis due to MV is a benign syndrome characterized by headache, fever, vomiting, meningeal signs, and lymphocytic pleocytosis in the CSF. The patient recovers in 1–2 weeks with supportive and symptomatic treatment (Bakir 1989; Valassina et al. 2000). Chronic infiltrative meningitis has been reported in an immunocompetent adult (Luzi et al. 1997).

1.6 Subacute Measles Encephalitis

This infection develops 1–10 months after primary measles infection in immunosuppressed hosts such as leukemia or acquired immune deficiency patients, and rarely in immunocompetent subjects. One of the largest series has recently been reported from South Africa in a group of eight HIV-positive adolescents and adults, four of whom had no history of rash (Albertyn et al. 2011).

The clinical manifestations start with myoclonia and mental changes. These symptoms resemble SSPE as described below; however, the course of SME is more rapidly progressive over weeks and months. Cognitive function declines, myoclonus, and refractory focal seizures, or, less commonly, generalized seizures are observed. Vision loss, hearing loss, and focal motor and sensory signs progress over weeks and usually end with a fatal outcome within months. Probably, the absence of an antiviral immune response allows more rapid spread of MV in the brain compared to SSPE (see below). EEG demonstrates slowing of the background rhythm and epileptiform discharges. MRI shows multifocal areas of increased T2-signal intensity more prominently in the cortex than the white matter, unlike SSPE where white matter changes predominate. The cortical involvement explains seizures being a frequent symptom in SME.

CSF may show normal findings or mild lymphocytic pleocytosis and oligoclonal bands; however, because of the underlying immune deficiency and the intracellular location of MV, measles virus IgG is normal or mildly elevated. Although PCR studies from various samples such as urine and brain tissue may be required for definite diagnosis, these can also be negative because SME is a nonproductive MV infection. In the Albertyn et al. (2011) series, only 2/6 CSF samples had detectable MV IgG and 2/8 contained MV RNA. On brain biopsy, neuronal loss, mononuclear infiltration, glial proliferation, and more specifically, eosinophilic inclusion bodies in neuronal and glial cells were observed. These bodies consist of paramyxovirus nucleocapsids and MV antigen as detected by immunohistochemistry. Such inclusion bodies are also observed in SSPE, in which pathological findings differ from SME by more marked inflammatory infiltration and demyelination.

Recovery from SME is rare and sequelae are frequent. There is no definite treatment but ribavirin might increase the chance of survival (Mustafa et al. 1993). Among the cases reported by Albertyn et al. (2011), two survived, including one with normal mental function. Higher $CD4^{+}$ cell counts were a good prognostic factor. Intravenous immunoglobulin appears to be a safe and reasonable therapeutic option based on previous experience in acute measles and postinfectious encephalomyelitis. However, there are no reports of SME cases treated with intravenous immunoglobulin.

1.7 Subacute Sclerosing Panencephalitis

SSPE is among the chronic neurological disorders where the discovery of a specific etiological agent came as an exciting scientific advance in our understanding of the disease, although isolation of MV was initially unsuccessful. Filamentous particles in the brain were described (Bouteille et al. 1965), then MV was recovered from the brain and established as the cause of the disease (Adels et al. 1968; Horta-Barbosa et al. 1969). Experimental evidence of the pathogenicity of the virus was obtained by transmission of the disease with brain extracts from patients to ferrets and between ferrets (Katz et al. 1968).

1.7.1 Epidemiology

SSPE is a result of natural measles infection especially when the latter is experienced at a young age, particularly before age of 2 years. The risk of developing

SSPE varies between series, probably due to environmental factors: in USA it was calculated as 22/100,000 measles cases (Bellini et al. 2005), and is probably closer to 1/1,000 after infantile measles. However, these are statistical estimates because of under- and over-reporting of measles and the absence of serological proof in most cases. The fact is that measles vaccine coverage in over 95 % of the population stops the transmission of measles and virtually eliminates SSPE. In addition to measles occurring at a young age, rural residence, crowded households, and adverse socioeconomical conditions have been associated with SSPE.

Boys are more frequently affected with a male/female ratio of 1.5–1.8 in different series. Mortality of acute measles is slightly higher in girls, however, not to a degree to explain the sex difference in SSPE. The majority of cases have an onset between 6 and 14 years. The mean age of onset varied from a median of 13 years before 1994 to a median of 7.6 years after 1995, possibly related to age at primary measles infection (Anlar et al. 2001a, b). The youngest case reported was 5 months old, and the oldest was 49 years old.

The interval between primary measles infection and neurological symptoms of SSPE is 1–10 years; it tends to be longer in adults and shorter (e.g., several months) in young children.

1.7.2 Etiology

Molecular studies confirm that wild-type MV is responsible for producing SSPE, as all strains amplified from brain tissue indicate MV circulating at the time of primary measles infection. Mutations occurring during persistence allow the virus to bypass the host's immune system and reproduce inside the cell in a less cytopathic fashion (Reuter and Schneider-Schaulies 2010). SSPE has been observed in previously immunized children, but most of these children had natural MV infection before immunization (Miki et al. 2002). Alternatively, vaccine failure occurs with an inability to seroconvert in a minority of vaccinated individuals, especially when vaccine is given before 12 months of age.

1.7.3 Pathogenesis

The route of entry of the MV to the brain and the cellular receptor for MV are unclear. CD46 is expressed on brain endothelial cells, ependyma, choroid plexus, neurons, and oligodendrocytes; however, the wild-type MV does not use this receptor. CD150 (Piskin et al. 2007) and possibly CD147 are more likely to act as receptors in the central nervous system. Incorporation of cyclophylin-B (cellular ligand for CD147) into MV particles is a prerequisite for cellular infection in vitro (Watanabe et al. 2010).

After entry, the site of MV persistence for years is unclear: it can be the nervous or lymphoid tissues. Antigens and genome of MV have been shown in tissues of individuals with no relevant symptoms (Anlar et al. 2002a, b; Katayama et al. 1998).

In the persistent state the viral RNA is complete but membrane proteins are missing, which impedes the production of infective viral particles. On the other hand, viral antigens are expressed and induce an antibody response. MV persistence is probably facilitated by the immaturity of the immune system in infants and the transient immunosuppression caused by MV mediated by blockade of IFNa/b-induced STAT signaling. Many studies investigated host predisposition to MV persistence. However, these were patients already diagnosed with SSPE, therefore, past the persistent state, when it is impossible to distinguish changes resulting from or causing the disease. Studies at gene level included single nucleotide polymorphisms in the Toll-like receptor (TLR)-2, interleukin (IL)-2 and IL-4, MxA (a type 1 IFN-inducible protein), PD1 (co-inhibitory for T cells), and CD46 produced variable results (Pipo-Deveza et al. 2010; Torisu and Hara 2006; Yilmaz et al. 2007; Torisu et al 2004).

Whatever the factor initiating persistence is, limited expression of MV antigens on neural cells, and hypermutations in MV matrix (M) and/or fusion (F) genes allow the MV to replicate while escaping host antibodies. The mutations probably result from long persistence rather than being a primary event. The enzymes adenosine deaminase acting on RNA (ADAR1 and ADAR2) catalyzing A to I mutations are expressed in brain and can take part in these mutations (Maas et al. 2006). As a result MV spreads cell to cell without releasing infectious viral particles.

The appearance of clinical symptoms can be due to MV becoming reactivated, or the intracellular infection reaching a threshold level. Alterations in the immune or hormonal systems, minor head trauma, or infections might contribute to transition from persistent to clinically manifest state.

1.7.4 Pathology

Brain biopsy and autopsy material show infiltration of mononuclear cells into the meninges and brain tissues. $CD4^+$ and B cells tend to gather in perivascular areas and $CD8^+$ cells in parenchymatous areas. Gliosis, astrocytic proliferation, neuronal degeneration, and demyelination are observed in various degrees. Inflammatory molecules, including IFN-gamma, HLA Class II, and tumor necrosis factor (TNF)-alpha are expressed on endothelial and glial cells (Anlar et al. 2001a). MxA, the type I interferon inducible protein, is perivascularly expressed (Anlar et al. 2001b). These findings indicate chronic encephalitis, but the diagnostic finding for SSPE is the presence of inclusion bodies in the cytoplasm or nucleus of neurons and glial cells. Nuclear inclusion bodies contain viral ribonucleoproteins or protein complexes. In early stages inflammation is more prominent; in late stages, astrocytic gliosis, demyelination, necrosis, neuronophagia, neurofibrillary tangles, and apoptosis of neuronal, oligodendroglial and microglial cells are observed. MV gene expression has no direct correlation with clinical course, as patients with rapid progression may have low levels of infection (Kühne Simmonds et al. 2006).

1.7.5 Clinical Findings

The diagnosis of SSPE cases with typical presentation is straightforward, but atypical presentations are not rare (Erturk et al. 2011). Initial symptoms frequently consist of behavioral changes, forgetfulness, and decreasing school performance in the child, which are often overlooked or attributed to psychological reasons. Myoclonic–atonic episodes begin as brief head dropping, unilateral arm or facial twitching, or falling without loss of consciousness. These episodes last only a few seconds and can occur repeatedly every few hours or minutes. At this stage the neurological examination can be normal or reveal hypertonicity of one or more extremities, tremor, ataxia, and apraxia. Less frequent manifestations are seizures, encephalopathy, increased intracranial pressure, hemiparesis, hemidystonia, and vision loss. The latter may result from occipital cortical lesions, optic atrophy, or retinopathy (Yüksel et al. 2011) Symptoms worsen over weeks and months, followed by loss of speech and ambulation in most patients. The course stabilizes in certain cases with the patient remaining in a mentally impaired but ambulatory state.

Clinical staging systems have been useful, but of limited value, due to lack of standardization. Briefly, mental and behavioral changes are the symptoms of stage 1; myoclonia characterize stage 2, and loss of independent ambulation, stage 3. In the latter stage myoclonia diminish or disappear and spastic quadriparesis, autonomic disturbances, tonic spasms, and fever are observed, although the patient shows some response to environmental stimuli. After stage 3 is reached, progression into coma and death occur in a few months to few years in most (about 60 %) cases. Other cases stabilize, sometimes with regain of motor functions and even independent walking for several years. About 20 % show acute or fulminant course from the beginning, presenting with altered consciousness with progression to death within a few months. The cause of death in SSPE patients is respiratory infections or disease affecting vital functions in the brainstem.

Besides clinical staging, standardized scoring systems for neurological and intellectual deficits such as the Neurological Disability Index (NDI) and cognitive assessment scales (Dyken et al. 1982; Öktem et al. 1997) are useful for research and follow-up. The authors use a modified version of the NDI, the SSPE Scoring System (SSS), based on their observations on large numbers of patients over years (Table 1).

1.7.6 Diagnosis

Seizures or myoclonia accompanied by mental and behavioral changes in a previously normal school-age child or adolescent should raise the suspicion of SSPE. The diagnosis is supported by EEG findings and confirmed by CSF examination for measles virus IgG and measles virus-specific IgG index.

Table 1 SSPE Scoring System (SSS)

Behavioral and mental		Myoclonia (before carbamazepine)	
Irritability: absent	0	Location: no myoclonia	0
Mild hyperactivity, restlesness	1	Focal, mild	1
Moderate restlesness	2	Focal 2 body parts, moderate amplitude	2
Marked irritability or delirium, lethargy	3	More than 2 body parts	3
Stupor, coma	4	Immobility	4
Personality		**Repetition**	
Normal	0	No myoclonia	0
Mild changes (excessive talking, apathy, etc.)	1	Irregular, less than once a day	1
Oppositional behavior, aggressive	2	Irregular, less than once per hour	2
Defiant or lethargic	3	Regular, more than once per hour	3
Stupor, coma	4	Immobility	4
Introversion or autism		**Convulsions (other than myoclonia)**	
None	0	None	0
Shy or withdrawn	1	Less than once a week	1
Limited interaction, stereotypies	2	Once a month /once a week	2
Marked autistic behavior/lethargy	3	Once a week/once a day	3
Stupor, coma	4	More than once a day	4
Mental-perceptive		**Daily functions**	
Normal	0	Dresses and feeds himself/herself	0
Dull (1 year difference with peers)	1	Can feed but not dress himself/herself	1
Borderline (2–3 years difference with peers)	2	Needs help while eating	2
Marked mental deficiency or lethargy	3	Expresses hunger/thirst, cannot feed him/herself	3
Stupor, coma	4	Totally dependent	4
Speech		**Following commands**	
Normal	0	Normal	0
Mild speech disturbance (talks in sentences, mild dysarthria)	1	Mild impairment (More than 2 of 4 commands)	1
Moderate speech disturbance (single words)	2	Moderate impairment (1–2 commands)	2
Severe speech disturbance (vocalizes, incomprehensible)	3	Hears commands, does not comply	3
Stupor, coma	4	Stupor, coma	4
Motor and sensory		**Vegetative and systemic**	
Reflex-tone		Vision	
Normal	0	Normal (counts/imitates fingers from 6 m.)	0
Mild hyperreflexia or hypertonia	1	Mild impairment (counts/imitates fingers from 2 m.)	1
Mild hyperreflexia and hypertonia	2	Moderate impairment (sees moving objects)	2
Moderate hyperreflexia and hypertonia	3	Marked impairment (sees light)	3
Severe hyperreflexia and hypertonia	4	Total loss of vision	4

Strength		Hearing	
Normal	0	Normal (hears whisper 30 cm)	0
Mild weakness (4/5) or atrophy	1	Mild impairment (hears voice 30 cm)	1
Mild weakness (4/5) and atrophy	2	Moderate impairment (hears loud voice)	2
Moderate weakness (3/5) and atrophy	3	Marked impairment (reacts to loud noise)	3
Marked weakness (0–2/5) and atrophy	4	No hearing	4
Posture/movement		**Sensory (touch, pressure, pain)**	
Normal	0	Normal	0
Mild chorea/athetosis	1	Does not feel touch, feels pressure	1
Mild dystonia, moderate chorea/athetosis	2	Does not feel touch and pressure, feels pain	2
Moderate dystonia, choreoathetosis, mild rigidity	3	Feels only deep pain	3
Severe extrapyramidal signs	4	Does not feel deep pain	4
Coordination		**Autonomic functions**	
Normal	0	Normal	0
Mild impairment (can walk)	1	Mild impairment (some urine incontinence)	1
Moderate impairment (walks with assistance)	2	Moderate impairment (some urine/bowel incontinence)	2
Marked impairment (cannot walk, sits without support)	3	Marked impairment (urine/bowel incontinence or sometimes fever, sweating)	
Severe incoordination (bedridden)	4	Severe impairment (continuous incontinence and sweating episodes)	4
Upper limb movements		**Nutrition**	
Uses objects appropriately	0	Normal	0
Uses some objects appropriately	1	Mild impairment (occasional choking)	1
Reaches, holds, may put in mouth	2	Moderate impairment (soft food)	2
Reaches, cannot hold	3	Marked impairment (only puree)	3
Does not reach for objects	4	Severe impairment (tube feeding)	4

Total score (maximum 80)
Modified from Dyken et al. (1982) by Anlar, Altunbasak, Köse, and Yüksel

The typical EEG pattern, bilateral, high amplitude slow, or sharp-slow wave complexes, is seen particularly in stage 2. The discharges are not always synchronous with myoclonia (Fig. 1a). The background rhythm is normal in early stages and becomes slower and suppressed over months/years. Sometimes EEG abnormalities can be asymmetrical or focal (Fig. 2). The typical periodic pattern becomes accentuated after administration of 5 mg diazepam intravenously, or, when routine EEG is normal in early stages, this pattern may appear only after diazepam (Fig. 1b). This feature of SSPE contrasts with epilepsy and other degenerative disorders, in which diazepam usually has a suppressive effect on epileptiform discharges.

CSF analysis for anti-measles virus IgG titers and intrathecal synthesis (measles virus-specific IgG index) is diagnostic. CSF protein, glucose, and cell count are normal. Pressure can be elevated in up to 25 % of the cases, sometimes associated with symptoms of increased intracranial pressure (Ölmez et al. 2007). The IgG index is markedly elevated and oligoclonal bands are observed. Most of

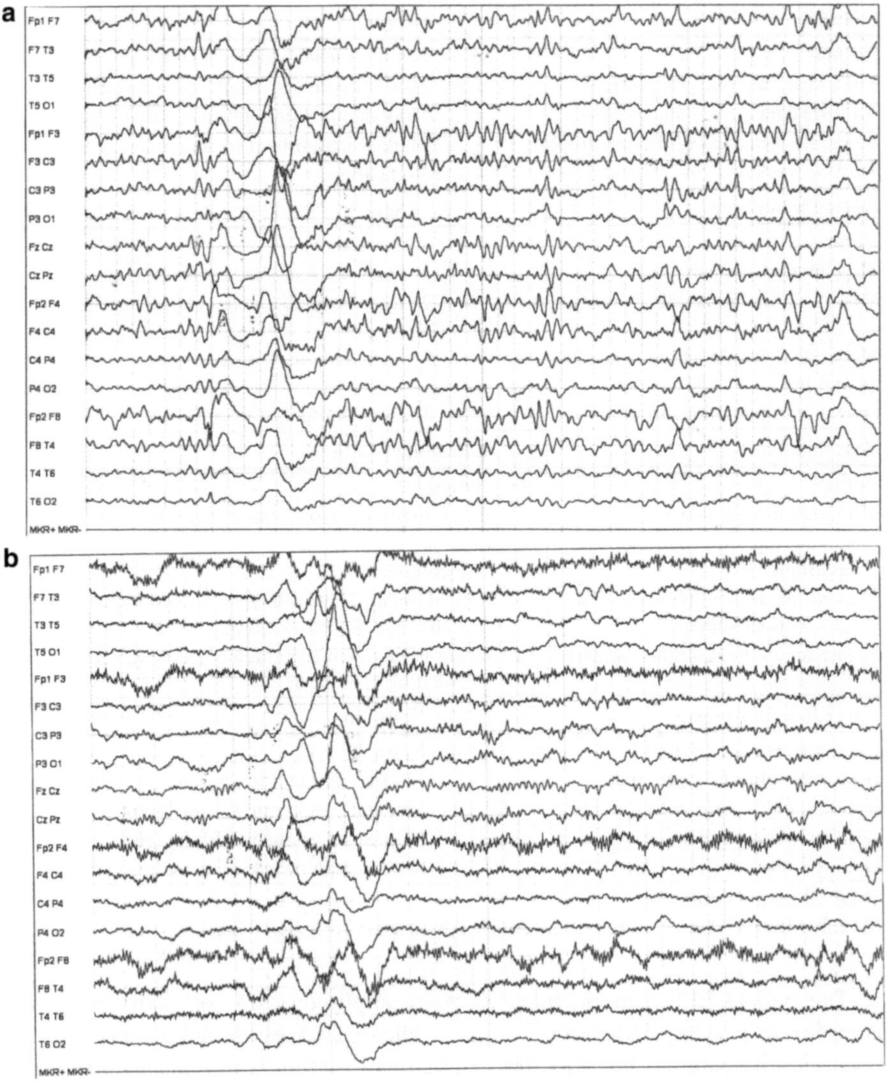

Fig. 1 (**a**) EEG in early stage: normal background rhythm and a slow-wave complex at the onset of the disease, (**b**) EEG recording of the same patient after diazepam showing slower background rhythm and persistence of paroxysmal activity

intrathecally synthesized IgG is not only against MV, but also against some other viruses at low titers (Anlar et al. 2002a, b). If the CSF protein and cells are increased or the IgG index is normal, alternative diagnoses should be considered. The presence of measles virus-specific IgM in the CSF might be associated with a more protracted clinical course (Connolly et al. 1971). The detection of MV RNA is diagnostic but may be difficult to demonstrate due to low copy numbers in the CSF (Nakayama et al. 1995).

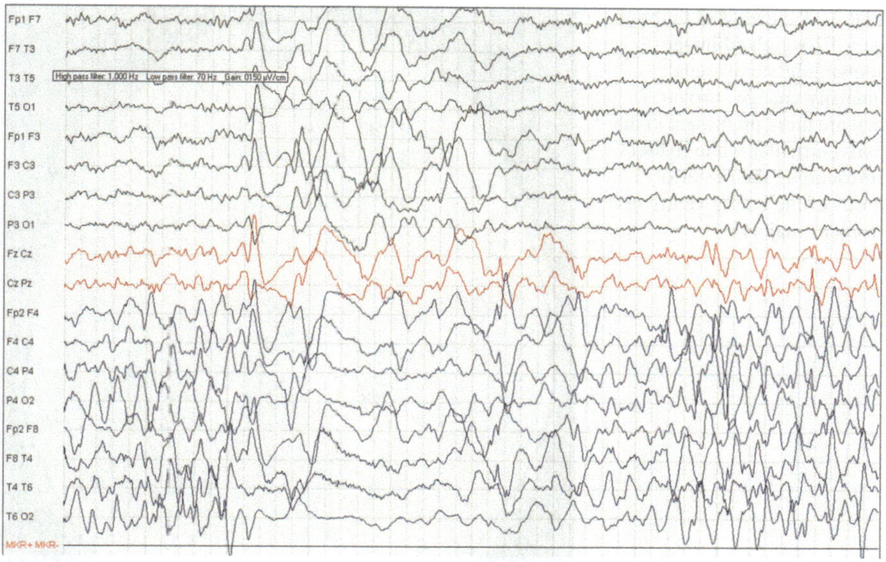

Fig. 2 Asymmetric periodic complexes in the two hemispheres

MRI is not diagnostic or even typical for SSPE, but is useful in excluding other disorders in the differential diagnosis. It can be normal in the initial stages. The most frequent finding is signal intensity changes on T2-weighted images, located in the periventricular or subcortical white matter (Fig. 3). Lesions usually progress to the midline and from the posterior to anterior regions, involving the basal ganglia and brainstem in more advanced disease. Pial or parenchymal contrast enhancement may be observed. Diffuse atrophy develops over years, more correlated with the duration of the disease rather than clinical stage (Anlar et al. 1996). The lesions have an inflammatory or demyelinating nature. Diffusion-weighted MRI may show lesions in the parenchyma appearing normal on routine MRI. MR spectroscopy may suggest neuronal loss, gliosis, demyelination, and inflammation by decreased *N*-acetylaspartate and increased choline and myo-inositol.

1.7.7 Treatment

Antiviral agents have no effect on SSPE and do not penetrate into the brain tissue when given from oral or parenteral route. Ribavirin has been administered intraventricularly to ten patients in combination with α-IFN and resulted in the reduction of MV IgG titers in the CSF (Tomoda et al. 2003). Clinically, progression slowed down in 5/10 treated patients compared to retrospective controls. However, some cases had slowed progression even before treatment was initiated.

The main drug used in the treatment of SSPE is inosine pranobex (inosiplex), a synthetic purine compound with antiviral and immunomodulatory actions. Its effect

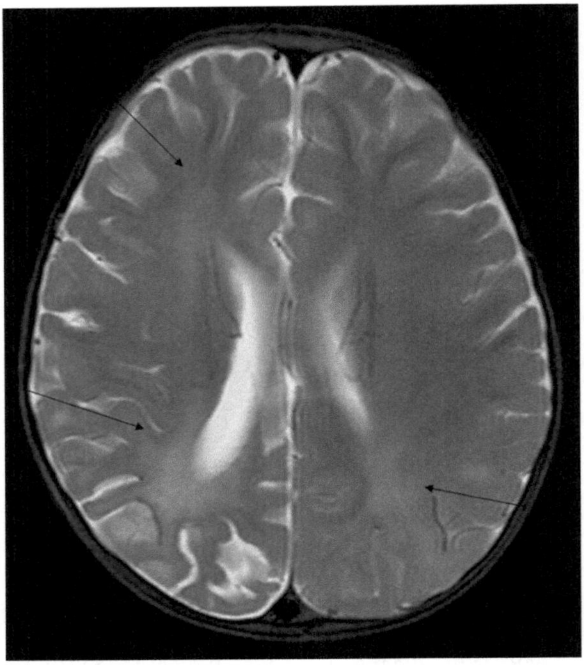

Fig. 3 Typical MRI findings on a T2-weighted image showing bilateral signal intensity changes (*arrows*) that are more prominent in the periventricular white matter surrounding the posterior horns of the lateral ventricles (forceps major)

is modest, but its safety and administration via oral route allow life-long usage. Side effects include gastric disturbance, hyperuricemia, and, rarely, renal stones. The dose is 50–100 mg/kg/day p.o. in divided doses. Inosiplex restores IFN-gamma synthesis from peripheral blood mononuclear cells in vitro and in vivo (Gadoth et al. 1989). Clinical studies comparing treated and non-treated children showed better survival, more frequent prolonged remission, and improved survival in treated patients (Jones et al. 1982, $n = 98$, Fukuyama et al. 1987, $n = 89$). On the other hand, some researchers reported no difference in outcome (Haddad and Risk 1980, $n = 18$) compared to controls ($n = 96$). Our data suggest better outcome in treated patients (Anlar and Yalaz 2011).

Interferons (IFN) have strong antiviral effects. Parenteral alpha-IFN was first given in small patient series or single cases. Intraventricular human lymphoblastoid alpha-IFN in combination with oral inosiplex resulted in improved or stable disease in about 50 % of the cases, and longer survival with slower progression compared to inosiplex alone (Yalaz et al. 1992; Anlar et al. 1997). Intrathecal application may induce meningeal inflammation, seizures, and neuropathy, and may be administered via lumbar puncture or intrathecal pump (Thurner et al. 2007). Another type of IFN, IFN beta1a might also prolong survival and delay progression when given three times per week subcutaneously with oral inosiplex (Anlar et al. 2004).

Amantadine is an antiviral medication that inhibits RNA replication. Certain series reported higher rates of remission with amantadine, a relatively safe drug given by oral route (Robertson et al. 1980).

Intravenous immunoglobulin treatment has been reported in single cases only (Gürer et al. 1996). We observe temporary improvement in some cases with acute deterioration during febrile illnesses, which is usually associated with nonspecific infections.

Corticosteroids may fasten the progression of the disease and are not indicated.

In vitro studies using small interfering RNA showed inhibition of MV replication (Otaki et al. 2007), but no in vivo studies have been published. Various natural or synthetic compounds have been studied in vitro.

Treatment approaches using antiviral and immunomodulatory agents can result in partial remission or stabilization at rates higher than expected in the natural course of SSPE. In general, patients with slow progression are more likely to benefit, perhaps because of their longer time window for drug effects. SSPE being a rare disease with a variable course among patients, and randomized controlled trials are difficult to execute. Special ethical and methodological regulations applied to rare or orphan diseases should be considered for SSPE.

The myoclonia of SSPE respond to carbamazepine, unlike other myoclonic attacks. This is a typical finding which supports the diagnosis of SSPE. Clonazepam, anti-spasticity agents, and other anticonvulsants can also be used as required. Supportive treatment is most influential on the outcome. In some cases, the persistent infection appears to stabilize and some functional repair takes place if the host is given enough time. Physical therapy, even in bedridden patients, and nutritional assistance sometimes requiring N/G tubing or gastrostomy in late stages, are important measures.

1.7.8 Prognosis

Age, sex, clinical, serological, or imaging features have not been found predictive of clinical course and outcome. To some extent, a very young age of onset appears to be associated with more rapid progression. The effects of treatment on outcome are illustrated by longer survival and higher rates of remission in treated patients. Beside clinical follow-up, EEG (and not IgG titers or imaging findings) is the best indicator of progression, stabilization, or remission. Changes in the background rhythm and the frequency of discharges correlate with, and may even precede, clinical changes.

1.7.9 Prevention

Measles vaccine prevents measles and SSPE when an immunization rate over 90 % is reached in a population. Recently the safety of the MMR vaccine has been questioned, which led to reduced immunization rates, accumulation of susceptible populations, and epidemics. Some previously immunized individuals can be susceptible due to low seroconversion or waning antibody levels in young adult age, termed as secondary vaccine failure (Paunio et al. 2000).

1.7.10 Differential Diagnosis

Patients are likely to be misdiagnosed initially, especially in populations in which SSPE is rare.

When presenting with typical symptoms and EEG findings, SSPE can be suspected and confirmed with CSF analysis. On the other hand myoclonia and mental deterioration also constitute the features of epilepsy, particularly progressive myoclonic epilepsies and mitochondrial disorders. Among neurometabolic disorders, Wilson's disease and leukodystrophies can resemble SSPE because they also present with progressive gait or movement disturbance and ataxia. Sydenham's chorea presents with irregular movements of the extremities, incoordination, and dysarthria. The absence of mental deterioration and progression are in favor of this type of chorea rather than SSPE.

Variant Creutzfeldt–Jakob disease (CJD) is a disease of young adults with myoclonus, cognitive decline, and seizures. Myoclonia in CJD are markedly related to auditory or tactile stimuli, whereas they are spontaneous in SSPE. CSF studies distinguish between the two diseases. Mass lesions and stroke are considered when SSPE starts with asymmetrical signs, and encephalitis or ADEM is considered when SSPE presents an acute fulminant course. SME's clinical features are very similar to SSPE, but the host is usually immunocompromised and CSF is negative for MV IgG. Progressive multifocal leukoencephalopathy is another slow infection of the immunocompromised host due to JC virus where cognitive and motor symptoms and white matter lesions on MRI are observed; virological studies allow differentiation. MR imaging rapidly rules out brain tumor in an SSPE patient presenting with focal signs. On the other hand, MR imaging of SSPE may be reminiscent of multiple sclerosis, especially in adult neurology clinics. However mental and behavioral symptoms are rare as initial manifestations of MS.

A recent review ($n = 307$) illustrates the wide list of differential diagnoses (Prashanth et al. 2007). In many cases, the first diagnosis was epilepsy, leukodystrophy, Schilder's disease, cerebral palsy, parkinsonism, Wilson's disease, vasculitis, spinocerebellar ataxia, motor neuron disease, nutritional amblyopia, retinitis, schizophrenia, and malingering. Clearly, a high index of suspicion is needed in order to prevent diagnostic delays and avoid unnecessary diagnostic and therapeutic interventions.

References

Adels BR, Gajdusek C, Gibbs CJ Jr, Albrecht P, Rogers NG (1968) Attempts to transmit subacute sclerosing panencephalitis and isolate a measles virus related agent, with a study of the immune response in patients and experimental animals. Neurology 18:30–51

Albertyn C, van der Plas H, Hardie D, Candy S, Tomoka T, Leepan EB, Heckmann JM (2011) Silent casualties from the measles outbreak in South Africa. S Afr Med J 101:313–314, 316–317

Anlar B, Yalaz K (2011) Prognosis in subacute sclerosing panencephalitis. Dev Med Child Neurol 53:965

Anlar B, Saatçi I, Köse G, Yalaz K (1996) MRI findings in subacute sclerosing panencephalitis. Neurology 47:1278–1283

Anlar B, Yalaz K, Oktem F, Köse G (1997) Long-term follow-up of patients with subacute sclerosing panencephalitis treated with intraventricular alpha-interferon. Neurology 48:526–528

Anlar B, Köse G, Gürer Y, Altunbaşak S, Haspolat S, Okan M (2001a) Changing epidemiological features of subacute sclerosing panencephalitis. Infection 29:192–195

Anlar B, Söylemezoğlu F, Aysun S, Köse G, Belen D, Yalaz K (2001b) Tissue inflammatory response in subacute sclerosing panencephalitis (SSPE). J Child Neurol 16:895–900

Anlar B, Ayhan A, Hotta H, Itoh M, Engin D, Barun S, Köseoglu O (2002a) Measles virus RNA in tonsils of asymptomatic children. J Paediatr Child Health 38:424–425

Anlar B, Pinar A, Yaşar Anlar F, Engin D, Ustaçelebi S, Kocagöz T, Us D, Akduman D, Yalaz K (2002b) Viral studies in the cerebrospinal fluid in subacute sclerosing panencephalitis. J Infect 44:176–180

Anlar B, Aydin OF, Guven A, Sonmez FM, Kose G, Herguner O (2004) Retrospective evaluation of interferon-beta treatment in subacute sclerosing panencephalitis. Clin Ther 26:1890–1894

Bakir TF (1989) Viral meningitis and encephalitis in Saudi Arabia. J Trop Pediatr 35:211–213

Bellini WJ, Rota JS, Lowe LE, Katz RS, Dyken PR, Zaki SR, Shieh WJ, Rota PA (2005) Subacute sclerosing panencephalitis: more cases of this fatal disease are prevented by measles immunization than was previously recognized. J Infect Dis 192:1686–1693

Bouteille M, Fontaine C, Vedrenne C, Delarue J (1965) Sur un cas d'encbphalite subaigue a inclusions. Etude anatomoclinique et ultrastructurale. Rev Neurol 113:454

Cherian T, Joseph A, John TJ (1984) Low antibody response in infants with measles and children with subclinical measles virus infection. J Trop Med Hyg 87:27–31

Committee on Infectious Diseases; American Academy of Pediatrics (2011) Policy statement–recommended childhood and adolescent immunization schedules–United States, 2011. Pediatrics 127:387–388

Connolly JH, Haire M, Hadden DS (1971) Measles immunoglobulins in subacute sclerosing panencephalitis. Br Med J 1:23–25

Dittmar S, Harms H, Runkler N, Maisner A, Kim KS, Schneider-Schaulies J (2008) Measles virus-induced block of transendothelial migration of T lymphocytes and infection-mediated virus spread across endothelial cell barriers. J Virol 82:11273–11282

Dyken PR, Swift A, DuRant RH (1982) Long-term follow-up of patients with subacute sclerosing panencephalitis treated with inosiplex. Ann Neurol 11:359–363

Erturk O, Karslıgil B, Cokar O, Yapıcı Z, Demirbilek V, Gurses C, Yalcinkaya C, Gokyigit A, Direskeneli GS, Yentur S, Onal E, Yılmaz G, Dervent A (2011) Challenges in diagnosing SSPE. Childs Nerv Syst 27:2041–2044

European Centre for Disease Prevention and Control (ECDC). European monthly measles monitoring (EMMO) 2011. Stockholm. http://ecdc.europa.eu/en/publications/Publications/1109_European_monthly_measles_monitoring_August_2011.pdf.

Filia A, Tavilla A, Bella A, Magurano F, Ansaldi F, Chironna M, Nicoletti L, Palù G, Iannazzo S, Declich S, Rota MC (2011). Measles in Italy, July 2009 to September 2010. Euro Surveill 16: pii: 19925.

Fukuyama Y, Nihei K, Matsumoto S, Ebina T, Kamoshita S, Sato T, Arima M, Yabuuchi H, Ueda S, Ohtahara S et al (1987) Clinical effects of MND-19 (Inosiplex) on subacute sclerosing panencephalitis–a multi-institutional collaborative study–The Inosiplex-SSPE Research Committee. Brain Dev 9:270–282

Gadoth N, Kott E, Levin S, Hahn T (1989) The interferon system in subacute sclerosing panencephalitis and its response to isoprinosine. Brain Dev 11:308–312

Gibbs FA, Gibbs EL, Spies HW, Carpenter PR (1964) Common types of childhood encephalitis: electroencephalographic and clinical relationships. Arch Neurol 10:1–11

Gunaratne PS, Rajendran T, Tilakaratne S (2001) Neurological complications of measles. Ceylon Med J 46:48–50

Gürer YK, Kükner S, Sarica B (1996) Intravenous gamma-globulin treatment in a patient with subacute sclerosing panencephalitis. Pediatr Neurol 14:72–74

Haddad FS, Risk WS (1980) Isoprinosine treatment in 18 patients with subacute sclerosing panencephalitis: a controlled study. Ann Neurol 7:185–188

Hirayama T, Ikeda K, Hidaka T, Nagata R, Yoshii Y, Kawabe K, Iwasaki Y (2010) Unilateral measles-associated retrobulbar optic neuritis without encephalitis: a case report and literature review. Case Rep Neurol 2:128–132

Horta-Barbosa L, Fuccillo DA, London WT, Jabbour JT, Zeman W, Sever JL (1969) Isolation of measles virus from brain cell cultures of two patients with subacute sclerosing panencephalitis. Proc Soc Exp Biol Med 132:272–277

Jones CE, Dyken PR, Huttenlocher PR, Jabbour JT, Maxwell K (1982) Inosiplex therapy in subacute sclerosing panencephalitis. A multicentre, non-randomised study in 98 patients. Lancet 1:1034–1037

Katayama Y, Kohso K, Nishimura A, Tatsuno Y, Homma M, Hotta H (1998) Detection of measles virus mRNA from autopsied human tissues. J Clin Microbiol 36:299–301

Katz M, Rorke LB, Masland WS, Koprowski H, Tucker SH (1968) Transmission of an encephalitogenic agent from brains of patients with subacute sclerosing panencephalitis to ferrets. Preliminary report. N Engl J Med 279:793–798

Kim SJ, Kim JS, Lee DY (2003) Neurologic outcome of acute measles encephalitis according to the MRI patterns. Pediatr Neurol 28:281–284

Kühne Simmonds M, Brown DW, Jin L (2006) Measles viral load may reflect SSPE disease progression. Virol J 3:49

Laboccetta AC, Tornay AS (1964) Measles encephalitis. Report of 61 cases. Am J Dis Child 107:247–255

Lee KY, Cho WH, Kim SH, Kim HD, Kim IO (2003) Acute encephalitis associated with measles: MRI features. Neuroradiology 45:100–106

Lisse I, Samb B, Whittle H, Jensen H, Soumare M, Simondon F, Aaby P (1998) Acute and long-term changes in T-lymphocyte subsets in response to clinical and subclinical measles. A community study from rural Senegal. Scand J Infect Dis 30:17–21

Luzi P, Leoncini L, Valassina M, Federico A, Palma L (1997) Chronic progressive leptomeningitis associated with measles virus. Lancet 350:338–339

Maas S, Kawahara Y, Tamburro KM, Nishikura K (2006) A-to-I RNA Editing and Human Disease. RNA Biol 3:1–9

Meyer E, Byers RK (1952) Measles encephalitis; a follow up study of sixteen patients. AMA Am J Dis Child 84:543–579

Miki K, Komase K, Mgone CS, Kawanishi R, Iijima M, Mgone JM, Asuo PG, Alpers MP, Takasu T, Mizutani T (2002) Molecular analysis of measles virus genome derived from SSPE and acute measles patients in Papua, New Guinea. J Med Virol 68:105–112

Moss WJ, Griffin DE (2012) Measles. Lancet 2012 379:153–164

Mustafa MM, Weitman SD, Winick NJ, Bellini WJ, Timmons CF, Siegel JD (1993) Subacute measles encephalitis in the young immunocompromised host: report of two cases diagnosed by polymerase chain reaction and treated with ribavirin and review of the literature. Clin Infect Dis 16:654–660

Nakajima M, Sakuishi K, Fukuda S, Fujioka S, Hashida H (2008) Recovery from adult measles encephalitis immediately after early immunomodulation. Clin Infect Dis 47:148–149

Nakayama T, Mori T, Yamaguchi S, Sonoda S, Asamura S, Yamashita R, Takeuchi Y, Urano T (1995) Detection of measles virus genome directly from clinical samples by reverse transcriptase-polymerase chain reaction and genetic variability. Virus Res 35:1–16

Naniche D, Varior-Krishnan G, Cervoni F, Wild TF, Rossi B, Rabourdin-Combe C, Gerlier D (1993) Human membrane cofactor protein (CD46) acts as a cellular receptor for measles virus. J Virol 67:6025–6032

Öktem F, Nester MJ, Anlar B (1997) Mental assessment in subacute sclerosing panencephalitis: Hacettepe Cognitive Short Assessment Scale. J Child Neurol 12:398–402

Ölmez A, Yilmaz D, Tan H, Duman O, Güngör S, Okuyaz C, Anlar B (2007) Cerebrospinal fluid pressures in subacute sclerosing panencephalitis. Brain Dev 29:409–412

Otaki M, Jiang DP, Sasayama M, Nagano-Fujii M, Hotta H (2007) Generation of recombinant adenovirus expressing siRNA against the L mRNA of measles virus and subacute sclerosing panencephalitis virus. Microbiol Immunol 51:985–991

Paunio M, Hedman K, Davidkin I, Valle M, Heinonen OP, Leinikki P, Salmi A, Peltola H (2000) Secondary measles vaccine failures identified by measurement of IgG avidity: high occurrence among teenagers vaccinated at a young age. Epidemiol Infect 124:263–271

Pipo-Deveza JR, Kusuhara K, Silao CL, Lukban MB, Salonga AM, Sanchez BC, Kira R, Takemoto M, Ishizaki Y, Yukaya N, Kusuhara K, Kira R, Torisu H, Ihara K, Sakai Y, Sanefuji M, Pipo-Deveza JR, Silao CL, Sanchez BC, Lukban MB, Salonga AM, Hara T (2010) PD1 as a common candidate susceptibility gene of subacute sclerosing panencephalitis. Hum Genet 127:411–419

Piskin AK, Akpinar P, Muftuoglu S, Anlar B (2007) Signaling lymphocyte activating molecule (SLAM) expression in subacute sclerosing panencephalitis. Brain Dev 29:439–442

Plaza JA, Nuovo GJ (2005) Histologic and molecular correlates of fatal measles infection in children. Diagn Mol Pathol 14:97–102

Prasad SR, Shaikh NJ, Verma S, Banerjee K (1995) IgG & IgM antibodies against measles virus in unvaccinated infants from Pune: evidence for subclinical infections. Indian J Med Res 101:1–5

Prashanth LK, Taly AB, Sinha S, Ravi V (2007) Subacute sclerosing panencephalitis (SSPE): an insight into the diagnostic errors from a tertiary care university hospital. J Child Neurol 22:683–688

Reuter D, Schneider-Schaulies J (2010) Measles virus infection of the CNS: human disease, animal models, and approaches to therapy. Med Microbiol Immunol 199:261–271

Robertson WC Jr, Clark DB, Markesbery WR (1980) Review of 38 cases of subacute sclerosing panencephalitis: effect of amantadine on the natural course of the disease. Ann Neurol 8:422–425

Sever AE, Rainey JJ, Zell ER, Hennessey K, Uzicanin A, Castillo-Solórzano C, Dietz V (2011) Measles elimination in the Americas: a comparison between countries with a one-dose and two-dose routine vaccination schedule. J Infect Dis 204(Suppl 2):S748–755

Shu M, Liu Q, Wang J, Ao R, Yang C, Fang G, Wan C, Guo W (2011) Measles vaccine adverse events reported in the mass vaccination campaign of Sichuan province, China from 2007 to 2008. Vaccine 29:3507–3510

Srivastava SP, Nema HV (1963) Optic neuritis in measles. Brit J Ophthal 47:180–181

Stanescu A, Janta D, Lupulescu E, Necula G, Lazar M, Molnar G, Pistol A (2011) Ongoing measles outbreak in Romania, 2011. Euro Surveill 16: pii:19932.

Tatsuo H, Ono N, Tanaka K, Yanagi Y (2000) SLAM (CDw150) is a cellular receptor for measles virus. Nature 406:893–897

Thurner B, Spangenberg P, Kleines M, Blaum M, Scheithauer S, Häusler M (2007) Continuous intrathecal interferon alpha application in subacute sclerosing panencephalitis. Pediatr Infect Dis J 26:863

Tomiyasu K, Ishiyama M, Kato K, Komura M, Ohnuma E, Inamasu J, Takahashi T (2009) Bilateral retrobulbar optic neuritis, Guillain-Barré syndrome and asymptomatic central white matter lesions following adult measles infection. Intern Med 48:377–381

Tomoda A, Nomura K, Shiraishi S, Hamada A, Ohmura T, Hosoya M, Miike T, Sawaishi Y, Kimura H, Takashima H, Tohda Y, Mori K, Kato Z, Fukushima A, Nishio H, Nezu A, Nihei K (2003) Trial of intraventricular ribavirin therapy for subacute sclerosing panencephalitis in Japan. Brain Dev 25:514–517

Torisu H, Hara T (2006) Analysis of MxA, IL-4, and IRF-1 genes in Filipino patients with subacute sclerosing panencephalitis. Neuropediatrics 37:222–228

Torisu H, Kusuhara K, Kira R, Bassuny WM, Sakai Y, Sanefuji M, Takemoto M, Hara T (2004) Functional MxA promoter polymorphism associated with subacute sclerosing panencephalitis. Neurology 62:457–460

Tyler HR (1957) Neurological complications of rubeola (measles). Medicine (Baltimore) 36:147–167

Valassina M, Meacci F, Valensin PE, Cusi MG (2000) Detection of neurotropic viruses circulating in Tuscany: the incisive role of Toscana virus. J Med Virol 60:86–90

Watanabe A, Yoneda M, Ikeda F, Terao-Muto Y, Sato H, Kai C (2010) CD147/EMMPRIN acts as a functional entry receptor for measles virus on epithelial cells. J Virol 84:4183–93

Yalaz K, Anlar B, Oktem F, Aysun S, Ustacelebi S, Gurcay O, Gucuyener K, Renda Y (1992) Intraventricular interferon and oral inosiplex in the treatment of subacute sclerosing panencephalitis. Neurology 42:488–491

Yilmaz V, Demirbilek V, Gürses C, Yentür SP, Uysal S, Yapici Z, Yilmaz G, Muncey A, Cokar O, Onal E, Gökyiğit A, Saruhan-Direskeneli G (2007) Interleukin (IL)-12, IL-2, interferon-gamma gene polymorphisms in subacute sclerosing panencephalitis patients. J Neurovirol 13:410–415

Yüksel D, Sonmez PA, Yilmaz D, Senbil N, Gurer Y (2011) Ocular findings in subacute sclerosing panencephalitis. Ocul Immunol Inflamm 19:135–138

Epstein–Barr Virus and Cytomegalovirus Infections

Alex Tselis

Abstract Epstein–Barr virus and cytomegalovirus are members of the human herpesviruses that have an extremely high seroprevalence in all populations studied. The initial infection is usually asymptomatic, or causes a febrile illness, but can rarely manifest itself neurologically. These viruses are increasingly important in the modern era of immunosuppression, whether due to AIDS or in the transplant or cancer chemotherapy population, and their reactivation gives rise to a wide spectrum of neurological diseases. The pathogenesis of these infections is not completely understood, but certainly multifaceted. In CMV lytic infection damages systemic tissues directly, whereas EBV involves an activated and distorted immune system. These diseases are treatable, but need to be recognized early in their course so that antiviral intervention can be effected promptly. The choice of therapeutic strategy can be counterintuitive: while CMV infections are conventionally managed with antiviral medications, EBV infections may demand a neoplastic treatment paradigm as an addition to (or alternative to) antiviral treatment.

Keywords Cytomegalovirus • Diagnostic virology • Encephalitis • Epstein-Barr virus • Immunosuppression • Lymphoproliferative disorder • Myelitis • Opportunistic infections • Primary CNS lymphoma

1 Introduction

Epstein–Barr Virus (EBV) and cytomegalovirus (CMV) are two herpesviruses occasionally associated with neurologic disease. They share with other herpesviruses the property of initial infection of young hosts, establishment of latency, and "reactivation" later in life, with variable consequences. While most initial infections with these viruses

A. Tselis (✉)
Department of Neurology, Wayne State University, Detroit, MI, USA
e-mail: atselis@med.wayne.edu

A.C. Jackson (ed.), *Viral Infections of the Human Nervous System*,
Birkhäuser Advances in Infectious Diseases,
DOI 10.1007/978-3-0348-0425-7_2,

are clinically self-limited, some have prominent neurological manifestations. In the modern era of immunocompromised patients who have had a transplant, cancer chemotherapy, autoimmune disease, or AIDS, reactivation of these viruses can have devastating consequences. These reactivations can have quite novel manifestations and reflect unusual pathogenetic mechanisms.

2 Epstein–Barr Virus

2.1 A Brief History

The history of the discovery of EBV is one of the great medical detective stories of the twentieth century. A febrile pharyngitis with cervical lymphadenopathy was described late in the nineteenth century. While a number of illnesses can have this presentation, a subset with very high peripheral mononuclear cell counts was defined in 1920 by Sprunt and Evans (1920) and called "infectious mononucleosis (IM)." The observation by Haganutziu and Deicher that serum sickness was associated with a sheep red cell agglutinin was confirmed by Paul and Bunnell (1932). They attempted to define the specificity of this observation by examining control sera. One of these showed a very high titer of such agglutinins, and was found to be from an IM patient. This led to the discovery of the so-called "heterophile antibodies (HA)," which evolved into a diagnostic test for IM. Attempts to transmit the disease to other humans or animals were inconsistently successful and further advances had to wait several decades.

In 1946, a British colonial surgeon, Denis Burkitt, was assigned to a post in Uganda, where he took care of a population of 250,000 people. In 1957, he was asked to see a child with a peculiar mass in the jaw, which rendered him "totally baffled." He saw other such cases and reviewed the hospital records for other cases. These showed that the tumor, a lymphoma, often affected the internal organs and the nervous system, rather than lymph nodes. He sent questionnaires to clinics around the continent using mails, and was able to establish the geographic distribution of this tumor, and noted that it overlapped the distribution of malaria and yellow fever, as well as an epidemic of o'nyong nyong fever. The fact that the geographical distribution of Burkitt's lymphoma (BL) overlapped that of several mosquito-borne diseases suggested the possibility that the disease was transmittable. Burkitt gave several talks about his findings on a visit to London, and Anthony Epstein, a virologist interested in tumor viruses, was present. He had Burkitt send him samples of the tumors and was able to detect a herpes-like virus by electron microscopy. However, the virus could not be cultured. For more accurate characterization of the virus, samples were sent to the laboratory of Werner and Gertrude Henle. They were able to show that antibodies to the Epstein–Barr virus (EBV) were present not only in pediatric oncology patients, but also were common in the general population. The first connection between EBV and a specific disease was made when a technician in the Henles' laboratory, who was seronegative, developed IM. Her serum, previously used as a negative control, became strongly seropositive (Henle et al. 1968). This observation provided the impetus for the studies of college students by Niederman et al. (1968) in which the etiologic role of EBV in IM was established. The role of EBV was then established in a number of tumors. This includes BL, a number of B and T cell lymphomas, Hodgkin's lymphoma, and leiomyosarcoma. Further, systemic "opportunistic lymphomas" in the context of transplantation, AIDS, and chemotherapy are often caused by EBV. These include posttransplant lymphoproliferative disorder (PTLD) and the experiment-of-nature X-linked lymphoproliferative disorder (XLPD), in which there is an uncontrolled proliferation of EBV-infected B cells because of a novel immune defect.

2.2 Basic Virology

The virus consists of a nucleocapsid containing a 184 kbp double stranded (ds) DNA molecule surrounded by 162 capsomers. The nucleocapsid is surrounded by a protein-rich tegument, which in turn is surrounded by an envelope.

The genome of the virus is structured similar to other herpesviruses, in which there are unique long and short regions, separated by a long run of internal repeats, and flanked by terminal repeats. There are about 190 genes per genome.

There are overall two types of genes in the EBV genome. When the virus infects its target cells, it replicates in two different ways, latent and lytic replication. In latent replication in EBV-infected B cells, the EBV genome replicates along with the cellular DNA, using the cell's own DNA polymerase. Thus, cellular and viral DNA are replicated by cellular DNA polymerase in latent replication. In the latent state, there is minimal expression of viral genes. In lytic replication, which occurs in epithelial cells and plasma cells, the viral DNA is replicated by viral DNA polymerase, and assembled into full virions that are released by lysis of the infected cell. It is important to note that antiviral drugs such as acyclovir and ganciclovir will inhibit the viral but not the cellular DNA polymerase. Thus, these drugs decrease lytic but not latent replication. The spectrum of disease depends on the type of replication as will be seen later.

2.3 Spectrum of Systemic Disease Associated with EBV

Primary EBV infection is often asymptomatic, especially in children. In young adults, the infection causes a febrile pharyngitis with prominent cervical lymphadenopathy and significant fatigue and malaise. This illness is called EBV-associated infectious mononucleosis (EBV IM). Usually, recovery is complete within a few weeks, although cases lasting several months have been reported. Interestingly, many patients develop a rash when treated for their pharyngitis with ampicillin, in order to cover a possible bacterial infection. The disease can be diagnosed by one of the slide tests to screen for it or more definitively by an EBV panel (see below). Other mimics of EBV IM include primary CMV disease, human herpes virus 6 disease (HHV6), acute retroviral syndrome, secondary disseminated syphilis, and acute toxoplasmosis (Hurt and Tammaro 2007).

Other manifestations of EBV IM include severe tonsillitis (which can potentially interfere with swallowing), splenomegaly (with a small risk of splenic rupture), hepatitis, myocarditis, pneumonitis, interstitial nephritis, and hemolytic anemia. These are uncommon, but point to the diversity of clinical manifestations of acute EBV infection.

EBV-infected B cells are transformed and tend to proliferate spontaneously. This proliferation, if uncontrolled, can result in serious disease. Therefore, EBV infection does not cause illness by causing lysis of tissues, but by the immune suppression of these proliferating B cells. Thus, rarely, IM can be severe, with poorly controlled proliferation of the infected B cells, and fatal results. This is a rare entity known as fatal IM (FIM) and can be seen in X-linked lymphoproliferative disorder, and it may be seen in other more subtle immune deficiencies. Acute EBV can cause a hemophagocytic syndrome, a sepsis-like syndrome caused by EBV triggering widespread macrophage activation and histiocytosis leading to a cytokine storm with multiple organ failure. In a few cases, EBV-driven lymphoproliferative syndrome can affect the central nervous system, as part of the systemic disease.

EBV can also result in a broad spectrum of neoplasms and lymphoproliferative states. One of the first to be characterized, as discussed above, is Burkitt's lymphoma, in which there is systemic lymphomatous involvement, particularly with visceral involvement. A high proportion of the original patients with Burkitt's lymphoma has central nervous system involvement. Others, as mentioned above, include Hodgkin's lymphoma (HL), posttransplant lymphoproliferative disorder (PTLD), X-linked lymphoproliferative disorder (XLPD), primary CNS lymphoma (especially in AIDS patients), nasopharyngeal carcinomas of Southeast Asia, T cell and NK cell lymphomas, and leiomyosarcomas. These generally involve latent infection of the neoplastic cells. Oral hairy leukoplakia, an infection of the tongue epithelium, is a lytic infection.

2.4 Pathology and Pathogenesis

EBV is transmitted by intimate oral contact, with virus shed asymptomatically in the saliva. The initial infection is of B cells in the oral mucosa. These cells are immortalized and proliferate, with latent replication of the virus within the B cells. The latently infected B cells express a very limited set of proteins and latency-associated RNA molecules. These sets (or latency types) depend on the stage of the illness (Table 1). These antigens are recognized by the immune system and a T cell response is generated. The infection is thereby controlled, but not eliminated. In some cases, the manifestations of the disease tend to be focal, with a clinical picture of hepatitis, meningitis, or encephalitis. It is not clear why this occurs in an otherwise systemic disease.

Table 1 Latency antigens and types

Latency type	Latency antigens						
	EBER	EBNA-1	EBNA-2	EBNA-3	LMP-1	LMP-2	BARTs
1	+	+	–	–	–	–	+
2	+	+	–	–	+	+	+
3	+	+	+	+	+	+	+
Other	+	+/–	–	–	–	+	+/–

Latency types	
Latency 1	Burkitt's lymphoma
Latency 2	Nasopharyngeal carcinoma, Hodgkin's disease
Latency 3	Infectious mononucleosis, lymphoproliferative disease
Other	Perpheral blood B lymphocytes

EBER Epstein–Barr virus-encoded RNA, *EBNA* Epstein–Barr nuclear antigen, *LMP* Latent membrane protein, *BART* BamHI A rightward transcripts

The pathogenesis of encephalitis (or meningitis or hepatitis or other focal visceral involvement) is not completely clear and there are several possibilities, which are not mutually exclusive. First, EBV may affect neurons (or other neural cells or endothelium) directly (Jones et al. 1995). There have been a few scattered reports of neurons and glial cells staining with EBV antigens, although there is not much detail (Biebl et al. 2009). In some patients with EBV encephalitis, as well as some with primary CNS lymphoma, lytic EBV mRNA was detected in the CSF, suggesting lytic replication of EBV in the brain in addition to latent replication (Weinberg et al. 2002a). Secondly, EBV-infected B cells are in an activated state and elaborate several proinflammatory cytokines, which can cause injury of the surrounding parenchyma (Foss et al. 1994). This injury is not necessarily irreversible. Third, EBV-infected B cells are actively attacked by EBV-specific cytotoxic T cells, and this can also injure the surrounding parenchyma. Finally, an acute disseminated encephalomyelitis can be triggered as in other viral infections.

Normally, EBV-infected B cells are suppressed (though not eliminated) by the immune system and lymphoproliferation can result during immunosuppression. In tissue culture in which T cells have been eliminated, B cells are immortalized and proliferate. In vivo, the B cell lymphoproliferation proceeds sequentially from polyclonal to oligoclonal to monoclonal, and evolves into a lymphoma. This can occur under circumstances of immunosuppression in transplant, chemotherapy, and AIDS patients as mentioned above. The lymphoproliferation can be accompanied by the elaboration of various cytokines, and a severe systemic illness resembling sepsis can result.

2.5 *Spectrum of Neurologic Disease Associated with EBV*

The spectrum of neurologic disease caused by EBV is very broad, and encompasses all of the neurological syndromes, pure or mixed: meningitis, encephalitis, myelitis, radiculopathy, plexitis, psychosis, and behavioral abnormalities. These syndromes may precede, follow, or occur independent of IM.

2.5.1 Aseptic Meningitis

Aseptic meningitis was one of the first reported complications of acute EBV infection, reported by Johansen (1931). Headaches are not rare in IM, and it is likely that some of these are due to aseptic meningitis. The early appreciation of aseptic meningitis is illustrated by a 1950 review of neurological complications of IM in which it was found in 41 % of the cases (Bernstein and Wolfe 1950). It is self-limiting.

2.5.2 Encephalitis

Encephalitis is an uncommon manifestation of IM with a broad clinical spectrum, but most cases have the usual presentation of fever, headache, confusion, seizures, and focal features. EBV encephalitis can precede, coincide with, or follow typical IM, and IM may be absent altogether (Silverstein et al. 1972; Friedland and Yahr 1977; Greenberg et al. 1982; Russell et al. 1985; Leavell et al. 1986; McKendall et al. 1990).

Brainstem encephalitis due to EBV has been reported in three cases, with one complete recovery, one with a residual ataxic gait, and one death (Shian and Chi 1994; North et al. 1993; Angelini et al. 2000). The syndrome of opsoclonus–myoclonus has been described in several cases of acute EBV infection. In one case, the patient had opsoclonus–myoclonus with ataxic gait. EBV was detected in the cerebrospinal fluid (CSF) by polymerase chain reaction (PCR) amplification. He was treated with intravenous methylprednisolone followed by intravenous immunoglobulin and returned to work 5 months later (Verma and Brozman 2002). Other cases of EBV-associated opsoclonus–myoclonus have a similar benign outcome.

Movement disorders have been reported in EBV encephalitis cases. In one case which resembled encephalitis lethargica, the patient developed an akinetic-rigid syndrome with tremor and sialorrhea. The MRI showed strongly abnormal signal in the striatum. Corticosteroids and antiparkinson drugs were given and the symptoms resolved over 2 months (Dimova et al. 2006). In another parkinsonian syndrome developed coincident with EBV encephalitis, antineuronal antibodies were detected in the serum of the patient but not three controls. Brain MRI was normal. Acyclovir, dexamethasone, and antiparkinsonian medications were given and the patient returned to normal over the next 2 months (Roselli et al. 2006).

2.5.3 Cranial Nerve Palsy

The most common cranial nerve palsy associated with acute EBV infection is Bell's palsy, which may be unilateral or bilateral (Grose et al. 1973; Egan 1960).

Sometimes several cranial nerves can be affected. A case of unilateral Bell's palsy with ipsilateral deafness and facial numbness has been reported to follow IM (Taylor and Parsons-Smith 1969). Optic neuritis and retinal involvement, which can be bilateral, has rarely occurred with IM (Ashworth and Motto 1947; Blaustein and Caccavo 1950; Bonynge and Van Hagen 1952).

2.5.4 Transverse Myelitis

Transverse myelitis has occasionally coincided with acute EBV infection. Several cases of TM have been reported in the literature, in which lower extremity paresthesias followed clinical IM, and progressed rapidly to flaccid paraplegia within a few days. Sensory levels and upgoing toes were seen (Cotton and Webb-Peploe 1966; Grose and Feorino 1973; Clevenbergh et al. 1997). One patient had a transient tetraparesis but normal gait on examination. Spinal sensory level was noted. Diagnosis was made by serology in two cases and PCR detection of EBV DNA in CSF in one (Clevenbergh et al. 1997). In all cases there was slow recovery over months. One of the patients received ACTH.

2.5.5 Cerebellar Ataxia

Acute cerebellar ataxia occurs in some patients with acute EBV infection, often following mild disease. Classically this has been attributed to varicella zoster virus (VZV) infection, especially in children. However, a significant number of cases are associated with EBV both in children and adults (Bergen and Grossman 1975; Cleary et al. 1980; Bennett and Peters 1961; Gilbert and Culebras 1972; Lascelles et al. 1973). The patients have gait ataxia and dysarthric speech, with mild pleocytosis and modestly increased CSF protein. Some have responded to ACTH, prednisone, and plasmapheresis (Schmahmann 2004). Recurrent cerebellitis was reported, in which a patient with dysarthria, dysmetria, and gait ataxia had a positive EBV VCA IgM, and resolved with prednisone. A year later, the symptoms recurred and resolved again with another course of prednisone (Shoji et al. 1983).

2.5.6 Alice-in-Wonderland Syndrome

Alice-in-Wonderland syndrome is a peculiar neuropsychiatric entity in which the patient develops metamorphopsia or distortion of spatial perception in which objects around the patient are perceived to be distorted in size, shape, and orientation. These episodes last about half an hour, and are understandably anxiety provoking. Neurologic examination is usually normal and EEGs are normal or minimally abnormal. Single patients were treated with prednisone and phenytoin, without clear effect. The symptoms resolve spontaneously over a few weeks (Copperman 1977; Eshel et al. 1987). Visual evoked potentials have an increased

P100-N145 wave complex, and hexamethylpropylene amine oxime single-photon emission computed tomography showed decreased perfusion in the visual tracts and visual cortex (Lahat et al. 1999; Kuo et al. 1998).

2.5.7 Acute Hemiplegia

Occasionally acute EBV infection can be associated with a rapidly developing hemiplegia, which can resemble a stroke. Some cases of so-called "acute hemiplegia of childhood" may well be due to acute EBV, and there are detailed reports of such cases. A 14-year-old girl had a left hemiplegia and left-sided numbness that evolved over several days along with right-sided headache, vomiting, and photophobia. She had two seizures and cervical lymphadenopathy. A fever prompted a CSF examination which showed moderate pleocytosis. She became confused and ataxic. Acute EBV infection was demonstrated by serology. She recovered completely in a few months (Leavell et al. 1986). Two other similar cases with unilateral headache and contralateral hemiplegia were reported in a 9-year-old girl and a 32-year-old man (Baker et al. 1983; Adamson and Gordon 1992). The former patient's hemplegia spontaneously improved to normal over a few days. The latter, who had a normal brain CT, resolved completely within a day of starting on dexamethasone.

2.5.8 Neurological Lymphoproliferative Disorder

As discussed above, EBV-infected B cells have a tendency to proliferate. This is stopped by the immune system, but if immunity is ineffective, then proliferation proceeds relatively unchecked, leading to polyclonal expansion and eventually oligoclonal and finally monoclonal lymphomas. Such lymphoproliferative disorders can affect the nervous system in the course of systemic disease. For many of these the distinction between infection, inflammation, and neoplasm is obscured. In one case of a 14-year-old girl with a chronic febrile illness, ataxia and hemiparesis led to an MRI of the brain which showed multifocal white matter lesions. Acute EBV was diagnosed by serology. These resolved with steroids, which needed to be used several times over the next few years, when she had relapses. Several years later she developed pneumonitis and a biopsy found lymphomatoid granulomatosis. A few years after that she developed disseminated intravascular coagulation with hemophagocytic syndrome. In patients with lymphomatoid granulomatosis, there is both pulmonary and CNS involvement. Often, biopsy of the lesions show scattered lymphocytes that stain positively for EBV antigens. Various treatments have been used, including chemotherapy and radiation, rituximab, and cyclophosphamide, with some success (Mizuno et al. 2003; Zaidi et al. 2004). In another case of lymphoproliferative disorder, a 17-year-old boy developed EBV-IM which in a few weeks evolved into a sepsis-like syndrome with encephalopathy. He was found to have hemophagocytic syndrome on bone marrow biopsy and a very high EBV load in the blood. He was treated with methylprednisolone, intravenous immunoglobulin, rituximab (B cell depleting antibody), etanercept (anti-TNFalpha antibody), and etoposide. His medical condition improved, but he showed no cognitive improvement and an MRI showed scattered nonenhancing frontal white matter disease. Intrathecal chemotherapy was instituted with both cognitive and imaging improvement (Mischler et al. 2006).

In patients with severe immunosuppression, especially in advanced HIV disease, primary CNS lymphoma (PCNSL) is not uncommon. In the AIDS population, this is almost 100 % driven by EBV, whereas PCNSL is only rarely EBV-related in those not infected with HIV (Larocca et al. 1998; Hochberg et al. 1983).

2.6 *Diagnosis*

The strategy of the diagnosis of EBV-related neurologic disease depends upon the patient's age, history, and degree of immunosuppression, in addition to the clinical presentation. The demonstration of the appropriate serologic findings, viral antigens, and DNA supports the clinical impression and may confirm the diagnosis. There are, of course, subtleties which will be mentioned below.

In the case of an adolescent patient with fever, headache, sore throat, enlarged cervical lymph nodes, and splenomegaly, leukocytosis with atypical lymphocytes in the peripheral smear, the diagnosis of EBV meningitis can be confirmed by a CSF examination to rule out other etiologies, and either a heterophile slide test or an EBV panel in the serum. Other neurological syndromes, especially in the past, have been attributed to EBV because of the coincidence of the symptoms and serology demonstrating acute EBV infection. More recently, the acute EBV panel is used to confirm disease, since the heterophil slide tests can be falsely negative (uncommon). The heterophil test continues to be relevant, however, since occasionally the EBV panel is difficult to interpret.

2.6.1 Serological Tests for EBV

Heterophile Slide Tests

It may be recalled that early in the twentieth century IM was noted to be associated with a sheep red cell agglutinin. This antibody is specific for but not directed at EBV antigens and is known as a heterophile antibody (HA), since it is elicited by one type of antigen and is directed to a separate, unrelated one. A positive serum HA test conclusively establishes an acute EBV infection. Before the EBV panel became available, neurologic disease was related to EBV by the coincidence of the clinical illness with a positive HA test.

Table 2 Serology in EBV infection

EBV status	VCA IgM	VCA IgG	EA	EBNA
Seronegative	−	−	−	−
Recent primary	+	+	+/−	−
Seropositive (remote infection)	−	+	+/−	+
Infectious mononucleosis	+	+	+	−
Reactivated infection	+/−	+++	+++	+

VCA IgG Viral capsid antigen immunoglobulin G, *VCA IgM* Viral capsid antigen immunoglobulin M, *EA* Early antigen (antibody to), *EBNA* Epstein–Barr nuclear antigen (antibody to)
(−) No antibody
(+/−) Either positive or negative
(+) Detectable antibody
(+++) High titer antibody

EBV Panel

The EBV panel tests for antibodies to specific EBV antigens. Different patterns of antibodies appear at different stages of EBV infection. These antigens are comprised of the viral capsid antigen (VCA), which is a structural protein, early antigen (EA), which is a complex expressed during viral lytic replication, and Epstein–Barr nuclear antigen (EBNA), which is a group of proteins confined to the nucleus and expressed during latent infection in B cells. It was found by Henle et al. (1974) that in acute EBV infection, the first antibody to appear is against EBV VCA, IgM followed by IgG, the second is to EA, and, finally, the third, to EBNA after the acute infection has resolved. Thus, a positive EBV VCA IgM and negative EBNA IgG indicate acute EBV infection while a positive EBV VCA IgG and positive EBNA IgG would be compatible with a remote infection. A guide to interpretation of the EBV panel is given in Table 2.

PCR Detection of EBV DNA

The detection of EBV DNA by PCR in the CSF has become the gold standard for the demonstration of EBV disease in the CNS, although few systematic studies have been done. There have been reports of acute neurologic syndromes in which EBV serology indicated acute infection, and EBV was detected in the CSF by PCR, which suggests the strategy of using both PCR and serology. In a series of 39 patients with acute neurologic disease, and PCR detection of EBV DNA in CSF, three categories of disease were noted: acute EBV encephalitis, PCNSL, and postinfectious EBV complications (such as acute disseminated encephalomyelitis, Guillain–Barre syndrome (GBS), and transverse myelitis). The quantity of EBV and degree of inflammation (as measured by pleocytosis) were both high in encephalitis. In PCNSL, the quantity of virus was high, but there was little inflammatory pleocytosis, as would be expected of a virally driven neoplasm. In postinfectious complications, the viral burden was low, and the inflammatory pleocytosis high. These patterns are as expected, and underline that detection of EBV DNA is not specific for EBV encephalitis (Weinberg et al. 2002a).

Furthermore, some patients with acute neurologic infections have been found to have EBV and another pathogen detected in the CSF (Weinberg et al. 2005). It was estimated that in 25 % of the patients (both immunocompetent and immunosuppressed) with EBV detected in the CSF, a second pathogen may be present. Some of the co-pathogens included CMV, VZV virus, JC polyomavirus, West Nile virus, pneumococcus, Cryptococcus, ehrlichiosis, and mycoplasma. These results may be due to "reactivation" of EBV because of another infection, or to dual, independent infections. The significance is unclear, and underscores the utility of EBV panels and heterophile testing to provide independent information.

Viral Antigen Detection

Viral antigen detection is not commonly used in the diagnosis of neurologic EBV disease, but is used mostly in systemic disease, particularly in transplants. Thus, the differentiation between lymphoproliferative disorder (PTLD) in a transplanted liver and rejection may be difficult. A biopsy that detects lymphocytes bearing latency antigens would suggest PTLD. The diagnosis cannot be made on morphology alone, since there is great variability and not all neoplasms have a monomorphic appearance. Similarly, the diagnosis of PCNSL in AIDS patients often relies upon the detection of latent antigens in lymphocytes.

2.6.2 Magnetic Resonance Imaging

There are no characteristic imaging findings that specifically suggest EBV encephalitis. Brain MRI can be normal, or show abnormal signal in the hemispheres (with gyral pattern or diffuse edema), basal ganglia, cerebellum, brainstem, thalamus, and limbic system (Tselis et al. 1997; Abul-Kasim et al. 2009). The abnormal signal may involve white matter as well as the deep gray structures, such as the basal ganglia and thalamus (Caruso et al. 2000; Garamendi et al. 2002; Phowthongkum et al. 2007). There are examples of simultaneous gray and white matter involvement (Fujimoto et al. 2003). There may be pathogenetic implications of the imaging findings. Thus, pure cortical or deep gray involvement may imply a "pure EBV encephalitis," whereas pure white matter involvement may be due to parainfectious demyelination.

Imaging findings may also have some prognostic value. In the Abul-Kasim et al. (2009) study, it was found that of those with normal imaging, 92.5 % had a good outcome, while of those with abnormal imaging, only 60.7 % did.

2.7 Management

The management of neurologic EBV disease depends upon the pathogenesis of the illness and there is no clear consensus on how to treat the diseases this virus causes. Therapeutic modalities would have to be exceptionally safe, since neurologic EBV disease tends to have a very benign course, even if it were very severe during the acute phase. Thus neurologic EBV disease tends to improve whether patients are treated with antivirals or not, and whether the patient is immunodeficient (e.g., HIV positive) or not (Weinberg et al. 2002a).

EBV encephalitis illustrates these issues well. If the major pathogenesis of the disease is direct lytic infection of neurons or endothelial cells in the brain (as in herpes simplex encephalitis), then antiviral drugs such as acyclovir or ganciclovir should be used since they inhibit viral DNA polymerase and prevent lytic infection. However, there is no much evidence for lytic infection in EBV encephalitis. In one autopsy, viral antigens were found in neurons and astrocytes (Biebl et al. 2009). In the CSF of EBV encephalitis and PCNSL, lytic EBV mRNAs were found but the source (neurons, glia, endothelial cells, lymphocytes, or plasma cells) is unknown (Weinberg et al. 2002b). In EBV IM, acyclovir reduces viral shedding, but has no effect on symptoms. It is not recommended to use acyclovir for EBV encephalitis by the Infectious Diseases Society of America (IDSA) guidelines, although corticosteroids can be given consideration (Tunkel et al. 2008).

On the other hand, if EBV encephalitis were due to the accumulation of activated EBV-infected B cells secreting inflammatory cytokines, which caused the damage, a strategy to eliminate such B cells would be considered, using a drug such as rituximab, which specifically depletes B cells. Of course, such a drug would have to have access to the CNS in order to remove parenchymally placed B cells. However, since the disease seems to have a relatively benign course, such treatment may not be especially useful. Other immunomodulatory or immunosuppressive drugs, such as corticosteroids or intravenous immunoglobulin, often seem to be followed by improvement and are relatively safe to use.

For neurological EBV disease that is part of an EBV lymphoproliferative syndrome (LPD), the disease has a systemic neoplastic character and chemotherapy and radiation, possibly combined with rituximab (to deplete B cells) should be considered.

3 Cytomegalovirus

3.1 A Brief History

In contrast to the dramatic history of the discovery of the nature of EBV, the elucidation of the pathogenesis of CMV disease came about by an almost logical accumulation of discrete steps of important observations and discoveries (Ho 2008; Riley 1997; Weller 1970, 2000).

The characteristic cytomegalic cells of CMV disease were first noted by Ribbert in 1881 in the kidney and parotic glands of a syphilitic neonate, and confirmed by Jesionek and Kiolemenoglu (1904). They interpreted these cells as protozoa. Others took up the search and found similar cells in other infants. The similarity of these cells to those seen in herpes zoster and herpes genitalis was remarked by Goodpasture and Talbot (1921) and by Von Glahn and Pappenheimer (1925). The prominence of these cells in salivary glands prompted the term "salivary gland virus." In 1926, a guinea pig model of salivary gland virus disease bolstered the case for the viral nature of the agent as salivary gland disease was shown to be transmissible by a filterable agent. As experience accumulated, a neonatal illness with petechiae, hepatosplenomegaly, and brain calcifications was characterized and correlated with the presence of cytomegalic cells. Wyatt et al. (1950) coined the term "generalized cytomegalic inclusion disease." When it was found that kidney tubule cells had viral inclusions, the idea of detecting cytomegalic cells in urine was used to make the diagnosis antenatally by Fetterman in 1952. The virus was isolated by three independent groups, those of Smith (1956), Weller et al. (1957), and Rowe et al. (1956). The latter developed a complement-fixation test that was used to show that the seroprevalence in human populations was very high with an increase in age prevalence. From the mid-1950s to the mid-1980s, more disease associations were established. These include the connection between congenital CMV infection, defined by CMV viruria, and deafness and cognitive difficulties later in life; the connection between CMV and CMV mononucleosis; transmission of CMV by transfused blood during cardiac surgery known as the "postperfusion syndrome"; and CMV disease in transplant and AIDS patients (Ho 2008; Riley 1997; Weller 2000).

3.2 Basic Virology

The structure of the CMV virion is similar to that of other herpesviruses with a double-stranded DNA viral genome enclosed in a capsid, which is surrounded by a protein-rich tegument, enveloped within a viral membrane. The genome codes for about 230–250 proteins, depending on the isolate (clinical vs laboratory), and is composed of a unique long (UL) and a unique short (US) region, flanked by terminal repeats. The proteins encoded by the open reading frames (ORFs) are labeled according to their position on the genome, following a common descriptive name. Thus, a phosphoprotein of molecular weight 65 coded by the 83rd ORF in the UL region would be labeled as pp65 (UL83).

CMV genes consist of latent and lytic types. The former are not as well characterized as those of EBV, but generate RNA transcripts that are reminiscent of the latency-associated transcripts (LATs) in herpes simplex infection or the EBERs of EBV infection. The lytic genes are grouped into three categories:

immediate early (or alpha) genes (IE), early (or beta) genes (E) and late (or gamma) genes (L). These permit viral takeover of macromolecular synthesis, synthesis of products necessary for DNA replication (e.g., viral DNA polymerase), and synthesis of structural components of the virion (e.g., capsid proteins), respectively.

3.3 Spectrum of Systemic CMV Disease

Initial infection is usually asymptomatic or results in a self-limited mononucleosis-like syndrome with fever, malaise, and sweats (Klemola and Kaariainen 1965). Signs of hepatitis are noted in about a third of the patients and there is less pharyngitis and only minimal cervical adenopathy. The heterophile antibody test is always negative and helps to differentiate CMV-associated IM (CMV IM) from EBV IM. Lymphocytosis with atypical cells is seen in both. Severe end organ involvement is rare in primary CMV infection in otherwise healthy hosts.

Serious CMV disease is mostly confined to immunosuppressed patients, especially AIDS, transplant, and chemotherapy patients. The disease is usually organ specific in solid organ transplants, but is often systemic in bone marrow or stem cell transplants (SCT). Active CMV infection after a transplant resembles CMV mononucleosis with evolution to involve specific organs, especially pneumonitis, hepatitis, colitis, esophagitis, gastritits, colitis, adrenalitis, and rarely encephalitis. Often the organ infected is the transplanted one, and in AIDS patients, multiple organs are often involved.

3.4 Pathology and Pathogenesis

In contrast to the multiple pathogenic processes by which EBV causes disease, the pathogenesis of direct CMV infection is much simpler, in that it mainly causes lytic infection of different types of cells. The typical CMV infected cell has a characteristic appearance (see Fig. 1), but CMV antigens can be detected in normal-appearing cells.

The initial infection occurs when virus, shed in secretions such as saliva, urine, and genital secretions, infects the naïve host. It attaches to and initially infects epithelial cells. A cell-associated viremia then ensues and the virus is deposited systemically, infecting fibroblasts, epithelial cells, endothelial cells, and smooth muscle cells (Sinzger et al. 1995). Viral antigen can be detected in multiple organs, including the brain, even in asymptomatic patients (Toorkey and Carrigan 1989). The virus latently infects myeloid precursor cells, from $CD34^+$ pluripotent stem cells to $CD14^+$ monocytes. When the latter enter visceral parenchyma and differentiate into macrophages and myeloid dendritic cells, the latent infection reactivates into a lytic one, with lytic infection of and damage to the surrounding

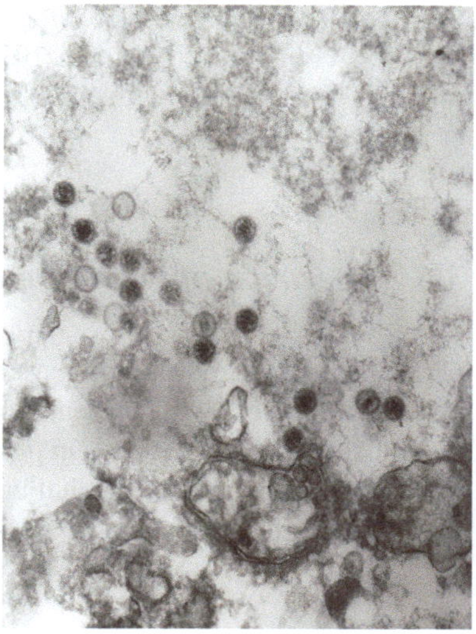

Fig. 1 Epstein–Barr virions seen in this transmission electron micrograph. Courtesy of Dr. Fred Murphy, CDC, CDC Public Health Image Library

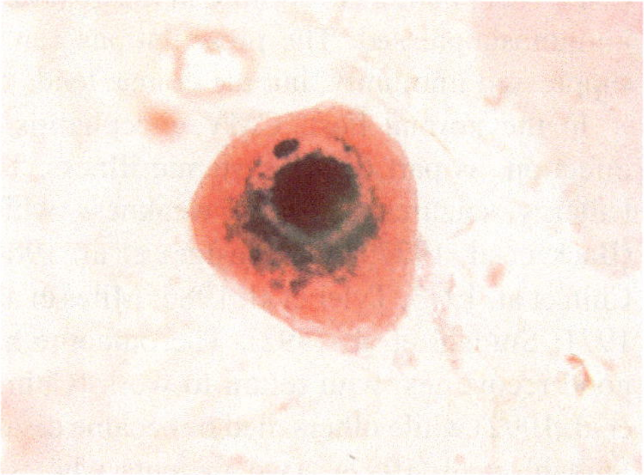

Fig. 2 Cytomegalic cell in urine. Courtesy of Dr. Haraszti, CDC, CDC Public Health Image Library

parenchyma. However, T cell immunity develops and active infection is suppressed.

CMV can "reactivate" periodically with nonspecific changes in CMV antibody titers and shedding of virus in saliva, urine, genital secretions, or even in the circulation. Thus, the virus can potentially spread through day care centers, caregivers, organ and blood recipients, and sexual partners. Known specific triggers of reactivation include radiation, allogeneic stimulation, TNFalpha, and cytotoxic drugs. In a murine model, CMV was reactivated in an allogeneic but not in a syngeneic kidney transplant (Hummel and Abecassis 2002). This was also noted in bone marrow transplant patients. In a study of 100 bone marrow transplants (BMT) between syngeneic identical twins, no CMV pneumonia was noted, whereas this occurred in 20 % of allogeneic pairs (Applebaum et al. 1982).

In the early transplant patients, pathologic examination of the brain showed scattered microglial nodules that were attributed to CMV encephalitis (Schober and

Herman 1973; Schneck 1965; Hotson and Pedley 1976). Inclusion-bearing cells are seen less commonly (Dorfman 1973). In patients with more severe immune suppression, for example with AIDS or transplants, ventriculitis was seen (Morgello et al. 1987).

3.5 Spectrum of Neurologic CMV Disease

CMV can affect the nervous system at all levels, from the hemispheres to the peripheral nerves, with presentations reflecting the pattern of anatomic involvement. Clinically, the patient can present with a febrile encephalopathy, myelopathy, optic neuropathy, psychosis, hallucinations, hemiplegia with headache, brainstem involvement, locked-in syndrome—the entire panoply of neurologic syndromes.

3.5.1 Encephalitis

CMV encephalitis is very rare in the general population and uncommon even in the immunosuppressed. The presentations can be similar in patients with intact and suppressed immunity, but the course tends to be more severe in the latter.

In the normal host, CMV encephalitis usually occurs during primary CMV infection, as part of the systemic illness. The illness consists of headache, fever, lethargy, seizures, and focal weakness, which is typical for any viral encephalitis (Back et al. 1977; Siegman-Igra et al. 1984; Dorfman 1973; Philips et al. 1977; Chin et al. 1973; Tyler et al. 1986; Miles et al. 1993; Waris et al. 1972; Perham et al. 1971; Studahl et al. 1992). The outcome has been variable. Several patients had good recoveries, with return to work (Chin et al. 1973; Back et al. 1977; Studahl et al. 1992) while others died or became disabled (Waris et al. 1972; Dorfman 1973; Studahl et al. 1992). Two patients who were treated with vidarabine recovered (Philips et al. 1977). A pregnant patient with CMV encephalitis made a complete recovery after treatment with acyclovir. A case of systemic primary CMV infection with multiple end organ involvement, including encephalitis, resolved completely after acyclovir therapy (Khattab et al. 2009).

Other unusual presentations of CMV encephalitis have been reported in the immunocompetent population. A rare form of CMV encephalitis with opsoclonus–myoclonus, treated with ganciclovir, steroids, and immunoglobulin has been reported. The patient recovered (Zaganas et al. 2007). Recently, a "paroxysmal" form of CMV encephalitis has been reported in the literature. In this condition, neurologic deficits lasting a few hours occur and then resolve, to be repeated over a week or so. The outcome appears to be benign, irrespective of whether patients are treated with antiviral drugs (Chalaupka Devetag and Boscariolo 2000; Richert et al. 1987).

In the AIDS patient, CMV encephalitis tends to present somewhat more indolently, with the first symptoms often noted only in retrospect (Arribas et al. 1996).

There are two recognizable presentations, mirroring to some extent the pathological findings. In the first, there is a syndrome of a flat affect, confusion and disorientation, lethargy, withdrawal, and apathy, which can be difficult to distinguish from HIV dementia (Holland et al. 1994). The pathology in these cases is that of diffuse microglial nodules in the brain parenchyma. The second type of CMV encephalitis begins in the same way, but multiple cranial nerves become involved, especially with nystagmus and facial palsy (Kalayjian et al. 1993). Often the patients have hypo- or hypernatremia (probably reflecting a concurrent CMV adrenalitis or possibly diencephalic involvement). Such patients have ventriculitis on MRI, and the CSF characteristically has a neutrophlic pleocytosis with hypoglycorrhachia. I have personally seen a case of AIDS-associated CMV encephalitis in which the CSF glucose was 0 mg/dL (confirmed on repeat testing). The prognosis appears to be rather poor, with a median survival of 42 days, irrespective of whether the patients were treated with antiviral drugs (Arribas et al. 1996). More recently, an open label study of a combination of both ganciclovir and foscarnet showed a median survival of 94 days in the participants, and when two patients were put on highly active antiretroviral therapy (HAART), they were able to survive beyond the study, off anti-CMV drugs (Anduze-Fafri et al. 2000). Finally, a case of AIDS-associated CMV encephalitis appearing after HAART was instituted was reported. The CD4 T cell count was low and the HIV viral load high. Ten days later, he had a headache and the CSF showed a mild pleocytosis with a high proportion of neutrophils. CMV PCR was positive. An MRI showed enhancement of the ependyma, typical of CMV ventriculitis. He was treated with ganciclovir and foscarnet with improvement. The CSF CMV PCR became negative and his symptoms resolved. He was given valganciclovir for maintenance therapy until there was complete immune recovery, and then discontinued. He had no recurrence to a follow-up 16 months later. This was most likely an immune reconstitution inflammatory syndrome (IRIS) causing a flare up of CMV ventriculitis (Janowicz et al. 2005).

A study of the natural history of AIDS-associated CMV encephalitis in the HAART era would be very valuable.

CMV encephalitis was reported early in the transplant era and had a poor prognosis (Dorfman 1973; Schober and Herman 1973; Hotson and Pedley 1976; Schneck 1965). In transplant patients, CMV is an important cause of systemic disease and patients are often put on prophylactic or preemptive antiviral drugs such as acyclovir or ganciclovir for several months after the transplant. This has reduced systemic CMV considerably but did not completely eliminate it (Ljungman 2002). Indeed, CMV encephalitis can occur in patients already on both ganciclovir and foscarnet for CMV viremia ("preemptive" treatment) (Seo et al. 2001). This is true especially for stem cell transplant recipients, who may develop CMV encephalitis late after transplant, and seem to have a poor prognosis despite treatment with various combinations of ganciclovir, foscarnet, and cidofovir (Reddy et al. 2010). This may be in part due to the emergence of resistance mutations during prolonged prophylactic or preemptive treatment.

3.5.2 Polyradiculopathy and Mononeuropathy Multiplex

CMV has been implicated as a potential cause of GBS, characterized by rapidly progressively ascending flaccid weakness. In a survey of the etiologies of inflammatory neurologic disorders, two patients with GBS were shown to be linked to CMV by CMV complement fixation seroconversion and in one patient, isolation of CMV from the urine, followed by the detection of cytomegalic cells in the urine (Klemola et al. 1967). In a similar study, ten patients with GBS (one of whom had Miller-Fisher variant) were found to have CMV IgM seroconversion (Schmitz and Enders 1977).

A superficially similar syndrome has been seen in patients with advanced AIDS except that it is due to direct infection of nerve roots and peripheral nerves. It is characterized by subacutely progressive lower extremity pain and paresthesias, flaccid weakness, and urinary retention with ascending weakness, reflecting progression from polyradiculopathy to necrotizing myelopathy. CSF often shows a neutrophilic pleocytosis with hypoglycorrhachia and is positive for CMV by PCR. EMG shows denervation changes and MRI demonstrates enhancing nerve roots (Bazan et al. 1991; Talpos et al. 1991).

CMV mononeuropathy multiplex is a rare complication seen in AIDS patients, in which there is multifocal sensory and motor loss, with progression to severe painful sensorimotor neuropathy. CSF is usually positive for CMV by PCR and EMG demonstrates the typical findings of a mononeuropathy multiplex. Sometimes, demyelination is prominent (Roullet et al. 1994; Morgello and Simpson 1994).

3.5.3 Pathogenetic Model of CMV Infection of the Nervous System

A pathogenetic model of CMV infection of the nervous system has been proposed as a way of summarizing the evolution of the disease (Tselis and Lavi 2000). The pattern of disease involvement in the CSN is combined with the severity of infection and summarized as follows:

1. Diffuse multifocal CMV encephalitis (CVE)
   a. Isolated inclusion-bearing cells
   b. Microglial nodule encephalitis
   c. Focal parenchymal necrosis
2. CMV ventriculoencephalitis
   a. Ependymitis
   b. Ependymitis and subependymitis
   c. CVE with necrotizing periventricular lesions
3. CMV radiculomyelitis
   a. CMV polyradiculitis
   b. Necrotizing radiculomyelitis

Inspection of this pattern suggests routes of access of virus to the nervous system: through the blood–brain barrier in parenchymal blood vessels, choroid plexus, and nerve roots, respectively, with the degree of infection depending on the viral inoculum.

3.6 Diagnosis

Diagnosis of CMV encephalitis is made on the basis of a compatible clinical picture and demonstration of CMV in the CSF. This has been validated in the HIV population, and is commonly used in other immunosuppressed patients such as in transplantation. In the AIDS population, CSF viral loads correlate to some extent with the extent and severity of encephalitis (Arribas et al. 1995). In the critically ill patient, it is important to consider other diagnostic possibilities such as seizures, septic encephalopathy, and effects of medications such as cyclosporine. Serologic methods, such as increase in titers of CMV antibody, are not useful.

3.7 Management

The currently available antiviral drugs that act against CMV are ganciclovir, foscarnet, and cidofovir. These have been shown to treat CMV retinitis in AIDS patients and their use in CMV encephalitis and radiculomyelitis has been an extrapolation.

Monotherapy seems not to affect the course of AIDS-associated CMV encephalitis (Arribas et al. 1996). The use of combination therapy with ganciclovir and foscarnet is probably more effective, although not ultimately curative (Anduze-Faris et al. 2000). The dose of ganciclovir was 5 mg/kg twice a day and foscarnet 90 mg/kg twice a day for an induction period of 3–6 weeks, followed by a maintenance phase of once daily dosing for both drugs. However, both drugs are rather toxic and the patient needs to be followed closely for bone marrow suppression (ganciclovir) and nephrotoxicity (foscarnet). Cidofovir has unreliable CNS penetration, and is not recommended in the IDSA guidelines (Tunkel et al. 2008). There is preliminary evidence that immune reconstitution from HAART therapy may allow long-term survival off anti-CMV drugs. There is even less data to guide the use of these drugs in the non-AIDS population. In the normal host, CMV encephalitis is often followed by disability, although a number of patients seem to recover well without anti-CMV medications. It is reasonable to treat with these drugs and follow the patients very closely for toxicity.

3.8 Summary and Conclusions

EBV and CMV are human gamma and beta herpesviruses that cause universal infection, usually self-limited. However, they are occasionally the cause of severe neurological syndromes. Despite the similarity of these viruses their effects are due to very different pathogeneses, EBV is primarily immunopathogenic and thus indirectly damaging whereas CMV causes more direct lytic infection. These viruses are more dangerous in the immunosuppressed, and are of increasing interest given the use of strongly immunosuppressing and immunomodulating agents. Despite a great deal of research and knowledge, we must still turn to clinical research to understand the natural history of the disease and test therapeutic modalities.

References

Abul-Kasim K, Palm L, Maly P, Sundgren P (2009) Neuroanatomic localization of Epstein-Barr virus encephalitis may be a predictive factor for its clinical outcome: a case report and a review of 108 cases in 28 reports. J Child Neurol 24:720–726

Adamson D, Gordon P (1992) Hemiplegia: a rare complication of acute Epstein-Barr virus (EBV) infection. Scand J Infect Dis 24:379–380

Anduze-Fafri B, Fillet A, Gozlan J, Lancar R, Boukli N, Gasnault J et al (2000) Induction and maintenance therapy of CMV central nervous system infection in HIV infected patients. AIDS 14:517–524

Anduze-Faris B, Fillet A, Gozlan J, Lancar R, Boukli N, Gasnault J et al (2000) Induction and maintenance therapy of CMV central nervous system infection in HIV infected patients. AIDS 14:517–524

Angelini L, Bugiani M, Zibordi F, Cinque P, Bizzi A (2000) Brainstem encephalitis resulting from Epstein-Barr virus mimicking an infiltrating tumor in a child. Pediatr Neurol 22(2):130–132

Applebaum F, Meyers J, Fefer A, Flourney N et al (1982) Nonbacterial nonfungal pneumonia following bone marrow transplantation in 100 identical twins. Transplantation 33:265–268

Arribas J, Clifford D, Fichtenbaum C, Commons D et al (1995) Level of cytomegalovirus (CMV) DNA in cerebrospinal fluid of subjects with AIDS and CMV infection of the central nervous system. J Infect Dis 172:527–531

Arribas J, Storch G, Clifford D, Tselis A (1996) Cytomegalovirus encephalitis. Ann Int Med 125:577–587

Ashworth J, Motto S (1947) Infectious mononucleosis complicated by papilloretinal edema. N Engl J Med 237:544–545

Back E, Hoglund C, Malmlund H (1977) Cytomegalovirus infection associated with severe encephalitis. Scand J Infect Dis 9:141–143

Baker F, Kotchmar G, Foshee W, Sumaya C (1983) Acute hemiplegia of childhood associated with Epstein-Barr virus infection. Pediatr Infect Dis J 2:136–138

Bazan C, Jackson C, Jinkins J, Barohn R (1991) Gadolinium enhanced MRI in a case of cytomegalovirus polyradiculopathy. Neurology 41:1522–1523

Bennett D, Peters H (1961) Acute cerebellar syndrome secondary to infectious mononucleosis in a fifty-two year old man. Ann Intern Med 55(1):147–149

Bergen D, Grossman H (1975) Acute cerebellar ataxia of childhood associated with infectious mononucleosis. J Pediatr 87:832–833

Bernstein T, Wolfe H (1950) Involvement of the nervous system in infectious mononucleosis. Ann Intern Med 33:1120–1138

Biebl A, Webersinke C, Traxler B, Povysil B, Furthner D, Schmitt K (2009) Fatal Epstein-Barr virus encephalitis in a 12 year old child: an underappreciated neurological complication? Nature Clin Pract Neurol 5:171–174

Blaustein A, Caccavo A (1950) Infectious mononucleosis complicated by bilateral papilloretinal edema. Arch Ophthalmol 43:853–856

Bonynge T, Van Hagen K (1952) Severe optic neuritis in infectious mononucleosis. J Am Med Assoc 145:933–934

Caruso J, Tung G, Gascon G, Rogg J, Davis L, Brown W (2000) Persistent preceding focal neurologic deficits in children with Epstein-Barr encephalitis. J Child Neurol 15:791–796

Chalaupka Devetag F, Boscariolo L (2000) Cytomegalovirus meningoencephalitis with paroxysmal course in immunocompetent adults: a new nosographical entity. Clinical, diagnostic, and therapeutic correlations and pathogenetic hpothesis. Eur Neurol 44:242–247

Chin W, Magoffin R, Frierson J, Lennette E (1973) Cytomegalovirus infection: a case with meningoencephalitis. JAMA 225:740–741

Cleary T, Henle W, Pickering L (1980) Acute cerebellar ataxia associated with Epstein-Barr virus infection. J Am Med Assoc 243:148–149

Clevenbergh R, Brohee P, Velu T, Jacobs E, Liesnard C, Deneft E, Thys J (1997) Infectious mononucleosis complicated by transverse myelitis: detection of the viral genome by polymerase chain reaction in the cerebrospinal fluid. J Neurol 244:592–594

Copperman S (1977) "Alice in Wonderland" syndrome as a presenting symptom of infectious mononucleosis in children. Clin Pediatr 16:143–146

Cotton P, Webb-Peploe M (1966) Acute transverse myelitis as a complication of glandular fever. Br Med J 1:654–655

Dimova P, Bojinova V, Georgiev D, Milanov I (2006) Acute reversible parkinsonism in Epstein-Barr virus-related encephalitis lethargica-like illness. Mov Disord 21:564–566

Dorfman L (1973) Cytomegalovirus encephalitis in adults. Neurology 23:136–144

Egan R (1960) Facial diplegia in infectious mononucleosis in the absence of Landry–Guillain–Barre syndrome. N Engl J Med 262(23):1178–1179

Eshel G, Eyov A, Lahat E, Brauman A (1987) Alice in Wonderland syndrome, a manifestation of acute Epstein-Barr virus infection. Pediatr Infect Dis J 6:68

Foss H, Herbst H, Hummel M, Araujo I, Latza U, Rancso C, Dallenbach F, Stein H (1994) Patterns of cytokine gene expression in infectious mononucleosis. Blood 83:707–712

Friedland R, Yahr M (1977) Meningoencephalopathy secondary to infectious mononucleosis unusual presentation with stupor and chorea. Arch Neurol 34:186–188

Fujimoto J, Asaoka J, Imaizumi T, Ayabe M, Shoji H, Kaji M (2003) Epstein-Barr virus infection of the central nervous system. Intern Med 42:33–40

Garamendi I, Montejo M, Cancelo L, Lopez L et al (2002) Encephalitis caused by Epstein-Barr virus in a renal transplant recipient. Clin Infect Dis 34:287–288

Gilbert J, Culebras A (1972) Cerebellitis in infectious mononucleosis. J Am Med Assoc 220(5):727

Greenberg D, Weinkle D, Aminoff M (1982) Periodic EEG complexes in infectious mononucleosis encephalitis. J Neurol Neurosurg Psychiatry 45:648–651

Goodpasture E, Talbot F (1921) Concerning the nature of "protozoan-like" cells in certain lesions of infancy. Am J Dis Child 21:415–421

Grose C, Feorino P (1973) Epstein-Barr virus and transverse myelitis. Lancet 1:892

Grose C, Feorino P, Dye L, Rand J (1973) Bell's palsy and infectious mononucleosis. Lancet 2:231–232

Henle G, Henle W, Diehl V et al (1968) Relation of Burkitt's tumor-associated herpes-type virus to infectious mononucleosis. Proc Natl Acad Sci 59:94–101

Henle W, Henle G, Horwitz C (1974) Epstein-Barr virus specific diagnostic tests in infectious mononucleosis. Hum Pathol 5:551–565

Ho M (2008) The history of cytomegalovirus and its diseases. Med Microbiol Immunol 197:65–73

Hochberg F, Miller G, Schooley R, Hirsch M, Feorino P, Henle G (1983) Central nervous system lymphoma related to Epstein-Barr virus. N Engl J Med 309:745–748

Holland N, Power C, Mathews V, Glass J et al (1994) CMV encephalitis in acquired immunodeficiency syndrome. Neurology 44:507–514
Hotson J, Pedley T (1976) The neurological complications of cardiac transplantation. Brain 99:673–694
Hummel M, Abecassis M (2002) A model for reactivation of CMV from latency. J Clin Virol 25:s123–s136
Hurt C, Tammaro D (2007) Diagnostic evaluation of mononucleosis-like illness. Amer J Med 120:911.e911–911.e918
Janowicz D, Johnson R, Gupta S (2005) Successful treatment of CMV ventriculitis immune reconstitution syndrome. J Neurol Neurosurg Psychiatry 76:691–692
Jesionek A, Kiolemenoglou R (1904) Uber einen Befund von protoenartigan…. Muench Med Wochenschr 51:1905–1907
Jones K, Rivera C, Sgadari C, Franklin J, Max E, Bhatia KG, Tosato G (1995) Infection of human endothelial cells with Epstein-Barr virus. J Exp Med 182:1213–1221
Johansen A (1931) Serous meningitis and infectious mononucleosis. Acta Med Scand 76:269
Kalayjian R, Cohen M, Bonomo R, Flanigan T (1993) Cytomegalovirus ventriculoencephalitis in AIDS: a syndrome with distinct clinical and pathological features. Medicine 72:67–77
Khattab M, Eslam M, Abd-eelfattah E (2009) Encephalitis, acute renal failure, and acute hepatitis caused by a viral infection in an immunocompetent young adult: a case report. J Med Case Rep 3:9289
Klemola E, Kaariainen L (1965) Cytomegalovirus as a possible cause of a disease resembling infectious mononucleosis. Br Med J 2:1099–1102
Klemola E, Weckman N, Haltia K, Kaariainen L (1967) The Guillain-Barre syndrome associated with acquired cytomegalovirus infection. Acta Med Scand 181:603–607
Kuo Y, Chiu N, Shen C et al (1998) Cerebral perfusion in children with Alice in Wonderland syndrome. J Child Neurol 19:105–108
Lahat E, Berkovitch M, Barr J, Peret G, Barzilai A (1999) Abnormal visual evoked potentials in children in "Alice in Wonderland" syndrome due to infectious mononucleosis. J Child Neurol 14:732–735
Larocca L, Capello D, Rinelli A, Nori S, Antinori A et al (1998) The molecular and phenotypic profile of primary central nervous system lymphoma identifies distinct categories of the disease and is consistent with histogenetic derivation from germinal center-related B-cells. Blood 92:1011–1019
Lascelles R, Johnson P, Longson M, Chiang A (1973) Infectious mononucleosis presenting as acute cerebellar syndrome. Lancet 2:707–709
Leavell R, Ray CG, Ferry P, Minnich LL (1986) Unusual acute neurological presentations with Epstein-Barr virus infection. Arch Neurol 43:186–188
Ljungman P (2002) Beta-Herpesvirus challenges in the transplant recipient. J Infect Dis 186:S99–S109
McKendall R, Sadiq S, Calverley J (1990) Unusual manifestations of Epstein-Barr virus encephalomyelitis. Infection 18(1):33–35
Miles C, Hoffman W, Lai C-W, Freeman J (1993) Cytomegalovirus-associated transverse myelitis. Neurology 43:2143–2145
Mischler M, Fleming G, Shanley T, Madden L, Levine J, Castle V, Filipovich A, Cornell T (2006) Epstein-Barr virus-induced hemophagocytic lymphohistiocytosis and X-linked lymphoproliferative disease: a mimicker of sepsis in the pediatric intensive care unit. Pediatrics 119:e1212–e1218
Mizuno T, Takanashi Y, Onodera H et al (2003) A case of lymphomatoid granulomatosis/angiocentric immunoproliferative lesion with long clinical course and diffuse brain involvement. J Neurol Sci 213:67–76
Morgello S, Simpson D (1994) Multifocal cytomegalovirus demyelinative polyneuropathy associated with AIDS. Muscle Nerve 17:176–182

Morgello S, Cho E-S, Nielsen S, Devinsky O, Petito C (1987) Cytomegalovirus encephalitis in patients with acquired immunodeficiency syndrome: an autopsy study of 30 cases and a review of the literature. Hum Pathol 18:289–297

Niederman J, McCollum R, Henle G, Henle W (1968) Infectious mononucleosis clinical manifestations in relation to EB virus antibodies. JAMA 203:139–142

North K, de Silva L, Procopis P (1993) Brain-stem encephalitis caused by Epstein-Barr virus. J Child Neurol 8:40–42

Paul J, Bunnell W (1932) The presence of heterophile antibodies in infectious mononucleosis. Am J Med Sci 183:191–194

Perham T, Caul E, Clarke S, Gibson A (1971) Cytomegalovirus meningoencephalitis. Br Med J 2:50

Philips C, Fanning L, Gump D, Phillips C (1977) Cytomegalovirus encephalitis in immunologically normal adults: successful treatment with vidarabine. JAMA 238:2299–2300

Phowthongkum P, Phantumchinda K, Jutivorakool K, Suankratay C (2007) Basal ganglia and brainstem encephalitis, optic neuritis, and radiculomyelitis in Epstein-Barr virus infection. J Infect 54:e141–e144

Reddy S et al (2010) CMV central nervous system disease in stem-cell transplant recipients: an increasing complication of drug-resistant CMV infection and protracted immunodeficiency. Bone Marrow Transplant 45:979–984

Richert J, Potolicchio S, Garagusi V, Manz H, Cohan S, Hartmann D, Johnson R (1987) Cytomegalovirus encephalitis associated with episodic neurologic deficits and OKT-8+ pleocytosis. Neurology 37:149–152

Riley H Jr (1997) History of the cytomegalovirus. South Med J 90:184–190

Roselli F, Russo L, Fraddosio A, Aniello M, De Mari M, Lamberti P, Livrea P, Defazio G (2006) Reversible parkinsonian syndrome associated with anti-neuronal antibodies in acute EBV encephalitis: a case report. Parkinsonism Relat Disord 12:257–260

Roullet E, Assuerus V, Gozlan J, Ropert A, Said G, Baudriment M et al (1994) Cytomegalovirus multifocal neuropathy in AIDS: analysis of 15 consecutive cases. Neurology 44:2174–2182

Russell J, Fisher M, Zivin J, Sullivan J, Drachman D (1985) Status epilepticus and Epstein-Barr virus encephalopathy diagnosis by modern serologic techniques. Arch Neurol 42:789–792

Rowe W, Hartley J, Waterman S, Al E (1956) Cytopathogenic agent resembling human salivary gland virus recovered from tissue cultures of human adenoids. Proc Soc Exp Biol Med 92:418–424

Schmahmann J (2004) Plasmapheresis improves outcome in postinfectious cerebellitis induced by Epstein-Barr virus. Neurology 62:1443

Schmitz H, Enders G (1977) Cytomegalovirus as a frequent cause of Guillain Barre syndrome. J Med Virol 1:21–27

Schneck S (1965) Neuropathologic features of human organ transplantation I. Probable cytomegalovirus infection. J Neuropathol Exp Neurol 24:415–429

Schober R, Herman M (1973) Neuropathology of cardiac transplantation. Survey of 31 cases. Lancet 1:962–967

Seo S, Regan A, Cihlar T, Lin D, Boulad F, George D, Prasad V, Kiehn T, Polsky B (2001) Cytomegalovirus ventriculoencephalitis in a bone marrow transplant recipient receiving antiviral maintenance: clinical and molecular evidence of drug resistance. Clin Infect Dis 33:e105–e109

Shian W, Chi C (1994) Fatal brainstem encephalitis caused by Epstein-Barr virus. Pediatr Radiol 24:596–597

Shoji H, Goto Y, Yanase Y, Sato Y, Nakashima K, Natori H, Kaji M (1983) Recurrent cerebellitis. A case report of a possible relationship with Epstein-Barr virus infection. Kurume Med J 30:23–26

Siegman-Igra Y, Michaeli D, Doron A, Weinberg M, Heilbrun Y (1984) Cytomegalovirus encephalitis in a noncompromised host. Isr J Med Sci 20:163–166

Silverstein A, Steinberg G, Nathanson M (1972) Nervous system involvement in infectious mononucleosis. The heralding and/or major manifestation. Arch Neurol 26:353–358
Sinzger C, Grefte A, Plachter B, Gouw A, The T, Jahn G (1995) Fibroblasts, epithelial cells and smooth muscle cells are major targets of human cytomegalovirus infection in lung and gastrointestinal tissues. J Gen Virol 76:741–750
Smith M (1956) Propagation in tissue cultures of a cytopathic virus from human salivary gland virus (SGV) disease. Proc Soc Exp Biol Med 92:242–430
Sprunt T, Evans F (1920) Mononuclear leucocytosis in reaction to acute infections ("infectious mononucleosis"). Johns Hopkins Hospital Bulletin 357:410–416
Studahl M, Ricksten A, Sandburg T, Burgstrom T (1992) Cytomegalovirus encephalitis in four immunocompetent patients. Lancet 340:1045–1046
Talpos D, Tien R, Hesselink J (1991) Magnetic resonance imaging of AIDS-related polyradiculopathy. Neurology 41:1996–1997
Taylor L, Parsons-Smith G (1969) Infectious mononucleosis, deafness and facial paralysis. J Laryngol Otol 83:613–616
Toorkey C, Carrigan D (1989) Immunohistochemical detection of an immediate early antigen of human cytomegalovirus in normal tissues. J Infect Dis 160:741–751
Tselis A, Lavi E (2000) Cytomegalovirus in the adult nervous system. Infectious diseases of the nervous System LE Davis and PGE Kennedy. Butteworth–Heinemann, Boston
Tselis A, Duman R, Storch G, Lisak R (1997) Epstein-Barr virus encephalomyelitis diagnosed by polymerase chain reaction: detection of the genome in the CSF. Neurology 48:1351–1355
Tunkel A, Glaser C, Bloch K, Sejvar J, Marra C, Roos K, Hartman H, Kaplan S, Scheld W, Whitley R (2008) The management of encephalitis: clinical practice guidelines by the Infectious Diseases Society of America. Clin Infect Dis 47:303–327
Tyler K, Gross R, Cascino G (1986) Unusual viral causes of transverse myelitis: hepatitis A and cytomegalovirus. Neurology 36:855–858
Verma A, Brozman B (2002) Opsoclonus-myoclonus syndrome following Epstein-Barr virus infection. Neurology 58:1131–1132
Von Glahn W, Pappenheimer A (1925) Intranuclear inclusions in visceral disease. Am J Pathol 1:445–465
Waris E, Rasanen O, Kreus K-E, Kreus R (1972) Fatal cytomegalovirus disease in a previously healthy adult. Scand J Infect Dis 4:61–67
Weinberg A, Li S, Palmer M, Tyler K (2002) Quantitative CSF PCR in Epstein-Barr virus infections of the central nervous system. Ann Neurol 52:543–548
Weinberg A, Bloch K, Li S, Tang Y, Palmer M, Tyler K (2005) Dual infections of the central nervous system with Epstein-Barr virus. J Infect Dis 191:234–237
Weller T, Macauley J, Craig J, Wirth P (1957) Isolation of intranuclearinclusion producing agents from infants with illnesses resembling cytomegalic inclusion disease. Proc Soc Exp Biol Med 94:4–12
Weller T (1970) Cytomegalovirus. The difficult years. J Infect Dis 122:532–539
Weller T (2000) Cytomegalovirus: a historical perspective. Herpes 7:66–69
Wyatt J, Saxton J, Lee R, Al E (1950) Generalized cytomegalic inclusion disease. J Pediat 36:271–294
Zaganas I, Prinianakis G, Xirouchaki N, Mavridis M (2007) Opsoclonus-myoclonus syndrome associated with cytomegalovirus encephalitis. Neurology 68:1636
Zaidi A, Kampalath B, Peltier W, Viesole D (2004) Successful treatment of systemic and central nervous system lymphomatoid granulomatosis with rituximab. Leuk Lymphoma 45:777–780

Herpes Simplex Virus Meningoencephalitis

Israel Steiner

Abstract Herpes simplex virus types 1 and 2 (HSV-1 and 2) are human neutropic viruses that establish latent infection in dorsal root ganglia for the entire life of the host. From this reservoir they can reactivate to cause human morbidity and mortality. Herpes simplex encephalitis (HSE) is one of the most devastating disorders caused by these viruses. The biology of their ability to establish latency, maintain it for the entire life of the host, reactivate, and cause primary and recurrent disease is being studied in animal models and in humans. Of special interest is the question whether HSE is the result of primary infection or is it the outcome of reactivation? The present review covers the biological, medical, and neurological aspects HSE, focusing on recent molecular findings of gene expression during latent infection of HSV-1. Despite accumulating knowledge, there are still several issues regarding both pathogenesis and therapy of HSV-1 that currently defy understanding.

Keywords Encephalitis • Gene therapy • Herpes simplex virus • Latency • Nervous system • Reactivation

1 Introduction

Herpes simplex virus types 1 and 2 (HSV-1 and 2) are important human pathogens. The name, herpes, is derived from the ancient Greek where it means to creep or crawl (Roizman and Whitley 2001). Herpetic disease might have first been described by Hippocrates (Cumston 1926), and genital herpes was reported in the eighteenth century in France of Louis XV, in prostitutes who were, at that time, under medical supervision (Astruc 1736).

I. Steiner (✉)
Department of Neurology, Rabin Medical Center, Beilinson Campus, 49100 Petach Tikva, Israel
e-mail: isteiner@cc.huji.ac.il

A.C. Jackson (ed.), *Viral Infections of the Human Nervous System*, Birkhäuser Advances in Infectious Diseases,
DOI 10.1007/978-3-0348-0425-7_3,

The nature of the disorders caused by these viruses has been elucidated in part. Research milestones, such as proving the infectious nature of the pathogens, understanding the biological and molecular concepts of latency and reactivation, devising specific antiviral agents, and sequencing the entire viral genomes, were all products of the biological and molecular revolution of the last century. Though herpes infections do not attract the same level of public attention that they did prior to the human immunodeficiency virus (HIV) and acquired immune deficiency syndrome (AIDS) era (Time 1982), they are still of major medical importance in the realms of neurology, dermatology, ophthalmology, pediatrics, infectious diseases, and obstetrics and gynecology as well as gene therapy to the human nervous system.

2 Structure

HSV-1 and HSV-2 are members of the herpesviridae family of DNA viruses. The genome is a double-stranded DNA molecule located within an icosapentahedral capsid which consists of 162 capsomers. The capsid is surrounded by an amorphous material called the tegument which is surrounded by an envelope that consists of polyamines, lipids, and glycoproteins. These glycoproteins confer distinctive properties to each virus and provide the antigens to which the host is capable of responding. HSV-1 and HSV-2 are closely related, with nearly 70 % genomic homology. These two viruses can be distinguished most reliably by DNA composition.

3 Primary Infection

HSV infects the human host via mucosal surfaces or damaged skin. Primary infection is in most instances asymptomatic and depends on the immunological condition of the host. HSV-1 and -2 tend to be transmitted via different routes. HSV-2 is usually the cause of genital herpes (Nahmias et al. 1990), whereas HSV-1 is typically transmitted during childhood via the orolabial mucocutaneous surfaces. There are, however, several exceptions. HSV-1 has become a principal causative agent of genital herpes in some developed countries (Nilsen and Myremel 2000; Tran et al. 2004), including the United States (Roberts et al. 2003), and HSV-2 can also be the cause of recurrent herpes labialis (Buxbaum et al. 2003). In less developed countries, HSV-1 primary infection with seroconversion usually takes place within the first two decades of life. In developed countries about half of seroconversion occurs at 20–40 years of age. HSV-1 seropositivity has reached 80–90 % worldwide (Smith and Robinson 2002).

4 Latency

Latent infection means that the genome of the pathogen is present in the host tissue without production of infective particles (Steiner and Kennedy 1995). During latency the pathogen maintains the potential to reactivate and to resume replication causing recurrent disease.

The neurotropic herpes viruses establish latent infection in dorsal root ganglia (DRG), and both latent HSV and latent varicella zoster virus (VZV) reside primarily in the trigeminal ganglia and in DRG. Although there are conspicuous clinical, cellular, and molecular differences between HSV-1 and -2 and VZV latency and reactivation, they share several features that govern the biology of their infection in the human nervous system and have important bearing upon latent infection in the human nervous system: (1) primary infection involves mucocutaneous surfaces acting as the portal of entry of viral particles into the peripheral nervous system; (2) primary and recurrent diseases caused by the same virus usually occur within the same cutaneous distribution; (3) under immunocompetent conditions, reactivation usually does not spread beyond the anatomic distribution and the vicinity of a single dorsal root ganglion, e.g., a dermatome; (4) although primary infection usually takes place during the first two to three decades of life, reactivation may occur at any time in the patient's life, though HSV-1 reactivations tend to decrease with age, attributed to depletion of viral genome copy number over the years.

These features can be grouped under a unifying hypothesis in herpes virology (Goodpasture 1929; Hope-Simpson 1965). Following primary infection, the virus gains access to axon endings within the mucocutaneous surfaces and is transported to the DRG where it is maintained in a latent state for the entire life of the human host. Under certain circumstances the virus can reactivate and travel to regions innervated by the respective DRG, causing recurrent disease.

During HSV-1 replication in cells in culture, the infected cell is destroyed, but replication in the individual neuron at the peripheral site of primary infection or in the DRG is not mandatory for the establishment of latency (Steiner et al. 1990). On the contrary, lack of replication within neurons carries obvious advantages for the virus because destruction of the host cell would prevent its ability to reside within this cell and establish a latent infection there (Steiner and Kennedy 1991). Viral particles are transported from the peripheral site of primary infection by retrograde axonal transport to the DRG, where the viral DNA genome establishes latent infection in neuronal cells in a nonintegrated (episomal) form (Rock and Fraser 1983). During latency in human DRG, restricted HSV-1 gene expression takes place (Steiner et al. 1988) to produce two colinear latency-associated transcripts (LATs), which are 2.0 and 1.5 kb in size, that accumulate in ganglion cells in DRG and so far have not been shown to be translated although they bind to polyribosomes (Goldenberg et al. 1997). The 2.0-kb LAT is a stable intron and the 1.5 kb is its splicing product. The structure and function of these transcripts have been studied using latency models in rodents and HSV-1 mutants that are unable to express the latency-associated gene. Theoretically latency requires the following functions: (1) prevention of host cell destruction during primary infection as lytic infection will not leave a viable host cell to maintain latency; (2) diverting viral replication cycle into a latent state; (3) maintenance of latency for the entire life of the host; and (4) reactivation. By and large these requirements have been mapped to the region on the HSV-1 genome that codes for the LATs including reactivation efficacy (Steiner et al. 1989). This phenotype has been attributed to improved ability of LAT-expressing viruses to establish latent infection. The LATs also have anti-apoptotic features and maintain latency by promoting the survival of infected neurons (Perng and Jones 2010). Recently it was shown that LAT function is mediated, at least in part, via microRNA (Shen et al. 2009; Tang et al. 2011), noncoding RNAs that participate in gene regulation. Like HSV-1, HSV-2 is also transcriptionally active during latency, and a mutant lacking HSV-2 LATs has a defective reactivation phenotype (Krause et al. 1995).

5 Reactivation

HSV reactivations, even when recurrent, are not accompanied by permanent sensory deficit. This suggests that reactivation is not associated with neuronal cell death, the usual outcome of HSV-1 replication. We have therefore suggested (Steiner and Kennedy 1991) that HSV-1 does not replicate in DRG during reactivation in a similar way to that characterized during its replication in cultured cells.

Reactivations can either be symptomatic (termed recrudescence) or asymptomatic with inadvertent transmission. In the oral cavity asymptomatic HSV-1 reactivation may exceed clinical recrudescence (Knaup et al. 2000), and asymptomatic HSV-2 shedding in seropositive individuals can occur in more than two-thirds of seropositive individuals (Wald et al. 2000).

6 Herpes Simplex Encephalitis

6.1 Epidemiology

Herpes simplex encephalitis (HSE), the most common cause of sporadic fatal viral encephalitis (Whitley 1997) has an incidence of 1–3 per million without seasonal or sex-related variability. It was associated with 70 % mortality in untreated patients and with up to 30 % mortality and a high incidence of severe and permanent neurological sequelae in treated cases (Whitley 1991). However, a recent study suggests a lower mortality rate of around 11 % and significant morbidity in about half the patients (Granerod et al. 2010). There is a bimodal age distribution of the disease: the majority of patients are either below 20 years of age or above 50 (Koskiniemi et al. 1996) with the peak between 60 and 64 years. In immunocompetent adults more than 90 % of HSE cases are due to HSV-1 (Aurelius et al. 1993). HSV-2 is responsible for 1.6–6.5 % of all HSE in adults and is typically observed in immunosuppressed individuals.

6.2 Pathogenesis

It is uncertain whether HSE is due to viral reactivation or primary infection. Theoretically, the mechanism of HSV's entry into the brain could be due to: (1) reactivation of the viral genome in the trigeminal ganglion, a natural reservoir of HSV-1 latent infection (Steiner et al. 1988), with resultant axonal spread via the trigeminal nerve into the frontal and temporal lobes (Davis and Johnson 1979); (2) in situ reactivation of the latent virus from CNS tissue where it can occasionally be identified (Fraser et al. 1981); (3) primary infection of the nervous system. Pathways for entry of HSV to the brain include both the olfactory and the trigeminal nerves and hematogenous spread. Arguments in favor of the third possibility include:

1. The finding that in at least half of the cases of HSE the encephalitis is due to a different viral strain from the one responsible for cold sores in the same individual (Whitley et al. 1982).
2. The observation that recurrent herpes labialis, which is due to reactivation from the trigeminal ganglia, rarely, if at all, results in HSE (Whitley 2004a).
3. The finding that neuronal cells expressing the latency-associated gene are protected from HSV-1 superinfection. This suggests that at least one of the hypothetical portals of entry of HSV-1, the trigeminal system, when latently infected with HSV, endows local immunity against HSV-1 reinfection (Mador et al. 2002).
4. The observation that in many cases HSE is not limited to fronto-temporal regions on one side but is rather present in both hemispheres in a bilateral fashion (Hirai et al. 2005; Baskin and Hedlund 2007).
5. Animal models show that the olfactory tract is a viable avenue for viral entry leading to focal infection in an area analogous to the medial temporal lobe in humans (Stroop and Schaefer 1986).
6. Evidence using electron microscopy that HSV-1 particles are present within nerve tracts in some human cases (Twomey et al. 1979).

It should be emphasized that the above three options are not mutually exclusive, namely that more than one mechanism may be responsible for HSV's entry into the brain resulting in encephalitis.

Despite anecdotal reports (Kumaravelu et al. 1998; Ejima et al. 1994), HSE is generally not a disorder of the immunocompromised host, except in the context of bone marrow transplantation (Darville et al. 1998) and in patients with AIDS (Kumaravelu et al. 1998; Kennedy 2004). However, deficiency of an intracellular protein, UNC-93B, causing impaired cellular interferon α/β and λ antiviral responses was associated with HSE (Casrouge et al. 2006).

The mechanisms that facilitate HSV-1 ability to penetrate the nervous system, evade the immune response, and cause encephalitis are also not fully understood. Under experimental conditions, the immune system can influence HSV gene expression during the acute phase of infection. However, infected neurons fail to express major histocompatability complex (MHC) antigens (Oldstone 1991), and it has been shown that an HSV-specific protein can bind to a cellular transporter associated with antigen processing. This leads to the retention of MHC class I molecules within the cell and enables the virus to evade the host immune response (Hill et al. 1995; Fruh et al. 1995).

ICP34.5 is an HSV-1-encoded protein that is required for neurovirulence (Chou et al. 1990; Harrow et al. 2004). This viral protein interacts with a host cellular protein, Beclin 1, which is needed for autophagy (Orvedahl et al. 2007). A virus lacking the Beclin 1-binding domain of ICP34.5 failed to inhibit autophagy and was attenuated in its ability to cause lethal encephalitis in mice. Thus, autophagy inhibition may be another molecular mechanism by which viruses evade innate immunity to cause disease.

Last, the rule of the immune response in tissue damage is undetermined. While many studies implicate the immune response to HSV-1 and its various cell populations (e.g., microglia and $CD8^+$ T cells) in causing widespread CNS pathology [reviewed in Conrady et al. (2010)], the exact cause of the extensive destruction of the CNS is unknown. Elucidating the mechanisms responsible for such damage may pave the way to clinical trials that will examine adjunct therapy besides antiviral agents such as corticosteroids (see below).

6.3 Clinical Features

Several studies delineated the clinical presentation and the characteristic features of HSE in the era prior to the routine use of polymerase chain reaction (PCR) for diagnosis, when, in many, the ultimate confirmation of the clinical diagnosis by brain histology was mandatory. PCR technology as a major diagnostic tool enlarged and modified our understanding of the clinical spectrum of HSE (Domingues et al. 1997; Studahl et al. 1998; Fodor et al. 1998). The symptoms and signs of HSE are related to nonspecific meningoencephalitis: headache, fever, and sometimes neck stiffness associated with signs of brain dysfunction (alterations of consciousness, personality and behavior, focal neurological signs, cognitive disturbances) and seizures. More specific to HSE are prodromal symptoms of upper respiratory tract infection and neurological findings related to dysfunction of the fronto-temporal lobes, sometimes mimicking acute psychiatric conditions.

Nevertheless, with PCR technology as the ultimate diagnostic tool, the spectrum of HSE seems to be much wider to include symptoms and signs of meningitis with minimal, if at all, evidence for brain parenchyma involvement.

Several points should be emphasized: (1) milder, less severe disease forms of encephalitis that do not conform with the classical fronto-temporal syndrome occur (Fodor et al. 1998); (2) fever is one of the most frequent features at presentation, and its absence should cast doubt upon the diagnosis; (3) headache is present in up to 90 % of HSE cases; (4) the disease is of acute onset, usually less than a week; (5) in the clinical series of the pre-PCR era, gray matter dysfunction was a dominant feature: personality changes, confusion, and disorientation were present in about three quarters of the patients and seizures in half; focal neurological signs were less frequent and were present in about a third of all patients.

In AIDS patients the presentation of HSE might vary and become a clinical challenge. Patients may present with behavioral abnormalities and without fever or even headache (Grover et al. 2004).

6.4 Diagnosis

HSE is a medical emergency because the prognosis is mainly dependent on early initiation of therapy. The differential diagnosis should usually be limited to infectious, parainfectious, and acute inflammatory conditions, and the final diagnosis is eventually based on CSF analysis and neuroimaging.

In the presence of focal neurological signs, neuroimaging prior to lumbar puncture might sometimes be indicated, provided that it can be obtained immediately. Lumbar puncture should be postponed only when strict contraindications are present (Hasbun et al. 2001). Then, anti-herpetic therapy should be introduced and eventually be amended according to neuroimaging and CSF findings when those become available. The CSF is abnormal in more than 95 % of HSE patients and contains moderate pleocytosis, usually mononuclear white and red blood cells. A moderate rise in CSF protein is present in more than 80 % of cases, and hypoglycorrhachia is the exception, detected in less than 5 % of biopsy-proven HSE patients (Sawyer et al. 1988). Demonstration of intrathecal production of anti-HSV antibodies, for many years the main noninvasive diagnostic tool, has been replaced by PCR for HSV-1 DNA that is very sensitive (98 %) and specific (94 %) when compared to the gold standard of histology, obtained by brain biopsy (Lakeman and Whitley 1995). False negative results might be present during the first days of disease, and when in doubt, a repeat lumbar puncture after 1–2 days is indicated. The optimal chance of obtaining a positive CSF PCR in HSE is probably between 2 and 10 days after the onset of the illness (Davis and Tyler 2005). CT scanning is usually normal within the first 4–6 days of disease and EEG though sensitive is a nonspecific diagnostic aid (Dutt and Johnston 1982). It may enable localization of the pathology to fronto-temporal brain regions before any abnormality can be visualized on imaging (mainly CT) studies. The findings are usually those of periodic sharp and slow wave complexes. Both have therefore been substituted by MRI which is much more sensitive demonstrating high signal intensity lesions on T2-weighted, diffusion weighted and FLAIR images earlier in the course (Fig. 1). Rarely, the MRI may be normal in HSE. Brain biopsy is now

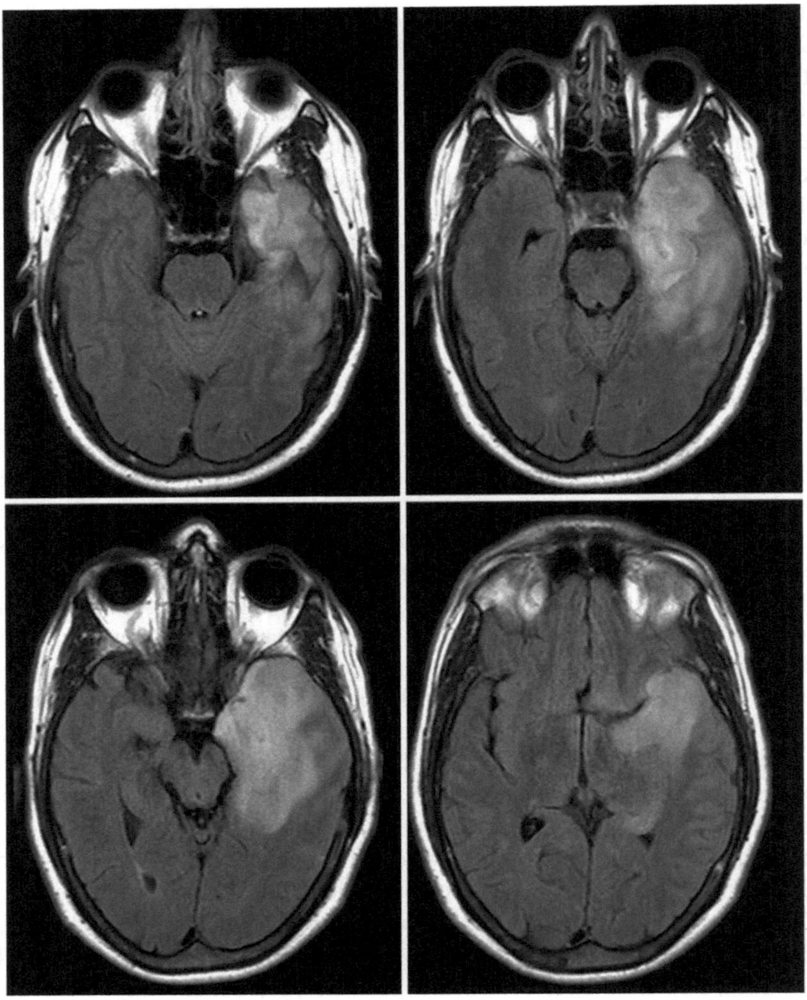

Fig. 1 FLAIR MRI Image of a 52-year-old patient with PCR-proven herpes encephalitis demonstrating signal in the left temporal lobe

seldom performed, indicated in cases unresponsive to anti-herpetic therapy, in the investigation of a relapsing encephalitic illness and where there is serious diagnostic doubt.

6.5 Therapy

To reduce permanent sequelae, treatment should be introduced as soon as possible. Therefore, when the clinical setting is suspicious, therapy should be initiated immediately and modified according to radiological, serological, and molecular results when they become available. The mainstay of treatment is still acyclovir (Whitley et al. 1986; James et al. 2009; Steiner et al. 2010). It prevents viral replication by inhibiting the viral DNA polymerase in infected cells and is activated only in cells that contain replicating virus. HSE is treated with acyclovir 10 mg/kg IV every 8 h. Dosage should be adjusted in cases of renal malfunction. Following relapses that were observed after a 10-day course of acyclovir therapy (VanLandigham et al. 1988), the current recommended protocol has increased from 10 days of therapy to 14–21 days. Although it is unclear for how long HSV DNA can be detected in the CSF after initiation of therapy, a consensus report (Cinque et al. 1996) as well as retrospective assessment (Ito et al. 2000) suggest that, when in doubt, identification of viral DNA by PCR on reexamination of the CSF may indicate the need for an additional 1–2 weeks of acyclovir therapy or reconsider the initial diagnosis. Some centers practice CSF PCR at the end of acyclovir treatment and advocate a further course of acyclovir if it remains HSV-positive (Cinque et al. 1996). Studies underway by the NIH collaborative antiviral study group are currently evaluating the potential benefits of prolonged oral agents in reducing the neurological sequelae of HSE.

The emergence of drug-resistant HSV and varicella-zoster virus (VZV) that carry alterations in the viral thymidine kinase gene that is required to metabolize acyclovir into its active compound affects an estimated 5–25 % of immunocompromised patients receiving long-term prophylactic treatment with acyclovir (Chen et al. 2000). For such patients, foscarnet, which requires no metabolic activation (Safrin et al. 1991) and cidofovir, a nucleoside analog that is phosphorylated to its active compound by cellular enzymes (Cundy 1999), have become the second-line drugs (Naesens and De Clercq 2001). With the long intracellular half-life of its metabolites, cidofovir can be administered once weekly. However, it is nephrotoxic and it should be administered with probenecid to decrease nephrotoxicity and its use is restricted to the treatment of acyclovir-resistant HSV isolates.

The use of corticosteroids as an adjunct treatment for HSE is controversial (Fitch and van de Beek 2008). The rationale for their use includes both anti-edematous and anti-inflammatory effects. However, their effect on HSE outcome is currently unknown. When encephalitis is complicated by severe, vasogenic cerebral edema with neuroimaging evidence of midline shift, high-dose steroids (dexamethasone) may have a role, even when currently evidence is lacking. A randomized clinical trial is underway to investigate the use of adjunctive dexamethasone in HSE (GACHE trial) (Hacke 2006).

6.6 Prognosis

Survival is dramatically improved by acyclovir therapy, but the quality of survival is still unsatisfactory (McGrath et al. 1997). While acyclovir has reduced mortality from HSE, most of survivors have persistent neurological symptoms, signs, or both. Cognitive impairment remains the main problem despite early diagnosis/treatment and a promising early outcome (Gordon et al. 1990; Kimberlin 2007).

7 Neonatal Encephalitis

About 80 % of neonatal encephalitis cases are due to HSV-2, and most are acquired from a mother who has active genital herpes infection at the time of delivery (Whitley et al. 1991). Neonates present with systemic findings (alterations in body temperature, lethargy, respiratory distress, anorexia, vomiting, cyanosis) and neurological signs (irritability, bulging fontanels, seizures, opisthotonus, and coma) (Overall 1994).

The infection can take one of three patterns: disseminated infection, an isolated CNS disease, or a focal infection confined to the skin, eye, or mouth. The skin, eye, and mouth findings are present in about 80 % of all cases, but although they are highly suggestive of the diagnosis, the condition may also resemble bacterial sepsis or meningitis, and therefore laboratory diagnosis is mandatory. This can be achieved rapidly by isolation of the virus from maternal genital lesions and secretions or from vesicles, peripheral blood, and CSF of the newborn. Staining of samples for viral antigen and PCR analysis will establish diagnosis within several hours. Both acyclovir and vidarabine have been shown to reduce the morbidity and mortality of HSV infection in the neonate, but acyclovir is preferred because of its safety profile and convenient dosing regimen. Acyclovir is given at 30 mg/kg/day intravenously in divided doses every 8 h for 14 days in infants with disease localized to skin, eyes, and mouth and for 21 days when the infection is disseminated or involves the CNS (Whitley 2004a, b). Since acyclovir can cause neutropenia and nephrotoxicity, the available data do not support the routine use of oral suppressive acyclovir therapy following treatment of acute neonatal HSV disease (Kimberlin 2004).

A prophylactic and therapeutic anti-HSV-2 vaccine has the potential to reduce the cases and the risk of neonatal herpetic disease. Unfortunately, the major determinants of effective immunity to HSV have yet to be identified (Stanberry 2004).

Even with appropriate treatment, prognosis of neonatal encephalitis is still very poor. One-third to half of all treated babies with disseminated disease die, and about two-thirds of survivors have neurological sequelae. This raises two unresolved issues:

1. Should a neonate who was discovered postnatally to have been delivered via an HSV-lesioned or culture-positive birth canal be treated prophylactically? In view of the serious consequences of untreated neonatal HSV infection we and others (Overall 1994) recommend prophylactic intravenous acyclovir (60 mg/kg per day in three divided doses) for 10 days.
2. Is cesarean section recommended in all cases of mothers with a history of HSV genital infection? Since asymptomatic HSV-2 shedding is a frequent occurrence in seropositive individuals (Wald et al. 2000), and cesarean section appears to decrease the risk of HSV transmission, our policy has been to avoid vaginal delivery in women with history of genital HSV infection. This approach might however be too cautious. Thus, the Royal College of Obstetricians and Gynecologists (RCOG) recommends cesarean section only for women

presenting with first-episode of genital herpes lesions at the time of delivery, but not for women who develop first episode genital herpes lesions during the first or second trimesters. For women who present with first-episode of genital herpes lesions within 6 weeks of the expected date of delivery or onset of preterm labor, elective cesarean section may be considered at term (Royal College of Obstetricians and Gynecologists 2002).

8 Mollaret's Meningitis

Mollaret's meningitis is characterized by recurrent self-limited lymphocytic meningitis in otherwise healthy individuals (Mollaret 1944). Recurrence takes place at intervals of several weeks to months and has been documented after up to 28 years (Tyler and Adler 1983). CSF contains from 200 to several thousand lymphocytes per ml^3, and large endothelial cells termed "Mollaret cells" may be present. CSF protein levels are elevated, and glucose may sometimes be low. Complete recovery occurs within several days. Diagnosis is established after other causes of lymphocytic meningitis have been ruled out. The causative agent is usually HSV-2 (Berger 1991; Picard et al. 1993; Tedder et al. 1994). Other pathogens, HSV-1 included (Yamamoto et al. 1991), have also been reported. While some reports suggested either shorter episodes or resolution of the syndrome with anti-herpetic treatment, one could argue that the therapy does not affect the viral reservoir in DRG and is not associated with prevention of future mucocutaneous disease (Berger and Houff 2008).

9 The Association of HSV with Other Neurological Diseases

Several neurological conditions have been associated with acute, chronic, or reactivated HSV infections. In some, there is presently insufficient evidence to implicate the virus in the etiology and pathogenesis of the disease. However, in some other conditions, such as Alzheimer's disease (AD), intractable focal epilepsy (Jay et al. 1998), multiple sclerosis, and acute disseminated encephalomyelitis (Steiner et al. 2001), the data appear to be more intriguing and may support the possibility that in a subgroup of patients HSV-1 has a causative role in pathogenesis.

Of special note is the possibility that in some sporadic cases of AD, HSV-1 might be the trigger that initiates or contributes to the pathogenetic cascade. The characteristic pathological features of AD favor the limbic system, the same region that takes the brunt of infection during HSE, leading to the speculation that there might be a cause-and-effect relationship between herpes infection of the CNS and sporadic AD (Ball 1982). Besides anecdotal reports of evidence for HSV-1 infection in brains of patients with dementia, more substantial data came from the demonstration that the APOE-4 allele of the APOE gene is a susceptibility factor for development of AD in patients harboring HSV-1 in their brain parenchyma

(Itzhaki et al. 1997). One explanation might be that prior damage, such as occurs in HSE and a possible defect in repair ability present in APOE-4 carriers (Danik et al. 1999), may eventually pave the way to the development of AD. HSV-1 neuroinvasiveness was shown to depend on the overall ApoE dosage and especially on the presence of isoform ApoE4, and ApoE4 was also shown to facilitate HSV-1 latency in the brain (Burgos et al. 2006).

Based on the assumption that HSV-1 resides in the geniculate ganglion in a similar fashion to colonization of other DRG, McCormick hypothesized in 1972 that HSV might be the causative agent of idiopathic peripheral facial nerve paralysis (Bell's palsy). Indeed, HSV nucleic acids were identified in the geniculate ganglion (Schulz et al. 1988) and PCR identified HSV DNA in endoneural fluids of the facial nerve or the auricular muscle of some patients with idiopathic facial palsy (Murakami et al. 1996) and the addition of acyclovir to prednisone for the treatment of this condition resulted in a very moderate but statistically significant clinical improvement compared to prednisone therapy alone (Adour et al. 1996). However, in humans, HSV-1 reactivation is not associated with motor impairment and HSV-1 causes recurrent reactivation, whereas Bell's palsy in the overwhelming majority of cases is a single episode and a rare condition compared to with the incidence of cold sores. Moreover, even if HSV-1 has an etiological role in Bell's palsy in a subgroup of patients, this does not imply a straightforward infective pathogenesis. The nerve may be damaged by edema and pressure within a narrow bony canal, by ischemia due to vascular congestion, or by a dysimmune-mediated condition. The indication to treat Bell's palsy with anti-HSV-1 therapy is therefore questionable (Steiner and Mattan 1999; Allen and Dunn 2004).

10 Conclusions

HSE is a medical emergency requiring immediate diagnosis and therapy. Prognosis of HSE is dependent on early diagnosis and immediate initiation of anti-herpetic therapy. The impact of adjuvant use of corticosteroids in HSE on outcome is currently unknown. While impressive progress has been made in understanding the biology of HSV human infection and in diagnosis and therapy of HSE, there are thus far many issues regarding both pathogenesis and effective therapy that are unknown. Whether HSE is due to viral reactivation or primary infection is unknown, but the possibility that the encephalitis is caused by primary HSV infection seems more plausible. During latent HSV infection restricted viral transcription takes place and may provide some, or even all of the requirements for the establishment, maintenance, and reactivation of the virus. Despite the ability to dramatically reduce mortality, many patients, however, remain with neurological incapacitation. Further knowledge on the biology, early diagnosis, therapy, and prevention of disease should be attained to reduce the morbidity associated with HSE.

Acknowledgments The research reported in this review was supported by grants from the Israel Science Foundation and the Israeli Ministry of Health, Chief Scientist.

References

Adour KK, Ruboyianes JM, Von Doersten PG, Byl FM, Trent CS, Quesenberry CP Jr, Hitchcock T (1996) Bell's palsy treatment with acyclovir and prednisone compared with prednisone alone: a double-blind, randomized, controlled trial. Ann Otol Rhinol Laryngol 105:371–378

Allen D, Dunn L. Aciclovir or valaciclovir for Bell's palsy (idiopathic facial paralysis). Cochrane Database Syst Rev 2004;(3):CD001869

Astruc J (1736) De morbis venereis libri sex. G Cavelier, Paris

Aurelius E, Johansson B, Skoldenberg B, Forsgren M (1993) Encephalitis in immunocompetent patients due to herpes simplex virus type 1 or 2 as determined by type-specific polymerase chain reaction and antibody assays of cerebrospinal fluid. J Med Virol 39:179–186

Ball MJ (1982) Limbic predilection in Alzheimer dementia: is reactivated herpesvirus involved? Can J Neurol Sci 9:303–306

Baskin HJ, Hedlund G (2007) Neuroimaging of herpesvirus infections in children. Pediatr Radiol 37:949–963

Berger JR (1991) Benign aseptic (Mollaret's) meningitis after genital herpes. Lancet 337:1360–1361

Berger JR, Houff S (2008) Neurological complications of herpes simplex virus type 2 infection. Arch Neurol 65:596–600

Burgos JS, Ramirez C, Sastre I, Valdivieso F (2006) Effect of apolipoprotein E on the cerebral load of latent herpes simplex virus type 1 DNA. J Virol 80:5383–5387

Buxbaum S, Geers M, Gross G, Schofer H, Rabenau HF, Doerr HW (2003) Epidemiology of herpes simplex virus types 1 and 2 in Germany: what has changed? Med Microbiol Immunol 192:177–181

Casrouge A, Zhang SY, Eidenschenk C, Jouanguy E, Puel A, Yang K, Alcais A, Picard C, Mahfoufi N, Nicolas N, Lorenzo L, Plancoulaine S, Senechal B, Geissmann F, Tabeta K, Hoebe K, Du X, Miller RL, Heron B, Mignot C, de Villemeur TB, Lebon P, Dulac O, Rozenberg F, Beutler B, Tardieu M, Abel L, Casanova JL (2006) Herpes simplex virus encephalitis in human UNC-93B deficiency. Science 314:308–312

Chen Y, Scieux C, Garrait V, Socie G, Rocha V, Molina JM et al (2000) Resistant herpes simplex virus type 1 infection: an emerging concern after allogeneic stem cell transplantation. Clin Infect Dis 31:927–935

Chou J, Kern ER, Whitley RJ, Roizman B (1990) Mapping of herpes simplex virus-1 neurovirulence to g134.5, a gene nonessentialfor growth in culture. Science 250

Cinque P, Cleator GM, Weber T, Monteyne P, Sindic CJ, van Loon AM (1996) The role of laboratory investigation in the diagnosis and management of patients with suspected herpes simplex encephalitis: a consensus report. The EU concerted action on virus meningitis and encephalitis. J Neurol Neurosurg Psychiatry 61:339–345

Conrady CD, Drevets DA, Carr DJ (2010) Herpes simplex type I (HSV-1) infection of the nervous system: is an immune response a good thing? J Neuroimmunol 220:1–9

Cumston CG (1926) The history of herpes from the earliest times to the nineteenth century. Ann Med Hist 8:284–291

Cundy KC (1999) Clinical pharmacokinetics of the antiviral nucleotide analogues cidofovir and adefovir. Clin Pharmacokinet 36:127–143

Danik M, Champagne D, Petit-Turcotte C, Beffert U, Poirier J (1999) Brain lipoprotein metabolism and its relation to neurodegenerative disease. Crit Rev Neurobiol 13:357–407

Darville JM, Ley BE, Roome AP, Foot AB (1998) Acyclovir-resistant herpes simplex virus infections in a bone marrow transplant population. Bone Marrow Transplant 22:587–589

Davis LE, Johnson RT (1979) An explanation for the localization of herpes simplex encephalitis? Ann Neurol 5:2–5

Davis LE, Tyler KL (2005) Molecular diagnosis of CNS viral infections. J Neurol Neurosurg Psychiatry 76:1–4

Domingues RB, Tsanaclis AM, Pannuti CS, Mayo MS, Lakeman FD (1997) Evaluation of the range of clinical presentations of herpes simplex encephalitis by using polymerase chain reaction assay of cerebrospinal fluid samples. Clin Infect Dis 25:86–91

Dutt MK, Johnston ID (1982) Computed tomography and EEG in herpes simplex encephalitis. Their value in diagnosis and prognosis. Arch Neurol 39:99–102

Ejima M, Tanaka H, Ueda M, Ota K, Maruyama S (1994) A case of chronic herpes encephalitis with myelodysplastic syndrome (MDS). Rinsho Shinkeigaku 34:246–249

Fitch MT, van de Beek D (2008) Drug insight: steroids in CNS infectious diseases—new indications for an old therapy. Nat Clin Pract Neurol 4:97–104

Fodor PA, Levin MJ, Weinberg A, Sandberg E, Sylman J, Tyler KL (1998) Atypical herpes simplex virus encephalitis diagnosed by PCR amplification of viral DNA from CSF. Neurology 51:554–559

Fraser NW, Lawrence WC, Wroblewska Z, Gilden DH, Koprowski H (1981) Herpes simplex type 1 DNA in human brain tissue. Proc Natl Acad Sci USA 78:6461–6465

Fruh K, Ahn K, Djaballah H, Sempe P, van Endert PM, Tampe R, Peterson PA, Yang Y (1995) A viral inhibitor of peptide transporters for antigen presentation. Nature 375:415–418

Goldenberg D, Mador N, Ball MJ, Panet A, Steiner I (1997) The abundant latency-associated transcripts of herpes simplex virus type 1 are bound to polyribosomes in cultured neuronal cells and during latent infection in mouse trigeminal ganglia. J Virol 71:2897–2904

Goodpasture EW (1929) Herpetic infections with special reference to involvement of the nervous system. Medicine 8:223–243

Gordon B, Selnes OA, Hart J Jr, Hanley DF, Whitley RJ (1990) Long-term cognitive sequelae of acyclovir-treated herpes simplex encephalitis. Arch Neurol 47:646–647

Granerod J, Ambrose HE, Davies NW, Clewley JP, Walsh AL, Morgan D, Cunningham R, Zuckerman M, Mutton KJ, Solomon T, Ward KN, Lunn MP, Irani SR, Vincent A, Brown DW, Crowcroft NS, UK Health Protection Agency (HPA). Aetiology of Encephalitis Study Group (2010) Causes of encephalitis and differences in their clinical presentations in England: a multicentre, population-based prospective study. Lancet Infect Dis 10:835–844

Grover D, Newsholme W, Brink N, Manji H, Miller R (2004) Herpes simplex virus infection of the central nervous system in human immunodeficiency virus-type 1-infected patients. Int J STD AIDS 15:597–600

Hacke W (2006) German trial of acyclovir and corticosteroids in herpes-simplex-virus-encephalitis (GACHE). ISRCTN Register (http://www.controlled-trials.com/ISRCTN45122933)

Harrow S, Papanastassiou V, Harland J, Mabbs R, Petty R, Fraser M, Hadley D, Patterson J, Brown SM, Rampling R (2004) HSV1716 injection into the brain adjacent to tumour following surgical resection of high-grade glioma: safety data and long-term survival. Gene Ther 11:1648–1658

Hasbun R, Abrahams J, Jekel J, Quagliarello VJ (2001) Computed tomography of the head before lumbar puncture in adults with suspected meningitis. N Engl J Med 345:1727–1733

Hill A, Jugovic P, York I, Russ G, Bennink J, Yewdell J, Ploegh H, Johnson D (1995) Herpes simplex virus turns off the TAP to evade host immunity. Nature 375:411–415

Hirai R, Ayabe M, Shoji H, Kaji M, Ichiyama T, Sakai K (2005) Herpes simplex encephalitis presenting with bilateral hippocampal lesions on magnetic resonance imaging, simultaneously complicated by small cell lung carcinoma. Intern Med 44:1006–1008

Hope-Simpson RE (1965) The nature of herpes zoster: a long term study and a new hypothesis. Proc R Soc Med 58:9–20

Ito Y, Kimura H, Yabuta Y, Ando Y, Murakami T, Shiomi M et al (2000) Exacerbation of herpes simplex encephalitis after successful treatment with acyclovir. Clin Infect Dis 30:185–187

Itzhaki RF, Lin WR, Shang D, Wilcock GK, Faragher B, Jamieson GA (1997) Herpes simplex virus type 1 in brain and risk of Alzheimer's disease. Lancet 349:241–244

James SH, Kimberlin DW, Whitley RJ (2009) Antiviral therapy for herpesvirus central nervous system infections: neonatal herpes simplex virus infection, herpes simplex encephalitis, and congenital cytomegalovirus infection. Antiviral Res 83:207–213

Jay V, Hwang P, Hoffman HJ, Becker LE, Zielenska M (1998) Intractable seizure disorder associated with chronic herpes infection. HSV1 detection in tissue by the polymerase chain reaction. Childs Nerv Syst 14:15–20

Kennedy PGE (2004) Viral encephalitis-causes, differential diagnosis and management. J Neurol Neurosurg Psychiatry 75(1):i10–i15

Kimberlin D (2004) Herpes simplex virus, meningitis and encephalitis in neonates. Herpes 11 (Suppl 2):65A–76A

Kimberlin DW (2007) Management of HSV encephalitis in adults and neonates: diagnosis, prognosis and treatment. Herpes 14:11–16

Knaup B, Schunemann S, Wolff MH (2000) Subclinical reactivation of herpes simplex virus type 1 in the oral cavity. Oral Microbiol Immunol 15:281–283

Koskiniemi M, Piiparinen H, Mannonen L, Rantalaiho T, Vaheri A (1996) Herpes encephalitis is a disease of middle aged and elderly people: polymerase chain reaction for detection of herpes simplex virus in the CSF of 516 patients with encephalitis the study group. J Neurol Neurosurg Psychiatry 60:174–178

Krause PR, Stanberry LR, Bourne N, Connelly B, Kurawadwala JF, Patel A, Straus SE (1995) Expression of the herpes simplex virus type 2 latency-associated transcript enhances spontaneous reactivation of genital herpes in latently infected guinea pigs. J Exp Med 181:297–306

Kumaravelu S, Duraiswamy R, Kalghatgi AT, Dhand VP (1998) Atypical presentation of herpes simplex encephalitis in acquired immune deficiency syndrome. J Assoc Physicians India 46:316–317

Lakeman FD, Whitley RJ (1995) Diagnosis of herpes simplex encephalitis: application of polymerase chain reaction to cerebrospinal fluid from brain-biopsied patients and correlation with disease. National Institute of Allergy and Infectious Diseases Collaborative Antiviral Study Group. J Infect Dis 171:857–863

Mador N, Panet A, Steiner I (2002) The latency-associated gene of herpes simplex virus type 1 (HSV-1) interferes with superinfection by HSV-1. J Neurovirol 8(Suppl 2):97–102

McCormick DP (1972) Herpes simplex virus as a cause of Bell's palsy. Lancet 1:937

McGrath N, Anderson NE, Croxson MC, Powell KF (1997) Herpes simplex encephalitis treated with acyclovir: diagnosis and long term outcome. J Neurol Neurosurg Psychiatry 63:321–326

Mollaret P (1944) La meningite endothelio-nucleocytaire multirecurrente benigne: syndrome nouveau ou maladie nouvelle? Rev Neurol 76:57–76

Murakami S, Mizobuchi M, Nakashiro Y, Doi T, Hat N, Yanagihara N (1996) Bell palsy and herpes simplex virus: identification of viral DNA in endoneural fluid and muscle. Ann Intern Med 127:27–30

Naesens L, De Clercq E (2001) Recent developments in herpesvirus therapy. Herpes 8:12–16

Nahmias AJ, Lee FK, Beckman-Nahmias S (1990) Sero-epidemiological and -sociological patterns of herpes simplex virus infection in the world. Scand J Infect Dis Suppl 69:19–36

Nilsen A, Myrmel H (2000) Changing trends in genital herpes simplex virus infection in Bergen Norway. Acta Obstet Gynecol Scand 79:693–696

Oldstone MB (1991) Molecular anatomy of viral persistence. J Virol 65:6381–6386

Orvedahl A, Alexander D, Talloczy Z, Sun Q, Wei Y, Zhang W, Burns D, Leib DA, Levine B (2007) HSV-1 ICP34.5 confers neurovirulence by targeting the Beclin 1 autophagy protein. Cell Host Microbe 1:23–35

Overall JC Jr (1994) Herpes simplex virus infection of the fetus and newborn. Pediatr Ann 23:131–136

Perng GC, Jones C (2010) Towards an understanding of the herpes simplex virus type 1 latency-reactivation cycle. Interdiscip Perspect Infect Dis 2010:262415

Picard FJ, Dekaban GA, Silva J, Rice GP (1993) Mollaret's meningitis associated with herpes simplex type 2 infection. Neurology 43:1722–1727

Roberts CM, Pfister JR, Spear SJ (2003) Increasing proportion of herpes simplex virus type 1 as a cause of genital herpes infection in college students. Sex Transm Dis 30:797–800

Rock DL, Fraser NW (1983) Detection of HSV-1 genome in the central nervous system of latently infected mice. Nature 302:523–525

Roizman B, Whitley RJ (2001) The nine ages of herpes simplex virus. Herpes 8:23–27

Royal College of Obstetricians and Gynecologists: Clinical Guidelines nomer 30: management of genital herpes in pregnancy. March 2002

Safrin S, Crumpacker C, Chatis P, Davis R, Hafner R, Rush J et al (1991) A controlled trial comparing foscarnet with vidarabine for acyclovir- resistant mucocutaneous herpes simplex in the acquired immunodeficiency syndrome. The AIDS clinical trials group. N Engl J Med 325:551–555

Sawyer J, Ellner J, Ransohoff DF (1988) To biopsy or not to biopsy in suspected herpes simplex encephalitis: a quantitative analysis. Med Decis Making 8:95–101

Schulz P, Arbusow V, Strupp M, Dieterich M, Rauch E, Brandt T (1988) Highly variable distribution of HSV-1 specific DNA in human geniculate, vestibular and spiral ganglia. Neurosci Lett 252:139–142

Shen W, Sa e Silva M, Jaber T, Vitvitskaia O, Li S, Henderson G, Jones C (2009) Two small RNAs encoded within the first 1.5 kilobases of the herpes simplex virus type 1 latency-associated transcript can inhibit productive infection and cooperate to inhibit apoptosis. J Virol 83:9131–9139

Smith JS, Robinson NJ (2002) Age-specific prevalence of infection with herpes simplex virus types 2 and 1: a global review. J Infect Dis 186(Suppl 1):S3–S28

Stanberry LR (2004) Clinical trials of prophylactic and therapeutic herpes simplex virus vaccines. Herpes 11(Suppl 3):161A–169A

Steiner I, Budka H, Chaudhuri A, Koskiniemi M, Sainio K, Salonen O, Kennedy PG (2010) Viral meningoencephalitis: a review of diagnostic methods and guidelines for management. Eur J Neurol 17:999–e57

Steiner I, Kennedy PG (1991) Herpes simplex virus latency in the nervous system–a new model. Neuropathol Appl Neurobiol 17:433–440

Steiner I, Kennedy PGE (1995) Herpes simplex virus type 1 latent infection in the nervous system. J Neurovirol 1:19–29

Steiner I, Mattan Y (1999) Bell's palsy and herpes viruses: to (acyclo)vir or not to (acyclo)vir? J Neurol Sci 170:19–23

Steiner I, Nisipiano P, Wirguin I (2001) Infections and the etiology of multiple sclerosis. Curr Neurol Neurosci Rep 1:271–276

Steiner I, Spivack JG, Deshmane SL, Ace CI, Preston CM, Fraser NW (1990) A herpes simplex virus type 1 mutant containing a non-transinducing Vmw65 protein establishes latent infection in vivo in the absence of viral replication and reactivates efficiently from explanted trigeminal ganglia. J Virol 64:1630–1638

Steiner I, Spivack JG, Lirette RP, Brown SM, MacLean AR, Subak-Sharpe J, Fraser NW (1989) Herpes simplex virus type 1 latency-associated transcripts are evidently not essential for latent infection. EMBO J 8:505–511

Steiner I, Spivack JG, O'Boyle DR, Lavi E, Fraser NW (1988) Latent herpes virus type 1 transcription in human trigeminal ganglia. J Virol 62:3493–3496

Stroop WG, Schaefer DC (1986) Production of encephalitis restricted to the temporal lobes by experimental reactivation of herpes simplex virus. J Infect Dis 153:721–731

Studahl M, Bergstrom T, Hagberg L (1998) Acute viral encephalitis in adults—a prospective study. Scand J Infect Dis 30:215–220

Tang S, Bertke AS, Patel A, Margolis TP, Krause PR (2011) Herpes Simplex Virus-2 miR-H6 is a Novel LAT-associated microRNA, but reduction of its expression does not influence viral latency establishment or recurrence phenotype. J Virol 85:4501–4509

Tedder DG, Ashley R, Tyler KL, Levin MJ (1994) Herpes simplex virus infection as a cause of benign recurrent lymphocytic meningitis. Ann Intern Med 121:334–338

Time. Today's Scarlet Letter, Herpes. 2nd Aug 1982

Tran T, Druce JD, Catton MC, Kelly H, Birch CJ (2004) Changing epidemiology of genital herpes simplex virus infection in Melbourne, Australia, between 1980 and 2003. Sex Transm Infect 80:277–279

Twomey JA, Barker CM, Robinson G, Howell DA (1979) Olfactory mucosa in herpes simplex encephalitis. J Neurol Neurosurg Psychiatry 42:983–987

Tyler KL, Adler D (1983) Twenty-eight years of benign recurring Mollaret meningitis. Arch Neurol 40:42–43

Wald A, Zeh J, Selke S, Warren T, Ryncarz AJ, Ashley R, Krieger JN, Corey L (2000) Reactivation of genital herpes simplex virus type 2 infection in asymptomatic seropositive persons. N Engl J Med 342:844–850

VanLandigham KE, Marsteller HB, Ross GW, Hayden FG (1988) Relapse of herpes simplex encephalitis after conventional acyclovir therapy. JAMA 259:1051–1053

Whitley RJ (2004a) Herpes simplex virus. In: Scheld MW, Whitley RJ, Marra CM (eds) Infections in the central nervous system, 3rd edn. Lippincott Williams and Wilkins, Philadelphia, PA, pp 123–144

Whitley R (2004b) Neonatal herpes simplex virus infection. Curr Opin Infect Dis 17:243–246

Whitley RJ (1997) Herpes simplex virus. In: Scheld WM, Whitley RJ, Durack DT (eds) Infections of the central nervous system, vol 2. Lippincott-Raven, Philadelphia, PA, pp 73–89

Whitley RJ (1991) Herpes simplex virus infections of the central nervous system encephalitis and neonatal herpes. Drugs 42:406–427

Whitley R, Arvin A, Prober C, Corey L, Burchett S, Plotkin S, Starr S, Jacobs R, Powell D, Nahmias A (1991) Predictors of morbidity and mortality in neonates with herpes simplex virus infections The National Institute of Allergy and Infectious Diseases Collaborative Antiviral Study Group. N Engl J Med 324:450–454

Whitley RJ, Alford CA, Hirsch MS, Schooley RT, Luby JP, Aoki FY et al (1986) Vidarabine versus acyclovir therapy in herpes simplex encephalitis. N Engl J Med 314:144–149

Whitley R, Lakeman AD, Nahmias A, Roizman B (1982) DNA restriction-enzyme analysis of herpes simplex virus isolates obtained from patients with encephalitis. N Engl J Med 307:1060–1062

Yamamoto LJ, Tedder DG, Ashley R, Levin MJ (1991) Herpes simplex virus type 1 DNA in cerebrospinal fluid of a patient with Mollaret's meningitis. N Engl J Med 325:1082–1085

Progressive Multifocal Leukoencephalopathy

Allen J. Aksamit

Abstract Progressive multifocal leukoencephalopathy (PML) is an opportunistic viral infection of the human central nervous system. It destroys oligodendrocytes leading to neurological deficits associated with demyelination. It most commonly occurs in HIV-infected patients, but increasing numbers of patients are being recognized in the context of immunosuppressive therapies for autoimmune diseases. The precise pathogenesis of infection by JC virus, a human papovavirus remains elusive, but much has been learned since the original description of the pathologic entity PML in 1958. Detection and diagnosis of this disorder has become more sophisticated with MR imaging of the brain and spinal fluid analysis using polymerase chain reaction (PCR) techniques. Immune reconstitution inflammatory syndrome complicates reversal of immunosuppression when PML has established a foothold in the brain. No effective therapy exists, but there is hope for both better management of patients diagnosed with exogenous immunosuppression and future prospects for antiviral therapy.

Keywords Progressive multifocal leukoencephalopathy • JC virus • PML • Immunosuppressive opportunistic infection • Inclusion bearing oligodendrocytes • Bizarre astrocytes • IRIS

1 Introduction

Progressive multifocal leukoencephalopathy (PML) is a JC virus infection of oligodendrocytes of the central nervous system white matter, leading to neurological deficits associated with demyelination. Death of oligodendrocytes leads to focal

A.J. Aksamit (✉)
Department of Neurology, Mayo College of Medicine, Mayo clinic, 200 First Street SW, Rochester, MN 55905, USA
e-mail: aksamit@mayo.edu

A.C. Jackson (ed.), *Viral Infections of the Human Nervous System*,
Birkhäuser Advances in Infectious Diseases,
DOI 10.1007/978-3-0348-0425-7_4,

loss of myelin, and dysfunction of associated myelinated tracts involving the cerebral hemispheres, cerebellum, or brainstem. For reasons that are unknown, the optic nerves or spinal cord oligodendrocytes are not clinically affected in PML, distinguishing it from multiple sclerosis (MS).

PML is regarded as an opportunistic infection by JC virus of the human nervous system. In recent years, the most common associated underlying immunosuppressive illness has been AIDS. However, a variety of non-AIDS immunosuppressive illnesses have been associated with occurrence of PML. These include lymphoreticular malignancy, most commonly (CLL) or non-Hodgkin's lymphoma. Of interest, the lymphoreticular malignancies associated with PML are considered B-cell neoplasms. The exact role of B cells in PML pathogenesis is controversial, but some authors have considered them a requirement for carrying JC virus to the brain (Gallia et al. 1997). However, controversy aside, the role of B cells in the pathogenesis of PML is not fully defined, and the selective role of B-cell neoplasms causing a tendency towards developing PML is uncertain. Organ transplantation, immunosuppression associated with rheumatoid arthritis, sarcoidosis, systemic lupus erythematosus, and dermatomyositis have all been associated with PML as predisposing immunosuppressive illnesses. It is somewhat unclear whether connective tissue disorders or sarcoidosis intrinsically cause an immunosuppressive risk for PML or whether the exogenous treatments of these disorders are required for the associated immunosuppressed state. In general, prolonged immunosuppression for 6 months or longer is required for PML to occur.

2 History

PML was first described as a neuropathologic entity in 1958 (Astrom et al. 1958). It was noted to be a microscopically multifocal demyelinating disease, with coalescence forming macroscopic lesions. Histologically, oligodendrocyte nuclear enlargement and bizarre astrocyte formation established the disease as unique. From early on it was suspected to be a viral illness, based on the pathologic appearance of the inclusion bearing oligodendrocytes and its occurrence in immunosuppressed populations (Richardson 1961). At the advent of electron microscopic investigation, electron micrographs revealed polyoma virus-size particles in the nuclei of the infected oligodendrocytes (Zurhein and Chou 1965). However, it took until 1971 when brain from a patient with PML was cultured with human fetal glial cells by Billie Padgett and Duard Walker (Padgett et al. 1971), who isolated a replicating human DNA polyoma virus. The patient's name from whom this virus was identified was John Cunningham, and therefore led to the designation as JC virus. Within 2 years, the group at Wisconsin did further investigations to show that much of the population carried antibodies to the newly described JC virus (Walker and Frisque 1986). Early on it was shown that the virus exposure occurred as a consequence of early adolescent or adulthood exposure, and persistent serological reactivity in 60–70 % of the population.

Another group isolated another polyoma virus called SV40 virus, from another PML patient (Weiner et al. 1972), but all subsequent PML cases studied by molecular means have been caused by JC virus.

Part of the unique PML histopathology was the finding of bizarre astrocytes. These are cytologically bizarre astrocytes that have the appearance of tumor-like glial cells. A case of glioma in a human PML patient was reported as early as 1974 (Castaigne et al. 1974). Subsequently however, this has been extraordinarily rare, and controversy about the association between the JC virus and human glioma remains (Del Valle et al. 2000; Major 2000; Major et al. 1992).

By 1982, PML was reported to be associated with AIDS (Berenguer et al. 2003; Berger et al. 1998; von Einsiedel et al. 1993). A review in 1984 showed only a few hundred cases of known PML, some of which were from the early aspects of the AIDS epidemic, although the majority had been associated with other immunosuppressive illness (Brooks and Walker 1984).

By 1984, the complete nucleotide sequence of the double-stranded DNA JC virus was published (Frisque et al. 1984). It was found to be closely linked to the SV40 virus and another human polyoma virus, designated BK virus. By 1985, nonradioactive means of in situ hybridization provided evidence that oligodendrocytes and bizarre astrocytes in the brain of PML patients were infected by JC virus (Aksamit et al. 1985). These tools became useful for confirming the presence of JC virus in cases of PML requiring pathological confirmation by brain biopsy. Immunohistochemical staining for JC virus antigen has also provided means for pathologic confirmation of this infection (Aksamit et al. 1986).

Although the pathogenesis of JC virus entry into the brain remains somewhat unclear, in 1988 B cells were reported to contain JC virus in PML patients (Houff et al. 1988). Other authors subsequently showed that JC virus was detected in lymphocytes of PML and HIV-positive patients by polymerase chain reaction (PCR) (Berger and Major 1999; Koralnik 2006).

The PCR revolution in virology also affected diagnosis of PML. For the first time, in 1992, JC virus was detected in the spinal fluid of PML patients (Fong et al. 1995; Gibson et al. 1993; McGuire et al. 1995; Telenti et al. 1992). This has become the noninvasive standard for diagnosing PML, and has now become accepted as surrogate marker for histologic proof of JC virus replication in the brain (Ryschkewitsch et al. 2010).

Despite the severe and usually fatal nature of PML, no good therapy has been defined. A landmark paper testing cytarabine in AIDS-related PML was published and found to be ineffective in AIDS patients (Hall et al. 1998). A variety of agents have been tried, but none with any reliable success (Aksamit 2008).

New interest developed in PML as an opportunistic infection because of the occurrence of PML in multiple sclerosis patients treated with the immunomodulatory monoclonal antibody natalizumab. The drug natalizumab was removed from the United States drug market for a period of time because of this association with PML infection (Aksamit 2006; Clifford et al. 2010; Kleinschmidt-DeMasters and Tyler 2005; Langer-Gould et al. 2005), but later returned to the market under stricter prescribing guidelines (Berger and Houff 2009). This, however, became a harbinger of PML occurring with other commonly used immune suppressive agents including rituximab and mycophenolate mofetil (Berger and Houff 2009; Clifford et al. 2011b). It is anticipated that additional immunosuppressive therapies may likewise continue to predispose to PML as an opportunistic infection.

It has been recognized in opportunistic infections, particularly associated with AIDS patients, that immune reconstitution after initiation of antiretroviral therapy creates a rebound inflammatory response that is damaging to the infected organ. This complication was named the immune reconstitution inflammatory syndrome (IRIS) (Berger 2009; Tan et al. 2009). IRIS may become manifest in PML AIDS patients treated with antiretroviral therapy, and subsequently was recognized in multiple sclerosis patients in whom natalizumab was stopped after recognition of PML infection, leading to an immune response that is damaging to the brain. Treating PML in this complex context of immunosuppressive disease has created uncertainty about treatment and challenges for the future.

3 Virology

JC virus is a double-stranded DNA icosahedral human polyomavirus. Sequencing of the DNA from the first isolate, referred to as the Mad-1 strain, revealed a circular supercoiled DNA genome of 5,130 bp (Frisque et al. 1984). The genome is divided into an early region and a late region separated by a tandem repeat regulatory region DNA sequence at the origin of DNA replication site. The DNA codes for at least two early and four late genes. Gene expression occurring before (early) and after (late) viral DNA replication gives the names to the respective regions of the viral genome (Frisque et al. 1984; Major et al. 1992; Weber and Major 1997). The early genes are designated as large T-antigen (T-Ag) and small T-antigen, respectively. T-Ag has primary importance in initiating viral DNA replication and interfacing with genes of the host cell. It interfaces with host DNA polymerases and topoisomerases, and cell DNA binding proteins, to bind in a complex to the viral DNA, particularly in the enhancer–promotor region near the origin of replication. T-Ag is thought to be important for growth induction. The presence of replicated viral DNA shifts the protein synthesis to the late region, shutting off early region transcription. Viral capsid proteins from the late region are designated VP1, VP2, and VP3. The fourth protein is designated the agnoprotein. The late proteins share significant sequence homology with other polyoma viruses SV40 and BK viruses. This leads to antigenic overlap between JC virus and these other viruses.

JC virus is part of the polyomavirus group which takes its name from a known propensity for some of these viruses to induce tumors in their infected hosts. Murine polyoma virus is the prototype in this regard. SV40, originally isolated from monkey species, has been extensively studied as an oncogenic virus in many species. Other human polyoma viruses, principally isolated from immunosuppressed patients, include BK virus, Merkel cell virus, KI virus, and WU virus.

In vitro, JC virus only infects human glial cells and select lymphoreticular cells. In humans, oligodendrocytes and astrocytes are the principal cells infected in the brain. Interestingly, the virus does not grow in glial cells derived from other species. The principal cells used for culture of the virus have been human fetal glial cells, or immortalized cells derived from these cells called SVG cells (Major et al. 1992).

In vivo, JC virus is shed in the urine during states of immunosuppression, so renal or uroepithelial cells must be infected. Precisely which cells in the urogenital system harbor infection is unclear. JC virus has been demonstrated in human lymphocytes, B cells, bone morrow cells, and dendritic cells of the lymphoreticular system (Gallia et al. 1997; Houff et al. 1988). On this basis, one or more of these cells have been regarded important in the spread of virus to the brain.

The JC virus receptor expressed on host cell membranes that binds JC virus has been suggested to be a 2 (2–6) sialic acid–*N*-linked glycoprotein (Liu et al. 1998). Also thought to be important is the cell membrane 5HT-2A receptor that binds JC virus allowing for cell entry (Elphick et al. 2004). This serotonin receptor has been targeted in therapeutic attempts by serotonin antagonist drugs such as mirtazipine.

Cell nuclear DNA-binding proteins selectively bind to JC virus DNA sequences explaining partly the neurotropism for oligodendrocytes and astrocytes in the human nervous system. A variety of nuclear factors have been studied to work in this role including SP1 and Pur alpha.

JC virus DNA isolates have been classified as “archetype” or “neurotropic” based on the DNA sequence of the regulator–promotor region near the origin of DNA replication (Major et al. 1992; Walker and Frisque 1986). The archetype viruses are relatively conserved in the tandem repeat sequences and are principally isolated from the kidney, whereas deletions or DNA rearrangements that are quite varied have been found in this sequence from the brain isolates of JC virus. It has been therefore postulated that a rearrangement of this viral sequence must occur in order to produce brain infection.

4 Pathogenesis

JC virus is a double-stranded DNA human polyomavirus acquired principally in childhood or young adulthood (Walker and Frisque 1986). The virus is thought to be acquired by a respiratory or oral route. After infection it remains latent in the body. Fifty to seventy percent of the adult population is seropositive. Recent interest in stratifying multiple sclerosis patients who are candidates for natalizumab therapy reconfirmed a seroprevalence in this patient population of adults as 50–60 % (Bozic et al. 2011). Presumably, all seropositive individuals harbor latent virus in kidney, lymphoreticular tissue, or brain after primary infection, but suffer no ill consequences as a result. Periodic asymptomatic systemic reactivation also occurs without consequence, principally detected as asymptomatic shedding of virus in the urine. PML is considered a JC virus reactivation infection in which a second event occurs. Whether the reactivation occurs systemically with

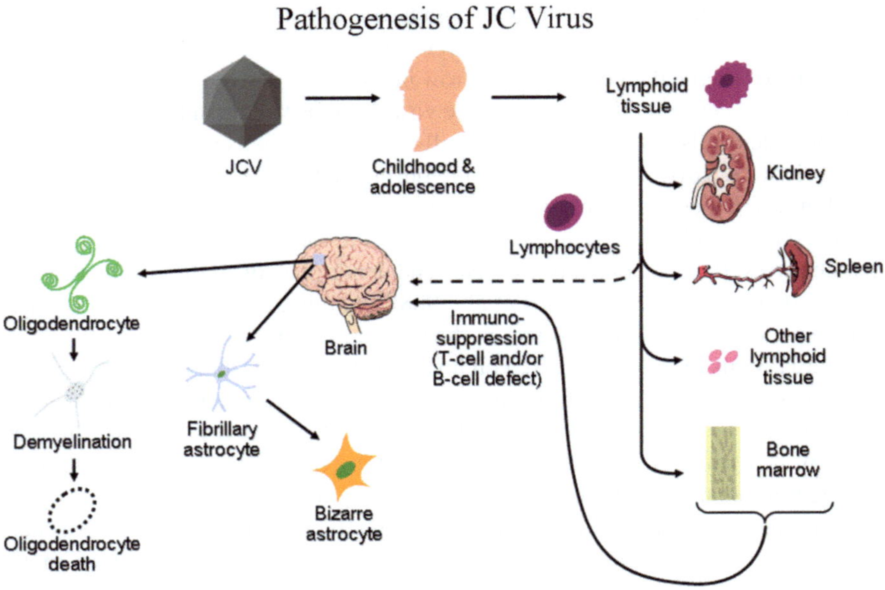

Fig. 1 Pathogenesis of JC virus infection causing PML: the virus is acquired in childhood or young adulthood, and becomes latent in lymphocytes, spleen, kidney, bone marrow, and other lymphoid tissue. It also may establish latency in the brain. With immunosuppression, JC virus replicates in oligodendrocytes, killing them causing demyelination, and non-productively infects astrocytes causing bizarre histologic changes

immunosuppression causing dissemination to the brain at that time, or the reactivation occurs from latent virus in the brain is unsettled (Fig. 1).

It has long been suspected that JC virus is carried into the nervous system via white blood cells. Clinically, this has been suspected because of the subcortical gray–white junction localization typical of the demyelinating lesions of PML. This is an end arteriole location for hematogenously disseminated deposits to occur, as occurs in brain abscess and metastatic brain cancer. However, which cells carry JC virus into the nervous system remain somewhat controversial. B cells and glia share a common DNA binding factor that may tie them in pathogenesis (Major et al. 1990). One hypothesis has been that B cells are the primary cells of importance because in some cases JC virus can be found in B cells of the bone marrow, and are detectable in cells defined by flow cytometry as B cells in the blood of immunosuppressed patients (Gallia et al. 1997; Major et al. 1992). One experience with double-labeling techniques of evaluating inflammatory lesions with B cells in PML lesions has been that the B cells do not harbor replicating JC virus (Aksamit 1995). Others have suggested that JC virus is present there in at least some cases (Major et al. 1992).

Still of controversial nature, but important for pathogenesis, is the timing of when JC virus enters the nervous system. The first hypothesis is that JC virus occurs as a reactivation infection in non-brain organs, in lymphoreticular cells or in the kidney, and is disseminated at the time of immune suppression via the blood stream to the nervous system, leading to infection of oligodendrocytes during an immunosuppressed state. A second hypothesis, however, is that JC virus infection occurs in childhood or adolescence. Latency in this model is established not only in lymphoreticular cells and kidney cells, but also in the brain at the time of primary infection. Then, at a time of immune suppression, virus is reactivated from a latent state in oligodendrocytes in the brain. This question of pathogenesis is relevant to the issues of immune surveillance of the nervous system for JC virus infection, and the role drugs such as natalizumab plays in preventing immune cells from access to the nervous system for containing JC virus infection. This second hypothesis is supported by the absence of JC viremia preceding PML in two of three initially reported PML patients treated with natalizumab where this was carefully studied (Kleinschmidt-DeMasters and Tyler 2005; Langer-Gould et al. 2005; Van Assche et al. 2005). PML in MS patients treated with this alpha-4 integrin-binding monoclonal antibody (natalizumab) raises the possibility that the alpha-4 integrin system is a critical mechanism for lymphocyte immune surveillance, preventing JC virus infection in immunosuppressed hosts.

JC virus behaves differently in oligodendrocytes and astrocytes in the brain, but the mechanism for the distinct pathology in the glial cells of the same patient is unclear. In oligodendrocytes, JC virus infection is a lytic one, where virus infects the cell and undergoes replication of viral DNA and synthesis of viral capsid proteins. Virus spreads in a centrifugal way from a central nidus of infection, to neighboring cells, leading to a circumferential expansion of demyelination (Aksamit 1995).

Astrocytes on the other hand take on a bizarre morphologic appearance with marked enlargement of the cells and distortion of the nuclei with enlargement or multiple nuclei. These cells are similar to those seen in giant cell astrocytomas unrelated to JC virus infection. Electron microscopic examination of these cells shows that no virions are present. In situ hybridization and to a lesser extent immunohistochemistry for viral proteins do show that these cells are infected in a nonproductive fashion (Aksamit et al. 1986). They have a "transformed" appearance in an oncogenic sense. The role of JC virus in human astrocytomas is uncertain and controversial (Del Valle et al. 2000; Major 2000).

5 Clinical

The clinical presentation of PML is either a focal or multifocal neurological disease. The name of the disorder, PML, is derived from neuropathological descriptions of *microscopic multifocal* abnormalities involving the brain white matter (Astrom et al. 1958; Aksamit 1995). In contrast, clinically, the typical presentation is of a unifocal cerebral or brainstem disorder. There is some disagreement about the frequency of focal versus multifocal imaging abnormalities at the time of initial clinical evaluation. Some authors regard PML to be associated with multifocal changes most commonly on MRI imaging, even if the clinical presentation is unifocal (Berger and Major 1999). Others believe that the majority of patients may present with a unifocal MRI scan as well as clinical unifocal syndrome (Aksamit 1995). In either case, either multifocal or unifocal clinical presentation is possible.

The variability in the neurological presentation of PML is a reflection of varying locations of the affected brain. Motor involvement with corticospinal tract findings, sensory involvement, cerebellar deficits, and visual field defects is common. Some syndromes regarded as "cortical" such as aphasia or visual–spatial disorientation occur commonly with PML because the pathology of PML is often immediately subcortical in the white matter, typically undermining the cortex referable to the clinical syndrome. This contrasts with the periventricular propensity of MS lesions.

Clinically, in the majority of non-AIDS-related PML patients, the neurological focal abnormalities are referable to cerebral hemispheric abnormalities. The ratio of cerebral to brainstem/cerebellar involvement is estimated to be approximately 10:1. For reasons that are unclear, brainstem involvement is more common in AIDS PML patients with a ratio of cerebral to brainstem involvement of approximately 4:1.

Consideration for the diagnosis of PML should be triggered by the clinical circumstances of a patient with immune suppression with a cell-mediated immunity defect and a subacute focal progressive neurological syndrome. Although PML has been regarded as a "slow virus" infection, it is really a subacute illness that typically evolves with focal neurological symptoms evolving over days to weeks. At times, the focal neurological syndrome can be acute and be mistaken for stroke. However, serial neurological examinations typically demonstrate that the patient does progress over ensuing days. This is parallel to the changes that occur involving the MRI of the brain. Patients with more immunopreserved status may have a slower clinical course, mimicking brain tumors like central nervous system lymphoma or glioma.

Central nervous system weakness or paralysis occurs in 60 % of PML patients (Berenguer et al. 2003; Berger et al. 1998; Brooks and Walker 1984). Gait abnormalities are common and occur in up to 65 % of patients at presentation, and cognitive disorders are the presenting manifestation in 30 %. Presenting manifestations are most commonly memory complaints or a behavioral disorder. Cognitive disturbances are present in the majority of cases with clinical progression in the cerebral hemispheres. Aphasia occurs in 20 %. Visual field defects are the presenting manifestation in 20 %. Focal cortical sensory loss (proprioception loss, astereognosis) is common, but poorly quantitated as to frequency of occurrence. Cortical limb monoparesis, limb apraxia, unilateral ataxia, or focal brainstem signs can all occur. Seizures are infrequent but are estimated to occur in 10 %.

6 Associated Illnesses

6.1 AIDS

AIDS is the most common mechanism of immunosuppression leading to PML. It is estimated that PML affects 1–4 % of symptomatic AIDS cases. PML can be the presenting manifestation of AIDS. However, more commonly, PML occurs with low CD4 blood counts (<200 cells/μL) and occurs later in the course of AIDS disease. As an opportunistic infection, PML has been less impacted by combination antiretroviral therapy (cART) than other opportunistic infections of the nervous system like toxoplasmosis or cryptococcal meningitis.

6.2 *Lymphoreticular Malignancy*

The lymphoreticular malignancies are the most common non-AIDS-related cause of immune suppression predisposing to PML. The most common disorders are chronic lymphocytic leukemia (CLL), Hodgkin's disease, and non-Hodgkin's lymphoma (Aksamit 2006). It is interesting that CLL and most non-Hodgkin's lymphomas are generally regarded as neoplasms of B-cell lineage. And yet all of these disorders are well known to be associated with opportunistic infections that manifest as disorders of T-cell immune deficiency, such as toxoplasmosis or cryptococcal meningitis. Therefore the precise immune defects that predispose to PML in these malignancies remain ill defined.

6.3 *Rheumatologic Disorders*

Virtually any form of rheumatological disorder has been reported associated with PML. Since these autoimmune disorders are commonly treated with immunosuppressive agents, it is somewhat unclear whether the underlying disease or the treatment is primarily responsible for the predisposition to PML. In one study looking at underlying disorders at non-AIDS PML, the frequency of connective tissue diseases present in PML patients was 15 %, with rheumatoid arthritis composing 5 %, systemic lupus erythematosus 7 %, dermatomyositis 2 %, and vasculitis 2 % (Aksamit 2006). Estimates of PML incidence in rheumatoid arthritis patients suggest the frequency is low, approximately 0.4 per 100,000 discharges (Calabrese and Molloy 2008).

6.4 *Transplantation*

PML is an important but rare cause of neurological disease in transplant recipients. A recent study found the estimated incidence of PML in heart and/or lung transplant patients occurring approximately 1.24 per every 1,000 transplant person years (Mateen et al. 2011). Mean onset of PML symptoms after transplantation was 17 months. PML is most often fatal (84 % case fatality) in this population but is compatible with recipient survival measured in years (3 of 50 patients). The risk of PML likely exists throughout the entire post-transplant period and should be suspected and quickly diagnosed because temporary reduction of immunosuppression may be compatible with PML improvement and long-term patient survival.

The case fatality among patients who develop PML post-transplantation is high with death occurring within 18 months in most cases. There are exceptions to this. In one series, three patients were still alive at 13, 44, and 155 months following PML symptom onset (Mateen et al. 2011). Among survivors, all patients had their immunosuppressive drug regimen significantly reduced or withdrawn at the time of diagnosis of PML. No patient survived without immunosuppressive medication reduction. Graft rejection is of significant concern with immunosuppressive drug reduction but is consistent with survival in some cases of PML.

6.5 PML in Multiple Sclerosis Patients Treated with Natalizumab

In November 2004, natalizumab, a monoclonal antibody directed against the alpha-4 integrin molecule expressed on lymphocytes and thought to be important for lymphocyte trafficking in immune surveillance, was approved by the FDA for treatment of MS patients after a controlled trial in MS patients (Miller et al. 2003; Rudick 2006). The mechanism of how anti-alpha-4 integrin antibodies alter lymphocyte function and the rationale for use in multiple sclerosis was the subject of a review (Rice et al. 2005). In February 2005, reports of two patients with multiple sclerosis who had been treated with natalizumab therapy in the prospective study were diagnosed with PML, leading to withdrawal of natalizumab from the market. The first patient had a fatal course (Kleinschmidt-DeMasters and Tyler 2005). The second patient's PML was proven by brain biopsy. He had a rocky clinical course with severe clinical deficits and immune reconstitution disease in the brain when natalizumab was stopped, but survived (Langer-Gould et al. 2005). A third case of natalizumab-related PML was subsequently recognized in a patient being treated for Crohn's disease with natalizumab, who died of a progressive neurological deficit, initially diagnosed as fatal astrocytoma. Further investigation of his brain pathology from cerebral biopsy revealed PML (Van Assche et al. 2005). In both MS cases, the patients were treated with natalizumab for over 2 years, and the third Crohn's disease case had been chronically immunosuppressed with other agents before starting natalizumab approximately 5 months prior to developing PML. This long duration of therapy is consistent with other cases of PML occurring late (usually longer than a year) after immunosuppressants introduced for transplantation, in which the date of onset of immune suppression can be accurately determined.

Since that time, natalizumab has been reintroduced to the market with a close monitoring system and a registry of patients administered by Biogen Idec, and with data available to physicians online for assessment of risk to patients (https://medinfo.biogenidec.com/medinfo). Several aspects have emerged from the data about these patients (Clifford et al. 2010). The incidence of PML has increased with the duration of natalizumab use and is proportional to the duration of exposure in the first 3 years of use. Incidence of confirmed PML cases in patients who have had

24 or more infusions of natalizumab is now estimated to be nearly one case of PML per 1,000 patients (Clifford et al. 2010). As of December 1, 2011, Biogen Idec reports 193 cases of natalizumab-associated PML. Of these, 39 people died. Those who have survived experienced varying degrees of disability.

6.6 Other Drugs

Other specific immunosuppressant drugs that can predispose to PML are prednisone, methotrexate, cyclophosphamide, and cyclosporine (Berger and Houff 2009). It is also apparent that newer immunosuppressives like leflunomide are a predisposing factor for developing PML (Rahmlow et al. 2008). The exact risk of use of these drugs such as monoclonal antibodies targeting lymphocyte trafficking or antagonizing specific cytokines remains to be defined on an individual basis.

Mycophenolate mofetil is used for a variety of B- and T-cell disorders. It was initially developed as a post-transplant antirejection drug. However, the drug is used off label for many other autoimmune disorders because of its favorable side effect profile. The presence of PML in patients treated with mycophenolate mofetil has been reported. There are online resources that help with risk assessment (http://www.gene.com/gene/products/information/cellcept/), and this subject has been covered in a recent review (Berger and Houff 2009).

Rituximab is a therapeutic monoclonal anti-CD20 antibody that targets B cells selectively and removes them from the circulation. This monoclonal antibody has been used to treat a variety of B-cell neoplasms (CLL, lymphoma) and autoimmune diseases, specifically rheumatoid arthritis, and off label for systemic lupus erythematosus. PML has been shown to occur in rituximab-treated patients (Carson et al. 2009; Clifford et al. 2011b). It is estimated that the risk is approximately 1 per 25,000 patients with rheumatoid arthritis treated with rituximab (Clifford et al. 2011b). Systemic lupus erythematosus may have a higher propensity to PML (Clifford et al. 2011b).

Another monoclonal anti-CD11a antibody named efalizumab was marketed for the treatment of refractory psoriasis. It was withdrawn from the United States market as of June 2009, because of PML occurring in patients treated with this drug. The frequency was estimated to be 4 of 46,000 patients treated (Korman et al. 2009). However, whether the rare occurrence of a usually fatal neurological infection should prohibit drugs such as this from use in all cases is uncertain.

Screening by serologic means allows one to determine whether a patient carries latent JC virus (Bozic et al. 2011). Since PML is regarded as a reactivation infection, this serologic testing has been proposed as a screening tool to use these above-mentioned immunosuppressives only with great caution in seropositive individuals. There is no reliable presymptomatic way to detect PML or JC virus infection of the brain. Imaging surveillance or PCR techniques applied to the CSF currently are the only way to detect PML occurrence. Withdrawal of drug, possibly in combination with antiviral therapy, promotes survival.

6.7 *Sarcoidosis*

Systemic sarcoidosis is associated with defects in cell and humoral immunity, with a granulomatous pathologic reaction. There are many reports of PML occurring in sarcoidosis patients. However, whether the predisposition to PML is part of the primary disease, or a result of the immunosuppressive treatment often used chronically to treat the disease, such as prednisone or methotrexate, is unclear. Some sarcoid patients who develop PML are on no active treatment. In either case, PML is an infrequent opportunistic infection in sarcoidosis patients, though incidence data are lacking. In one series of non-AIDS PML cases, sarcoidosis was present in 9 % of the patients (Aksamit 2006).

7 Radiologic Features

An MRI scan of the head is far superior to a CT scan in visualizing abnormalities related to PML (Whiteman et al. 1993). A normal MRI head scan with focal neurological deficits would strongly argue against PML as the cause of neurological deficits in most clinical circumstances. It is rare for patients to have clinically detectable neurological deficits and no MRI abnormalities. This rare occurrence occurs occasionally because of the subcortical nature of the demyelination that occurs with PML, which may undermine the cortex and not be easily visualized on the MRI scan early in the course. However, the general rule is that MRI brain scan is almost always abnormal in association with PML despite the fact that the findings can be nonspecific, and therefore difficult to recognize. The typical MRI abnormalities are localized to the subcortical white matter at the gray–white junction, with increased T2 signal, little in the way of associated mass effect and little contrast enhancement after gadolinium administration (Fig. 2). Each of these rules has exceptions with documented PML, but generally these findings correlate well with PML.

Some authors have suggested that occipital PML is more common than frontal PML (Whiteman et al. 1993). Others' experience is that MRI abnormalities are equally common in the frontal lobes as compared to the occipital lobes. Indeed, any cerebral lobe is potentially vulnerable to disease and presumably some of the variability of the presenting syndrome represents the impact of the neurological deficit on the patient's functions of daily living. Likewise, any area of the white matter areas of the brainstem can be affected, but most commonly affected is the cerebellum. Again, the clinical deficit is referable to the location of the white matter affected. A focal abnormality on MRI scan with little mass effect and little contrast enhancement in the brainstem would suggest PML.

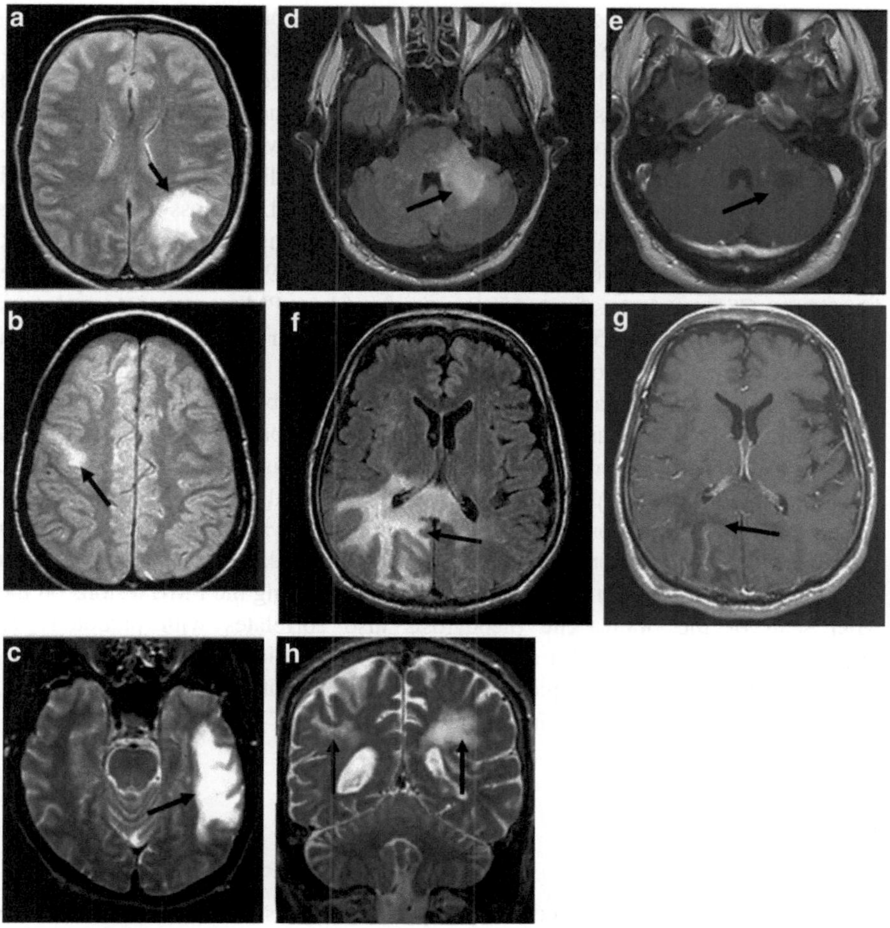

Fig. 2 MRI head scans of PML: (**a–c**) Axial images of bright T2 signal (*arrows*) of unifocal PML lesions undermining the cerebral gray–white junction of the occipital (**a**), frontal (**b**), and temporal (**c**) cortex; (**d**, **e**) Axial images of bright T2 FLAIR (**d**) and dark T1 minimally contrast enhancing (**e**) cerebellar and pontine PML lesions (*arrows*); (**f**, **g**) Axial images of bright T2 FLAIR (**f**) and dark T1 minimally contrast enhancing (**g**) occipital lobe and transcallosal PML lesions mimicking butterfly glioma or lymphoma; (**h**) Coronal image of bright T2 signal (*arrows*) of multifocal PML lesions undermining bilaterally the cerebral gray–white junction

8 Diagnostic Tests

PCR has been used to amplify JC virus DNA from the spinal fluid of PML patients. Current studies show that spinal fluid detection of JC virus is specific for pathologically significant PML (Bossolasco et al. 2005; Fong et al. 1995; Gibson et al. 1993; McGuire et al. 1995; Telenti et al. 1992; Weber et al. 1994). All of these studies have been retrospective analyses of spinal fluid specimens of patients with either pathologically confirmed or clinically suspected PML, based on MRI and coincident underlying immunosuppressive illness. Sensitivity of detection is not 100 %. Specifically, combining the data from studies published show that 76 % were correctly diagnosed by this assay (Aksamit 1997). It is unclear why false negatives occur. The most likely explanation is low abundance of virus DNA in spinal fluid. Other possible explanations include storage and handling of specimens, small volume of spinal fluid assay, inhibitors in the spinal fluid, and loss of DNA during CSF concentration. Combining data from studies (Aksamit 1997; Fong et al. 1995; Gibson et al. 1993; McGuire et al. 1995; Weber et al. 1994), the false-positive rate identifying CSF specimens from AIDS patients without PML or other controls was only 2 %. Sensitivity of 75 % has been born out in larger routine experiences with PML in AIDS patients (Bossolasco et al. 2005).

PCR analysis of spinal fluid for JC virus is the best noninvasive test for confirmation of PML caused by JC virus. The other parameters of spinal fluid analysis are typically normal. Minimal spinal pleocytosis may be evident, but the usual finding is <20 cells/μL. Greater than 20 cells would suggest the possibility of either a different infection or relative immune preservation of the host with a prominent cellular reaction against the virus. In general, a pleocytosis correlates with increased gadolinium enhancement or IRIS involving the PML lesions on the MRI scan of the brain. The pleocytosis also correlates with pathology of perivascular lymphocytic infiltrates involving the brain in relatively immune preserved hosts. CSF protein is typically only modestly elevated, usually <100 mg/dL. Other microbiological studies, including culturing for JC virus, is unrevealing.

9 Pathology

Brain biopsy of the advancing edges of suspected PML lesions is advised to confirm diagnosis when spinal fluid PCR is negative for JC virus. In situ hybridization or immunohistochemistry is among the best confirmatory techniques for identifying JC virus in the biopsy specimen (Aksamit 1993; Aksamit et al. 1985, 1986). Cells (oligodendrocytes) at the gray–white matter junction are the most common site for initial infection, and are the best candidate location when searching for pathologic evidence of infection.

The spectrum of neuropathology in the brain of JC virus infection in PML patients has been widened recently. JC virus behaves differently in oligodendrocytes, astrocytes, and granule cell neurons of the cerebellum.

The demyelination of PML has long been known to be secondary to oligodendrocyte infection and lysis (Astrom et al. 1958). The pathology of PML includes associated nuclear enlargement of infected oligodendrocytes (Fig. 3). Virion accumulations in the nucleus produce the light microscopic "inclusion bearing" oligodendrocytes. These oligodendrocyte nuclei are markedly enlarged, with ground glass, amphophilic, basophilic, or sometimes eosinophilic staining characteristics after hematoxylin and eosin staining. The nuclear enlargement of

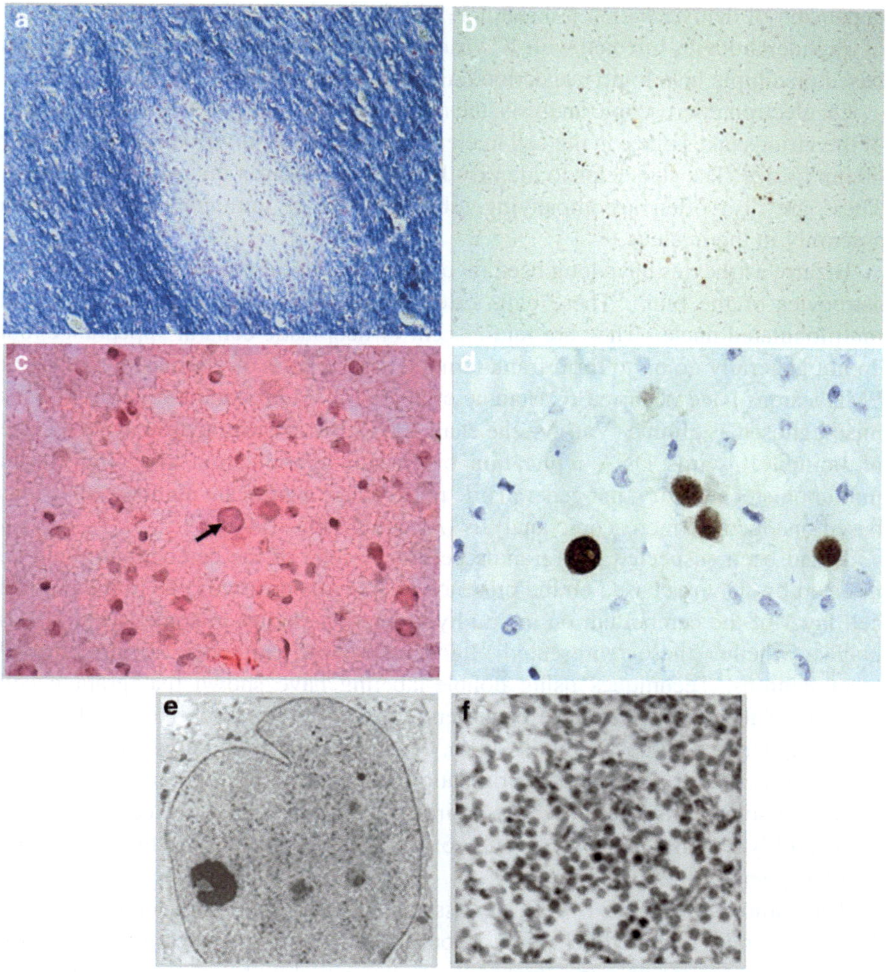

Fig. 3 Oligodendrocyte involvement in PML: (**a**) PML microscopic area of demyelination (*center*) surrounded by normal myelin stain (Luxol fast blue, original magnification 100×); (**b**) Same lesion as shown in (**a**) stained by in situ hybridization for JC virus DNA, *brown-stained* nuclei identify infected oligodendrocyte nuclei spreading out from the demyelinated center (hematoxylin counterstain, original magnification 100×); (**c**) Enlarged oligodendrocyte nucleus (*arrow*, hematoxylin and eosin stain, original magnification 630×); (**d**) In situ hybridization for JC virus DNA, *brown-stained* nuclei identifies infected oligodendrocyte nuclei (hematoxylin counterstain, original magnification 400×); (**e**) Electron micrograph of an enlarged oligodendrocyte nucleus filled with JC virus virions (*small dots*) expanding the nucleus (original magnification 6,300×); (**f**) Electron micrograph of JC virus 40-nm virions packing the nucleus of an infected oligodendrocyte (original magnification 75,000×)

the oligodendrocytes proceeds to lytic death of the oligodendrocyte, and the release of virus to infect neighboring cells. Virus spreads in a centrifugal way from a central nidus of infection to neighboring cells, leading to a circumferential expansion of demyelination (Aksamit 1995). By in situ hybridization, the number of oligodendrocytes infected with JC virus is more numerous than would suspected based on simple histologic inspection (Aksamit et al. 1986).

On electron microscopic analysis, the nuclei are filled with JC virions. Assembly of the virions takes place in the cell nucleus, into icosahedral virions approximately 40 nm in size. The size helps to identify JC virus on electron microscopic analysis. There are also often accompanying tubular forms of immature capsid protein assembly in the nucleus.

Bizarre astrocytes have long been recognized as a non-lytic JC virus infection of astrocytes of the brain. These cells often look bizarre with giant enlarged and multilobulated nuclei. They are reminiscent of neoplastic cells in appearance, but do not generally go on to form frank tumors. Rare reports of gliomas in the bed of PML lesions have occurred (Castaigne et al. 1974). More commonly, the PML is misdiagnosed as glioma (Van Assche et al. 2005). Bizarre astrocytes have evidence of limited JC virus DNA replication by in situ hybridization and viral capsid immunohistochemistry, but generally do not produce significant numbers of virions based on electron microscopic analysis (Aksamit 1993, 1995).

It had been suspected that granule cell layer cells of the cerebellum could be infected by JC virus based on the presence of JC virus positive cells in the granule cell layer of the cerebellum on in situ hybridization studies in some cases. It was unclear whether these represented oligodendrocytes, astrocytes, or granule cell layer neurons. Techniques using double-labeling have shown that granule cell neurons of the cerebellum have been demonstrated to be infectible with JC virus (Du Pasquier et al. 2003; Koralnik 2006). Granule cell neurons show lytic infection by JC virus, based on in situ hybridization studies, the presence of viral capsid protein in some cells, the presence of virions within the cells by electron microscopy, and lytic cell death with atrophy of the granule cell layer of the cerebellum (Du Pasquier et al. 2003).

The pathology of PML is typically associated with limited or no inflammation. This is generally regarded to be proportional to the immunosuppression of the host. That is, more severe immunosuppression is associated with little or no inflammation. By contrast, AIDS PML patients who have been treated with cART have a better chance of clearing the virus (Clifford et al. 1999). Patients who have had immune reconstitution in AIDS have developed IRIS in the nervous system (Du Pasquier and Koralnik 2003).

10 Treatment

There are no good treatments for PML. General principles about treatment include improvement in immune status, possibly in combination with antiviral therapy, promotes survival. Also, antiviral therapies can be offered if neurologic stabilization satisfies the quality-of-life goals for the patient. Although the prognosis of PML is generally dismal, removal of the immune suppression influence of an external drug allows the patient's own immune system to clear JC virus from the brain. This is an effective approach, but can also lead to IRIS which, when it occurs in the brain after immune restoration, may need treatment. IRIS should be treated if accompanied by neurologic deterioration with short-term corticosteroids.

If the patient's immunosuppression is AIDS-related, cART should be initiated if it has not previously been used. If the AIDS patient is already receiving cART, the therapy should be changed to optimize treatment, with the goals of normalization or near normalization of the CD4 count. Cytosine arabinoside has failed in AIDS-related PML patients (Hall et al. 1998). For AIDS patients with PML, or failing non-AIDS patients with neurologic deterioration, cidofovir can be considered (De Luca et al. 2001; Razonable et al. 2001; Viallard et al. 2007), although several other studies have suggested it is ineffective (De Luca et al. 2008; Houston et al. 2001; Marra et al. 2002).

For non-AIDS PML patients, there is no effective therapy. However if the goal to stabilize neurologic deterioration is acceptable in the clinical context of the systemic disease, one may consider intravenous cytosine arabinoside 2 mg/kg/day daily for 5 days. A single non-blinded study showed approximately a 30 % response rate in patients where 85 % mortality was expected with 1 year (Aksamit 2001).

Mirtazapine or risperidone have also been suggested as options for treatment, but their effectiveness is not yet been proven. The receptor for JC virus entry into the cell has been reported to be a subtype of the serotonin receptors 5-HT2A (Elphick et al. 2004), combined with a sialic acid–*N*-linked glycoprotein on the cell membrane. These findings have led to proposals to treat PML with a variety of psychotropic medications known to block the 5-HT2A receptor. Mirtazapine, an antidepressant, and risperidone, an antipsychotic medication, have been proposed to be most specific for potential blockade of this receptor. Mirtazapine has been used anecdotally at 15–30 mg/day, and successes have been reported in patients with non-AIDS-related PML (Owczarczyk et al. 2007; Vulliemoz et al. 2006). This result cannot be fully attributed to the 5-HT receptor blockade, however, because these patients also had discontinuation or alteration of their immunosuppressive regimen and treatment with cytosine arabinoside or cidofovir. Risperidone has been suggested to be a high-affinity receptor blocker and therefore potentially more potent (Kast et al. 2007).

Mefloquine has been suggested to be potentially helpful based on in vitro screening of compounds with activity against JCV. It has in vitro activity against JC virus. Mefloquine hydrochloride was used in several cases of rituximab-associated PML (Clifford et al. 2011b). However, a recent clinical trial was stopped for lack of demonstrable efficacy (Clifford et al. 2011a).

Alpha-interferon has been anecdotally reported to have some success in treating PML; however, some of these patients were treated with multiple agents making these reports difficult to interpret. Currently, the consensus (largely from anecdotal experience) is that α-interferon is not helpful in the treatment of PML (Geschwind et al. 2001; Koralnik 2006). β-Interferon also seems to be of no help (Nath et al. 2006). Two DNA topoisomerase inhibitor drugs, camptothecin and topotecan, have been shown to have antiviral activity in vitro against JC virus. These drugs are available as antineoplastic agents, and a few case reports of their use in treating PML have been published (Royal et al. 2003; Vollmer-Haase et al. 1997). No series of PML patients treated in this fashion exists to suggest that these drugs are successful in patients in whom other therapies have failed.

11 IRIS

IRIS can occur in the brain after PML diagnosis and withdrawal of the immunosuppressive agent. The immune response is important in clearing JC virus from the brain. Therefore at least limited inflammation is probably important in neurologic survival (Berger 2009). IRIS, however, is thought to be injurious to the brain, and many cases need additional treatment. Treatment is typically short courses of high-dose intravenous methylprednisolone, usually 1 g/day for 5 days (Tan et al. 2009). However there is no universal consensus on route, type, dose, or duration of corticosteroid therapy. Despite treatment success and survival, the PML-associated deficits are expected to be permanent.

In the circumstance of natalizumab-associated PML, management of the PML has routinely used plasma exchange (PLEX) or immunoabsorption to hasten clearance of the drug and shorten the period in which natalizumab remains active (usually several months). Exacerbation of symptoms and enlargement of lesions on MRI have occurred within a few days to a few weeks after PLEX, indicative of IRIS. This syndrome seems to be more common and more severe in patients with natalizumab-associated PML than it is in patients with HIV-associated PML treated with cART.

12 Prognosis

In general, PML has been regarded as nearly a universally fatal disease. However more recent experience with PML suggests it can be survived. A pre-AIDS era study showed the 4-month survival to be 30 %, and 12-month survival 15 % (Brooks and Walker 1984). In the pre-cART AIDS era, PML was fatal in 95 % in 6 months (Clifford et al. 1999). Institution of optimized cART therapy has produced 50 % survival at 1 year (Marra et al. 2002). Natalizumab survivors have had a higher survival rate of 80 % (https://medinfo.biogenidec.com/medinfo); although PML-associated deficits are expected to be permanent. These patients usually also require the treatment of IRIS.

References

Aksamit AJ Jr (1993) Nonradioactive in situ hybridization in progressive multifocal leukoencephalopathy. Mayo Clin Proc 68(9):899–910

Aksamit AJ Jr (1995) Progressive multifocal leukoencephalopathy: a review of the pathology and pathogenesis. Microsc Res Tech 32(4):302–311. doi:10.1002/jemt.1070320405

Aksamit AJ (1997) Cerebrospinal fluid in the diagnosis of central nervous system infections. In: Roos KL (ed) Central nervous system infectious diseases and therapy. Marcel Dekker, New York, pp 731–745

Aksamit AJ (2001) Treatment of non-AIDS progressive multifocal leukoencephalopathy with cytosine arabinoside. J Neurovirol 7(4):386–390. doi:10.1080/13550280152537292

Aksamit AJ (2006) Review of progressive multifocal leukoencephalopathy and natalizumab. Neurologist 12(6):293–298. doi:10.1097/01.nrl.0000250948.04681.96

Aksamit AJ (2008) Progressive multifocal leukoencephalopathy. Curr Treat Options Neurol 10 (3):178–185

Aksamit AJ, Mourrain P, Sever JL, Major EO (1985) Progressive multifocal leukoencephalopathy: investigation of three cases using in situ hybridization with JC virus biotinylated DNA probe. Ann Neurol 18(4):490–496. doi:10.1002/ana.410180412

Aksamit AJ, Sever JL, Major EO (1986) Progressive multifocal leukoencephalopathy: JC virus detection by in situ hybridization compared with immunohistochemistry. Neurology 36 (4):499–504

Astrom KE, Mancall EL, Richardson EP Jr (1958) Progressive multifocal leuko-encephalopathy; a hitherto unrecognized complication of chronic lymphatic leukaemia and Hodgkin's disease. Brain 81(1):93–111

Berenguer J, Miralles P, Arrizabalaga J, Ribera E, Dronda F, Baraia-Etxaburu J, Domingo P, Marquez M, Rodriguez-Arrondo FJ, Laguna F, Rubio R, Lacruz Rodrigo J, Mallolas J, de Miguel V (2003) Clinical course and prognostic factors of progressive multifocal leukoencephalopathy in patients treated with highly active antiretroviral therapy. Clin Infect Dis 36 (8):1047–1052. doi:10.1086/374048

Berger JR (2009) Steroids for PML-IRIS: a double-edged sword? Neurology 72(17):1454–1455. doi:10.1212/01.wnl.0000343735.44983.5e

Berger JR, Houff S (2009) Opportunistic infections and other risks with newer multiple sclerosis therapies. Ann Neurol 65(4):367–377. doi:10.1002/ana.21630

Berger JR, Major EO (1999) Progressive multifocal leukoencephalopathy. Semin Neurol 19(2):193–200. doi:10.1055/s-2008-1040837

Berger JR, Pall L, Lanska D, Whiteman M (1998) Progressive multifocal leukoencephalopathy in patients with HIV infection. J Neurovirol 4(1):59–68. doi:10.3109/13550289809113482

Bossolasco S, Calori G, Moretti F, Boschini A, Bertelli D, Mena M, Gerevini S, Bestetti A, Pedale R, Sala S, Lazzarin A, Cinque P (2005) Prognostic significance of JC virus DNA levels in cerebrospinal fluid of patients with HIV-associated progressive multifocal leukoencephalopathy. Clin Infect Dis 40(5):738–744. doi:10.1086/427698

Bozic C, Richman S, Plavina T, Natarajan A, Scanlon J, Subramanyam M, Sandrock A, Bloomgren G (2011) Anti-John Cunnigham virus antibody prevalence in multiple sclerosis patients: baseline results of STRATIFY-1. Ann Neurol 70(5):713–721

Brooks BR, Walker DL (1984) Progressive multifocal leukoencephalopathy. Neurol Clin 2 (2):299–313

Calabrese LH, Molloy ES (2008) Progressive multifocal leucoencephalopathy in the rheumatic diseases: assessing the risks of biological immunosuppressive therapies. Ann Rheum Dis 67 (Suppl 3):iii64–65. doi:10.1136/ard.2008.097972

Carson KR, Evens AM, Richey EA, Habermann TM, Focosi D, Seymour JF, Laubach J, Bawn SD, Gordon LI, Winter JN, Furman RR, Vose JM, Zelenetz AD, Mamtani R, Raisch DW, Dorshimer GW, Rosen ST, Muro K, Gottardi-Littell NR, Talley RL, Sartor O, Green D, Major EO, Bennett CL (2009) Progressive multifocal leukoencephalopathy after rituximab therapy in HIV-negative patients: a report of 57 cases from the research on adverse drug events and reports project. Blood 113(20):4834–4840. doi:10.1182/blood-2008-10-186999

Castaigne P, Rondot P, Escourolle R, Ribadeau dumas J, Cathala F, Hauw J (1974) Progressive multifocal leukoencephalopathy and multiple gliomas. Rev Neurol 130(9–10):379–392

Clifford DB, Yiannoutsos C, Glicksman M, Simpson DM, Singer EJ, Piliero PJ, Marra CM, Francis GS, McArthur JC, Tyler KL, Tselis AC, Hyslop NE (1999) HAART improves prognosis in HIV-associated progressive multifocal leukoencephalopathy. Neurology 52 (3):623–625

Clifford DB, De Luca A, Simpson DM, Arendt G, Giovannoni G, Nath A (2010) Natalizumab-associated progressive multifocal leukoencephalopathy in patients with multiple sclerosis: lessons from 28 cases. Lancet Neurol 9(4):438–446. doi:10.1016/S1474-4422(10)70028-4

Clifford D, Nath A, Cinque P, Brew BJ, Workman A, Gorelik L, Zhao J, Duda P (2011a) Mefloquine treatment in patients with progressive multifocal leukoencephalopathy. Neurology 9(4 Suppl):A28

Clifford DB, Ances B, Costello C, Rosen-Schmidt S, Andersson M, Parks D, Perry A, Yerra R, Schmidt R, Alvarez E, Tyler KL (2011b) Rituximab-associated progressive multifocal leukoencephalopathy in rheumatoid arthritis. Arch Neurol 68(9):1156–1164. doi:10.1001/archneurol.2011.103

De Luca A, Giancola ML, Ammassari A, Grisetti S, Cingolani A, Larussa D, Alba L, Murri R, Ippolito G, Cauda R, Monforte A, Antinori A (2001) Potent anti-retroviral therapy with or without cidofovir for AIDS-associated progressive multifocal leukoencephalopathy: extended follow-up of an observational study. J Neurovirol 7(4):364–368

De Luca A, Ammassari A, Pezzotti P, Cinque P, Gasnault J, Berenguer J, Di Giambenedetto S, Cingolani A, Taoufik Y, Miralles P, Marra CM, Antinori A (2008) Cidofovir in addition to antiretroviral treatment is not effective for AIDS-associated progressive multifocal leukoencephalopathy: a multicohort analysis. AIDS 22(14):1759–1767. doi:10.1097/QAD.0b013e32830a5043

Del Valle L, Azizi SA, Krynska B, Enam S, Croul SE, Khalili K (2000) Reactivation of human neurotropic JC virus expressing oncogenic protein in a recurrent glioblastoma multiforme. Ann Neurol 48(6):932–936

Du Pasquier RA, Koralnik IJ (2003) Inflammatory reaction in progressive multifocal leukoencephalopathy: harmful or beneficial? J Neurovirol 9(Suppl 1):25–31. doi:10.1080/13550280390195315

Du Pasquier RA, Corey S, Margolin DH, Williams K, Pfister LA, De Girolami U, Mac Key JJ, Wuthrich C, Joseph JT, Koralnik IJ (2003) Productive infection of cerebellar granule cell neurons by JC virus in an HIV + individual. Neurology 61(6):775–782

Elphick GF, Querbes W, Jordan JA, Gee GV, Eash S, Manley K, Dugan A, Stanifer M, Bhatnagar A, Kroeze WK, Roth BL, Atwood WJ (2004) The human polyomavirus, JCV, uses serotonin receptors to infect cells. Science 306(5700):1380–1383. doi:10.1126/science.1103492

Fong IW, Britton CB, Luinstra KE, Toma E, Mahony JB (1995) Diagnostic value of detecting JC virus DNA in cerebrospinal fluid of patients with progressive multifocal leukoencephalopathy. J Clin Microbiol 33(2):484–486

Frisque RJ, Bream GL, Cannella MT (1984) Human polyomavirus JC virus genome. J Virol 51(2):458–469

Gallia GL, Houff SA, Major EO, Khalili K (1997) Review: JC virus infection of lymphocytes—revisited. J Infect Dis 176(6):1603–1609

Geschwind MD, Skolasky RI, Royal WS, McArthur JC (2001) The relative contributions of HAART and alpha-interferon for therapy of progressive multifocal leukoencephalopathy in AIDS. J Neurovirol 7(4):353–357. doi:10.1080/13550280152537238

Gibson PE, Knowles WA, Hand JF, Brown DW (1993) Detection of JC virus DNA in the cerebrospinal fluid of patients with progressive multifocal leukoencephalopathy. J Med Virol 39(4):278–281

Hall CD, Dafni U, Simpson D, Clifford D, Wetherill PE, Cohen B, McArthur J, Hollander H, Yainnoutsos C, Major E, Millar L, Timpone J (1998) Failure of cytarabine in progressive multifocal leukoencephalopathy associated with human immunodeficiency virus infection. AIDS Clinical Trials Group 243 Team. N Engl J Med 338(19):1345–1351. doi:10.1056/NEJM199805073381903

Houff SA, Major EO, Katz DA, Kufta CV, Sever JL, Pittaluga S, Roberts JR, Gitt J, Saini N, Lux W (1988) Involvement of JC virus-infected mononuclear cells from the bone marrow and spleen in the pathogenesis of progressive multifocal leukoencephalopathy. N Engl J Med 318(5):301–305. doi:10.1056/NEJM198802043180507

Houston S, Roberts N, Mashinter L (2001) Failure of cidofovir therapy in progressive multifocal leukoencephalopathy unrelated to human immunodeficiency virus. Clin Infect Dis 32(1):150–152. doi:10.1086/317539

Kast RE, Focosi D, Petrini M, Altschuler EL (2007) Treatment schedules for 5-HT2A blocking in progressive multifocal leukoencephalopathy using risperidone or ziprasidone. Bone Marrow Transplant 39(12):811–812. doi:10.1038/sj.bmt.1705682

Kleinschmidt-DeMasters BK, Tyler KL (2005) Progressive multifocal leukoencephalopathy complicating treatment with natalizumab and interferon beta-1a for multiple sclerosis. N Engl J Med 353(4):369–374. doi:10.1056/NEJMoa051782

Koralnik IJ (2006) Progressive multifocal leukoencephalopathy revisited: has the disease outgrown its name? Ann Neurol 60(2):162–173. doi:10.1002/ana.20933

Korman BD, Tyler KL, Korman NJ (2009) Progressive multifocal leukoencephalopathy, efalizumab, and immunosuppression: a cautionary tale for dermatologists. Arch Dermatol 145(8):937–942. doi:10.1001/archdermatol.2009.175

Langer-Gould A, Atlas SW, Green AJ, Bollen AW, Pelletier D (2005) Progressive multifocal leukoencephalopathy in a patient treated with natalizumab. N Engl J Med 353(4):375–381. doi:10.1056/NEJMoa051847

Liu CK, Wei G, Atwood WJ (1998) Infection of glial cells by the human polyomavirus JC is mediated by an N-linked glycoprotein containing terminal alpha(2–6)-linked sialic acids. J Virol 72(6):4643–4649

Major EO (2000) From telomeres to T-antigens: many roads…multiple pathways…novel associations in the search for the origins of human gliomas. Ann Neurol 48(6):823–825

Major EO, Amemiya K, Elder G, Houff SA (1990) Glial cells of the human developing brain and B cells of the immune system share a common DNA binding factor for recognition of the regulatory sequences of the human polyomavirus, JCV. J Neurosci Res 27(4):461–471. doi:10.1002/jnr.490270405

Major EO, Amemiya K, Tornatore CS, Houff SA, Berger JR (1992) Pathogenesis and molecular biology of progressive multifocal leukoencephalopathy, the JC virus-induced demyelinating disease of the human brain. Clin Microbiol Rev 5(1):49–73

Marra CM, Rajicic N, Barker DE, Cohen BA, Clifford D, Donovan Post MJ, Ruiz A, Bowen BC, Huang ML, Queen-Baker J, Andersen J, Kelly S, Shriver S (2002) A pilot study of cidofovir for progressive multifocal leukoencephalopathy in AIDS. AIDS 16(13):1791–1797

Mateen FJ, Muralidharan R, Carone M, van de Beek D, Harrison DM, Aksamit AJ, Gould MS, Clifford DB, Nath A (2011) Progressive multifocal leukoencephalopathy in transplant recipients. Ann Neurol 70(2):305–322. doi:10.1002/ana.22408

McGuire D, Barhite S, Hollander H, Miles M (1995) JC virus DNA in cerebrospinal fluid of human immunodeficiency virus-infected patients: predictive value for progressive multifocal leukoencephalopathy. Ann Neurol 37(3):395–399. doi:10.1002/ana.410370316

Miller DH, Khan OA, Sheremata WA, Blumhardt LD, Rice GP, Libonati MA, Willmer-Hulme AJ, Dalton CM, Miszkiel KA, O'Connor PW (2003) A controlled trial of natalizumab for relapsing multiple sclerosis. N Engl J Med 348(1):15–23. doi:10.1056/NEJMoa020696

Nath A, Venkataramana A, Reich DS, Cortese I, Major EO (2006) Progression of progressive multifocal leukoencephalopathy despite treatment with beta-interferon. Neurology 66 (1):149–150. doi:10.1212/01.wnl.0000191322.93310.a1

Owczarczyk K, Hilker R, Brunn A, Hallek M, Rubbert A (2007) Progressive multifocal leucoencephalopathy in a patient with sarcoidosis–successful treatment with cidofovir and mirtazapine. Rheumatology 46(5):888–890. doi:10.1093/rheumatology/kem049

Padgett BL, Walker DL, ZuRhein GM, Eckroade RJ, Dessel BH (1971) Cultivation of papova-like virus from human brain with progressive multifocal leucoencephalopathy. Lancet 1 (7712):1257–1260

Rahmlow M, Shuster EA, Dominik J, Deen HG Jr, Dickson DW, Aksamit AJ Jr, Robles HA, Freeman WD (2008) Leflunomide-associated progressive multifocal leukoencephalopathy. Arch Neurol 65(11):1538–1539. doi:10.1001/archneur.65.11.1538

Razonable RR, Aksamit AJ, Wright AJ, Wilson JW (2001) Cidofovir treatment of progressive multifocal leukoencephalopathy in a patient receiving highly active antiretroviral therapy. Mayo Clin Proc 76(11):1171–1175

Rice GP, Hartung HP, Calabresi PA (2005) Anti-alpha4 integrin therapy for multiple sclerosis: mechanisms and rationale. Neurology 64(8):1336–1342. doi:10.1212/01.WNL.0000158329.30470.D0
Richardson EP Jr (1961) Progressive multifocal leukoencephalopathy. N Engl J Med 265:815–823. doi:10.1056/NEJM196110262651701
Royal W 3rd, Dupont B, McGuire D, Chang L, Goodkin K, Ernst T, Post MJ, Fish D, Pailloux G, Poncelet H, Concha M, Apuzzo L, Singer E (2003) Topotecan in the treatment of acquired immunodeficiency syndrome-related progressive multifocal leukoencephalopathy. J Neurovirol 9(3):411–419. doi:10.1080/13550280390201740
Rudick R, Stuart W, Calabresi P, et al. (2006) Natalizumab plus Interferon Beta-1a for relapsing multiple sclerosis. New England Journal of Medicine 354:911–923
Ryschkewitsch CF, Jensen PN, Monaco MC, Major EO (2010) JC virus persistence following progressive multifocal leukoencephalopathy in multiple sclerosis patients treated with natalizumab. Ann Neurol 68(3):384–391. doi:10.1002/ana.22137
Tan K, Roda R, Ostrow L, McArthur J, Nath A (2009) PML-IRIS in patients with HIV infection: clinical manifestations and treatment with steroids. Neurology 72(17):1458–1464. doi:10.1212/01.wnl.0000343510.08643.74
Telenti A, Marshall WF, Aksamit AJ, Smilack JD, Smith TF (1992) Detection of JC virus by polymerase chain reaction in cerebrospinal fluid from two patients with progressive multifocal leukoencephalopathy. Eur J Clin Microbiol Infect Dis 11(3):253–254
Van Assche G, Van Ranst M, Sciot R, Dubois B, Vermeire S, Noman M, Verbeeck J, Geboes K, Robberecht W, Rutgeerts P (2005) Progressive multifocal leukoencephalopathy after natalizumab therapy for Crohn's disease. N Engl J Med 353(4):362–368. doi:10.1056/NEJMoa051586
Viallard JF, Lazaro E, Ellie E, Eimer S, Camou F, Caubet O, Lafon ME, Fleury H, Pellegrin JL (2007) Improvement of progressive multifocal leukoencephalopathy after cidofovir therapy in a patient with a destructive polyarthritis. Infection 35(1):33–36. doi:10.1007/s15010-006-5103-y
Vollmer-Haase J, Young P, Ringelstein EB (1997) Efficacy of camptothecin in progressive multifocal leucoencephalopathy. Lancet 349(9062):1366. doi:10.1016/S0140-6736(05)63201-1
von Einsiedel RW, Fife TD, Aksamit AJ, Cornford ME, Secor DL, Tomiyasu U, Itabashi HH, Vinters HV (1993) Progressive multifocal leukoencephalopathy in AIDS: a clinicopathologic study and review of the literature. J Neurol 240(7):391–406
Vulliemoz S, Lurati-Ruiz F, Borruat FX, Delavelle J, Koralnik IJ, Kuntzer T, Bogousslavsky J, Picard F, Landis T, Du Pasquier RA (2006) Favourable outcome of progressive multifocal leucoencephalopathy in two patients with dermatomyositis. J Neurol Neurosurg Psychiatry 77(9):1079–1082. doi:10.1136/jnnp. 2006.092353
Walker D, Frisque R (1986) The biology and molecular biology of JC virus. In: Salzman NP (ed) The Papovaviridae. Plenum, New York
Weber T, Major EO (1997) Progressive multifocal leukoencephalopathy: molecular biology, pathogenesis and clinical impact. Intervirology 40(2–3):98–111
Weber T, Turner RW, Frye S, Luke W, Kretzschmar HA, Luer W, Hunsmann G (1994) Progressive multifocal leukoencephalopathy diagnosed by amplification of JC virus-specific DNA from cerebrospinal fluid. AIDS 8(1):49–57
Weiner LP, Herndon RM, Narayan O, Johnson RT, Shah K, Rubinstein LJ, Preziosi TJ, Conley FK (1972) Isolation of virus related to SV40 from patients with progressive multifocal leukoencephalopathy. N Engl J Med 286(8):385–390. doi:10.1056/NEJM197202242860801
Whiteman ML, Post MJ, Berger JR, Tate LG, Bell MD, Limonte LP (1993) Progressive multifocal leukoencephalopathy in 47 HIV-seropositive patients: neuroimaging with clinical and pathologic correlation. Radiology 187(1):233–240
Zurhein G, Chou SM (1965) Particles resembling papova viruses in human cerebral demyelinating disease. Science 148(3676):1477–1479

Varicella Zoster Virus Infections

Maria A. Nagel, Randall J. Cohrs, and Don Gilden

Abstract Varicella zoster virus (VZV) is an exclusively human neurotropic alphaherpesvirus. Primary infection causes varicella (chickenpox), after which the virus becomes latent in ganglionic neurons along the entire neuraxis. With advancing age or immunosuppression, cell-mediated immunity to VZV declines and virus reactivates to cause zoster (shingles), dermatomal-distribution pain and rash on an erythematous base in 1–3 dermatomes. Zoster may develop anywhere on the body. Skin lesions resolve within 1–2 weeks, while complete cessation of pain usually takes 4–6 weeks. Unfortunately, zoster can be followed by chronic pain (postherpetic neuralgia), cranial nerve palsies, zoster paresis, vasculopathy, meningoencephalitis, cerebellitis, myelopathy, and multiple ocular disorders. VZV reactivation also produces chronic radicular pain without rash (zoster sine herpete). In fact, all of the neurological and ocular disorders listed above may also develop without rash. This review covers clinical, laboratory and pathological features of neurological complications of VZV reactivation, including diagnostic testing to verify active VZV infection in the nervous system and the potential value of examining saliva for VZV DNA in patients with neurological disease without rash. Additional perspectives are provided by discussions of the VZV genome, VZV latency, viral pathology, pathogenesis and immunity, and of the value of vaccination of elderly individuals to boost cell-mediated immunity to VZV and prevent VZV reactivation.

Keywords Latency • Neurologic disease • Pathogenesis • Reactivation • VZV

M.A. Nagel (✉) • R.J. Cohrs • D. Gilden
Departments of Neurology and Microbiology, University of Colorado School of Medicine, 12700 E. 19th Avenue, Box B182, Aurora, CO 80045, USA
e-mail: maria.nagel@ucdenver.edu; randall.cohrs@ucdenver.edu; don.gilden@ucdenver.edu

A.C. Jackson (ed.), *Viral Infections of the Human Nervous System*,
Birkhäuser Advances in Infectious Diseases,
DOI 10.1007/978-3-0348-0425-7_5,

1 Varicella Zoster Virus Genome

The 124,884-bp varicella zoster virus (VZV) DNA was the first herpesvirus genome to be completely sequenced (Davison and Scott 1986). To date, the DNA sequence of 24 VZV isolates has been determined. Genomic variation among VZV isolates is extremely low: 0.2% in total DNA size range, >95% sequence identity and only 557 single nucleotide polymorphisms (Tyler et al. 2007). While five phylogenetic clades of VZV have been identified, no phenotypic changes associated with the differences in DNA sequence have been found (Breuer et al. 2010). Of the human alphaherpesviruses [herpes simplex virus (HSV)-1, -2 and VZV], the VZV genome is the smallest and is extraordinarily stable after propagation multiple times in tissue culture (Sauerbrei et al. 2007). As in all neurotropic alphaherpesviruses, the VZV genome is composed of two unique segments, each bound by smaller inverted repeated segments resulting in four different DNA isomers. Thus, viral genes contained within the repeated DNA segments are present twice per genome. However, while all four orientations of the VZV genome are found in the virus particle, only two predominate.

The VZV genome is highly compact. The virus DNA molecule contains 71 open reading frames (ORFs) >300 bp in length. Three ORFs are duplicated because of their location within inverted repeated DNA segments (Davison and Scott 1986). All 68 VZV ORFs (Cohrs et al. 2003; Kennedy et al. 2005) as well as three additional VZV ORFs (McMillan et al. 1997; Ross et al. 1997; Kemble et al. 2000) have been detected in RNA extracted from productively infected cells in tissue culture. The 71 VZV ORFs are separated on average by 187 nucleotides, indicating that the promoter regions for these genes are highly compact, a notion supported by experimental analysis of the transcriptional regulatory region in various VZV genes (Inchauspe et al. 1989; Disney et al. 1990; Ling et al. 1992; Michael et al. 1998; Cohrs et al. 1998).

Herpesvirus genes are expressed in a coordinated fashion (Honess and Roizman 1974, 1975). Immediate-early (IE) virus genes, the first to be expressed in infected cells, are transcribed in the presence of cycloheximide, an inhibitor of translation, and encode proteins that modify the host cell transcription machinery to recognize virus promoters and inactivate the host response to foreign DNA. Initiation of early virus gene transcription is dependent on prior IE gene products and precedes virus DNA replication. Early herpesvirus genes are transcribed in the presence of acyclovir, an inhibitor of virus DNA replication, and encode proteins required to replicate the virus genome. Late herpesvirus genes are transcribed after initiation of virus DNA replication and encode structural proteins required for packaging of the replicated virus DNA and for assembly of infectious progeny virus particles. While the temporal expression of HSV-1 genes is well described, the lack of high-titer, cell-free VZV has hindered a similar description of VZV genes. Thus, the kinetic classification of VZV genes remains based on their homology to well-characterized HSV-1 genes.

2 VZV Pathogenesis

VZV is a highly contagious human pathogen. While VZV DNA can be found on surfaces such as door knobs, toys, air-conditioning filters and tables in rooms where children or adults have active varicella or zoster, the enveloped VZV is highly sensitive to desiccation such that infection is typically acquired through inhalation of aerosolized virus (Connelly et al. 1993; Asano et al. 1999; Yoshikawa et al. 2001). Since the virus has a host range restricted to humans, the exact mechanism of disease production remains unclear. The currently accepted scenario after exposure to the virus involves infection of Langerhans and plasmacytoid dendritic cells of the upper respiratory mucosa and nasal pharyngeal region (Abendroth et al. 2001; Morrow et al. 2003), with subsequent transfer of the virus by these antigen-presenting cells to CD4 T cells which, in turn, transmit the virus to dermal endothelial cells (Abendroth et al. 2000; Taylor and Moffat 2005; Huch et al. 2010). During primary infection, VZV gains access to ganglionic neurons, most likely via a hematogenous route as supported by the finding that ganglionic infection with simian varicella virus, a closely related neurotropic alphaherpesvirus, precedes the onset of rash (Mahalingam et al. 2001).

3 VZV Latency

VZV latency is characterized by the presence of the virus genome in an "endless" circular configuration associated with cellular histone complexes that function in part to regulate virus gene transcription (Clarke et al. 1995; Gary et al. 2006). PCR analysis of trigeminal ganglia from ten subjects by laser-capture microdissection detected latent VZV DNA in 71 (4.1%) of 1,722 neurons and in only 1 (<0.01%) of 14,200 satellite cells (Wang et al. 2005). Further, each latently infected neuron contained, on average, seven copies of VZV DNA. That study, in which individual cells were identified, isolated and analyzed, is in agreement with the results of the studies using whole ganglion homogenates or isolated neurons (Mahalingam et al. 1993; LaGuardia et al. 1999; Cohrs et al. 2000; Levin et al. 2003).

While the site of latency, virus genome configuration and DNA burden during latency are similar for HSV-1 and VZV, there are significant differences between the two neurotropic alpha herpesviruses in gene expression. During latency, a single HSV-1 gene is transcribed (Stevens et al. 1987). This latency-associated transcript (LAT) is an 8.5-kb primary transcript which, after splicing, results in a stable 2.0-kb intron (Natarajan et al. 1991; Zabolotny et al. 1997). The LAT transcript is processed into at least four microRNAs (Umbach et al. 2008) which, along with LAT, have been implicated in neuronal survival by inhibiting apoptotic cell death (Peng et al. 2003; Bloom 2004; Shen et al. 2009). VZV does not contain a homologous LAT region. During latency, at least 12 VZV gene transcripts have been detected (Cohrs et al. 1994; Kennedy et al. 1999; Nagel et al. 2011) but no

VZV encoded microRNA has been detected in latently infected human ganglia (Umbach et al. 2009). Of the VZV genes transcribed during latency, VZV ORF 63 transcripts are the most prevalent and abundant (Cohrs and Gilden 2007). VZV ORF 63 encodes a 278-amino acid phosphorylated protein found in the cytoplasm of latently infected neurons (Mahalingam et al. 1996). This protein has been associated with neuronal protection from apoptotic cell death, but its mode of action is yet to be determined (Hood et al. 2006).

4 VZV-Specific Immunity

Varicella is followed by the production of VZV-specific antibody and VZV-specific T cell-mediated immunity (Kumagai et al. 1980; Arvin et al. 1986). Antibodies to VZV glycoproteins I–IV and to three nonglycosylated proteins remain throughout life. VZV-specific antibodies are also present in some adults with no history of varicella or zoster, indicative of subclinical infection (Vafai et al. 1988).

T cell immunity to VZV is more important than the antibody response. For example, agammaglobulinemic humans who are unable to produce VZV-specific antibodies are protected against second episodes of varicella because of their ability to mount a VZV-specific T cell-mediated immune response (Good and Zak 1956). Furthermore, individuals with T cell immune deficiency disorders have more severe disease than normal hosts (Gershon and Steinberg 1979). In stem cell transplant recipients who received inactivated VZV vaccine, protection was correlated with VZV-specific T cell immunity, but not with anti-VZV antibody (Hata et al. 2002).

VZV-specific T cell-mediated immunity maintains latent VZV in ganglia. The immune response is also boosted by subclinical reactivation of latent virus or environmental exposure to virus. Importantly, the incidence of zoster increases with age as VZV-specific T cell-mediated immunity declines. The frequency of VZV-specific memory CD4 T cells is significantly influenced by age. CD4 T cells decrease during the first 3 years after varicella (Burke et al. 1982), and a comparison of the cell-mediated immune response to VZV antigen in vitro in young adults and individuals over age 60 years revealed fivefold fewer CD4 cells producing interferon-gamma or interleukins 4 and 5, as well as fewer CD4 early effectors and CD8 effector memory cells in the older group (Patterson-Bartlett et al. 2007).

While reduced cell-mediated immunity to VZV with age or after exposure to immunosuppressive regimens in cancer patients or bone marrow transplant recipients often results in VZV reactivation (Dolin et al. 1978), virus-specific T cells are rarely seen in human ganglia latently infected with VZV (Verjans et al. 2007). Analysis of human ganglia from donors who had zoster 1–5 months before death revealed VZV glycoprotein E in neurons and infiltration of non-cytolytic $CD8^+$ T cells, but neurons that were positive for VZV glycoprotein E were neither positive for MHC class I antigens nor surrounded by T cells, suggesting that immune control of virus reactivation may not depend on direct contact with T cells (Gowrishankar et al. 2010). Since VZV has been shown to downregulate MHC I surface expression, virus latency is probably regulated by an innate immune response involving cytokines or chemokines (Cohen 1998; Abendroth et al. 2001; Eisfeld et al. 2007). CXCL10 has been proposed as a potential driver of T cell recruitment, based on its detection and that of its receptor (CXCR3) in human ganglia from zoster patients (Steain et al. 2010). Recognition of the essential role of cell-mediated immunity to VZV for protection against and recovery from varicella and zoster has led to studies designed to boost the cell-mediated immune response to VZV by immunization of elderly adults (see below).

5 VZV Epidemiology

Based on the prevalence of anti-VZV IgG antibodies in 21,288 individuals in the US from 1988 to 1994 (Kilgore et al. 2003) and on the detection of VZV DNA in the trigeminal ganglia of 201 of 207 cadavers (Inoue et al. 2010), VZV infects more than 97% of the adult population. One in three individuals will develop zoster during their lifetime (Yawn et al. 2007). While a zoster vaccine is available, only 6.7% of individuals over age 60 years have received it (Lu et al. 2011); thus, more than 900,000 individuals in the US are still likely to develop zoster annually. In patients with zoster, two-thirds are older than 50 years and 90% are immunocompetent. Furthermore, 10% experience at least one zoster-related non-pain complication and one in four has zoster-related pain that persists 30 days or more (Yawn et al. 2007).

In the US, zoster, post-herpetic neuralgia (PHN) and other neurological diseases caused by VZV reactivation are a significant health care challenge. Currently, the annual medical care cost of treating zoster is estimated to be $1.1 billion, most of which is used to treat immunocompetent adults 50 years and older (Yawn et al. 2009). This cost is expected to rise since the incidence of zoster increases with age and the population aged 65 and older is expected to double from 36 million in 2003 to 72 million in 2030.

6 Neurological Complications of VZV Reactivation (Fig. 1)

6.1 Zoster

Herpes zoster is the most common manifestation of VZV reactivation. Zoster is characterized by a vesicular eruption on an erythematous base (Fig. 2) in one to three dermatomes, accompanied by severe, lancinating radicular pain and itching, and unpleasant sensations (dysesthesias) produced by touch (allodynia) as well as decreased sensation in the affected area. Rash and pain usually develop within a few days of each other, although pain can precede rash by weeks to months (Gilden et al. 1991). After reactivation from cranial nerve, dorsal root and autonomic

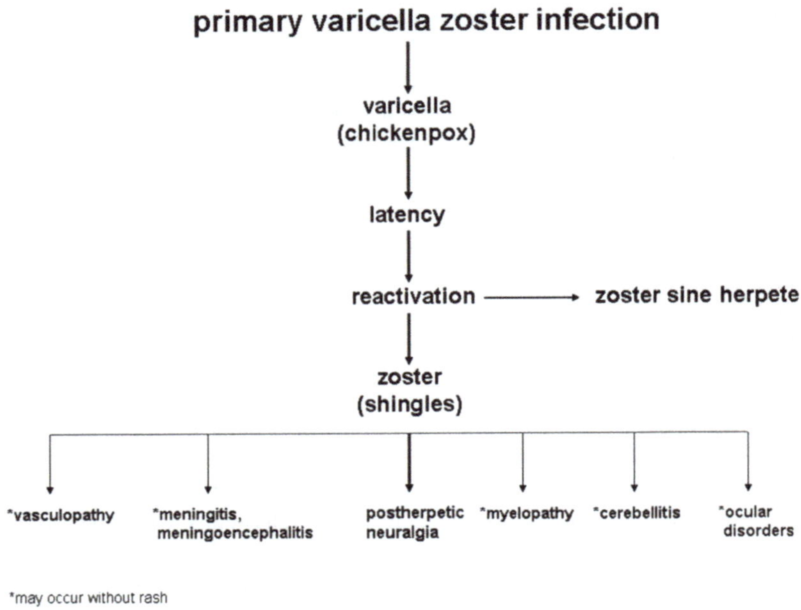

Fig. 1 Neurological complications of VZV reactivation

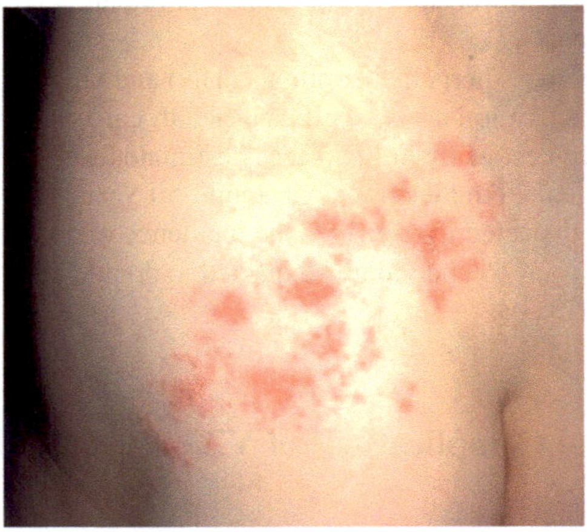

Fig. 2 Maculopapular-vesicular eruption in a L5 distribution produced by reactivation of VZV

ganglia, VZV can travel peripherally to the corresponding dermatome; thus, zoster can affect any level of the neuraxis. The thoracic region is most commonly affected (over 50%), followed by the face, cervical and lumbosacral regions.

Zoster most frequently occurs in the elderly (Harnisch 1984) as cell-mediated immunity to VZV declines. Other groups at risk for zoster are patients taking immunosuppressive drugs (such as cancer patients, organ transplant recipients, and patients with AIDS) (Gilden et al. 2003). Zoster in an otherwise healthy young person may be the first manifestation of HIV infection (Leppard and Naburi 1998; Tyndall et al. 1995). Varicella in infancy also predisposes to zoster in early adulthood (Kakourou et al. 1998).

Optic neuritis occurs rarely in association with herpes zoster (Carroll and Mastaglia 1979; Selbst et al. 1983; Miller et al. 1986; Meenken et al. 1998; Kurimoto et al. 2011) and may be bilateral (Gündüz and Ozdemir 1994). Ophthalmoplegia after zoster most often involves cranial nerve III, followed by cranial nerves VI and IV (Grimson and Glaser 1978; Carroll and Mastaglia 1979; Archambault et al. 1988; Sodhi and Goel 2001; Karmon and Gadoth 2005; Kurimoto et al. 2011). Herpes zoster ophthalmicus (HZO) is often accompanied by zoster keratitis, which can lead to blindness. Patients with HZO and visual symptoms should have an immediate slit-lamp examination by an ophthalmologist, particularly if skin lesions extend to the nose indicating involvement of the ophthalmic branch (V_1) of the trigeminal nerve (Hutchinson sign). Involvement of the maxillary (V_2) and mandibular (V_3) distribution of the trigeminal nerve can produce osteonecrosis and spontaneous tooth exfoliation (Manz et al. 1986; Garty et al. 1985; Volvoikar et al. 2002; Feller et al. 2008; Lambade et al. 2011).

Involvement of the geniculate ganglion (cranial nerve VII) causes weakness or paralysis of ipsilateral facial muscles. The combination of facial palsy and vesicles in the external auditory canal (zoster oticus) or on the tympanic membrane or on the ipsilateral anterior two-thirds of the tongue or hard palate (Payten and Dawes 1972) constitutes the Ramsay Hunt syndrome (RHS). This syndrome is often associated with tinnitus, hearing loss, nausea, vomiting, vertigo and nystagmus, indicating the involvement of cranial nerve VIII within the bony facial canal. Facial paralysis is often more severe than in Bell's palsy, and patients are less likely to recover completely (Robillard et al. 1986). Zoster is also followed by the involvement of cranial nerves IX, X, XI and XII (Steffen and Selby 1972; Asnis et al. 1996).

Cranial neuropathies frequently occur weeks after zoster, raising the possibility that the disease is due to micro-infarction of cranial nerves. Virus particles can potentially spread transaxonally along trigeminal and other ganglionic afferent fibers to cause occlusion of small vessels supplying cranial nerves in the same manner that produces VZV vasculopathy in larger arteries (see Sect. 6.3). The blood supply of cranial nerves III, IV, V_1 and VI comes from the internal carotid circulation, while V_2, V_3 and VII, IX, X, XI and XII are supplied by the external carotid circulation (Lapresle and Lasjaunias 1986), It is important to recognize that multiple forms of trigeminal (Easton 1970; Yamada et al. 2003; Hevner et al. 2003) and facial (Morgan and Nathwani 1992; Terada et al. 1998; Murakami et al. 1998) distribution zoster as well as polyneuritis cranialis due to VZV (Osaki et al. 1995; Murata et al. 2010) may occur in the absence of rash.

Zoster paresis (weakness) may present with arm weakness or diaphragmatic paralysis (Brostoff 1966; Stowasser et al. 1990) after cervical distribution zoster; leg weakness after lumbar or sacral distribution zoster; or urinary retention after sacral distribution zoster (Izumi and Edwards 1973; Jellinek and Tulloch 1976). Magnetic resonance imaging (MRI) of patients with zoster paresis reveals involvement of both anterior and posterior roots at the spinal level that corresponds to the patient's clinical deficit (Hanakawa et al. 1997; Umehara et al. 2011). Rarely, clinical deficits in cervical zoster paresis extend to the brachial plexus, confirmed by both electrodiagnostic testing and MRI (Choi et al. 2009). In 45 patients with zoster paresis, 67% had near-complete recovery (Gupta et al. 1969), and in another study of 61 cases, 55% had complete functional recovery (Thomas and Howard 1972). Rare cases of thoracic zoster have also been associated with abdominal muscle weakness, which can result in abdominal herniation (Tjandra and Mansel 1986; Molinero et al. 2002).

The cardinal pathological features of zoster are characterized by inflammation and hemorrhagic necrosis with associated neuritis, localized leptomeningitis, unilateral segmental poliomyelitis and degeneration of related motor and sensory roots (Head and Campbell 1900; Denny-Brown et al. 1944). Demyelination is seen in areas with mononuclear cell (MNC) infiltration and microglial proliferation. Intranuclear inclusions, viral antigen and herpesvirus particles have been found in acutely infected ganglia (Cheatham et al. 1956; Esiri and Tomlinson 1972; Ghatak and Zimmerman 1973).

Antiviral drugs, such as oral valacyclovir (1 g three times daily for 7–10 days) or acyclovir (800 mg five times daily for 7–10 days), speed the healing of rash and shorten the duration of acute pain. Immunocompromised patients require intravenous acyclovir (5–10 mg/kg three times daily for 5–7 days). Because zoster pain may be associated with an inflammatory response, many clinicians administer a short course of corticosteroids, e.g., oral prednisone, 1 mg/kg for 5–7 days, in addition to antiviral therapy.

6.2 Postherpetic Neuralgia

The most common neurological complication of zoster is PHN, defined as dermatomal-distribution pain that persists for more than 3 months after zoster. Age is the most important factor in predicting the development of PHN. More than 40% of zoster patients over 60 years of age experience chronic pain. Except for its longevity, the pain of PHN and associated allodynia are the same as in zoster. The incidence of PHN is slightly greater in women (Hope-Simpson 1975) and after trigeminal distribution zoster (de Moragas and Kierland 1957; Rogers and Tindall 1971; Hope-Simpson 1975).

The cause and pathogenesis of PHN are unknown. Two non-mutually exclusive theories are that: (1) excitability of ganglionic or even spinal cord neurons is altered; and (2) persistent productive virus infection exists in ganglia. Analysis of ganglia from an early case of PHN of 2.5 months duration revealed diffuse and focal infiltration by chronic inflammatory cells (Smith 1978), an observation confirmed by Watson et al. (1991) who found prominent collections of lymphocytes in ganglia from a patient with PHN of 2 years duration (Fig. 3). The inflammatory response in ganglia of these subjects raised the possibility of prolonged viral

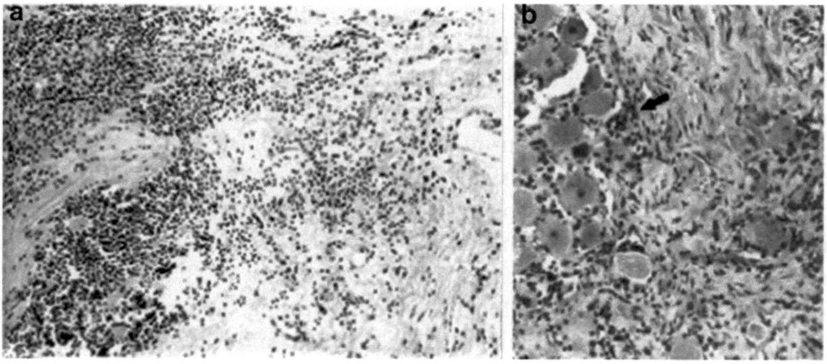

Fig. 3 Hematoxylin and eosin-stained sections of dorsal root ganglia from patients with postherpetic neuralgia. Note diffuse and focal infiltration by inflammatory cells (**a**). *Arrow* in *panel* (**b**) points to a prominent collection of lymphocytes. *The figure was previously published in Smith (1978). Reprinted with permission of Elsevier, Copyright 1978*

infection. Further evidence that PHN may be produced by low-level ganglionitis has come from the detection of VZV DNA and proteins in blood MNCs of many patients with PHN (Vafai et al. 1988; Devlin et al. 1992; Mahalingam et al. 1995) and from the favorable response of some PHN patients to antiviral treatment (Terada et al. 1998; Gilden et al. 2003; Quan et al. 2006).

PHN is difficult to manage and no universal treatment exists. First-line therapies include tricyclic antidepressants (TCAs), gabapentin and pregabalin, and topical lidocaine patches. Opioids, tramadol, capsaicin cream and the capsaicin 8% patch are recommended as second- or third-line therapies. TCAs such as amitriptyline are usually started at a dose of 10–25 mg orally at bedtime with a maximum dose of 150–200 mg/day. Secondary amine TCAs, such as nortriptyline and despramine, can also be used due to a superior safety profile compared to the tertiary amine amitryptyline (Dworkin et al. 2007; Attal et al. 2010). The calcium channel alpha (2)-delta ligands gabapentin and pregabalin are also used, with pregabalin providing equivalent efficacy to that of gabapentin, but at much lower doses due to its higher bioavailability and rapid absorption. Pregabalin is given at 75–150 mg orally twice daily or 50–100 mg orally three times daily (150–300 mg/day). If minimal relief is obtained at 300 mg daily for 2 weeks, the dose can be increased to a maximum of 600 mg/day in two or three divided doses. Opioids such as extended-release oxycodone, morphine and methadone have shown efficacy in patients with PHN. Tramadol is better tolerated but less effective than these stronger opioids. The lidocaine 5% patch has significant analgesic efficacy in patients with PHN (Baron et al. 2009; Hans et al. 2009). Capsaicin 0.075% cream is sometimes prescribed, but the American Academy of Neurology (AAN) guidelines state that the analgesia provided is below the threshold for a clinically important effect (Dubinsky et al. 2004). The new capsaicin 8% patch (Backonja et al. 2008, 2010), which delivers a high concentration of capsaicin in a single 60-min application after application of local anesthetic, is promising for the treatment of PHN but its use awaits long-term safety data. Combination therapy such as gabapentin and nortriptyline (Gilron et al. 2009), morphine and gabapentin (Gilron et al. 2005) or pregabalin and the lidocaine 5% patch (Rehm et al. 2010) may provide greater analgesic effects.

For patients refractory to non-invasive pharmacological intervention, invasive procedures such as botulinum toxin injection, sympathetic blockade, implantable spinal cord stimulators and peripheral nerve field stimulation have been used with limited success. Botulinum toxin is minimally invasive and has been successful in decreasing PHN pain in several cases (Ruiz and Bermejo 2008; Sotiriou et al. 2009; Xiao et al. 2010). Sympathetic blockade helps with acute zoster and early PHN but has no long-term benefit in established PHN (Winnie and Hartwell 1993). Epidural injection of steroids produced modest effects but the relief was short lived (van Wijck et al. 2006). Spinal cord stimulation has limited short-term success (Harke et al. 2002).

A newer potentially promising treatment for PHN is percutaneous peripheral nerve field stimulation. Under intravenous sedation or monitored anesthesia care, stimulating electrodes are placed subcutaneously over the area of maximal pain. The leads are connected to an external pulse generator for a 2–14-day trial period. If there is over 50% improvement of pain, then a permanent pulse generator is implanted. This can be done in an outpatient setting. After device placement, subjects became pain free with minimal to no medication needed for ophthalmic (Surjya et al. 2010), cervical (Lynch et al. 2011) and thoracic (Yakovlev and Peterson 2007; Kouroukli et al. 2009) distribution PHN.

6.3 VZV Vasculopathy

After VZV reactivation from ganglia, the virus can also travel centrally to infect cerebral arteries and cause ischemic and hemorrhagic stroke (VZV vasculopathy). The exact incidence of VZV vasculopathy is unknown although it is a significant stroke risk factor. In adults with zoster, the risk of stroke is increased by 30% within the following year (Kang et al. 2009) and by 4.5-fold when zoster is in the ophthalmic distribution of the trigeminal nerve (Lin et al. 2010). Furthermore, up to one-third of pediatric ischemic arteriopathies are associated with varicella (Braun et al. 2009; Amlie-Lefond et al. 2009; Ciccone et al. 2010). VZV vasculopathy affects both immunocompromised and immunocompetent individuals and can present as headache, mental status changes and focal neurological deficits. In a study of 30 virologically verified cases of VZV vasculopathy (Nagel et al. 2008), lesions at grey–white matter junctions were frequently seen on MRI (Fig. 4), and magnetic resonance angiography (MRA) revealed focal arterial stenosis and occlusion (Fig. 5) in more than two-thirds of patients. Both large and small arteries were involved in 50% of the patients, small arteries in 37%, and large arteries alone in only 13% of the 30 subjects. Importantly: (1) up to one-third of the patients did not have preceding zoster rash; (2) up to one-third did not have a CSF pleocytosis;

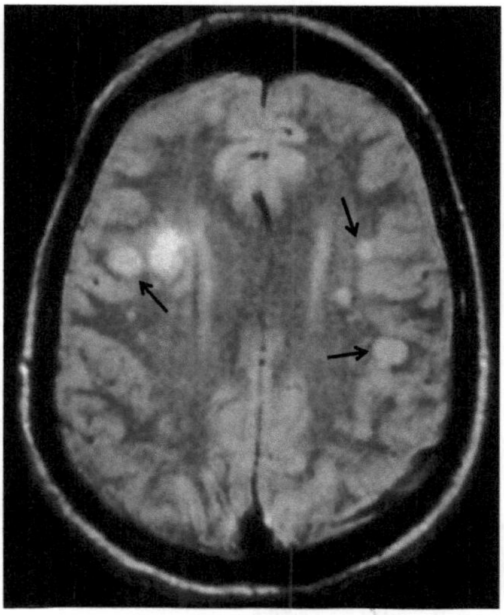

Fig. 4 Brain MRI of a patient with VZV multifocal vasculopathy. Proton-density brain MRI scan shows multiple areas of infarction, particularly involving white matter, in both hemispheres. *Arrows* point to gray–white matter junction lesions

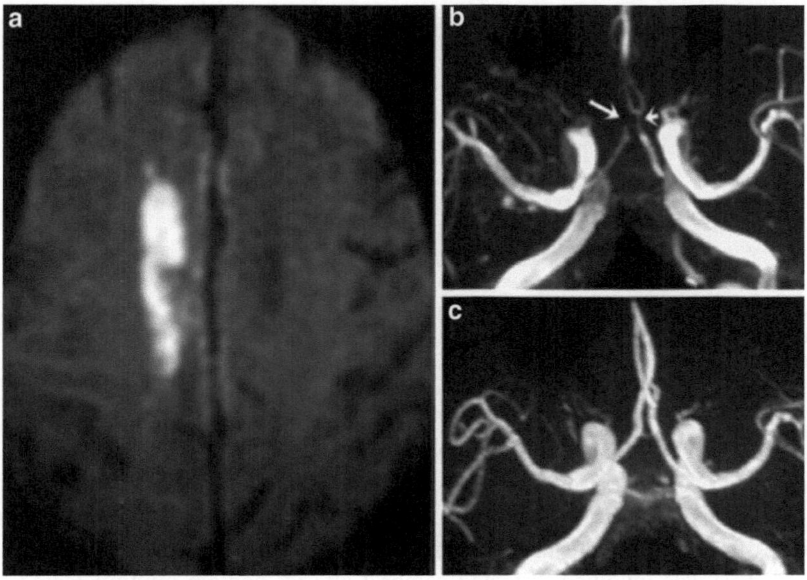

Fig. 5 Brain MRI and MRA images of a patient with VZV vasculopathy and infarction. (**a**) Diffusion-weighted image (*B* value 1,000) shows restricted diffusion in the right anterior cerebral artery territory, indicating acute infarction. (**b**) Three-dimensional time-of-flight MRA at the time of infarction shows marked narrowing of anterior cerebral arteries, with a new flow gap at the junction of A1 and A2 segments of the right anterior cerebral artery, indicating occlusion on the right (*long arrow*) and marked stenosis on the left (*short arrow*). (**c**) Three-dimensional time-of-flight MRA of the Circle of Willis 5 months before infarction shows normal anterior cerebral arteries

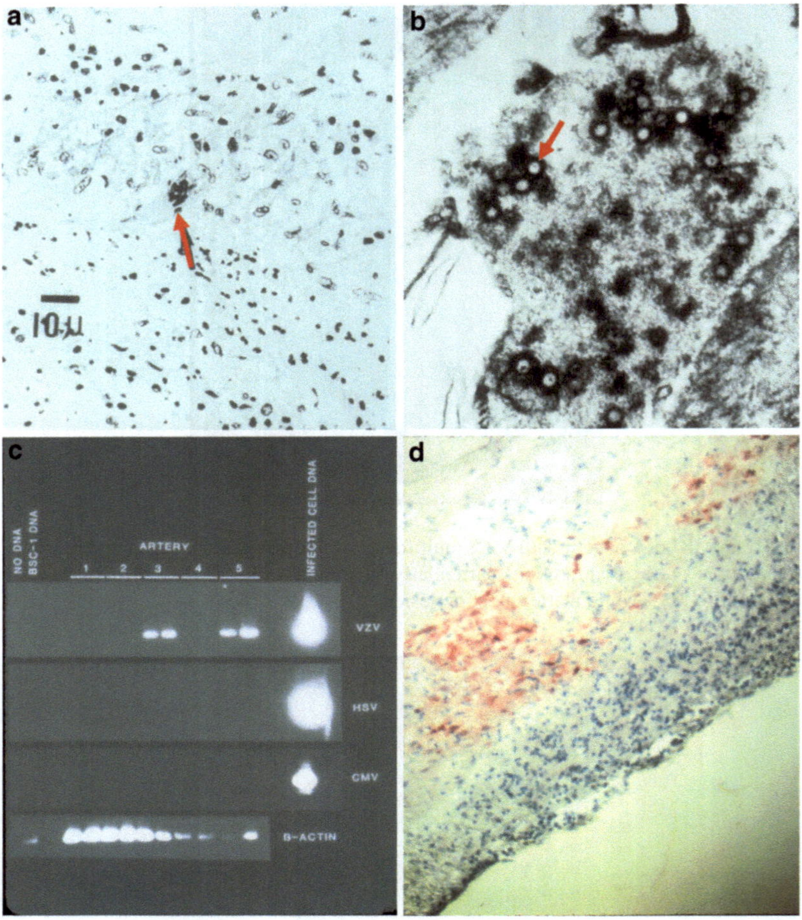

Fig. 6 Pathological and virological findings in the arteries of patients who died from VZV vasculopathy. (**a**) Cerebral artery with multinucleated giant cells (*arrow*). (**b**) Herpes virions (*arrow*) within a cerebral artery. (**c**) VZV DNA in two (lanes 3 and 5) of five cerebral arteries. (**d**) VZV antigen (*red*) in the media of a cerebral artery

(3) VZV DNA PCR analysis of CSF was only 30% sensitive; and (4) symptoms and signs often occurred months after zoster (Nagel et al. 2008).

Infarctions are mostly bland but can also be hemorrhagic. Deep white-matter lesions often predominate and are ischemic or demyelinative, depending on the size of blood vessels involved. Infected cerebral arteries contain multinucleated giant cells, Cowdry A inclusion bodies, herpesvirus particles detected by electron microscopy, as well as VZV DNA and VZV antigen (Fig. 6). A variety of vascular pathology has been reported, ranging from neointimal proliferation to necrosis with and without inflammation (Kleinschmidt-DeMasters and Gilden 2001). A recent study that examined the histological and immunohistochemical features of VZV-infected cerebral arteries from subjects with VZV vasculopathy revealed several distinct features: (1) the presence of VZV antigen in the arterial adventitia from early VZV vasculopathy and in the media and intima of protracted cases of VZV vasculopathy, supporting the notion of virus persistence in the artery and spread from the "outside-in"; (2) a thickened arterial intima composed of myofibroblasts and cells most likely of medial smooth muscle origin; (3) a disrupted internal elastic lamina; and (4) a disrupted medial layer with a significant loss of normal smooth muscle cells (Nagel et al. 2011). The morphological changes explain why there is arterial occlusion and loss of vascular contractility, contributing to stroke. Since inflammation also contributes to certain models of coronary and pulmonary vascular remodeling, future studies on the inflammatory response in VZV-infected arteries are needed.

After zoster, the increased risk of stroke from intracranial artery involvement rather than coronary, pulmonary or other systemic vascular complications most likely reflects inherent differences in the cellular and structural composition of intracranial versus systemic arteries. For example, intracranial cerebral arteries, unlike systemic arteries, contain neither an external elastic lamina that may affect transmural migration of virus and cells nor a vaso vasorum in the adventitia above the level of the medulla unless pathological angiogenesis occurs (Lee 1995). Like horseradish peroxidase, which travels along trigeminal ganglionic afferent fibers to trigeminal ganglia after application to the external surface of cerebral arteries (Mayberg et al. 1981, 1984), reactivated VZV in ganglia may also travel along ganglionic afferent fibers to the adventitia of cerebral arteries, a notion consisted with the presence of viral antigen predominantly in the adventitia of cerebral arteries in early VZV vasculopathy (Salazar et al. 2011) and in the media and intima of protracted cases (Nagel et al. 2011).

6.4 VZV Meningitis, Meningoencephalitis, Meningoradiculitis and Cerebellitis

Like VZV vasculopathy, these neurological complications of VZV reactivation can occur in the absence of zoster rash, as demonstrated by recent reports of VZV meningitis (Habib et al. 2009), meningoradiculitis (Gunson et al. 2011) and cerebellitis (Moses et al. 2006; Ratzka et al. 2006) in which diagnosis was confirmed by the detection of VZV DNA and anti-VZV antibody in CSF.

6.5 VZV Myelopathy

VZV myelopathy can present as a self-limiting, monophasic spastic paraparesis, with or without sensory features and sphincter problems. This so-called post-infectious myelitis usually occurs in immunocompetent patients days to weeks

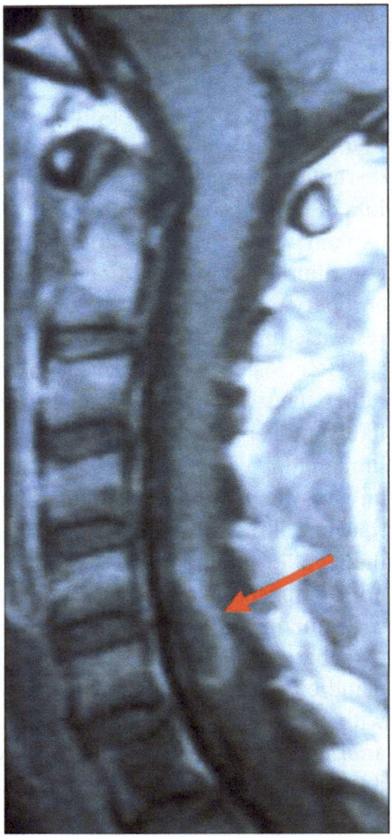

Fig. 7 MRI abnormality in a patient with varicella zoster virus myelitis. Note cervical, longitudinal, serpiginous enhancing lesion (*arrow*)

after acute varicella or zoster. Its pathogenesis is unknown. The CSF usually contains a mild mononuclear pleocytosis, with a normal or slightly elevated protein. Steroids are used to treat these patients (Pina et al. 1997), although some improve spontaneously (Celik et al. 2001).

VZV may directly invade the spinal cord or affect the spinal arteries to produce myelopathy. In such instances, VZV myelopathy may present as an insidious, progressive and sometimes fatal myelitis, mostly in immunocompromised individuals, such as patients with AIDS. MRI reveals longitudinal serpiginous enhancing lesions (Fig. 7). Diagnosis is confirmed by the presence of VZV DNA or anti-VZV IgG or both in CSF (Gilden et al. 1994a). Pathological and virological analyses of the spinal cord from fatal cases have revealed frank invasion of VZV in the parenchyma (Kleinschmidt-DeMasters and Gilden 2001) and, in some instances, spread of virus to adjacent nerve roots (Devinsky et al. 1991). Early diagnosis and aggressive treatment with intravenous acyclovir have been helpful, even in immunocompromised patients (de Silva et al. 1996). The benefit of steroids in addition to antiviral agents is unknown. Rarely, VZV myelitis recurs, even in immunocompetent patients (Gilden et al. 1994a). VZV myelitis may also occur in the absence of zoster rash.

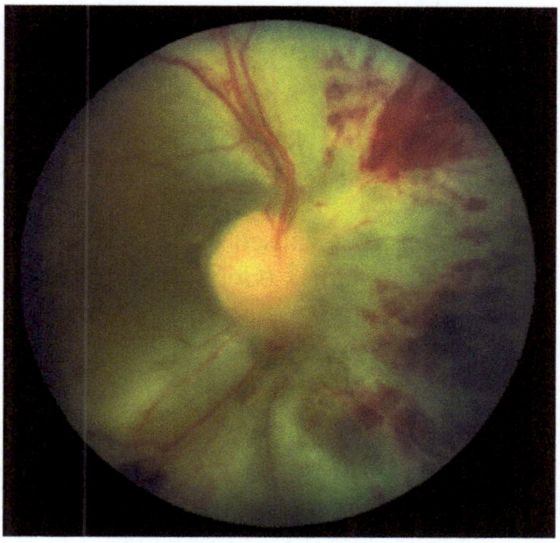

Fig. 8 Fundus photograph of a patient with VZV vasculopathy and progressive outer retinal necrosis. Note diffuse retinal hemorrhages and whitening with macular involvement

VZV can also produce spinal cord infarction identified by diffusion-weighted MRI and confirmed virologically (Orme et al. 2007). Thus, VZV vasculopathy can cause stroke in the spinal cord as well as in the brain.

6.6 *Ocular Disease*

VZV infection produces acute retinal necrosis (ARN) or progressive outer retinal necrosis (PORN). ARN in both immunocompetent and immunocompromised individuals presents with periorbital pain and floaters with hazy vision and loss of peripheral vision. Treatment is typically intravenous acyclovir, steroids and aspirin followed by oral acyclovir (Bonfioli and Eller 2005). Intravitreal injections of foscarnet and oral acyclovir have also been effective. PORN presents with painless loss of vision, floaters and constricted visual fields with resultant retinal detachment. Multifocal, discrete opacified lesions begin in the outer retinal layers peripherally and/or at the posterior pole; only late in disease are inner retinal layers involved. Diffuse retinal hemorrhages and whitening with macular involvement bilaterally are characteristic findings (Fig. 8). VZV is the most common cause of PORN, although HSV and cytomegalovirus can also cause this disease. Most cases are seen in AIDS patients with $CD4^+$ T cell counts less than 10 cells/mm^3 of blood (Guex-Crosier et al. 1997) as well as in other immunosuppressed individuals (Lewis et al. 1996). PORN may be preceded by retrobulbar optic neuritis and aseptic meningitis (Franco-Paredes et al. 2002), central retinal artery occlusion or ophthalmic-distribution zoster (Menerath et al. 1995) and may occur together with multifocal vasculopathy or myelitis. Treatment with intravenous acyclovir has given poor or inconsistent results (Johnston et al. 1993) and even when acyclovir helped, VZV retinopathy recurred when drug therapy was tapered or stopped. PORN patients treated with ganciclovir alone or in combination with foscarnet had a better final visual acuity than those treated with acyclovir or foscarnet (Moorthy et al. 1997). The best treatment for PORN in AIDS patients may be prevention with HAART (Austin 2000).

Like other neurological disorders caused by VZV, ocular disease caused by VZV can also occur in the absence of rash. Multiple cases of PORN (Friedman et al. 1993; Galindez et al. 1996) and a case of severe unremitting eye pain without rash were shown to be caused by VZV based on detection of VZV DNA in nasal and conjunctival samples (Goon et al. 2000). In addition, third cranial nerve palsies (Hon et al. 2005), retinal periphlebitis (Noda et al. 2001), uveitis (Akpel and Gottsch 2000; Hon et al. 2005), iridocyclitis (Yamamoto et al. 1995) and disciform keratitis (Silverstein et al. 1997) that occurred without rash were confirmed virologically to be caused by VZV.

6.7 Zoster Sine Herpete (Radicular Pain in the Absence of Rash)

Zoster sine herpete is recognized by clinicians as chronic radicular pain without rash caused by VZV. Zoster sine herpete was first described in a report of multiple patients with dermatomal distribution radicular pain in areas similar to pain with rash in zoster (Lewis 1958). The first two virologically confirmed cases of zoster sine herpete were verified by the detection of VZV DNA in CSF (Gilden et al. 1994b). A third case of thoracic-distribution zoster sine herpete, in which electromyography of paraspinal muscles demonstrated frequent fibrillation potentials restricted to chronically painful thoracic root segments was confirmed by the detection of VZV DNA in blood MNCs and anti-VZV IgG antibody in CSF (Amlie-Lefond et al. 1996). Blumenthal et al. (2011) recently described a patient with zoster sine herpete whose CSF did not contain amplifiable VZV DNA but did contain anti-VZV IgG with reduced serum/CSF ratios of anti-VZV IgG indicative of intrathecal synthesis. Perhaps the most compelling evidence that persistent radicular pain without rash can be caused by chronic active VZV ganglionitis came from analysis of a trigeminal ganglionic mass removed from an immunocompetent adult who had experienced relentless trigeminal-distribution pain for more than a year; pathological and virological analyses of the ganglionic mass revealed active VZV ganglionitis (Fig. **9**) (Hevner et al. 2003). The detection of VZV DNA and anti-VZV IgG and IgM antibody has expanded the spectrum of neurological diseases produced by VZV in the absence of rash to include VZV meningoencephalitis, vasculopathy, myelitis, cerebellar ataxia and polyneuritis cranialis.

6.8 Diagnostic Tests

The diagnosis of VZV-induced neurological disease is straightforward when the characteristic dermatomal distribution rash of zoster is present. When zoster rash is not present in a patient with neurological disease that can be caused by VZV

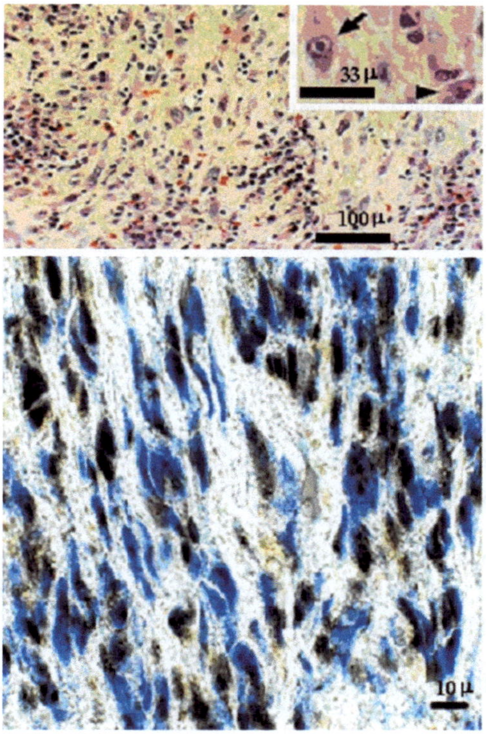

Fig. 9 Ganglionitis and intranuclear inclusions in zoster sine herpete. Hematoxylin and eosin staining of the trigeminal ganglion (*top panel*) shows widespread chronic inflammation with fibrosis and loss of neurons. Cells in some foci contain Cowdry type A intranuclear inclusions (*arrow, inset*) indicative of virus infection; inflammatory cells are mainly lymphocytes with some plasma cells (*arrowhead, inset*). Immunohistochemical staining of same ganglion (*bottom panel*) with mouse monoclonal antibody against VZV gene 63 protein reveals VZV antigen (*brown staining*) in multiple cells throughout the ganglion. Adjacent sections stained with antibody directed against HSV or with normal rabbit serum were negative (not shown)

(e.g., zoster sine herpete, vasculopathy, meningoencephalitis, myelopathy or retinal necrosis), examination of CSF and serum is necessary. The routine cell count can be helpful, since a mild lymphocytic pleocytosis is characteristically found in VZV vasculopathy, myelitis and meningoencephalitis. Furthermore, increased red blood cells and polymorphonuclear leukocytes in CSF may also be seen when VZV infects the nervous system.

In the absence of rash, the CSF should be examined virologically for VZV DNA by PCR and for anti-VZV IgG and IgM. Detection of VZV DNA in CSF or anti-VZV IgM in serum or CSF is a strong presumptive evidence of recent VZV infection. If anti-VZV IgG antibody is present in CSF, the antibody index should be calculated to determine whether anti-VZV antibody is being produced intrathecally. For molecules such as albumin and total IgG, the serum/CSF ratio is usually more than 100:1. A reduced ratio of anti-VZV IgG antibody compared to ratios for albumin or total IgG is seen in many neurological diseases produced by VZV. Importantly, many cases of VZV vasculopathy are protracted and VZV DNA is only found ~30% of the cases (Nagel et al. 2008). The detection of anti-VZV IgG antibody in CSF with intrathecal synthesis is superior to the detection of VZV DNA in CSF to diagnose VZV vasculopathy (Nagel et al. 2007), recurrent myelopathy and brainstem encephalitis produced by VZV (Haug et al. 2010).

6.9 VZV DNA in Human Saliva

In addition to the detection of VZV DNA in blood MNCs and CSF, VZV DNA may also be present in saliva. VZV DNA was present in the saliva of all of 54 patients with acute zoster involving the face, trunk, and upper and lower extremities (Mehta et al. 2008), as well as in the saliva of most patients with the Ramsay Hunt syndrome (Furuta et al. 2001; Lackner et al. 2010). Many patients with acute peripheral facial palsy without rash also had VZV DNA in saliva (Furuta et al. 2001, 2005), as did four of the eight patients who experienced delayed facial palsy after orofacial surgery (Furuta et al. 2000). Importantly, VZV DNA persists for years in the saliva of individuals who develop zoster after 60 years of age (Nagel et al. 2011). While it remains unclear why VZV DNA persists in the saliva of individuals with a history of zoster, the findings are consistent with the earlier studies showing that the presence of VZV is not restricted to the skin of the affected dermatome. For example, VZV DNA is also present in blood MNCs during acute zoster (Gilden et al. 1987) as well as in blood MNCs of some elderly individuals with no history of zoster (Devlin et al. 1992) and even in stressed healthy astronauts (Mehta et al. 2004; Cohrs et al. 2008). Overall, the detection of VZV DNA in saliva and blood indicates that after reactivation from ganglia, the virus does more than travel transaxonally retrograde to skin. The detection of VZV DNA in the saliva of some elderly individuals for many years after zoster may reflect their inability to drive the virus back to the latent state, just as a smoldering ganglionitis has been speculated to explain the development of PHN (Gilden et al. 2005).

Finally, the potential usefulness of saliva in the diagnosis of patients with neurological and ocular disease should be considered. Given that VZV may cause not only radicular pain in the absence of rash, but also cerebellitis, meningoencephalitis, vasculopathy, myelitis and multiple serious ocular disorders without rash, future studies are needed to establish whether VZV DNA can be detected in the saliva of such patients and confirmed by the presence of intrathecal synthesis of antibodies against VZV.

As mentioned in the sections above, once the diagnosis of neurological disease caused by VZV has been confirmed, patients are treated with at least a 2-week course of intravenous acyclovir with or without prednisone; however, the duration of treatment may be longer in cases of recurrent disease and in immunosuppressed patients.

6.10 Prevention of VZV Reactivation

There are two VZV vaccines, both of which contain live attenuated VZV. The first developed was the varicella (Oka strain) vaccine (Varivax, Merck), which has been used to vaccinate children to prevent varicella (chickenpox) in Japan since 1975 and in the US since 1996. The second vaccine (Zostavax, Merck) is used to prevent zoster in elderly individuals. The only difference between the two vaccines is that Zostavax contains 19,400 pfu per dose, 14-fold more virions than in the varicella vaccine. Varivax generates VZV-specific humoral and cell-mediated immune responses, particularly CD8 T cells (Frey et al. 2003). The memory cell response that occurs after vaccination protects from infection during re-exposure to VZV.

Zostavax is indicated for the prevention of herpes zoster in individuals aged 60 years and older. Zoster vaccine administered to people over 50 years of age resulted in increased numbers of CD4 and CD8 cells, CD4 and CD8 effector memory T cells, and CD8 early-effector T cells; the half-life of the boost in T cell immunity to VZV is at least 5 years (Levin et al. 1998). Zoster vaccine also boosts VZV-specific immunity in adults with a history of zoster before vaccination or with chronic illness.

The Shingles Prevention Study (SPS) of the licensed zoster vaccine was a placebo-controlled, double-blind study of more than 38,000 adults over age 60 years and randomized to receive either zoster vaccine or placebo. All subjects were monitored for zoster. Endpoints included the burden of illness due to zoster and zoster-associated pain as well as the incidence of clinically significant PHN. Subjects received a single dose of Zostavax ($n = 19{,}270$) or placebo ($n = 19{,}276$). Racial distribution across both vaccination groups was similar: White (95%), black (2%), Hispanic (1%) and other (1%). The gender distribution was 59% male and 41% female in both groups. The most common side effects reported by the participants after zoster vaccination were redness, pain, itching, swelling, warmth or bruising at the injection site and sometimes headache. Varicella-like rashes at the injection site were more common in zoster vaccine than in placebo recipients (0.1% vs. 6.4%; $p < 0.05$).

After a mean follow-up of 3 years, the SPS found that use of the Zostavax vaccine reduced the incidence of zoster by 51%. Subjects in the immunization group who developed zoster reported significantly less pain and discomfort than those in the placebo group and PHN was less frequent (an overall 61% lower burden of disease). While the vaccine group had a significantly greater risk of a serious adverse event (1.9% vs. 1.3%) and experienced more adverse events at the injection site (48.3% vs. 16.6%) than did the placebo group during the first 42 days after vaccination, no significant differences were seen between the groups in the incidence of vaccine-related serious adverse events (both <0.1%) at the end of the study.

In the US, the Center for Disease Control and Prevention Advisory Committee on Immunization Practices recently recommended zoster vaccine for all persons aged over 50 years with no prior indications and in persons reporting a previous episode of zoster or who have chronic medical conditions. By 2008, 3 years after zoster vaccine was licensed and recommended by the Advisory Committee on Immunization Practices for persons aged 60 and older, less than 7% of the age group in the US was vaccinated (Lu et al. 2011). This was due to a combination of lack of patient awareness regarding the availability of a vaccine, physicians' uncertainty about the duration of protection and different cost-sharing plans for immunization. This is disappointing because Zoster vaccine should be universally administered to all individuals over age 50.

Several important questions regarding zoster vaccines remain. How many years will the current zoster vaccine maintain immunity to prevent zoster? Is zoster vaccine safe for immunocompromised individuals? Will a killed VZV vaccine produce a significant increase in cell-mediated immunity to VZV? Should multiple vaccinations be considered in individuals for every decade of life after age 60? Should zoster vaccine be refined to include epitopes that induce cell-mediated immunity to VZV?

Acknowledgments This work was supported in part by NIH grants AG006127 to DG, AG032958 to DG and RJC, and NS067070 to MAN. We thank Marina Hoffman for editorial assistance and Lori DePriest for manuscript preparation.

References

Abendroth A, Morrow G, Cunningham AL et al (2001) Varicella-zoster virus infection of human dendritic cells and transmission to T cells: implications for virus dissemination in the host. J Virol 75:6183–6192

Abendroth A, Slobedman B, Lee E et al (2000) Modulation of major histocompatibility class II protein expression by varicella-zoster virus. J Virol 74:1900–1907

Akpel EK, Gottsch JD (2000) Herpes zoster sine herpete presenting with hyphema. Ocul Immunol Inflamm 8:115–118

Amlie-Lefond C, Bernard TJ, Sébire G et al (2009) Predictors of cerebral arteriopathy in children with arterial ischemic stroke: results of the International Pediatric Stroke Study. Circulation 119:1417–1423

Amlie-Lefond C, Mackin GA, Ferguson M et al (1996) Another case of virologically confirmed zoster sine herpete with electrophysiologic correlation. J Neurovirol 2:136–138

Archambault P, Wise JS, Rosen J et al (1988) Herpes zoster ophthalmoplegia. Report of six cases. J Clin Neuroophthalmol 8:185–193

Arvin AM, Koropchak CM, Williams BR et al (1986) Early immune response in healthy and immunocompromised subjects with primary VZV infection. J Infect Dis 154:422–429

Asano Y, Yoshikawa T, Ihira M et al (1999) Spread of varicella-zoster virus DNA to family members and environments from siblings with varicella in a household. Pediatrics 103:e61

Asnis DS, Micic L, Giaccio D (1996) Ramsay Hunt syndrome presenting as a cranial polyneuropathy. Cutis 57:421–424

Attal N, Cruccu G, Baron R et al (2010) EFNS guidelines on the pharmacological treatment of neuropathic pain. Eur J Neurol 17(9):e1113–e1188

Austin RB (2000) Progressive outer retinal necrosis syndrome: a comprehensive review of its clinical presentation, relationship to immune system status, and management. Clin Eye Vis Care 12:119–129

Backonja MM, Malan TP, Vanhove GF et al (2010) NGX-4010, a high-concentration capsaicin patch, for the treatment of postherpetic neuralgia: a randomized, double-blind, controlled study with an open-label extension. Pain Med 11:600–608

Backonja M, Wallace MS, Blonsky ER et al (2008) NGX-4010C116 Study Group. NGX-4010, a high-concentration capsaicin patch, for the treatment of postherpetic neuralgia: a randomised, double-blind study. Lancet Neurol 7:1106–1112

Baron R, Tolle TR, Gockel U et al (2009) A cross-sectional cohort survey in 2100 patients with painful diabetic neuropathy and postherpetic neuralgia: differences in demographic data and sensory symptoms. Pain 146:34–40

Bloom DC (2004) HSV LAT and neuronal survival. Int Rev Immunol 23:187–198

Blumenthal DT, Shacham-Shmueli E, Bokstein F et al (2011) Zoster sine herpete: virological verification by detection of anti-VZV IgG antibody in CSF. Neurology 76:484–485

Bonfioli AA, Eller AW (2005) Acute retinal necrosis. Semin Ophthalmol 20:155–160

Braun KP, Bulder MM, Chabrier S et al (2009) The course and outcome of unilateral intracranial arteriopathy in 79 children with ischaemic stroke. Brain 132:544–557

Breuer J, Grose C, Norberg P et al (2010) A proposal for a common nomenclature for viral clades that form the species varicella-zoster virus: summary of VZV Nomenclature Meeting 2008, Barts and the London School of Medicine and Dentistry 24-25. J Gen Virol 91:821–828

Brostoff J (1966) Diaphragmatic paralysis after herpes zoster. Br Med J 2:1571–1572

Burke BL, Steele RW, Beard OW et al (1982) Immune responses to varicella-zoster in the aged. Arch Intern Med 142:291–293

Carroll WM, Mastaglia FL (1979) Optic neuropathy and ophthalmoplegia in herpes zoster oticus. Neurology 29:726–729

Celik Y, Tabak F, Mert A et al (2001) Transverse myelitis caused by varicella. Clin Neurol Neurosurg 103:260–261

Cheatham WJ, Dolan TF Jr, Dower JC et al (1956) Varicella: report on two fatal cases with necropsy, virus isolation, and serologic studies. Am J Pathol 32:1015–1035

Choi JY, Kang CH, Kim BJ et al (2009) Brachial plexopathy following herpes zoster infection: two cases with MRI findings. J Neurol Sci 285:224–226

Ciccone S, Faggioli R, Calzolari F et al (2010) Stroke after varicella-zoster infection: report of a case and review of the literature. Pediatr Infect Dis J 29:864–867

Clarke P, Beer T, Cohrs R et al (1995) Configuration of latent varicella-zoster virus DNA. J Virol 69:8151–8154

Cohen JI (1998) Infection of cells with VZV down-regulates surface expression of class I major histocompatibility complex antigens. J Infect Dis 177:1390–1393

Cohrs RJ, Gilden DH (2007) Prevalence and abundance of latently transcribed varicella-zoster virus genes in human ganglia. J Virol 81:2950–2956

Cohrs RJ, Mehta SK, Schmid DS et al (2008) Asymptomatic reactivation and shed of infectious varicella zoster virus in astronauts. J Med Virol 80:1116–1122

Cohrs RJ, Hurley MP, Gilden DH (2003) Array analysis of viral gene transcription during lytic infection of cells in tissue culture with Varicella-Zoster virus. J Virol 77:11718–11732

Cohrs RJ, Randall J, Smith J et al (2000) Analysis of individual human trigeminal ganglia for latent herpes simplex virus type 1 and varicella-zoster virus nucleic acids using real-time PCR. J Virol 74:11464–11471

Cohrs RJ, Barbour M, Gilden DH (1998) Varicella-zoster virus gene 21: transcriptional start site and promoter region. J Virol 72:42–47

Cohrs RJ, Srock K, Barbour MB et al (1994) Varicella-zoster virus (VZV) transcription during latency in human ganglia: construction of a cDNA library from latently infected human trigeminal ganglia and detection of a VZV transcript. J Virol 68:7900–7908

Connelly BL, Stanberry LR, Bernstein DI (1993) Detection of varicella-zoster virus DNA in nasopharyngeal secretions of immune household contacts of varicella. J Infect Dis 168:1253–1255

Davison AJ, Scott JE (1986) The complete DNA sequence of varicella-zoster virus. J Gen Virol 66:1759–1816
de Moragas JM, Kierland RR (1957) The outcome of patients with herpes zoster. Arch Dermatol 75:193–196
Denny-Brown D, Adams RD, Fitzgerald PJ (1944) Pathologic features of herpes zoster: a note on "geniculate herpes". Arch Neurol Psychiatry 51:216–231
de Silva SM, Mark AS, Gilden DH et al (1996) Zoster myelitis: improvement with antiviral therapy in two cases. Neurology 47:929–931
Devinsky O, Cho ES, Petito CK et al (1991) Herpes zoster myelitis. Brain 114:1181–1196
Devlin ME, Gilden DH, Mahalingam R et al (1992) Peripheral blood mononuclear cells of the elderly contain varicella-zoster virus DNA. J Infect Dis 165:619–622
Disney GH, McKee TA, Preston CM et al (1990) The product of varicella-zoster virus gene 62 autoregulates its own promoter. J Gen Virol 71:2999–3003
Dolin R, Reichman RC, Mazur MH et al (1978) NIH conference. Herpes zoster-varicella infections in immunosuppressed patients. Ann Intern Med 89:375–388
Dubinsky RM, Kabbani H, El-Chami Z et al (2004) Practice parameter: treatment of postherpetic neuralgia: an evidence-based report of the Quality Standards Subcommittee of the American Academy of Neurology. Neurology 63:959–965
Dworkin RH, O'Connor AB, Backonja M et al (2007) Pharmacologic management of neuropathic pain: evidence-based recommendations. Pain 132:237–251
Easton HG (1970) Zoster sine herpete causing acute trigeminal neuralgia. Lancet 2:1065–1066
Eisfeld AJ, Yee MB, Erazo A et al (2007) Downregulation of class I major histocompatibility complex surface expression by VZV involves open reading frame 66 protein kinase-dependent and -independent mechanisms. J Virol 81:9034–9049
Esiri MM, Tomlinson AH (1972) Herpes zoster: demonstration of virus in trigeminal nerve and ganglion by immunofluorescence and electron microscopy. J Neurol Sci 15:35–48
Feller L, Wood NH, Raubenheimer EJ et al (2008) Alveolar bone necrosis and spontaneous tooth exfoliation in an HIV-seropositive subject with herpes zoster. SADJ 63:106–110
Franco-Paredes C, Bellehemeur T, Merchant A et al (2002) Aseptic meningitis and optic neuritis preceding varicella-zoster progressive outer retinal necrosis in a patient with AIDS. AIDS 16:1045–1049
Frey CR, Sharp MA, Min AS et al (2003) Identification of CD8+ T cell epitopes in the immediate early 62 protein (IE62) of varicella-zoster virus, and evaluation of frequency of CD8+ T cell response to IE62, by use of IE62 peptides after varicella vaccination. J Infect Dis 182:40–52
Friedman SM, Mames RN, Sleasman JW et al (1993) Acute retinal necrosis after chickenpox in a patient with acquired immunodeficiency syndrome. Arch Ophthalmol 111:1607–1608
Furuta Y, Ohtani F, Aizawa H et al (2005) Varicella-zoster virus reactivation is an important cause of acute peripheral facial paralysis in children. Pediatr Infect Dis J 24:97–101
Furuta Y, Ohtani F, Sawa H et al (2001) Quantitation of varicella-zoster virus DNA in patients with Ramsay Hunt syndrome and zoster sine herpete. J Clin Microbiol 39:2856–2859
Furuta Y, Ohtani F, Fukuda S et al (2000) Reactivation of varicella-zoster virus in delayed facial palsy after dental treatment and oro-facial surgery. J Med Virol 62:42–45
Galindez OA, Sabates NR, Whitacre MM et al (1996) Rapidly progressive outer retinal necrosis caused by varicella zoster virus in a patient infected with human immunodeficiency virus. Clin Infect Dis 22:149–151
Garty BZ, Dinari G, Sarnat H et al (1985) Tooth exfoliation and osteonecrosis of the maxilla after trigeminal herpes zoster. J Pediatr 106:71–73
Gary L, Gilden DH, Cohrs RJ (2006) Epigenetic regulation of varicella-zoster virus open reading frames 62 and 63 in latently infected human trigeminal ganglia. J Virol 80:4921–4926
Gershon AA, Steinberg SP (1979) Cellular and humoral immune responses to VZV in immunocompromised patients during and after VZV infections. Infect Immun 26:170–174
Ghatak NR, Zimmerman HM (1973) Spinal ganglion in herpes zoster. Arch Pathol 95:411–415

Gilden DH, Cohrs RJ, Mahalingam R (2005) VZV vasculopathy and postherpetic neuralgia: progress and perspective on antiviral therapy. Neurology 64:21–25
Gilden DH, Cohrs RJ, Mahalingam R (2003) Clinical and molecular pathogenesis of varicella virus infection. Viral Immunol 16:243–258
Gilden DH, Beinlich BR, Rubinstien EM et al (1994a) Varicella-zoster virus myelitis: an expanding spectrum. Neurology 44:1818–23
Gilden DH, Wright RR, Schneck SA et al (1994b) Zoster sine herpete, a clinical variant. Ann Neurol 35:530–533
Gilden DH, Dueland AN, Cohrs R et al (1991) Preherpetic neuralgia. Neurology 41:1215–1218
Gilden DH, Rozenman Y, Murray R et al (1987) Detection of varicella-zoster virus nucleic acid in neurons of normal human thoracic ganglia. Ann Neurol 22:377–380
Gilron I, Bailey JM, Tu D et al (2009) Nortriptyline and gabapentin, alone and in combination for neuropathic pain: a double-blind, randomised controlled crossover trial. Lancet 374:1252–1261
Gilron I, Bailey JM, Tu D et al (2005) Morphine, gabapentin, or their combination for neuropathic pain. N Engl J Med 352:132–1334
Good RA, Zak SJ (1956) Disturbances in gamma globulin synthesis as experiments of nature. Pediatrics 18:109–149
Goon P, Wright M, Fink C (2000) Ophthalmic zoster sine herpete. J R Soc Med 93:191–192
Gowrishankar K, Steain M, Cunningham AL et al (2010) Characterization of the host immune response in human ganglia after herpes zoster. J Virol 84:8861–8870
Grimson BS, Glaser JS (1978) Isolated trochlear nerve palsies in herpes zoster ophthalmicus. Arch Ophthalmol 96:1233–1235
Guex-Crosier Y, Rochat C, Herbort CP (1997) Necrotizing herpetic retinopathies. A spectrum of herpes virus-induced diseases determined by the immune state of the host. Ocul Immunol Inflamm 5:259–265
Gündüz K, Ozdemir O (1994) Bilateral retrobulbar neuritis following unilateral herpes zoster ophthalmicus. Ophthalmologica 208:61–64
Gunson RN, Aitken C, Gilden D (2011) A woman with acute headache and sacral dermatomal numbness. J Clin Virol 50:191–193
Gupta SK, Helal BH, Kiely P (1969) The prognosis in zoster paralysis. J Bone Joint Surg Br 51:593–603
Habib AA, Gilden D, Schmid DS et al (2009) Varicella zoster virus meningitis with hypoglycorrhachia in the absence of rash and in an immunocompetent woman. J Neurovirol 15:206–208
Hanakawa T, Hashimoto S, Kawamura J et al (1997) Magnetic resonance imaging in a patient with segmental zoster paresis. Neurology 49:631–632
Hans G, Sabatowski R, Binder A et al (2009) Efficacy and tolerability of a 5% lidocaine medicated plaster for the topical treatment of post-herpetic neuralgia: results of a long-term study. Curr Med Res Opin 25:1295–1305
Harke H, Gretenkort P, Ladleif HU et al (2002) Spinal cord stimulation in postherpetic neuralgia and in acute herpes zoster pain. Anesth Analg 94:694–700
Harnisch JP (1984) Zoster in the elderly: clinical, immunologic and therapeutic considerations. J Am Geriatr Soc 32:789–793
Hata A, Asanuma H, Rinki M et al (2002) Use of an inactivated varicella vaccine in recipients of hematopoietic-cell transplants. N Engl J Med 347:26–34
Haug A, Mahalingam R, Cohrs RJ et al (2010) Recurrent polymorphonuclear pleocytosis with increased red blood cells caused by varicella zoster virus infection of the central nervous system. J Neurol Sci 292:85–88
Head H, Campbell AW (1900) The pathology of herpes zoster and its bearing on sensory localization. Brain 23:353–523
Hevner R, Vilela M, Rostomily R et al (2003) An unusual cause of trigeminal-distribution pain and tumour. Lancet Neurol 2:567–572

Hon C, Au WY, Cheng VC (2005) Ophthalmic zoster sine herpete presenting as oculomotor palsy after marrow transplantation for acute myeloid leukemia. Haematologica 90(12suppl):EIM04
Honess RW, Roizman B (1975) Regulation of herpesvirus macromolecular synthesis: sequential transition of polypeptide synthesis requires functional viral polypeptides. Proc Natl Acad Sci USA 72:1276–1280
Honess RW, Roizman B (1974) Regulation of herpesvirus macromolecular synthesis. I. Cascade regulation of the synthesis of three groups of viral proteins. J Virol 14:8–19
Hood C, Cunningham AL, Slobedman B et al (2006) Varicella-zoster virus ORF63 inhibits apoptosis of primary human neurons. J Virol 80:1025–1031
Hope-Simpson RE (1975) Postherpetic neuralgia. J R Coll Gen Pract 25:571–575
Huch JH, Cunningham AL, Arvin AM et al (2010) Impact of varicella-zoster virus on dendritic cell subsets in human skin during natural infection. J Virol 84:4060–4072
Inchauspe G, Nagpal S, Ostrove JM (1989) Mapping of two varicella-zoster virus-encoded genes that activate the expression of viral early and late genes. Virology 173:700–709
Inoue H, Motani-Saitoh H, Sakurada K et al (2010) Detection of varicella-zoster virus DNA in 414 human trigeminal ganglia from cadavers by the polymerase chain reaction: a comparison of the detection rate of varicella-zoster virus and herpes simplex virus type 1. J Med Virol 82:345–349
Izumi AK, Edwards J Jr (1973) Herpes zoster and neurogenic bladder dysfunction. JAMA 224:1748–1749
Jellinek EH, Tulloch WS (1976) Herpes zoster with dysfunction of bladder and anus. Lancet 2:1219–1222
Johnston WH, Holland GN, Engstrom RE Jr et al (1993) Recurrence of presumed varicella-zoster virus retinopathy in patients with acquired immunodeficiency syndrome. Am J Ophthalmol 116:42–50
Kakourou T, Theodoridou M, Mostrou G et al (1998) Herpes zoster in children. J Am Acad Dermatol 39:207–210
Kang JH, Ho JD, Chen YH et al (2009) Increased risk of stroke after a herpes zoster attack: a population-based follow-up study. Stroke 40:3443–3448
Karmon Y, Gadoth N (2005) Delayed oculomotor nerve palsy after bilateral cervical zoster in an immunocompetent patient. Neurology 65:170
Kemble GW, Annunziato P, Lungu O et al (2000) Open reading frame S/L of varicella-zoster virus encodes a cytoplasmic protein expressed in infected cells. J Virol 74:11311–11321
Kennedy PG, Grinfeld E, Craigon M et al (2005) Transcriptomal analysis of varicella-zoster virus infection using long oligonucleotide-based microarrays. J Gen Virol 86:2673–2684
Kennedy PG, Grinfeld E, Gow JW (1999) Latent Varicella-zoster virus in human dorsal root ganglia. Virology 258:451–454
Kilgore PE, Kruszon-Moran D, Seward JF et al (2003) Varicella in Americans from NHANES III: implications for control through routine immunization. J Med Virol 70:S111–118
Kleinschmidt-DeMasters BK, Gilden DH (2001) Varicella-zoster virus infections of the nervous system: clinical and pathologic correlates. Arch Pathol Lab Med 125:770–780
Kouroukli I, Neofytos D, Panaretou V et al (2009) Peripheral subcutaneous stimulation for the treatment of intractable postherpetic neuralgia: two case reports and literature review. Pain Pract 9:225–229
Kumagai T, Chiba Y, Wataya Y et al (1980) Development and characteristics of the cellular immune response to infection with varicella-zoster virus. J Infect Dis 141:7–13
Kurimoto T, Tonari M, Ishizaki N et al (2011) Orbital apex syndrome associated with herpes zoster ophthalmicus. Clin Ophthalmol 5:1603–1608
Lackner A, Kessler HH, Walch C et al (2010) Early and reliable detection of herpes simplex virus type 1 and varicella zoster virus DNAs in oral fluid of patients with idiopathic peripheral facial nerve palsy: decision support regarding antiviral treatment? J Med Virol 82:1582–1585
LaGuardia JJ, Cohrs RJ, Gilden DH (1999) Prevalence of varicella-zoster virus DNA in dissociated human trigeminal ganglion neurons and nonneuronal cells. J Virol 73:8571–8577

Lambade P, Lambade D, Saha TK et al. (2011) Maxillary osteonecrosis and spontaneous teeth exfoliation following herpes zoster. Oral Maxillofac Surg doi:10.1007/s10006-011-03-08
Lapresle J, Lasjaunias P (1986) Cranial nerve ischaemic arterial syndromes. Brain 109:207–215
Lee RM (1995) Morphology of cerebral arteries. Pharmacol Ther 66:149–173
Leppard B, Naburi AE (1998) Herpes zoster: an early manifestation of HIV infection. Afr Health 21:5–6
Levin MJ, Cai GY, Manchak MD et al (2003) Varicella-zoster virus DNA in cells isolated from human trigeminal ganglia. J Virol 77:6979–6987
Levin MJ, Barber D, Goldblatt E et al (1998) Use of a live attenuated varicella vaccine to boost varicella-specific immune responses in seropositive people 55 years of age and older: duration of booster effect. J Infect Dis 178:S109–112
Lewis GW (1958) Zoster sine herpete. Br Med J 2:418–421
Lewis JM, Nagae Y, Tano Y (1996) Progressive outer retinal necrosis after bone marrow transplantation. Am J Ophthalmol 122:892–895
Lin HC, Chien CW, Ho JD (2010) Herpes zoster ophthalmicus and the risk of stroke: a population-based follow-up study. Neurology 74:792–797
Ling P, Kinchington PR, Sadeghi-Zadeh M et al (1992) Transcription from varicella-zoster virus gene 67 (glycoprotein IV). J Virol 66:3690–3698
Lu PJ, Euler GL, Harpaz R (2011) Herpes zoster vaccination among adults aged 60 years and older in the US 2008. Am J Prev Med 40:e1–6
Lynch PJ, McJunkin T, Eross E et al (2011) Case report: successful epiradicular peripheral nerve stimulation of the C2 dorsal root ganglion for postherpetic neuralgia. Neuromodulation 14:58–61. doi:10.1111/j.1525-1403.2010.00307x
Mahalingam R, Wellish M, Soike K et al (2001) Simian varicella virus infects ganglia before rash in experimentally infected monkeys. Virology 279:339–342
Mahalingam R, Wellish M, Cohrs R et al (1996) Expression of protein encoded by varicella-zoster virus open reading frame 63 in latently infected human ganglionic neurons. Proc Natl Acad Sci USA 93:2122–2124
Mahalingam R, Wellish M, Brucklier J et al (1995) Persistence of varicella-zoster virus DNA in elderly patients with postherpetic neuralgia. J Neurovirol 1:130–133
Mahalingam R, Wellish M, Lederer D et al (1993) Quantitation of latent varicella-zoster virus DNA in human trigeminal ganglia by polymerase chain reaction. J Virol 67:2381–2384
Manz HJ, Canter HG, Melton J (1986) Trigeminal herpes zoster causing mandibular osteonecrosis and spontaneous tooth exfoliation. South Med J 79:1026–1028
Mayberg MR, Zervas NT, Moscowitz MA (1984) Trigeminal projections to supratentorial pial and dural blood vessels in cats demonstrated by horseradish peroxidase histochemistry. J Comp Neurol 223:46–56
Mayberg MR, Langer RS, Zervas NT et al (1981) Perivascular meningeal projections from cat trigeminal ganglia: possible pathway for vascular headaches in man. Science 213:228–230
McMillan DJ, Kay J, Mills JS (1997) Characterization of the proteinase specified by varicella-zoster virus gene 33. J Gen Virol 78(Pt 9):2153–2157
Meenken C, van den Horn GJ, de Smet MD et al (1998) Optic neuritis heralding varicella zoster virus retinitis in a patient with acquired immunodeficiency syndrome. Ann Neurol 43:534–536
Mehta SK, Tyring SK, Gilden DH et al (2008) Varicella-zoster virus in the saliva of patients with herpes zoster. J Infect Dis 197:654–657
Mehta SK, Cohrs RJ, Forghani B et al (2004) Stress-induced subclinical reactivation of varicella zoster virus in astronauts. J Med Virol 72:174–179
Menerath JM, Gerard M, Laurichesse H et al (1995) Bilateral acute retinal necrosis in a patient with acquired immunodeficiency syndrome. J Fr Ophtalmol 18:625–633
Michael EJ, Kuck KM, Kinchington PR (1998) Anatomy of the varicella-zoster virus open-reading frame 4 promoter. J Infect Dis 178(suppl 1):S27–S33
Miller DH, Kay R, Schon F et al (1986) Optic neuritis following chickenpox in adults. J Neurol 233:182–184

Molinero J, Nagore E, Obón L et al (2002) Metameric motor paresis following abdominal herpes zoster. Cutis 69:143–144
Moorthy RS, Weinberg DV, Teich SA et al (1997) Management of varicella zoster virus retinitis in AIDS. Br J Ophthalmol 81:189–194
Morgan M, Nathwani D (1992) Facial palsy and infection: the unfolding story. Clin Infect Dis 14:263–271
Morrow GB, Slobedman B, Cunningham AL et al (2003) Varicella-zoster virus productively infects mature dendritic cells and alters their immune function. J Virol 77:4950–4959
Moses H, Nagel MA, Gilden DH (2006) Acute cerebellar ataxia in a 41 year old woman. Lancet Neurol 5:984–988
Murakami S, Honda N, Mizobuchi M et al (1998) Rapid diagnosis of varicella zoster virus infection in acute facial palsy. Neurology 51:1202–1205
Murata K, Miwa H, Kondo T (2010) Polyneuritis cranialis caused by varicella zoster virus in the absence of rash. Neurology 74:85–86
Nagel MA, Choe A, Traktinskiy I et al (2011) Varicella-zoster virus transcriptome in latently infected human ganglia. J Virol 85:2276–2287
Nagel MA, Cohrs RJ, Mahalingam R et al (2008) The varicella zoster vasculopathies: clinical, CSF, imaging, and virologic features. Neurology 70:853–860
Nagel MA, Forghani B, Mahalingam R et al (2007) The value of detecting anti-VZV IgG antibody in CSF to diagnose VZV vasculopathy. Neurology 68:1069–1073
Natarajan R, Deshmane S, Valyi-Nagy T et al (1991) A herpes simplex virus type 1 mutant lacking the ICP0 introns reactivates with normal efficiency. J Virol 65:5569–5573
Noda Y, Nakazawa M, Takahashi D et al (2001) Retinal periphlebitis as zoster sine herpete. Arch Ophthalmol 119:1550–1552
Orme HT, Smith G, Nagel MA et al (2007) VZV spinal cord infarction identified by diffusion-weighted magnetic resonance imaging (DWI). Neurology 69:398–400
Osaki Y, Matsubayashi K, Okumiya K et al (1995) Polyneuritis cranialis due to varicella-zoster virus in the absence of rash. Neurology 45:2293–2294
Patterson-Bartlett J, Levin MJ, Lang N et al (2007) Phenotypic and functional characterization of ex vivo T cell responses to the live attenuated herpes zoster vaccine. Vaccine 25:7087–7093
Payten RJ, Dawes JDK (1972) Herpes zoster of the head and neck. J Laryngol Otol 86:1031–1055
Peng W, Henderson G, Perng GC et al (2003) The gene that encodes the herpes simplex virus type 1 latency-associated transcript influences the accumulation of transcripts (Bcl-x(L) and Bcl-x (S)) that encode apoptotic regulatory proteins. J Virol 77:10714–10718
Pina MA, Ara JR, Capablo JL et al (1997) Myelitis and optic neuritis caused by varicella. Rev Neurol 25:1575–1576
Quan D, Hammack BN, Kittelson J et al (2006) Improvement of postherpetic neuralgia after treatment with intravenous acyclovir followed by oral valacyclovir. Arch Neurol 63:940–942
Ratzka P, Schlachetzki JC, Bähr M et al (2006) Varicella zoster virus cerebellitis in a 66-year-old patient without herpes zoster. Lancet 367:182
Rehm S, Binder A, Baron R (2010) Post-herpetic neuralgia: 5% lidocaine medicated plaster, pregabalin, or a combination of both? A randomized, open, clinical effectiveness study. Curr Med Res Opin 26:1607–1619
Robillard RB, Hilsinger RL Jr, Adour KK (1986) Ramsay Hunt facial paralysis: clinical analyses of 185 patients. Otolaryngol Head Neck Surg 95:292–297
Rogers RS III, Tindall JP (1971) Herpes zoster in the elderly. Postgrad Med 50:153–157
Ross J, Williams M, Cohen JI (1997) Disruption of the varicella-zoster virus dUTPase and the adjacent ORF9A gene results in impaired growth and reduced syncytia formation in vitro. Virology 234:186–195
Ruiz Huete C, Bermejo PE (2008) Botulinum toxin type A in the treatment of neuropathic pain in a case of postherpetic neuralgia [in Spanish]. Neurologia 23:259–262
Salazar R, Russman AN, Nagel MA et al (2011) VZV ischemic optic neuropathy and subclinical temporal artery involvement. Arch Neurol 68:517–520

Sauerbrei A, Philipps A, Zell R et al (2007) Genotyping of varicella-zoster virus strains after serial passages in cell culture. J Virol Methods 145:80–83
Selbst RG, Selhorst JB, Harbison JW et al (1983) Parainfectious optic neuritis report and review following varicella. Arch Neurol 40:347–350
Shen W, Silva Sa E, Jaber T et al (2009) Two small RNAs encoded within the first 1.5 kilobases of the herpes simplex virus type 1 latency-associated transcript can inhibit productive infection and cooperate to inhibit apoptosis. J Virol 83:9131–9139
Silverstein BE, Chandler D, Neger R et al (1997) Disciform keratitis: a case of herpes zoster sine herpete. Am J Ophthalmol 123:254–255
Smith FP (1978) Pathological studies of spinal nerve ganglia in relation to intractable intercostal pain. Surg Neurol 10:50–53
Sodhi PK, Goel JL (2001) Presentations of cranial nerve involvement in two patients with herpes zoster ophthalmicus. J Commun Dis 33:130–135
Sotiriou E, Apalla Z, Panagiotidou D et al (2009) Severe post-herpetic neuralgia successfully treated with botulinum toxin A: three case reports. Acta Derm Venereol 89:214–215
Steain M, Gowrishankar K, Rodriguez M et al (2010) Upregulation of CXCL10 in human dorsal root ganglia during experimental and natural varicella zoster virus infection. J Virol 85:626–663
Steffen R, Selby G (1972) 'Atypical' Ramsay Hunt syndrome. Med J Aust 1:227–230
Stevens JG, Wagner EK, vi-Rao GB et al (1987) RNA complementary to a herpesvirus alpha gene mRNA is prominent in latently infected neurons. Science 235:1056–1059
Stowasser M, Cameron J, Oliver WA (1990) Diaphragmatic paralysis following cervical herpes zoster. Med J Aust 153:555–556
Surjya PU, Shiv PR, Mishra S et al (2010) Successful treatment of an intractable postherpetic neuralgia (PHN) using peripheral nerve field stimulation (PNFS). Am J Hosp Palliat Care 27:59–62
Taylor SL, Moffat JF (2005) Replication of varicella-zoster virus in human skin organ culture. J Virol 79:11501–11506
Terada K, Niizuma T, Kawano S et al (1998) Detection of varicella-zoster virus DNA in peripheral mononuclear cells from patients with Ramsay Hunt syndrome or zoster sine herpete. J Med Virol 56:359–363
Thomas EJ, Howard FM Jr (1972) Segmental zoster paresis—a disease profile. Neurology 22:459–466
Tjandra J, Mansel RE (1986) Segmental abdominal herpes zoster paresis. Aust N Z J Surg 56:807–808
Tyler SD, Peters GA, Grose C et al (2007) Genomic cartography of varicella-zoster virus: a complete genome-based analysis of strain variability with implications for attenuation and phenotypic differences. Virology 359:447–458
Tyndall MW, Nasio J, Agoki E et al (1995) Herpes zoster as the initial presentation of human immunodeficiency virus type 1 infection in Kenya. Clin Infect Dis 21:1035–1037
Umbach JL, Nagel MA, Cohrs RJ et al (2009) Analysis of human alphaherpesvirus microRNA expression in latently infected human trigeminal ganglia. J Virol 83:10677–10683
Umbach JL, Kramer MF, Jurak I et al (2008) MicroRNAs expressed by herpes simplex virus 1 during latent infection regulate viral mRNAs. Nature 454:780–783
Umehara T, Sengoku R, Mitsumura H et al (2011) Findings of segmental zoster paresis on MRI. J Neurol Neurosurg Psychiatry. doi:10.1136/jnnp. 2010.235127
Vafai A, Wellish M, Gilden DH (1988) Expression of varicella-zoster virus in blood mononuclear cells of patients with postherpetic neuralgia. Proc Natl Acad Sci USA 85:2767–2770
van Wijck AJ, Opstelten W, Moons KG et al (2006) The PINE study of epidural steroids and local anaesthetics to prevent postherpetic neuralgia: a randomised controlled trial. Lancet 367:219–224

Verjans GM, Hintzen RQ, van Dun JM et al (2007) Selective retention of herpes simplex virus-specific T cells in latently infected human trigeminal ganglia. Proc Natl Acad Sci USA 104:3496–3501

Volvoikar P, Patil S, Dinkar A (2002) Tooth exfoliation, osteonecrosis and neuralgia following herpes zoster of trigeminal nerve. Indian J Dent Res 13:11-14

Wang K, Lau TY, Morales M et al (2005) Laser-capture microdissection: refining estimates of the quantity and distribution of latent herpes simplex virus 1 and varicella-zoster virus DNA in human trigeminal ganglia at the single-cell level. J Virol 79:14079–14087

Watson CPN, Deck JH, Morshead C et al (1991) Postherpetic neuralgia: further post-mortem studies of cases with and without pain. Pain 44:105–117

Winnie AP, Hartwell PW (1993) Relationship between time of treatment of acute herpes zoster with sympathetic blockade and prevention of post-herpetic neuralgia: clinical support for a new theory of the mechanism by which sympathetic blockade provides therapeutic benefit. Reg Anesth 18:277–282

Xiao L, Mackey S, Hui H et al (2010) Subcutaneous injection of botulinum toxin a is beneficial in postherpetic neuralgia. Pain Med 11:1827–1833

Yakovlev AE, Peterson AT (2007) Peripheral nerve stimulation in treatment of intractable postherpetic neuralgia. Neuromodulation 373–375. doi:10.1111/j.1525-1403.2007.00126.x

Yamada S, Atsuta N, Tokunaga S et al (2003) Ipsilateral truncal sensory deficit in a patient with ophthalmic zoster sine herpete. Neurology 60:1049–1050

Yamamoto S, Tada R, Shimomura Y et al (1995) Detecting varicella-zoster virus DNA in iridocyclitis using polymerase chain reaction: a case of zoster sine herpete. Arch Ophthalmol 113:1358–1359

Yawn BP, Itzler RF, Wollan PC et al (2009) Health care utilization and cost burden of herpes zoster in a community population. Mayo Clin Proc 84:787–794

Yawn BP, Saddier P, Wollan PC et al (2007) A population-based study of the incidence and complication rates of herpes zoster before zoster vaccine introduction. Mayo Clin Proc 82:1341–1349

Yoshikawa T, Ihira M, Suzuki K et al (2001) Rapid contamination of the environments with varicella-zoster virus DNA from a patient with herpes zoster. J Med Virol 63:64–66

Zabolotny JM, Krummenacher C, Fraser NW (1997) The herpes simplex virus type 1 2.0-kilobase latency-associated transcript is a stable intron which branches at a guanosine. J Virol 71:4199–4208

Part II
Meningitis/Encephalitis/Poliomyelitis

Enterovirus Infections

Burk Jubelt

Abstract The human enteroviruses (EV) comprise one group of the picornavirus family. The best known members are the polioviruses (PV), coxsackieviruses, and echoviruses. They replicate in the oropharynx and gastrointestinal (GI) tract and are primarily spread by fecal–hand–oral contamination. With systemic invasion nonspecific febrile illness occurs as well as specific syndromes (rashes, hand-foot-and-mouth disease, herpangina, pleurodynia, myocarditis/pericarditis, and conjunctivitis). With systemic replication a high level viremia may result in central nervous system (CNS) invasion. EV activity can be endemic in warm climates or epidemic in temperate climates. In temperate climates, because of improved hygiene, newborns were not exposed to EV until they were older, resulting in large epidemics of poliomyelitis, which were finally curtailed with the killed PV vaccines in the 1950s and the live oral PV vaccines in the 1960s. Today, "aseptic" meningitis is the most common neurologic syndrome caused by EV. EV are also the most common cause of viral meningitis. Other EV neurologic syndromes include encephalitis, rhombencephalitis, paralytic disease, persistent infections, and the severe group B coxsackievirus fatal systemic infections of neonates. Diagnostic clues can come from epidemics, systemic manifestations, household infections, the cerebrospinal picture, and the neurologic syndrome. However, definite diagnosis depends on laboratory methods, primarily nucleic acid amplification. Treatment of acute infections is supportive. Preventative methods include good hygiene and aggressive polio vaccination programs.

Keywords Enteroviruses • Coxsackieviruses • Echoviruses • Polioviruses • Aseptic meningitis • Encephalitis • Rhombencephalitis • Brainstem encephalitis • Paralytic disease • Acute cerebellar ataxia • Cranial nerve palsies • Chronic infections

B. Jubelt (✉)
Departments of Neurology, Microbiology/Immunology and Neuroscience, SUNY Upstate Medical University, 750 E. Adams Street, Syracuse, NY 13210, USA
e-mail: jubeltb@upstate.edu

A.C. Jackson (ed.), *Viral Infections of the Human Nervous System*,
Birkhäuser Advances in Infectious Diseases,
DOI 10.1007/978-3-0348-0425-7_6,

1 Introduction

The enteroviruses are the most common cause of viral meningitis (Chonmaitree et al. 1982; Nicolosi et al. 1986; Kupila et al. 2006). In the Centers for Disease Control aseptic meningitis surveillance report for 1976, enteroviruses accounted for 83 % of viral meningitis cases of known etiology (Centers for Disease Control 1979). Also, meningitis is the most common neurologic syndrome caused by the enteroviruses (Centers for Disease Control 1981; Moore 1982).

However, the viruses that put enteroviruses at center stage are the notorious polioviruses, which cause paralytic disease. Presumed cases of poliomyelitis were probably first reported during the eighteenth Egyptian dynasty (1580–1350 BC) (Paul 1955). Although epidemics of poliomyelitis were reported in modern times in the 1840s, it was not until 1908 that the viral cause of poliomyelitis was identified [reviewed by Jubelt and Lipton (1989)]. Over the next five decades numerous epidemics of poliomyelitis were recorded. The poliomyelitis epidemics were eventually controlled with killed vaccine in the 1950s and later with the live oral poliovirus vaccine in the 1960s.

The coxsackieviruses and echoviruses (ECHO) were identified around 1950. In the 1970s diseases associated with two new enteroviruses, EV70 and EV71, were characterized (Hung et al 1976; Schmidt et al 1974). Recently many additional enterovirus strains have been isolated, most of which have caused aseptic meningitis or have not been associated as yet with disease (Piconavirus Study Group 2012).

2 The Enteroviruses: Classification

The human enteroviruses comprise one group (until recently referred to as a genus or subfamily) in the family *Picornaviridae*. The piconaviruses are comprised of both human and animal subfamilies, which are generally species specific. The term "picornavirus" was derived from "pico" meaning very small and "RNA" for the type of genomic nucleic acid (Melnick et al 1963).

The enteroviruses share a number of physical (acid pH stability, virion shape, density in CsCl, sedimentation coefficient) and biochemical (lack of a lipid envelope, replication at 37 °C, similar replication schemes) properties (Matthews 1982) which allows them to survive and replicate in the gastrointestinal tract.

The virion is composed of a positive single strand of RNA of approximately 7,400 nucleotides and a 3′ poly-A tail (Pevear et al. 1990). The polyprotein is translated as one long single protein, which is then cleaved to form all the individual viral proteins (Kitamuri et al. 1981). The capsid is an icosahedron (spheroidal) that is 22–30 nm in diameter and composed of four proteins. Three of them, VP1, VP2, and VP3, are each repeated 60 times on the external surface of the capsid. Once the virus completes the replication cycle it is generally released from the host cell via cell lysis, thus killing the infected cell.

Table 1 Human enterovirus species and types

Enterovirus species	Types
Human enterovirus A	Human Coxsackievirus A2–8, 10, 12, 14, 16
	Human enterovirus 71, 76, 89–92, 114
Human enterovirus B	Human Coxsackievirus A9, B1–6[a]
	Human echovirus 1–9, 11–21, 24–27, 29–33[a,b,c]
	Human enterovirus 69, 73–75, 77–88, 93, 97, 98, 100, 101, 106, 107, 110
Human enterovirus C	Human poliovirus 1–3
	Human Coxsackievirus A1, 11, 13, 15, 17-22, 24
	Human enteroviruses 95, 96, 97, 99, 102, 104, 105, 109, 113, 116
Human enterovirus D	Human enterovirus 68, 70, 94, 111

Modified from: Picornaviridae Study Group 2012 and Wong et al. 2010

[a]Coxsackievirus A23 reclassified as echovirus 9

[b]Vacated numbers are not used

[c]Echoviruses 1 and 8 share antigens, Echovirus 10 reclassified as reovirus I, echovirus 28 as rhinovirus type IA, echoviruses 22 and 23 as parechoviruses, and echovirus 34 as coxsackievirus A24

Originally the various enteroviruses (polioviruses, coxsackieviruses, and echoviruses) were classified by host range (infection of newborn mice, growth in tissue culture), systemic manifestations, and neurological syndromes. Later it was recognized that many of these properties, especially mouse infectivity and tissue culture replication, overlapped between the coxsackieviruses, echoviruses, and the newer enteroviruses designated as EV with a number, e.g., EV 70, EV 71 (Matthews 1982; Melnick 1985).

A recently proposed new classification scheme would divide serotypes into different species partially on the basis of sequence similarity and genome organization (Stanway et al. 2005). This new classification divides human enteroviruses into four species, human enterovirus A (HEV-A), HEV-B, HEV-C, and HEV-D (Table 1). In this scheme, HEV species "share greater than 70 % amino acid (aa) identity in P1, share greater than 70 % aa identity in nonstructural proteins 2C + 3CD, share a limited range of host-cell receptors, share a limited natural host range, have a genome base composition (G + C) which varies by no more than 2.5 %, and share a significant degree of compatibility in proteolytic processing, replication, encapsidation, and genetic recombination" (Stanway et al. 2005). In this classification, some coxsackie A viruses are HEV-A species, one (A9) an HEV-B species, and many HEV-C species (Table 1). Coxsackie B viruses and echoviruses are HEV-B species, while the polioviruses are HEV-C species.

3 Pathogenesis and Pathology

Enteroviruses (EV) enter the host via fecal–hand–oral (most common), oral secretion, or airborne aerosol (rare) transmission. EV replicate in the mucosa of the oropharynx and gastrointestinal tract (Fig. 1). Within 24–48 h, virus can be detected

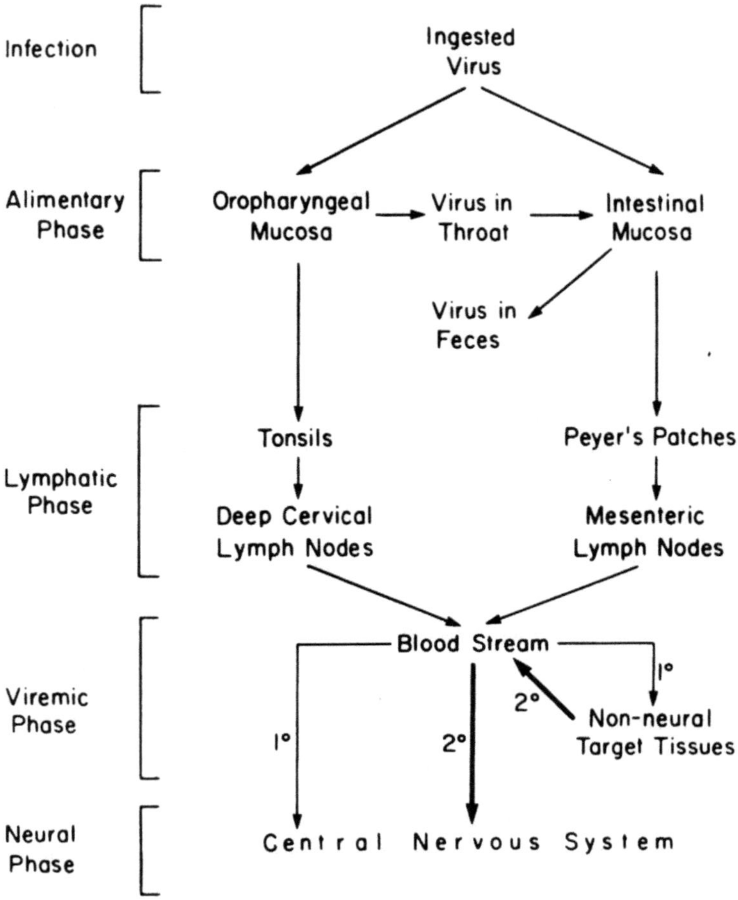

Fig. 1 Schematic representation of the pathogenesis of enteroviral infections. The importance of primary (1°) and secondary (2°) viremia in central nervous system invasion is shown. Reproduced with permission from Jubelt (1984), by courtesy of the Publisher

in the throat and stool (Horstman 1963). The incubation period for gastroenteritis and upper respiratory infections, if they occur, is usually 2–3 days. Virus excretion from the oropharynx usually lasts several weeks but may persist up to a month, whereas excretion in the stool may continue for 3–4 months (Horstman 1963). Virus may then spread through the lymphatics to lymph nodes and then to the bloodstream resulting in a primary viremia (Fig. 1). This initial primary viremia does not usually result in central nervous system (CNS) invasion, but allows EV to reach nonneural target tissues (brown fat, muscle, skin, myocardium, pericardium, and pancreas) where further amplification of viral titer occurs. A secondary viremia of greater magnitude ensues and is more likely to result in CNS invasion. This secondary viremia usually occurs during the first week of illness (days 3–7) and

Table 2 Extraneural clinical syndromes caused by enteroviruses

Syndrome	Causative enteroviruses
Rashes	Coxsackieviruses Groups A and B Echoviruses
Hand-foot-and-mouth disease	Coxsackieviruses Group A EV71
Herpangina	Coxsackieviruses Group A
Pleurodynia (epidemic myalgia, Bornholm disease)	Coxsackieviruses Group B
Myocarditis/pericarditis	Coxsackieviruses Group B
Conjunctivitis	EV70 (AHC*) Coxsackievirus A24

**AHC* acute hemorrhagic conjunctivitis

may result in the systemic manifestations of EV infections. Most EV infections are subclinical (inapparent) at this state (Horstman 1963).

The most common systemic illness due to EV infection is a nonspecific febrile illness lasting on average 3 days. Other nonspecific symptoms that may occur are sore throat, headache, malaise, and myalgias. During poliomyelitis epidemics, this was referred to as the minor illness and occurred in only 5–10 % of those infected (Paul 1955). Although the vast majority of EV infections are subclinical and usually benign, some specific nonneurologic syndromes may occur that can be more serious (carditis, conjunctivitis, and pleurodynia) or can provide clues to diagnosis (Table 2). Neonates can acquire coxsackie B virus or echovirus infections during birth or postnatally when there is a maternal EV infection during delivery (Modlin 1988). Infants younger than 10 days of age are at higher risk for severe disease because of their inability to mount a significant immune response and lack of specific antibody. Severe manifestations may be systemic (cardiac, liver) or neurologic (meningitis, encephalitis). Up to 20 % of infected infants will develop symptomatic disease (Modlin 1988). The infection may be aborted at this point or result in CNS invasion.

The precise route that EV take to invade the CNS is unclear. Most of the experimental data come from studies of poliovirus (PV) infections. Viremia precedes and appears to be required for CNS invasion (Nathanson and Bodian 1962). After the viremia, it is unclear if virus enters the CNS via areas where the blood–brain barrier is defective, such as the area postrema (Bodian 1954) or via seeding the neuromuscular junction followed by retrograde axonal transport to the neuronal cell body (Ren and Racaniello 1992; Ford et al. 2002). It is thought that coxsackieviruses and echoviruses enter the CSF through the choroid plexus from the viremia.

For EV that primarily infect neurons (PV, EV70, and EV71), viral spread is thought to occur along nerve fiber pathways (Bodian 1955), probably by fast axonal transport (Jubelt et al. 1986). Neurons other than spinal cord anterior horn cells may also be infected. In the brain, this includes large motor neurons in the precentral gyrus (Betz cells) and neurons of the thalamus and hypothalamus (Bodian 1972; Ford et al. 2002). In the brainstem, motor nuclei (facial, hypoglossal, and nucleus ambiguus), sensory nuclei (vestibular and trigeminal), and reticular formation nuclei may be involved. Infection within these areas may result in brainstem encephalitis and respiratory insufficiency. The cerebellar vermis and roof nuclei may also be involved (Bodian 1972; Ford et al. 2002). In the spinal cord, intermediate, posterior horn, and even dorsal root ganglia may be infected (Bodian 1972). Nervous system involvement with some of the other enteroviruses, especially EV70 and EV71, may resemble the poliovirus pattern with infection of anterior horn cells and paralytic disease (Chumakov et al. 1979; Hung and Kono 1979; Wadia et al. 1983; Hashimoto and Hagiwara 1982). More recently EV71 has caused outbreaks of brainstem encephalitis in Asian-Pacific countries (Huang et al. 1999; Ooi et al. 2010). The inflammatory response to these neuronal infections consists of meningeal, perivascular, and parenchymal infiltrates. Polymorphonuclear cells may be seen in the CSF early in the process, but within days, the inflammatory response becomes primarily mononuclear. Additional changes in the parenchyma include a microglial cell response and neuronophagia (Jubelt et al. 1980; Wolinsky et al. 1982). The viruses that cause meningitis (coxsackieviruses, echoviruses, and many other EV) probably replicate in meningeal and ependymal cells and disseminate through the CSF (Fenner et al. 1974; Johnson 1998).

Immunity against PV and other EV includes activation of both the innate and specific immune responses. The innate immune response is the earliest immune response and is nonspecific as compared to the adaptive immune system. Induction of innate immune responses is due to the recognition of viral components by host pattern recognition receptors, including Toll-like receptors and RNA helices, which recognize double-stranded RNA during virus replication. These two classes of pattern recognition receptors induce type I interferons (IFN-α and IFN-β) (Kawai and Akira 2006; Pichlmair et al. 2007). The cellular component of the innate immune response includes dentritic cells, $\gamma\delta$ T cells, NK T cells, and microglia. In vitro and mouse studies have demonstrated innate immune protection against PV and coxsackie B virus infections (Ida-Hosonuma et al. 2005; Flodstrom-Tullberg et al. 2005; Wang et al. 2010). Investigations into the role of the innate immune response in protection from EV infections are just in their infancy.

The vast majority of studies of immunity during PV and EV infections have focused on systemic antibody responses. Natural infection or immunization with poliovirus leads to a brisk serotype-specific antibody response, which is important for viral clearance and life-long immunity. Neutralizing antibodies of the IgM class are detectable in serum as early as 1–3 days, reach peak titers in 2–3 weeks, and fall to undetectable levels by 3 months (Svehag and Mandel 1964; Ogra et al. 1968) (Fig. 2). Neutralizing antibodies of the IgG class increase slowly, reach peak titers in 2–3 months, and decrease in titer over several years' time. Thus, IgG neutralizing antibodies are detectable for many years and provide life-long immunity (Rousseau et al. 1973; Roebauck and Chamberlain 1982). The induction of high serum neutralizing antibody titers appears to depend on the presence of viremia (Bodian 1954; Jubelt and Lipton 1987), although a low level antibody response may occur after only virus replication in the intestinal mucosa and local lymphatics (Sabin 1957). The presence of circulating antibody is crucial in preventing hematogenous

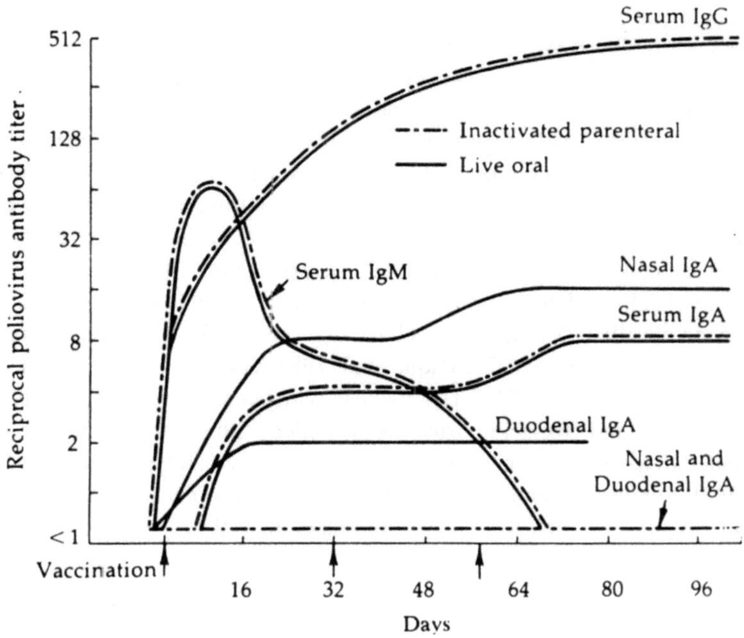

Fig. 2 Comparison of antibody responses to live oral and inactivated parenteral poliovirus vaccines. Serum IgG, IgM, and IgA responses were identical for the two vaccines. Nasal and duodenal IgA were produced only by live oral vaccination. Reproduced with permission from Fenner et al. (1974), by courtesy of the Publisher Modified from Ogra et al. (1968)

dissemination of virus and subsequent CNS virus invasion that leads to paralysis (Stevens 1959; Nathanson and Bodian 1962; Melnick et al. 1966). High serum antibody titers may also interdict intestinal virus replication and prolonged virus shedding in the feces (Bodian and Nathanson 1960; Marine et al. 1962). Serum serotype-specific neutralizing antibody responses to the coxsackieviruses and echoviruses resemble those seen after poliovirus infections (Ogra 1970; El-Hagrassy et al. 1980; Welliver et al. 1982).

Alimentary tract infections with vaccine and wild-type strains of poliovirus induce both a serum antibody response as well as a secretory IgA antibody response in the pharynx and intestinal tract (Keller et al. 1969; Ogra and Karzon 1969). This secretory antibody response in the gut appears to be a local phenomenon independent of the antibody response in the circulation. After immunization with the live, attenuated vaccines, an IgA response can be detected in the alimentary tract within 2 weeks and it persists for years (Ogra and Karzon 1971). However, parenteral immunization with inactivated poliovirus vaccine does not induce an IgA response in the gut, although the kinetics of the serum antibody response is similar to that with live oral vaccines (Ogra et al. 1968). The presence of gut immunity decreases the duration of virus replication and excretion in the stool, protects the individual from reinfection, and decreases transmission to others (Sabin et al. 1963; Welliver et al. 1982). Secretory IgA antibody responses similar to those observed following poliovirus infection have also been detected during coxsackievirus and echovirus infections (Ogra 1970; Welliver et al. 1982).

In the CNS, there also appears to be local antibody responses to PV that is independent of the serum antibody response and appears to be important for viral clearance (Morgan 1947; Ogra et al. 1973). In addition, the persistence of polioviruses and echoviruses in the CNS of agammaglobulinemic children as well as the positive response of these children to specific intravenous antibody supports a role for antibody in CNS viral clearance (Wyatt 1973; Wilfert et al. 1977; McKinney et al. 1987; Misbah et al. 1992; DeVries et al. 2011).

The role of cell-mediated immunity (CMI) in protection from or clearance of EV has not been clarified. Animal studies have demonstrated CMI during a number of EV infections (Wong et al. 1977; Paque et al. 1981; Gauntt et al. 1983). These CMI responses are mediated by $CD4^+$ Th1 T-cells (Wang et al. 1989; Mahon et al. 1992). Treatment of PV and coxsackievirus infected mice with antithymocyte sera (ATS) has revealed that T cells are necessary for viral clearance (Jubelt et al. 1989; Khatib et al. 1983). Studies of humans vaccinated with PV have demonstrated both $CD4^+$ and $CD8^+$ cytotoxic T-cell responses (Simons et al. 1992).

4 Epidemiology

Human enteroviruses have a worldwide distribution and are responsible for the majority of viral infections (Melnick 1997). In the United States, EV are responsible for 10–15 million (3–5 % of the population) symptomatic infections per year (Spigland et al. 1966; Strikas et al. 1986). EV transmission depends on a number of factors, the environment (crowded conditions, standards of hygiene, and sanitation), season of the year, geography, and host characteristics (especially age). In tropical regions, EV disease is endemic and present year round (Melnick 1997), whereas in temperate northern climates infections are epidemic occurring from May through October, resulting in the term "summer viruses." About 70 % of illness occurs from July through mid-December (Atkinson et al. 1998). In both tropical and temperate climates, several different EV may circulate at the same time (Chonmaitree et al. 1982; Moore et al. 1983). Because up to 50 % of EV infections are asymptomatic or cause only mild febrile illness, it is believed that many cases go undiagnosed.

Prior to the mid to late 1800s, infants were exposed to PV and other EV shortly after birth when they were less susceptible to severe disease (Paul et al. 1952). At that time sporadic cases of paralytic poliomyelitis did occur in infants as true "infantile paralysis," although most cases were subclinical (Paul 1955). The improved personal hygiene and social conditions in the late 1800s probably account for the severe epidemic form of poliomyelitis that developed in temperate zones and developed areas of the world (Nathanson and Martin 1979). Less crowding and improved sanitation reduced infant exposure and produced a growing pool of older children, adolescents, and adults susceptible to infection. When poliovirus infects such a virgin population, large epidemics occur with a high rate of paralysis and death (Horstman 1963; Peart 1949; Weinstein 1957). Since the introduction of the inactivated poliovirus vaccine (Salk vaccine) in 1955 and the oral poliovirus vaccine (Sabin vaccine) in the early 1960s there was a dramatic drop in paralytic poliomyelitis cases in developed countries (Melnick 1981). In the United States, the peak incidence of paralytic cases occurred in the 1952 epidemic with over 21,000 paralytic cases (Paul 1971). By the 1980s there were less than ten cases per year (Centers for Disease Control 1986).

As previously noted, most EV are primarily spread between individuals by fecal–hand–oral transmission, fecal contamination of food and utensils or rarely via oral secretions (Jubelt and Lipton 1989). Aerosol respiratory spread is rare (Couch et al. 1970). The viruses that cause acute hemorrhagic conjunctivitis (AHC), coxsackievirus A21 and EV71, are transmitted by hands contaminated with conjunctival secretions or by fomites (Hung and Kono 1979; Patriarca et al. 1983). EV are frequently found in both fresh and salt water, where contact with the contaminated water during recreational activities or land irrigation occasionally is the mode of infection. EV are usually introduced into the household by small children who are not toilet trained (Artenstein et al. 1964). Viral cultures of the skin of these children often reveals enteric flora, referred to as the "fecal veneer." Secondary transmission in the family is rapid and most members are infected in 4–5 days (Clemmer et al. 1966) depending upon household size, socioeconomic states, and prior immunity (Hall et al. 1970; Melnick 1997). Besides households, other crowded situations include school athletic teams, summer camps, hospital nurseries, and homes for the mentally challenged (Jenista et al. 1984; Alexander et al. 1993; Melnick 1997).

Age is an important determinant of susceptibility, clinical manifestations, severity, and outcome. Young children are the primary targets and most important transmitters of EV (Moore and Moresn 1984). However, older children, adolescents, and adults usually develop more severe disease (Weinstein 1957; Moore 1982; Jubelt and Lipton 1989). The severe and disseminated coxsackievirus B infections of newborns are exceptions (Grist et al. 1978). EV disease occurs more frequently in males than females and tends to be more severe in males (Moore and Moresn 1984; Gondo et al. 1995). Other predisposing factors for poliomyelitis include tonsillectomy years before or at the time of infection, injections in the month prior to infection, physical exertion, extremity trauma, and pregnancy [reviewed by Jubelt and Lipton (1989)]. Physical exertion also leads to a higher incidence and greater severity of echovirus aseptic meningitis (Baron et al. 1982).

5 Clinical Features of Human Disease: Enteroviral Neurologic Disease Syndromes

Involvement of the CNS is the most common complication of enteroviral infections (Moore 1982), and enteroviral CNS disease syndromes can involve most anatomical regions of the CNS. The syndromes associated with specific infecting viruses are shown in Table 3.

Table 3 Neurologic syndromes associated with enteroviruses

Syndrome	Virus type
Aseptic meningitis	Echoviruses
	Coxsackieviruses
	Enterovirus 71
	Parechoviruses 1, 2[a]
	Polioviruses
Encephalitis	Coxsackieviruses
	Echoviruses
	Enterovirus 71
	Polioviruses
Rhombencephalitis/Brainstem encephalitis	Enterovirus 71
	Coxsackievirus A16
	Coxsackievirus B1
	Echoviruses
Paralytic disease	Polioviruses
	Coxsackieviruses
	Echoviruses
	Enteroviruses 70, 71
Acute cerebellar ataxia	Polioviruses
	Coxsackieviruses group A
	Echoviruses 6, 9
	Enterovirus 71
Isolated cranial nerve palsies, especially facial	Poliovirus 1–3
	Coxsackieviruses A10, B5
	Echovirus 4
	Enterovirus 70
Chronic infections	Polioviruses (vaccine strains)
	Coxsackievirus A15, B3
	Echoviruses

[a]Parechoviruses are no longer classifies as EV (see Table 1)

5.1 *Aseptic Meningitis*

Aseptic meningitis is the most common neurologic syndrome caused by enteroviruses (Grist et al. 1978; CDC 1981; Moore 1982). Also, enteroviruses are the most common cause of viral meningitis (Chonmaitree et al. 1982; Ponka and Petierson 1982; Nicolosi et al. 1986; Kupila et al. 2006). In the 1976 CDC surveillance report, enteroviruses accounted for 83 % of viral meningitis of known etiology (CDC 1979). This disease syndrome is primarily caused by echovirus and coxsackievirus serotypes (Table 3) (CDC 1981, 2001, 2010; Chonmaitree et al. 1982; Morens et al. 1991; Rotbart et al. 1999) and EV 71 (Table 3) (Schmidt et al. 1974; Chonmaitree et al. 1981; Chang et al. 1999; Jeong et al. 2010). Enteroviral meningitis can be sporadic or occur as a neurological complication during systemic epidemic disease.

Clinical presentation of viral meningitis includes headache, nausea, photophobia, fever, a stiff neck, and general irritability. Onset can be abrupt. Headaches may be severe (Sawyer 1999). Although photophobia, nausea, and vomiting are common, the presentation of viral meningitis is strongly influenced by age. In older children and young adults, there is frequently a short prodromal period of 2–3 days with sore throat, fever, malaise, nausea, vomiting, diarrhea, and headache occurring before medical attention is sought. Upon presentation children frequently have fever, and up to 33 % of them will have Kernig and/or Brudzinski signs. In younger children and infants, signs are much less specific and include increased irritability and nonspecific rash. Other specific nonneurologic syndromes may also occur (see Table 2). Nuchal rigidity is not always apparent and in infants is rarely seen. Fifty percent of children older than one will develop nuchal rigidity (Zaoutis and Klein 1998). Generally children less than 3 years of age are most susceptible to viral meningitis, whether sporadic or epidemic. For all age groups disease can progress to involve additional organ systems (e.g., renal failure, carditis).

The *differential diagnosis* of enteroviral meningitis not only includes other viruses, but also some bacterial, fungal, and parasitic infections (Johnson 1998; Jubelt 2010a, b, c). Other viruses causing meningitis include herpes simplex 2, arboviruses, mumps, varicella, and lymphocytic choriomeningitis (LCM) virus. Often less invasive or subacute bacterial, fungal, and parasitic infections can present with a mononuclear pleocytosis similar to viral meningitis. These infections include bacterial infections (brucellosis, mycoplasma, listeria, partially treated bacterial meningitis, tuberculosis), fungal infections (cryptococcosis, coccidiomycosis, and candidiasis), spirochetes (leptospirosis, secondary syphilis, and Lyme disease), rickettsial infections (primarily Rocky Mountain spotted fever), and parasitic infections (cysticercosis and angiostrongylus eosinophilic meningitis). A number of noninfectious diseases (e.g., carcinomatous meningitis, collagen vascular meningitis, and sarcoidosis) may also present as aseptic meningitis (Johnson 1998).

Diagnosis is based on imaging studies to exclude other diagnoses and CSF analysis. The presenting neurologic manifestations of meningitis are not dissimilar enough between etiologies in order to make a specific diagnosis. Systemic manifestations (Table 2) as well as household disease and an epidemic pattern may provide clues to diagnosis. With progression of many of the nonviral infections, there is a decrease in the CSF glucose, which helps to clarify the diagnosis since hypoglycorrhachia is rarely seen in viral meningitis. Computerized tomography (CT), magnetic resonance imaging (MRI), and electroencephalography (EEG) are usually normal. Occasionally, the EEG may reveal diffuse slowing without clinical encephalitis. The CSF profile in viral meningitis usually consists of a mononuclear cell pleocytosis with a normal glucose. Occasionally during the first 24–48 h of the infection, polymorphonuclear cells may be seen, mimicking bacterial meningitis. Very rarely the CSF glucose may be low, as in fungal and tuberculous meningitis (Jubelt and Lipton 1989). Except for EV 70, virus can usually be cultured from the throat and stool. Except for PV and EV 70, virus may be cultured from the CSF. PCR techniques applied to the CSF are available for most EV and are replacing culture for diagnosis (Lee and Davies 2007; Archimbaud et al. 2009).

The *prognosis* of viral meningitis is good as this entity is usually benign, self-limited, and patients make a complete recovery. However, some studies suggest that children less than 3 months of age may suffer from speech and language delay (Zaoutis and Klein 1998). The duration of illness is 3–7 days, although adolescents and adults may be symptomatic for several weeks (Rotbart et al. 1999). Although many cases of viral meningitis result in hospitalization, this is primarily due to the time it takes to exclude the differential diagnoses. It is important to distinguish viral meningitis due to enteroviral infection from aseptic meningitis due to herpes simplex virus (treatable with acyclovir) or bacterial meningitis (life threatening without antibiotic treatment).

There is no specific *approved therapy* for EV meningitis. Pleconaril is a specific antiviral agent under study that has been available for compassionate use (Sawyer 1999). Again, good personal hygiene can be preventative.

5.2 *Encephalitis*

The term "encephalitis" is used when there is clinical evidence of brain involvement. Enterovirus encephalitis results from direct viral invasion of and damage to the parenchyma (primary encephalitis). Encephalitis is the second most common neurologic syndrome caused by enteroviruses (Centers for Disease Control 1981; Yang et al. 2005). In years of high enterovirus activity, these viruses have caused almost a fourth of all cases of encephalitis of known etiology (Centers for Disease Control 1982; Frantzidou et al. 2008). Encephalitis has primarily been caused by coxsackieviruses, echoviruses, and EV 71 (see Table 3) (Horstmann and Yamada 1968; Grist et al. 1978; Chonmaitree et al. 1981).

Clinical manifestations of encephalitis include those of meningitis (fever, headache, nuchal rigidity), plus those of encephalitis (altered mental status, seizures, and focal neurologic deficits); thus the term meningoencephalitis is more correct (Melnick 1965; Horstmann and Yamada 1968; Grist et al. 1978; Morens et al. 1991). Mental status changes are usually those of mild obtundation although bizarre behavior and coma may occur infrequently (Acharya et al. 2001; Cree et al. 2003). Seizures may occur, but are usually isolated rather than multiple. Focal deficits are usually mild and transient.

The *differential diagnosis* of enterovirus encephalitis includes many of the same viruses that are responsible for aseptic meningitis (measles, mumps, varicella, arboviruses, and herpes simplex) and other viruses, i.e., adenovirus, cytomegalovirus, Epstein–Barr virus, herpes zoster (Centers for Disease Control 1979; Koskiniemi and Vaheri 1982; Beghi et al. 1984). A wide variety of nonviral infections are included in the differential: mycoplasma, Legionnaires disease, Lyme disease, leptospirosis, syphilis, brucellosis, subacute bacterial endocarditis, brain abscess, Rocky Mountain spotted fever, toxoplasmosis, cysticercosis, amebiasis, schistosomiasis, malaria, trypanosomiasis, echinococcus, and trichinosis

(Ho 1978; Johnson 1998; Jubelt 2010a, b, c). Noninfectious diseases to be considered include vasculitis, sarcoid, and gliomatosis cerebri (Johnson 1998).

Again, *diagnosis* is based on imaging studies to exclude other diagnoses and CSF analysis. Similar to viral meningitis, the neurologic manifestations of viral encephalitis are not specific enough to be useful to make a diagnosis. Similar to EV meningitis, the systemic manifestations (Table 2), the pattern of disease in the family, and an epidemic occurrence are suggestive of the diagnosis of EV encephalitis. The CT and MRI are usually normal. Rarely focal parenchymal lesions may be seen (Modlin et al. 1991). The EEG usually reveals generalized slowing, but focal sharp waves or focal slowing may be seen. The CSF profile is similar to that of EV meningitis. Coxsackieviruses, echoviruses, and EV 71 may be isolated from the CSF, stool, and throat. PV is not isolated from the CSF, but usually from the stool and throat. Again PCR analysis of the CSF has become the diagnostic test of preference (Lee and Davies 2007; Archimbaud et al. 2009).

The *prognosis* of EV encephalitis is usually good as it is usually mild with good recovery. Exception to this is the fulminant encephalitis seen with neonatal group B coxsackievirus infections and the chronic encephalitis seen in agammaglobulinemic patients (see below). Also, when enterovirus encephalitis occurs in infants, subtle residual intellectual deficits may be seen (Sells et al. 1975).

There is no specific approved *therapy* for EV encephalitis. Treatment is supportive.

5.3 EV 71 Rhombencephalitis/Brainstem Encephalitis

Rhombencephalitis/brainstem encephalitis is a serious neurological complication of EV 71 infection. This complication first appeared during epidemics of EV 71 HFMD in Taiwan and Malaysia during the late 1990s (Huang et al. 1999; Ho et al. 1999; Lum et al. 1998). Over the last 10 years, regular epidemics have continued to occur in countries across the Asia-Pacific region (Ooi et al. 2010). More recently, cases have occurred in Europe (Vallet et al. 2009). During epidemics, the vast majority of cases that progressed to rhombencephalitis occurred in children under the age of 5 years (Dolin 1999). Fatality rates have ranged from 11 to 14 % (Huang et al. 1999; Lum et al. 1998). Similar to EV encephalitis, there is direct virus invasion of the brainstem (Hsueh et al. 2000).

Clinical manifestations of rhombencephalitis include initial symptoms of myoclonic jerks, tremors, ataxia, and cranial nerve palsies. As the disease progresses, coma and respiratory failure may occur (Huang et al. 1999). Polio-like acute flaccid paralysis and aseptic meningitis each occurred in about 10 % of cases (Huang et al. 1999).

The *differential diagnosis* is extensive and includes bacteria, especially listeria, other viruses (CMV, HSV, Cox A16, echovirus), autoimmune diseases, and paraneoplastic syndromes (Jubelt et al. 2011). When poliovirus causes encephalitis, it more often involves the brainstem, with extensive reticular formation damage, i.e., bulbar polio, rather than the cerebral cortex (Baker 1949; Howe and Wilson 1957; Plum and Swanson 1959). However, in a majority of these cases of bulbar poliomyelitis, extremity paralysis also occurs (see Sect. 5.4).

Diagnosis again relies on imaging studies and CSF analysis. Most patients (70–75 %) have MRI T2 hyperintense lesions in the brainstem and cerebellum (Huang et al. 1999; Shen et al. 1999). Lesions in the brainstem most often involve the pontine tegmentum. Occasionally, lesions extend to the thalamus and cervical cord. In the CSF, there is a mononuclear pleocytosis and normal glucose. In the report by Huang et al. (1999), the mean CSF white blood cell (WBC) count was 194 cells/m^3 (range 5–379). Virus may be isolated from the throat, stool, HFMD vesicles, and the CSF. Viral gene amplification is becoming more readily available (Ooi et al. 2010).

There is no specific *therapy* for EV 71 rhombencephalitis. There are case reports of improvement with IVIg, but no controlled studies (Ooi et al. 2010).

5.4 Paralytic Disease

As previously noted, poliomyelitis has become rare in developed countries since the introduction of the killed polio vaccine in 1955 and the live vaccine in 1961 (Paul 1971; Centers for Disease Control 1981). From 1975 through 1984, there were 118 cases of paralytic poliomyelitis in the United States, and a majority of these cases were caused by the live poliovirus vaccines (Centers for Disease Control 1986; Nokowane et al. 1987). In 2000, inactivated polio (Salk) vaccine (IPV) became the standard for vaccination in the United States. Since then there have been two cases of endogenous poliomyelitis, one in a patient with common variable immunodeficiency who was exposed prior to 2000 (DeVries et al. 2011), and a second in an under vaccinated community exposed to imported vaccine-derived poliovirus (Alexander et al. 2009). However, paralytic poliomyelitis remains a significant health problem in underdeveloped areas of the world (Afghanistan, Pakistan, India, Nigeria, Angola, Chad, Congo, Sudan) (CDC 2009; Aylward and Yamada 2011).

Once poliovirus epidemics were reduced it was found that infection with other EV could cause paralytic disease (Table 3). EV 71 caused flaccid paralysis in 11–14 % of cases as one of the neurological complications during HFMD epidemics (Huang et al. 1999; Lum et al. 1998). It also has caused isolated cases. EV 70 has caused severe flaccid paralysis during epidemics of AHC (Chopra et al. 1986). Approximately 1 in 10,000–15,000 cases of AHC develop paralytic disease. Paralytic cases caused by coxsackieviruses and echoviruses are isolated and rare. They are indistinguishable from poliomyelitis cases. In the period 1976–1979, 52 cases of paralytic disease caused by enteroviruses were recorded in the United States; 25 were caused by poliovirus, 18 by echoviruses, 7 by coxsackieviruses, and 2 by EV 71 (CDC 1981).

The *clinical manifestations* of acute paralytic disease begin with the symptoms of the prodromal or minor illness, fever, malaise, nausea, muscle aches, and sometimes diarrhea. Prodromal symptoms may not occur with EV 70 since it enters the host via the eye. Paralysis usually begins between days 3 and 7 of infection. Paralysis is usually asymmetrical, flaccid, more proximal than distal, and often patchy. The reflexes are lost as paralysis progresses. Over the next several days, paralysis may develop in other extremities and bulbar involvement with impaired respiration may occur. Extension of paralysis is unlikely to occur after the fifth or sixth day of paralysis. Paralysis caused by coxsackieviruses and echoviruses is usually mild compared to that seen with PV, EV 70, and EV 71. Generally, the weakened muscles regain some strength over the next several years.

The *differential diagnosis* of paralytic disease should include all of the EV, as well as other viruses that can cause acute lower motor neuron paralysis, rabies (Chopra et al. 1980), herpes zoster (Thomas and Howard 1972), and West Nile virus (Jeha et al. 2003). Other entities in the differential include acute inflammatory polyradiculitis (Guillain–Barré syndrome), botulism, acute toxic neuropathies, acute intermittent porphyria, acute transverse myelitis, and acute spinal cord compression from epidural abscess (Price and Plum 1978; Asbury 1981; Gear 1984).

The *diagnosis* depends on the clinical manifestations, CSF analysis, and neurophysiological studies. The clinical manifestations of acute lower motor neuron paralysis are usually easy to recognize. Usually imaging studies are negative but there have been a few reports of MRI T2 hyperintensities in the anterior horns (Rao and Bareman 1997). Neurophysiological studies may be especially useful in excluding other diagnoses. During the first week of paresis, stimulation of motor nerves may reveal reduced amplitude of compound muscle action potentials (cMAP) (David and Doyle 1997; Edwards et al. 2000). F waves are often unobtainable (So and Olney 1991). In the second week, signs of denervation (fibrillation and positive sharp waves) begin to appear (Weichers 1988). CSF analysis reveals a mononuclear pleocytosis with a normal glucose similar to other CNS viral infections. In the first 24–48 h polymorphonuclear cells may be found, converting to mononuclear cells thereafter (Jubelt and Lipton 1989). PV, coxsackieviruses, echoviruses, and EV 71 can be isolated from the stool and throat but rarely from the CSF when the presentation is paralytic. EV 70 is not found in the stool or throat and is usually not isolated from the CSF. Genomic amplification is now available for diagnosis for most of these EV (Archimbaud et al. 2009; Ooi et al. 2010).

Treatment for paralytic disease is supportive. Poliovaccine is used for prophylactic treatment to prevent the occurrence of poliovirus-induced paralysis. Both the IPV and the live Sabin oral poliovirus vaccine (OPV) are used throughout the world. Vaccination is not available for the other EV.

5.5 Persistent Infections

Persistent CNS enteroviral infections have primarily occurred in immunodeficient children with agammaglobulinemia. The viruses that cause these infections are of low virulence, primarily echoviruses (McKinney et al. 1987) and vaccine-like PV

(Wyatt 1973). Coxsackieviruses have caused several cases (McKinney et al. 1987), one of which was normal immunologically (Berger et al. 2006).

Clinical manifestations depend on the infecting virus. The clinical course is one of slow progression over months to years with intermittent plateaus. Persistent echovirus infections primarily involve the brain, with alterations in behavioral and mental status, headaches, seizures, pyramidal tract involvement, ataxia, and tremor (Wilfert et al. 1977; McKinney et al. 1987). About half of the patients with chronic echovirus infections develop a dermatomyositis-like syndrome, presumably from viral invasion of the muscle (Webster et al. 1978; Hays and Gamboa 1986). Virus has been isolated from muscle in only two instances (Asherson and Webster 1980; Mease et al. 1981). In the PV cases, there is a prolonged incubation period of several months to years from the time of vaccination until the onset of neurological disease (Wyatt 1973; DeVries et al. 2011). Some cases begin with lower motor neuron paralysis but the persistent CNS infection continues to cause progressive intellectual and cerebral dysfunction. In other cases, paralysis occurs later, after cerebral dysfunction (Davis et al. 1977).

The *differential diagnosis* would include those of chronic progressive inflammatory cerebral neurological dysfunction. Possibilities would include other viruses (rabies, CMV, EBV, adenovirus, HIV) and nonviral infections (mycoplasma, tertiary Lyme disease, syphilis, brucellosis, cysticercosis, malaria).

Diagnosis includes serum studies demonstrating deficient immunoglobulins on serum protein electrophoresis or when testing for individual immunoglobulins (IgM, IgG, IgA). Examination of the CSF reveals the viral picture of a mononuclear pleocytosis with a normal glucose. Echovirus but not poliovirus has been frequently isolated from the CSF. Either virus can at times be isolated from the stool because of intestinal infection.

Standard *treatment* involves the regular use of intravenous immunoglobulins for the treatment of agammaglobulinemia (Misbah et al. 1992). Unfortunately death is a frequent outcome even when specific antibody is given intravenously or even intrathecally (McKinney et al. 1987; DeVries et al. 2011). Live oral poliovaccine should not be given to these patients. Unfortunately, in many cases the immunodeficiency was not obvious at the time of vaccination.

5.6 Uncommon Neurologic Syndromes

Acute cerebellar ataxia has been infrequently caused by coxsackieviruses (Feldman and Larke 1972), echoviruses (McCallister et al. 1959), polioviruses (Mendez-Cushion et al. 1962), and frequently by EV 71 in Japan (Ishimaur et al. 1980). Many nonenteroviruses have caused this syndrome (Johnson 1998; Kanagarajan et al. 2003). There is no specific treatment. Prognosis for recovery is good.

Cranial nerve palsies are well documented to occur with enterovirus infections. EV 70 has caused cranial nerve palsies in as many as one-half of patients with neurologic complications, the majority also having limb paralysis (Katiyar et al.

1983; Wadia et al. 1983). Isolated cranial nerve palsies have occurred in as many as one-fourth of patients with neurologic involvement, although in some epidemics, the incidence was much lower (Hung and Kono 1979; Katiyar et al. 1983; Wadia et al. 1983). In those with isolated palsies, the facial nerve is affected most frequently followed by the motor trigeminal nerve. Less frequently reported isolated palsies include those of cranial nerves two to four, six, and eight to twelve. In EV 71 rhombencephalitis, involvement of cranial nerves is common (Huang et al. 1999). Most commonly involved cranial nerves are three, four, six, seven, nine, ten, and twelve (Ooi et al. 2010). Isolated facial nerve palsies have infrequently been reported to occur after poliovirus, coxsackievirus, and echovirus infections [reviewed by Weiner (1978)]. The prognosis for isolated palsies is relatively good. In those patients with extremity paralysis or rhombencephalitis, the prognosis is poor. There is no specific therapy.

Post-Polio Syndrome (PPS) is a term used to describe the late weakness and other problems that occur in patients who had poliomyelitis 30–40 years earlier (Mulder et al. 1972; Jubelt and Cashman 1987). PPS includes three major symptom components: systemic (fatigue, pain, cold intolerance), musculoskeletal (joint dysfunction, joint pains), and neurologic (weakness) (Jubelt and Cashman 1987). Most often new weakness develops in previously involved muscles (Codd et al. 1985; Halstead et al. 1985). Even though detailed virologic studies have not been performed in these patients, the limited evidence to date does not support an ongoing or chronic poliovirus infection [for review, see Jubelt and Cashman (1987)]. However, inflammatory changes have been found in spinal cord postmortem material (Pezeshkpour and Dalakas 1988), which suggests an infectious or an immune-mediated syndrome. Another leading hypothesis invokes an excessive metabolic demand (overwork) on the remaining motor units that have previously reinnervated denervated fibers [for review see Jubelt and Cashman (1987)]. This excessive demand of an increased motor unit territory would result in loss of the reinnervating terminal sprouts or the motor neuron itself. PPS weakness progresses at a rate of 1–2 % per year (Dalakas et al. 1986; Stalberg and Grimby 1995). Treatment is primarily supportive although nonfatiguing strengthening exercises may be beneficial (Jubelt 2004).

Both acute and chronic *myopathies* have infrequently been related to coxsackievirus and echovirus infections (Grist et al. 1978; Melnick 1982). Acute myositis, sometimes accompanied by myoglobinuria, appears to have been caused by coxsackieviruses in a handful of cases based upon serologic conversion and/or virus isolation (Hays and Gamboa 1986). The onset is abrupt with fever, myalgia, elevated creatinine phosphokinase, and occasionally visible myoglobinuria. Usually with acute myositis there is full recovery. Coxsackie B viruses have also been related to chronic myopathies and polymyositis/dermatomyositis in a handful of cases (Tang et al. 1975; Travers et al. 1977; Bowles et al. 1987).

The *group B coxsackieviruses also cause severe, often fatal systemic neonatal disease* with pneumonia, myocarditis, myositis, necrotic hepatitis, hemorrhagic manifestations, and meningoencephalitis (Grist et al. 1978; Kaplan et al. 1983).

Less frequently, echoviruses and, rarely, group A coxsackieviruses have caused this syndrome (Morens et al. 1991; Anonymous 1986; Modlin 1986).

Other neurologic syndromes reported to be due to enteroviruses have included acute infantile hemiplegia, opsiclonus-myoclonus, parkinsonism, hemichorea, transverse myelitis, and Guillain–Barré syndrome (Jubelt and Lipton 1989). These disease syndromes are very rare, and a cause-and-effect relationship is not always clear.

6 Diagnosis

Epidemiologic clues to the diagnosis of enteroviral infections include epidemics, extraneural manifestations (Table 2), household infections, and obviously paralytic disease and rhombencephalitis. The CSF picture (mononuclear pleocytosis with normal glucose) is suggestive of viral CNS infections.

Laboratory diagnosis has utilized three methods: serology, viral culture (virus isolation), and nucleic acid amplification by the polymerase chain reaction (PCR) (Hamilton et al. 1999). Serology is of limited value because it is slow, requires both acute and convalescent blood samples, and is difficult due to the large number of antigenically distinct serotypes (at least 60) that need to be included in the assay. Viral culture has been the gold standard for detecting enterovirus. During infection, enteroviruses are shed from the intestine and oropharynx; thus stool samples and oropharyngeal swabs can be grown in tissue culture and the identification made based on cytopathic effect. Although viral cultures take 4–8 days (up to 14 days depending on the strain), they are useful for the isolation and identification of serotypes. However, viral culture is limited because of low relative sensitivity (<75 % for diagnosis of aseptic meningitis) and the fact that enteroviruses can be shed for weeks after exposure and may be part of an asymptomatic infection and not the disease at hand (Sawyer 1999). Commercial PCR methods (e.g., Amplicon™) specific for enteroviruses have been developed and are now readily available in the clinical setting. PCR requires a small amount of clinical material and is rapid (12–24 h), very sensitive, and highly specific. PCR methods are now the diagnostic gold standard. Most EV can be detected by PCR from the oropharynx, stool, and CSF (Lee and Davies 2007; Noordhoek et al. 2008; Archimbaud et al. 2009; Ooi et al. 2010).

7 Treatment

As noted earlier, treatment of patients is primarily supportive. Preventative methods are used to prevent virus dissemination and control epidemics by good hygiene practices and aggressive polio vaccination programs. There are no FDA approved specific treatments for the various enterovirus infections.

8 Prognosis

Prognosis for recovery is good for most EV infections except paralytic disease, EV 71 rhombencephalitis, and systemic neonatal coxsackievirus infections (see specific syndromes above).

References

Acharya VZ, Talwar D, Elliott SP (2001) Enteroviral encephalitis leading to a locked-in state. J Child Neurol 15:864–866

Alexander JP, Chapman LE, Pallansch MA et al (1993) Coxsackievirus B2 infection and aseptic meningitis: a focal outbreak among members of a high school football team. J Infect Dis 167:1201–1205

Alexander JP, Ehresmann K, Seward J et al (2009) Transmission of imported vaccine-derived poliovirus in an undervaccinated community in Minnesota. J Infect Dis 199:391–397

Anonymous (1986) Avoiding the danger of enteroviruses to newborn infants (Editorial). Lancet 1:194–195

Archimbaud C, Chambon M, Bailly JL et al (2009) Impact of rapid enterovirus molecular diagnosis on the management of infants, children, and adults with aseptic meningitis. J Med Virol 81:42–48

Artenstein MS, Caoiganur FC, Buescher EL (1964) Epidemic coxsackievirus infection with mixed clinical manifestations. Ann Intern Med 60:196–203

Asbury AK (1981) Diagnostic considerations in Guillain-Barre syndrome. Ann Neurol 9:1–5

Asherson GL, Webster ADB (1980) Sex-linked hypogammaglobuJinaemia. In: Asherson GL, Webster ADM (eds) Diagnosis and treatment of immunodeficiency diseases. Blackwell, Oxford, pp 7–36

Atkinson PF, Sharland M, Maquire H (1998) Predominate enteroviral serotypes causing meningitis. Arch Dis Child 78:373–374

Aylward B, Yamada T (2011) The polio endgame. N Engl J Med 364:2273–2275

Baker AB (1949) Neurologic signs of bulbar poliomyelitis. In: Poliomyelitis: papers and discussion presented at the first international poliomyelitis conference. Lippincott, Philadelphia, pp 241–244

Baron RC, Hatch MH, Eeman KD, MacCormack JN (1982) Aseptic meningitis among members of a high school football team. An outbreak associated with echovirus 16 injection. J Am Med Assoc 248:1724–1727

Beghi E, Nicolosi A, Kurland LT, Mulder OW, Hauser WA, Shuster L (1984) Encephalitis and aseptic meningitis, Olmsted County, Minnesota, 1950-1981: I Epidemiology. Ann Neurol 16:283–294

Berger JR, Chumley W, Pittman T, Given C, Nuovo G (2006) Persistent coxsackie B encephalitis: report of a case and review of the literature. J Neurovirol 12:511–516

Bodian O (1954) viremia in experimental poliomyelitis I. General aspects of infection after intravascular inoculation with strains of high and low invasiveness. Am J Hyg 60:339–357

Bodian O (1955) Emerging concepts of poliomyelitis infection. Science 122:105–108

Bodian D (1972) Poliomyelitis. In: Minckler J (ed) Pathology of the nervous system. McGraw-Hill, New York, pp 2323–2394

Bodian O, Nathanson N (1960) Inhibitory effect of passive antibody on virulent poliovirus excretion and on immune response in chimpanzees. Bull Johns Hopkins Hosp 107:143–162

Bowles NE, Oubcwitz V, Sewry CA, Archard LC (1987) Dermatomyositis, polymyositis, and coxsackie-B-virus infection. Lancet 1:1004–1007

Centers for Disease Control (1979) Aseptic meningitis surveillance, annual summary 1976
Centers for Disease Control (1981) Enterovirus surveillance report, 1970–1979
Centers for Disease Control (1982) Poliomyelitis surveillance summary, 1980–1981
Centers for Disease Control (1986) Poliomyelitis-United States, 1975–1984. MMWR 35:180–182
Centers for Disease Control (2001) Echovirus type 13—United States, 2001. MMWR 50:777–780
Centers for Disease Control and Prevention (2009) Wild poliovirus type 1 and type 3 importations—15 countries, Africa, 2008–2009. MMWR 58:357–362
Centers for Disease Control and Prevention (2010) Nonpolio enterovirus and human parechovirus surveillance—United States, 2006–2008. MMWR 59:1577–1580
Chang LY, Lin TY, Huang YC (1999) Comparison of enterovirus 71 and coxsackievirus A16 clinical illnesses during the Taiwan enterovirus epidemic, 1998. Pediatr Infect Dis J 18:1092–1096
Chonmaitree T, Menegus MA, Schervish-Swierkosz EM, Schwalenstocker E (1981) Enterovirus 71 infection: report of an outbreak with two cases of paralysis and a review of the literature. Pediatrics 67:489–493
Chonmaitree T, Menegus MA, Powell KR (1982) The clinical relevance of 'CSF viral culture'. A two year experience with aseptic meningitis in Rochester, NY. J Am Med Assoc 247:1843–1847
Chopra JS, Banerjee AK, Murthy JM, Pal SR (1980) Paralytic rabies: a clinico-pathological study. Brain 103:789–802
Chopra JS, Sawhney IMS, Dhand UK, Prabhakar S, Maik S, Sehgal S (1986) Neurological complications of acute haemorrhagic conjunctivitis. J Neurol Sci 73:177–191
Chumakov M, Varoshilova M, Shindarov L et al (1979) Enterovirus 71 isolated from cases or epidemic poliomyelitis-like disease in Bulgaria. Arch Virol 60:329–340
Clemmer DI, Li F, Leblanc DR, Fox JP (1966) An outbreak of subclinical infection with coxsackievirus B3 in southern Louisiana. Am J Epidemiol 83:123–129
Codd MB, Mulder DW, Kuju LT, Beard CM, O'Fallon WM (1985) Poliomyelitis in Rochester, Minnesota, 1935-1955: epidemiology and long-term sequele: a preliminary report. In: Halstead FL, Wiechers DO (eds) Late effects of poliomyelitis. Symposia Foundation, Miami, pp 121–134
Couch RM, Douglas RG Jr, Lindgren KM, Gerone PJ, Knight V (1970) Airborne transmission of respiratory infection with coxsackievirus A type 21. Am J Epidemiol 91:78–86
Cree BC, Bernardini GL, Gays AP, Lowe G (2003) A fatal case of coxsackievirus B4 meningoencephalitis. Arch Neurol 60:107–112
Dalakas MC, Eldre G, Hallett M et al (1986) A long term follow-up study of patients with postpoliomyelitis neuromuscular symptoms. N Engl J Med 314:959–963
David WS, Doyle JJ (1997) Acute infantile weakness: a case of vaccine-associated poliomyelitis. Muscle Nerve 20:747–749
Davis LE, Bodian D, Price D, Butler I, Vickers J (1977) Chronic progressive poliomyelitis secondary to vaccination of an immunodeficient child. N Engl J Med 297:241–245
DeVries AS, Harper J, Murray A et al (2011) Vaccine-derived poliomyelitis 12 years after infection in Minnesota. N Engl J Med 364:2316–2323
Dolin R (1999) Enterovirus 71—emerging infections and emerging questions. N Engl J Med 341:984–985
Edwards EA, Grant CC, Huang QS, Powell KFH, Croxson MC (2000) A case of vaccine-associated paralytic poliomyelitis. J Paediatr Child Health 36:408–411
El-Hagrassy MMO, Banatvala JE, Coltart DJ (1980) Coxsackie-B-virus specific IgM responses in patients with cardiac and other diseases. Lancet 2:1160–1162
Feldman W, Larke RPB (1972) Acute cerebellar ataxia associated with the isolation of coxsackie type A9. Can Med Assoc J 106:1104–1107
Fenner R, McAuslan BR, Mims CS, Sambrook J, White DO (1974) The biology of animal viruses. Academic, New York

Flodstrom-Tullberg M, Hulcrantz M, Stotland A et al (2005) RNase and double-stranded RNA-dependent protein kinase exert complementary roles in islet cell defense during coxsackievirus infection. J Immunol 174:1171–1177

Ford DJ, Ropka SL, Collins GH, Jubelt B (2002) The neuropathology observed in wild-type mice inoculated with human poliovirus mirrors human paralytic poliomyelitis. Microb Pathog 33:97–107

Frantzidou F, Kamaria F, Dumaidi K, Skoura L, Antoniadis A, Papa A (2008) Aseptic meningitis and encephalitis because of herpesviruses and enteroviruses in an immunocompetent adult population. Eur J Neurol 15:995–997

Gauntt CJ, Paque RE, Trousdale MD et al (1983) Temperature-sensitive mutant of coxsackievirus B3 establishes resistance in neonatal mice that protects them during adolescence against coxsackievirus B3-induced myocarditis. Infect Immun 39:851–864

Gear JH (1984) Nonpolio causes of polio-like paralytic syndromes. Rev Infect Dis 6:S379–S384

Gondo K, Kusuhara K, Take H et al (1995) Echovirus type 9 epidemic in Kagoshima, southern Japan: seroepidemiology and clinical observation of aseptic meningitis. Pediatr Infect Dis J 14:787–791

Grist NR, Bell EJ, Asaad F (1978) Enteroviruses in human disease. Prog Med Virol 24:114–187

Hall CE, Cooney MIC, Fox JP (1970) The Seattle Virus Watch Program. I. Infection and illness experience of Virus Watch families during a community-wide epidemic of enterovirus type 30 aseptic meningitis. Am J Public Health 60:1456–1465

Halstead LS, Wiechers DO, Rossi CD (1985) Late effects of poliomyelitis: a national survey. In: Halstead LS, Wiechers DO (eds) Late effects of poliomyelitis. Syposia Foundation, Miami, pp 11–31

Hamilton MS, Jackson MA, Abel D (1999) Clinical utility of polymerase chain reaction testing for enteroviral meningitis. Pediatr Infect Dis J 18:533–537

Hashimoto I, Hagiwara A (1982) Studies on the pathogenesis of and propagation of enterovirus 71 in poliomyelitis-like disease in monkeys. Acta Neuropathol 58:125–132

Hays AP, Gamboa ET (1986) Acute viral myositis. In: England AG, Banker BQ (eds) Myology, vol 2. McGraw-Hill, New York, pp 1439–1466

Ho M (1978) Acute viral encephalitis. In: Vinken PJ, Bruyn GW, Klawans HL (eds) Handbook of clinical neurology: infections of the nervous system, Part II, vol 34. North-Holland, Amsterdam, pp 63–82

Ho M, Chen ER, Hsu KW (1999) An epidemic of enterovirus 71 infection in Taiwan. N Engl J Med 341:929–935

Howe HA, Wilson JL (1957) Poliomyelitis. In: Rivers TM, Horsfall FL (eds) Viral and rickettsial infections of man, 3rd edn. Lippincott, Philadelphia, PA, pp 432–478

Horstman DM (1963) Epidemiology of poliomyelitis and allied diseases—1963. Yale J Biol Med 36:5–26

Horstmann DM, Yamada N (1968) Enterovirus infections of the central nervous system. In: Zimmerman HM (ed) Association for research in nervous and mental disease: infections of the nervous system; Proceedings of the association, 1964, New York, NY (Research Publications Vol. XLIV). Williams and Wilkins, Baltimore, pp 236–253

Hsueh C, Jung SM, Shih SR (2000) Acute encephalomyelitis during an outbreak of enterovirus 71 infection in Taiwan: report of an autopsy case with pathologic, immunofluorescence, and molecular studies. Mod Pathol 13:1200–1205

Hung TP, Sung SM, Uang HC, Landsborough D, Green IJ (1976) Radiculomyelitis following acute haemorrhagic conjunctivitis. Brain 99:771–790

Hung T, Kono R (1979) Neurological complications of acute haemorrhagic conjunctivitis (a polio-like syndrome in adults). In: Winken PJ, Bruyn GW, Klawans HL (eds) Handbook of clinical neurology, vol 38, Neurological manifestations of systemic disease, part I. North-Holland, Amsterdam, pp 595–623

Huang CC, Liu CC, Chang YC, Chen CY, Wang ST, Yeh TF (1999) Neurologic complications in children with enterovirus 71 infection. N Engl J Med 341:936–942

Ida-Hosonuma M, Iwsaki T, Yoshikawa T et al (2005) The alpha/beta interferon response controls tissue tropism and pathogenicity of poliovirus. J Virol 79:4460–4469
Ishimaur Y, Nakano S, Yamaoka K, Takami S (1980) Outbreaks of hand, foot and mouth disease by enterovirus 71: high incidence of complicating disorders of central nervous system. Arch Dis Child 55:583–588
Jeha LE, Sila CA, Lederman RJ, Prayson RA, Isada CM, Gordon SM (2003) West Nile virus infection: a new acute paralytic illness. Neurology 61:55–59
Jenista JA, Powell KR, Menegus MA (1984) Epidemiology of neonatal enterovirus infection. J Pediatr 104:685–690
Jeong EJ, Lee JH, Kim MS et al (2010) Molecular characterization of enteroviruses detected in Gyeong-Ju and Po-Hang provinces of Korea in 2003. Arch Virol 155:1701–1712
Johnson RT (1998) Viral infections of the nervous system, 2nd edn. Raven, New York
Jubelt B (1984) Enterovirus and mumps virus infections of the nervous system. Neurol Clin 2:187–213
Jubelt B (2004) Post-polio syndrome. Curr Treat Options Neurol 6:87–93
Jubelt B (2010a) Viral infections and post viral syndromes. In: Rowland LP, Pedley TA (eds) Merritt's neurology, 12th edn. Lippincott Williams & Wilkins, Wolters Kluver, Philadelphia, PA, pp 156–185
Jubelt B (2010b) Parasitic infections. In: Rowland LP, Pedley TA (eds) Merritt's neurology, 12th edn. Lippincott Williams & Wilkins, Wolters Kluver, Philadelphia, PA, pp 228–238
Jubelt B (2010c) Bacterial infections. In: Rowland LP, Pedley TA (eds) Merritt's neurology, 12th edn. Lippincott Williams & Wilkins, Wolters Kluver, Philadelphia, PA, pp 125–147
Jubelt B, Cashman NR (1987) Neurological manifestations of post-polio syndrome. Crit Rev Neurobiol 3:199–200
Jubelt B, Gallez-Hawkins G, Narayan O, Johnson RT (1980) Pathogenesis of human poliovirus infection in mice: I. Clinical and pathological studies. J Neuropath Exp Neurol 39:138–148
Jubelt B, Goldfarb SJ, Paradise MJ, Close MG (1986) Fast axonal transport of human, poliovirus (HPV) (abstract). Neurology 36(Suppl 1):204
Jubelt B, Lipton HL (1987) Lansing poliovirus infection in mice: antibody demonstrable by enzyme linked immunosorbent assay (ELISA) and immunoprecipitations but not by neutralization. J Neuroimmunol 14:109–121
Jubelt B, Lipton H (1989) Enterovirus infections. In: Vinken PJ, Bruyn GW, Klawans HL, McKendall RR (eds) Viral disease, Handbook of clinical neurology, revised series, vol 12 (56), Elsevier North Holland, Amsterdam, Chapter 18, 307:347
Jubelt B, Ropka SL, Goldfarb S (1989) Anti-thymocyte serum delays clearance of human poliovirus from the mouse central nervous system. J Neuroimmunol 22:223–232
Jubelt B, Mihai C, Li TM, Veerapaneni P (2011) Rhombencephalitis/brainstem encephalitis. Curr Neurol Neurosci Rep 11:543–552
Kanagarajan K, Ganesh S, Alakhras M, Go ES, Recco RA, Zaman MM (2003) West Nile virus infection presenting as cerebellar ataxia and fever: case report. South Med J 96:600–601
Kaplan MH, Klein SW, McPhee J, Harper RG (1983) Group B coxsackievirus infections in infants younger than three months of age: a serious childhood illness. Rev Infect Dis 5:1019–1032
Katiyar BC, Misra S, Singh RB et al (1983) Adult polio-like syndrome following enterovirus 70 conjunctivitis (natural) history of the disease. Acta Neurol Scand 67:263–274
Kawai T, Akira S (2006) Innate immune recognition of viral infection. Nat Immunol 7:131–137
Keller R, Dwyer J, Oh W, Amodio MD (1969) Intestinal IgA neutralizing antibodies in newborn infants following poliovirus immunization. Pediatrics 43:330–338
Khatib R, Hatib GK, Chason JL, Lerner AM (1983) Alterations in coxsackievirus B4 heart muscle disease in ICR Swiss mice by anti-thymocyte serum. J Gen Virol 64:231–236
Kitamuri N, Semler B, Rothberg PG et al (1981) Primary structure, gene organization and polypeptide expression of poliovirus RNA. Nature 291:547–553
Koskiniemi M, Vaheri A (1982) Acute encephalitis of viral origin. Scand J Infect Dis 14:181–187

Kupila L, Vuorinen T, Vainionpaa R, Hukkanen V, Marttila RJ, Kotilainen P (2006) Eitology of aseptic meningitis and encephalitis in an adult population. Neurology 66:75–80
Lee BE, Davies HD (2007) Aseptic meningitis. Curr Opin Infect Dis 20:272–277
Lum LC, Wong KT, Lam SK (1998) Fatal enterovirus 71 encephalomyelitis. J Pediatr 133:795–798
Mahon BP, Katrak K, Mills KH (1992) Antigenic sequences of poliovirus recognized by T cells: serotype-specific epitopes on VP1 and VP2 and cross-reactive epitopes on VP4 defined by using CD4+ T-cell clones. J Virol 66:7012–7020
Marine WM, Chen TDY, Gravelle CR (1962) Limitation of fecal and pharyngeal poliovirus excretion in Salk vaccinated children. Am J Hyg 76:173–195
Matthews REF (1982) Classification and nomenclature of viruses. Intervirology 17:1–200
McCallister RM, Hummeler K, Coriell LL (1959) Acute cerebellar ataxia: report of a case with isolation of type 9 ECHO virus from the cerebrospinal fluid. N Engl J Med 261:1159–1162
McKinney RE Jr, Katz SL, Wilfert CM (1987) Chronic enteroviral meningoencephalitis in agammaglobulinemic patients. Rev Infect Dis 9:334–356
Mease PF, Ochs HD, Wedgewood RJ (1981) Successful treatment of echovirus meningoencephalitis and myositis – fascitis with intravenous immune globulin therapy in a patient with X-linked agammaglobulinemia. N Engl J Med 304:1278–1281
Melnick JL (1965) Echoviruses. In: Horsfall FL Jr, Tamm I (eds) Viral and rickettsial infections of man, 4th edn. Lippincott, Philadelphia, PA, pp 513–545
Melnick JL (1981) Towards the eradication of poliomyelitis. In: Peterson EM, deLaMasa LM (eds) Medical virology: proceedings of the international symposium on medical virology. Elsevier Biomedical Division, New York, pp 261–299
Melnick JL (1982) Enteroviruses. In: Evans AS (ed) Viral infections of humans: epidemiology and control, 2nd edn. Plenum, New York, pp 187–251
Melnick JL (1985) Enteroviruses: polioviruses, coxsackieviruses, echoviruses and newer enteroviruses. In: Fieldst BN, Knipe DM, Cbanock RM, Melnick JL, Roizman B, Shope RE (eds) Virology. Raven, New York, pp 739–794
Melnick JL (1997) Poliovirus and other enteroviruses. In: Evans AS, Kaslow RA (eds) Viral infections of humans. Plenum, New York, pp 583–663
Melnick JL, Cockburn WC, Dalldorf G et al (1963) Picornavirus group. Virology 19:114–116
Melnick JL, Proctor RO, Campo AR, Diwan AR, Ben-Porath E (1966) Free and bound virus in serum after administration of oral poliovirus vaccine. Am J Epidemiol 84:329–342
Mendez-Cushion P, Sanchez-Longo LF, Valcarcel M, Rosen L (1962) Acute cerebellar ataxia in children associated with infection by poliovirus I. Pediatrics 29:808–815
Misbah SA, Spickett GP, Ryba PCJ (1992) Chronic enteroviral meningitis in agammaglobulinemia: case report and literature review. J Clin Immunol 12:266–270
Modlin JF (1986) Perinatal echovirus infections: insights from a literature review of 61 cases of serious infection and 16 outbreaks in nurseries. Rev Infect Dis 8:918–926
Modlin JF, Dagan R, Berlin LE, Virshup DM, Yolken RN, Menegus M (1991) Focal encephalitis with enterovirus infections. Pediatrics 88:841–845
Moore M (1982) From the Centers for Disease Control: enteroviral disease in the U.S., 1970-1979. J Infect Dis 145:103–108
Moore M, Moresn DM (1984) Enteroviruses, including polioviruses. In: Belshe RM (ed) Textbook of human virology. PSG, Littleton, pp 407–483
Moore M, Baron RC, Filstein MR et al (1983) Aseptic meningitis and high school football players, 1978 and 1980. J Am Med Assoc 249:2039–2042
Morens DM, Pallansch MA, Moore M (1991) Polioviruses and other enteroviruses. In: Belshe RM (ed) Textbook of human virology, 2nd edn. Mosby Year Book, St. Luis, pp 427–297
Morgan IM (1947) The role of antibody in experimental poliomyelitis. 111: Distribution of antibody in and out of the central nervous system in paralyzed monkeys. Am J Hyg 45:390–400
Mulder DW, Rosenbaum RA, Layton DD Jr (1972) Late progression of poliomyelitis or forme fruste amyotrophic lateral sclerosis. Mayo Clin Proc 47:756–761

Nathanson N, Bodian D (1962) Experimental poliomyelitis following intramuscular virus infection III. The effect of passive antibody on paralysis and viremia. Bull Johns Hopkins Hosp III:198–220
Nathanson N, Martin JR (1979) The epidemiology of poliomyelitis: enigmas surrounding its appearance, epidemicity, and disappearance. Am J Epidemiol 110:672–692
Nicolosi A, Hauser WA, Beghi E, Kurland LT (1986) Epidemiology of the central nervous system infections in Olmsted County, Minnesota, 1950-1981. J Infect Dis 154:399–408
Nokowane BM, Wassilak SGF, Orenstein WA et al (1987) Vaccine associated paralytic poliomyelitis: United States: 1973 through 1984. J Am Med Assoc 257:1335–1340
Noordhoek GT, Weel JF, Poelstra E, Hooghiemstra M, Brandenburg AH (2008) Clinical validation of a new real-time PCR assay for detection of enteroviruses and parechoviruses, and implications for diagnostic procedures. J Clin Virol 41:75–80
Ogra PL (1970) Distribution of echovirus antibody in serum, nasopharynx, rectum and spinal fluid after natural infection with echovirus type 6. Infect Immun 2:150–155
Ogra PL, Karzon DT (1969) Distribution of poliovirus antibody in serum, nasopharynx and alimentary tract following segmental immunization of lower alimentary tract with polio vaccine. J Immunol 102:1423–1430
Ogra PL, Karzon DT (1971) Formation and function of poliovirus antibody in different tissues. Prog Med Virol 13:156–193
Ogra PL, Darzon DT, Righthand R, Macoillivray M (1968) Immunoglobulin response in serum and secretions after immunization with live and inactivated polio vaccine and natural infection. N Engl J Med 279:893–900
Ogra PL, Ogra SS, Al-Nakeed S (1973) Local antibody response to experimental poliovirus infection in the central nervous system of rhesus monkeys. Infect Immun 8:931–937
Ooi MH, Wong SC, Lewthwaite P, Cardosa MJ, Soloman T (2010) Clinical features, diagnosis, and management of enterovirus 71. Lancet Neurol 9:1097–1105
Paque RE, Gauntt CJ, Nealon TJ (1981) Assessment of cell-mediated immunity against coxsackievirus B3 induced myocarditis in a primate model (*Papto papto*). Infect Immun 31:470–479
Patriarca PA, Onorato IM, Sklar VE et al (1983) Acute hemorrhagic conjunctivitis: Investigation of a large-scale community outbreak in Dade county, FL. J Am Med Assoc 249:1283–1289
Paul JR (1955) Epidemiology of poliomyelitis. In: Poliomyelitis. Geneva, W.H.O. Monograph Series No. 26
Paul JR (1971) History of poliomyelitis. Yale University Press, New Haven
Paul JR, Melnick JL, Barnett VH, Coldblum N (1952) A survey of neutralizing antibodies to poliomyelitis virus in Cairo, Egypt. Am J Hyg 55:402–413
Peart AFW (1949) An outbreak of poliomyelitis in Canadian Eskimos in wintertime. Can J Public Health 10:405–417
Pevear DC, Oh CK, Cunningham LL, Calenoff MD, Jubelt B (1990) Localization of genomic region specific for the attenuated, mouse-adapted poliovirus type 2 strain W-2. J Gen Virol 71:43–52
Pezeshkpour GH, Dalakas MC (1988) Long-term changes in the spinal cords of patients with old poliomyelitis: signs of continuous disease activity. Arch Neurol 45:505–508
Pichlmair A, Reis E, Sousa C (2007) Innate recognition of viruses. Immunity 27:370–383
Picornaviridae Study Group (2012) http://www.picornastudygroup.com/types/index.html. Accessed 22 Jan 2012
Plum F, Swanson AG (1959) Central neurogenic hyperventilation in man. Arch Neurol Psychiatry 81:531–549
Ponka A, Petierson T (1982) The incidence and aetiology of central nervous system infections in Helsinki in 1980. Acta Neurol Scand 66:529–535
Price RW, Plum F (1978) Poliomyelitis. In: Vinken PJ, Bruyn GW, Klawans HL (eds) Handbook of clinical neurology, vol 34, Infections of the nervous system, part II. North-Holland, Amsterdam, pp 93–132

Rao DG, Bareman DE (1997) Hyperintensities of the anterior horn cells on MRI due to poliomyelitis. J Neurol Neurosurg Psychiatry 63:720
Ren R, Racaniello V (1992) Poliovirus spreads from muscle to the central nervous system by neural pathways. J Infect Dis 166:747–753
Roebauck M, Chamberlain R (1982) Prevalence of antibodies to poliovirus in 1978 among subjects aged 0-88years. Br Med J 284:697–700
Rousseau WE, Noble GR, Tegmeier GE, Jordan MC, Chin TDY (1973) Persistence of poliovirus neutralizing antibodies eight years after immunization with live, attenuated-virus vaccine. N Engl J Med 289:1357–1359
Rotbart HA, McCracken GH, Whitley RJ (1999) Clinical significance of enteroviruses in serious summer febrile illnesses of children. Pediatr Infect Dis J 18:869–874
Sabin AB (1957) Properties of attenuated polioviruses and their behavior in human beings. In: St. Whitelock OV (ed) Cellular biology, nucleic acids and viruses. New York Academy of Sciences, New York, pp 113–127
Sabin AB, Michaels RH, Krugman S, Eyer M, Berman P, Warren RJ (1963) Effect of oral poliovaccine in newborn children. II. Intestinal resistance and antibody response at 6 months in children fed type I vaccine at birth. Pediatrics 31:641–650
Sawyer MH (1999) Enterovirus infections: diagnosis and treatment. Pediatr Infect Dis J 18:1033–1040
Schmidt JJ, Lennette EH, Ho HH (1974) An apparently new enterovirus isolated from patients with disease of the central nervous system. J Infect Dis 129:304–309
Sells CJ, Carpenter RL, Gray C (1975) Sequelae of central nervous system enterovirus infections. N Engl J Med 293:1–4
Shen W-C, Chiu H-H, Chow K-C, Tsai C-H (1999) MR imaging findings of enteroviral encephalomyelitis: an outbreak in Taiwan. Am J Neuroradiol 20:1889–1895
Simons JM, Kutubuddin M, Chow M (1992) Characterization of poliovirus-specific T lymphocytes in the peripheral blood of Sabin-vaccinated humans. J Virol 67:1262–1268
So YT, Olney RK (1991) AAEM case report #23: acute paralytic poliomyelitis. Muscle Nerve 14:1159–1164
Spigland I, Fox JP, Elveback LR et al (1966) The Virus Watch program: a continuing surveillance of viral infections in metropolitan New York families II. Laboratory methods and preliminary report on infections revealed by virus isolation. Am J Epidemiol 83:413–435
Stalberg E, Grimby G (1995) Dynamic electromyography and muscle biopsy changes in a 4-year follow-up: study of patients with a history of polio. Muscle Nerve 18:699–707
Stanway G, Brown F, Christian P et al (2005) Family picornaviridae. In: La B, Fauquet CM, Mayo MA, Maniloff J, Desselberger U (eds) Virus taxonomy, Eight Report of the International Committee on the Taxonomy of Viruses. Elsevier Academic Press, London, pp 757–778
Stevens KM (1959) Estimate of molecular equivalent of antibody required for prophylaxis and therapy of poliomyelitis. J Hyg 57:198–200
Strikas RA, Anderson LJ, Parker RA (1986) Temporal and geographic patterns of isolates of nonpolio enterovirus in the United States, 1970-1983. J Infect Dis 153:346–351
Svehag SE, Mandel B (1964) The formation and properties of poliovirus-neutralizing antibody of rabbit origin. I. 19S and 7S antibody formation difference in kinetics and antigen dose requirement for induction. J Exp Med 119:1–19
Tang TT, Sedmak GV, Siegesmund KA, McCreadie SR (1975) Chronic myopathy associated with coxsackie virus type A9. N Engl J Med 292:608–611
Thomas JE, Howard FM (1972) Segmental zoster paresis, a disease profile. Neurology 22:459–466
Travers RL, Hughes GRV, Cambridge G, Sewell JR (1977) Coxsackie B neutralization titres in polymyositis/dermatomyositis. Lencet I:1268
Vallet S, Legrand-Quillien M-C, Dailland T et al (2009) Fatal case of enterovirus 71 infection, France, 2007. Emerg Infect Dis 15:1837–1840

Wadia NH, Katrak SM, Misra VP et al (1983) Polio-like motor paralysis associated with acute hemorrhagic conjunctivitis in an outbreak in 1981 in Bombay, India: Cinical and serologic studies. J Infect Dis 147:660–668

Wang K, Sun L, Jubelt B, Waltenbaugh C (1989) Cell-mediated immune responses to poliovirus. I. Conditions for induction, characterization of effector cells, and cross-reactivity between serotypes for delayed hypersensitivity and T cell proliferative responses. Cell Immunol 119:252–262

Wang JP, Cerny A, Asher DR, Kurt-Jones EA, Bronson RT, Finberg RW (2010) MDA5 and MAVS mediate type I interferon responses to coxsackie B virus. J Virol 84:254–260

Webster ADB, Tripp JH, Hayward AR et al (1978) Echo-virus encephalitis and myositis in primary immunoglobulin deficiency. Arch Dis Child 53:33–37

Weiner LP (1978) Enterovirus infections of the nervous system other than poliovirus. In: Vinken PJ, Bruyn GW, Klawans HL (eds) Handbook of cinical neurology, vol 24, Infections of the nervous system, part II. North-Holland, Amsterdam, pp 133–143

Weinstein L (1957) Influence of age and sex on susceptibility and clinical manifestations in poliomyelitis. N Engl J Med 257:47–52

Welliver RC, Drucker MM, Ogra PL (1982) Immunology of enteroviruses. In: Nahmias AJ, O'Reilly RJ (eds) Immunology of human infection. Comprehensive immunology, vol 9. Plenum, New York, pp 185–203

Weichers DO (1988) New concepts of the reinnervated motor unit revealed by vaccine-associated poliomyelitis. Muscle Nerve 11:356–364

Wilfert CM, Buckley RH, Mohanakumar T et al (1977) Persistent and fatal central nervous system ECHO virus infections in patients with agammaglobulinemia. N Engl J Med 296:1485–1489

Wolinsky JS, Jubelt B, Burke S, Narayan O (1982) The hematogenous origin of the inflammatory response in acute poliomyelitis. Ann Neurol 11:59–68

Wong CY, Woodruff JJ, Woodruff JR (1977) Generation of cytotoxic lymphocytes during coxsackievirus B-3 infection I. Model and viral specificity. J Immunol 118:1159–1164

Wong SSY, Yip CCY, Lau SKP, Yuen KY (2010) Human enterovirus 71 and hand, foot and mouth disease. Epidemiol Infect 138:1072–1089

Wyatt HV (1973) Poliomyelitis in hypogammaglobulinemics. J Infect Dis 128:802–806

Yang TT, Huang LM, Lu CY et al (2005) Clinical features and factors of unfavorable outcomes for non-polio enterovirus infection of the central nervous system in northern Taiwan, 1994–2003. J Microbiol Immunol Infect 38:417–424

Zaoutis T, Klein JD (1998) Enterovirus infections. Pediatr Rev 19:183–191

Part III
Retroviruses

Human Immunodeficiency Virus Infection/AIDS

Jeffrey Rumbaugh, Taylor Harrison, and William Tyor

Abstract For any practitioner treating patients with human immunodeficiency virus (HIV), this chapter provides a framework for considering the varied array of neurological complications. HIV can affect both central and peripheral nervous systems. Some of these complications, such as HIV-associated neurocognitive dysfunction, are the direct result of HIV itself. Other complications are due to opportunistic infections or are adverse effects of the treatments that are used against HIV. The clinical manifestations of these complications are reviewed, with emphasis on the key features useful for making proper and timely diagnoses. Up-to-date treatment recommendations, based on the most recent research and expert opinion, are included.

Keywords CNS lymphoma • HIV • HIV-associated neurocognitive disorders • IRIS • Neuromuscular disorders • Opportunistic infections

1 Introduction

As the human immunodeficiency virus (HIV) pandemic enters its fourth decade with over 25 million lives lost, HIV is firmly established as one of the most destructive pathogens known to man. Moreover, an estimated 7,000 new infections per day contributes to the 33.3 million people currently living with HIV infection (UNAIDS). Neurological complications affecting either the central nervous system

J. Rumbaugh • W. Tyor (✉)
Department of Neurology, Emory University School of Medicine, Neurology (GEC), Atlanta Veterans Administration Medical Center, 1670 Clairmont Road, Decatur, GA 30033, USA
e-mail: jrumbau@emory.edu; wtyor@emory.edu

T. Harrison
Department of Neurology, Emory University School of Medicine, Grady Memorial Hospital, 80 Jesse Hill Jr. Drive, Box 036, Atlanta, GA 30303, USA
e-mail: tharri4@emory.edu

A.C. Jackson (ed.), *Viral Infections of the Human Nervous System*,
Birkhäuser Advances in Infectious Diseases,
DOI 10.1007/978-3-0348-0425-7_7, © Springer Basel 2013

(CNS) or peripheral nervous system (PNS) are common in HIV-infected patients, particularly with more severe immunodeficiency, and such pathology is associated with significant morbidity and mortality. These neurological complications may represent end-organ manifestations of HIV infection, sequelae of associated opportunistic infections or neoplasm, or toxicities related to antiretroviral or other drug regimens. Familiarity with HIV-associated neurological disorders is key to managing this clinical population given that protean manifestations may be observed at all stages of infection (with the frequency and timing of onset variable among the different syndromes) and the physical examination findings are frequently confounded by diverse pathology affecting both the CNS and PNS.

2 HIV-Associated Neurocognitive Disorders

2.1 Introduction/Epidemiology

HIV-associated neurocognitive disorders (HAND) are the most common forms of cognitive dysfunction worldwide in people under the age of 40, affecting people in their prime, and thus having a large socioeconomic impact. The incidence of HIV-associated dementia (HAD) has declined dramatically since the introduction of combination antiretroviral therapy (cART) in 1996. However, despite the decreasing incidence of HAD, the prevalence of milder forms of HAND is actually increasing as patients live longer with their HIV infection, and may be as high as 35–40 %. While HAD usually occurs at end-stages of HIV infection, milder HAND can occur at any level of immunosuppression. The decline of HAD does not reflect a neuroprotective effect of cART per se, but, rather, improved immune status among those with HIV. Prevalence of HAD among those with advanced HIV infection is still as high as 5–10 %. These would include patients with poor compliance or poor access to antiretroviral therapies and/or multidrug resistant forms of HIV. Also, HIV-infected individuals over the age of 50 years are twice as likely to develop HAD compared to those under the age of 50 (Nath et al. 2008).

2.2 Clinical Manifestations

The clinical manifestations of HAD were defined early in the AIDS epidemic (Navia et al. 1986) and they largely apply today. Early, bradyphrenia, or slowness of mental functions, is commonly seen and patients can appear apathetic or depressed (Atkinson and Grant 1997). These patients need to be distinguished from primarily depressed patients exhibiting memory problems using neuropsychological testing. Components of HAD include memory impairment, impairment of executive functions, and mood abnormalities (Navia et al. 1986). They may have trouble reading and comprehending. There may be also gait disturbances with stumbling, tremor, and fine motor impairments (McArthur and Brew 2010). Other signs may include impaired rapid eye movements, hyperreflexia, and frontal release signs. These signs and memory impairment become more apparent as the disease progresses, eventually culminating in severe neurological sequelae—a bedridden, mute state. Prior to cART, death usually occurred over several months (Navia et al. 1986; Harrison et al. 1995).

There have been several attempts to standardize the nomenclature for HAND primarily for research purposes. Early, the Memorial Sloan Kettering (MSK) Rating Scale was employed to stage HAND (Price and Brew 1988). In 1991 the American Academy of Neurology (AAN) published criteria, which essentially split HAND into two major categories—HAD and minor cognitive motor disorder (MCMD) (Janssen 1991). Briefly, the AAN criteria for HAD were: (1) an acquired abnormality in at least two cognitive (nonmotor) areas causing impairment in work or activities of daily living (ADLs), and (2) either an abnormality of motor function or specified neuropsychiatric or psychosocial functions (e.g., motivation, emotional control, social behavior). In addition, the patient could not have other etiologies that might explain the disorder. MCMD is characterized by mild impairment in functioning and can be missed clinically if a careful mental status exam is not performed. In addition, patients are more likely to notice cognitive deficits if their activities require relatively sophisticated cognition. More recently, nosology for HAND has been revised (Antinori et al. 2007). MCMD now more closely corresponds to mild neurocognitive disorder (MND), although MND differs in that neuropsychological testing is technically required to define abnormalities in two cognitive domains. Otherwise MND requires mild impairment in ADLs, such as inefficiency at work, in the home, or problems in social settings. The third category of HAND is asymptomatic neurocognitive impairment (ANI). This also requires formal neuropsychological testing to diagnose. There must be acquired impairments in cognitive functioning of at least two test domains assessing language, attention, executive function, speed of memory recall, information processing speed, sensory, and motor skills.

It is beyond the scope of this chapter to provide a detailed discussion of the clinical presentation of pediatric NeuroAIDS. Common findings include impaired brain growth as assessed by head circumference, associated with loss of or failure to develop fine and gross motor control, spasticity, and extraparamidal syndromes (Mintz 2005). Delays in cognitive, speech, and language milestones have also been commonly reported (Van Rie et al. 2007).

2.3 Differential Diagnosis and Diagnostic Evaluation

Diagnosis of HAND is based on clinical grounds (Schouten et al. 2011). Ancillary testing is done only to support the diagnosis or exclude other diagnoses. Most commonly, HAND must be differentiated from opportunistic infections of the

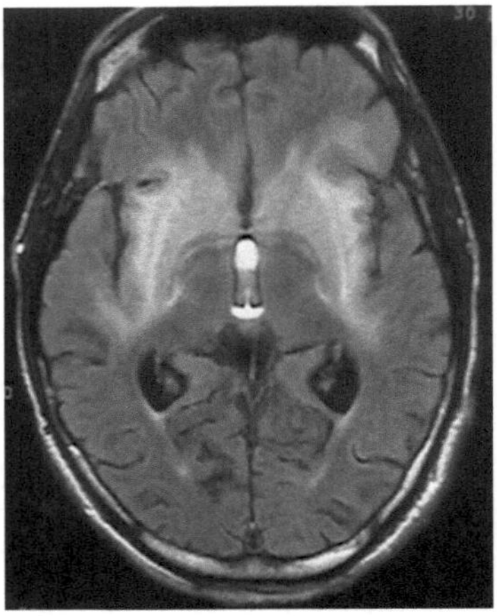

Fig. 1 Brain MRI with subcorticalT2 hyperintensity on FLAIR sequences in HAND

central nervous system (CNS) which can affect cognition. Drug abuse can also affect cognition (see below). Indigent individuals may be subject to nutritional deficiencies or other issues which could affect cognitive status. Older individuals may have neurocognitive impairment on the basis of vascular risk factors or Alzheimer's pathology, independent of their HIV infection.

Tests should potentially include: thyroid function tests, metabolic profile/liver function tests, ammonia, vitamin B12 level, toxicology, rapid plasma reagin (RPR)/fluorescent treponemal antibody absorbed (RPR/FTA-ABS), cytomegalovirus (CMV) polymerase chain reaction (PCR), cryptococcal antigen, and urine toxicology. The $CD4^{+}$ T lymphocyte count can be an indicator of opportunistic infections.

Cerebrospinal fluid (CSF) examination in HAND usually shows mild pleocytosis, mildly elevated protein concentration, and normal glucose (Schouten et al. 2011). This CSF pattern can often be in patients with HIV who are completely neurologically asymptomatic, so it is a nonspecific finding. To exclude other conditions it is prudent to test the CSF for the following: venereal disease research laboratory (VDRL)/FTA-ABS, cryptococcal antigen; Epstein–Barr virus (EBV), CMV, varicella zoster virus (VZV), herpes simplex virus (HSV), and/or JC-virus (JCV) PCRs; bacterial, fungal, and acid-fast bacilli cultures.

Some clinicians are concerned about the possibility of "CNS escape." This refers to a situation in which HIV may be well controlled systemically, but not in the CNS. This may suggest that a patient's cART be changed to one with higher CSF penetration (see below). This condition can be followed by measuring HIV RNA level (viral load) in serum and CSF.

Brain magnetic resonance imaging (MRI) is used to exclude opportunistic infection, malignancy, or stroke (Schouten et al. 2011). In HAND, brain MRI may show diffuse, patchy periventricular white matter T2-weighted hyperintensities and generalized atrophy (Fig. 1). There may be preferential atrophy in the basal ganglia. However, these findings are nonspecific.

Formal neurocognitive testing can be used to diagnose and/or quantify the cognitive impairment.

2.4 Management

Despite attempts to address specific pathogenic mechanisms of HAND in phase I trials, the mainstay of HAND treatment remains cART (Tyor 2009). There are additional factors that affect the expression and treatment of HAND, one being the determination of when to institute cART. cART should be strongly considered in any patient with HAND. Additional factors to consider are (1) the decline in the incidence of HAD, an AIDS defining diagnosis, since the advent of cART and (2) the emergence of highly prevalent but less severe forms (i.e., MND and ANI) (Tozzi et al. 2010). MND and ANI are not detected routinely and therefore cART is typically instituted after HAD has been discovered. Although cART can reverse cognitive dysfunction in many patients, cases of HAND are not infrequently discovered coincident with cART. Therefore, HAND occurs despite cART and one must consider whether specific agents penetrate the blood–brain barrier (BBB) and suppress CNS HIV. Suppressing CNS HIV load is probably important, although studies have given somewhat conflicting results (Tozzi et al. 2010). There have been attempts to estimate CNS penetration of specific antiretroviral agents by measuring them in the CSF. These levels may or may not reflect brain parenchymal concentrations (Tyor 2009). Nevertheless, a CNS penetration effectiveness (CPE) ranking system has been developed, which may be helpful in decision making (Table 1) (Letendre et al. 2008). However, the routine clinical applicability of the CPE system is debatable (Schouten et al. 2011). Regardless, there is substantial evidence suggesting zidovudine (AZT) is beneficial for patients with HAND. cART regimens with higher CPE scores (i.e., better BBB penetration) may be more effective. Recently the issue of potential antiretroviral drug neurotoxicity has been raised (Robertson et al. 2010). Efavirenz is associated with cognitive dysfunction. It remains to be seen whether agents associated with peripheral neuropathy (i.e., dideoxyinosine, stavudine, indinavir, and amprenavir) are also neurotoxic.

2.5 Prognosis

Prior to the advent of cART HAD was more commonly diagnosed in HIV-infected individuals (Heaton et al. 2011; Schouten et al. 2011). HAD, an AIDS-defining illness most commonly observed in immunocompromised patients (CD4 count less than 200/mm^3) (McArthur et al. 1993), was diagnosed in approximately 15 % of HIV+ patients, whereas in the cART era, the rate is less than 5 %. Since the advent of cART life expectancy for HAD patients has increased significantly. HAD is usually seen in cART-naïve patients or due to cART viral resistance. Not only has

Table 1 CNS Penetration of Selected Antiretroviral Drugs

Antiretroviral drug class	CNS penetration effectiveness score		
	0	0.5	1
Nucleoside reverse transcriptase inhibitor	Tenofovir Adefovir Didanosine Zalcitabine	Emtracitabine Lamivudine Stavudine	Abacavir Zidovudine
Nonnucleoside reverse transcriptase inhibitor		Efavirenz	Nevirapine Delaviridine
Protease inhibitor	Ritonavir Nelfinavir Saquinavir Saquinavir[a] Tipranavir[a]	Atazanavir[a] Ampresnavir Atazanavir Fosamprenavir Indivir	Amprenavir[a] Lopinavir[a] Indinavir[a] Fosamprenavir[a] Darunavir
Integrase inhibitor		Raltegravir Elvitegravir	
Entry inhibitor	Enfuritide T-1249		Maraviroc Vicriviroc

[a]Rotinavir-boosted protease inhibitor

the rate of HAD decreased, but HAD patients can improve following cART initiation. However, there are exceptions to this and there is evidence that HAD now occurs with higher CD4 counts in cART treated patients. Despite these seeming gains in the incidence of HAD and its treatment, it is increasingly recognized that a larger portion of HIV-infected individuals exhibit cognitive impairments that cannot be categorized as frank dementia. HAND is seen even in patients on cART and the incidence of HAD increases with age, suggesting milder forms will follow the same pattern (Valcour et al. 2004). Increasingly, evidence shows that factors associated with aging and cognitive dysfunction such as cerebrovascular disease significantly impact the development of HAND (Becker et al. 2009). Therefore, other factors that influence cognitive performance such as cerebrovascular disease, thyroid status, vitamin B12, and other comorbid conditions will affect prognosis, especially in older patients.

2.6 Pathogenesis

Pathologically, chronic HIV encephalitis is characterized by microglial nodules, multinucleated giant cells, astrocytosis, microgliosis, neuronal loss, and primarily infection of mononuclear phagocytes and not neurons. Damage to the nervous system is, therefore, believed to be indirect. HAND has been attributed to a variety of neurotoxins including HIV proteins, particularly Tat protein and gp120, as well as to the host response to the presence of this virus in the brain. Neurons are injured by various upregulated, inflammatory mediators, creating a toxic milieu. Astrocytes are infected by HIV, but the infection is nonproductive. Thus, astrocytes may not function properly, so the neurons may lack their support. Astrocytes may also serve as a continual source of viral proteins, like Tat and gp120 that are secreted into the extracellular milieu. HIV invades the CNS very rapidly after initial infection and before ART is started. Once in the brain, HIV integrates into the DNA of mononuclear phagocytes and astrocytes, then, even if ART is subsequently started, it does not target proteins like Tat and gp120 (Nath et al. 2008).

2.7 Impact of Comorbidities

2.7.1 Hepatitis C Virus

Hepatitis C virus (HCV) can be considered a rare cause of viral encephalitis in the absence of HIV infection (Seifert et al. 2008). In contrast, HCV is frequently detected in the brains of HIV- and HCV-seropositive individuals at autopsy (Letendre et al. 2007). However, whether HCV encephalitis as opposed to liver disease or other effects (e.g., drugs of abuse) cause more frequent cognitive dysfunction in HIV-infected patients is unclear (Ryan et al. 2004; Clifford et al. 2009). Recently Vivithanaporn et al. (2012) have shown that HIV and HCV coinfected individuals tend to have more severe HAND than HIV-infected patients alone. While *active* drug abusers were excluded from the study, there was nevertheless a high correlation between history of drug abuse and coinfection. Therefore, it still remains controversial as to whether HCV infection really increases cognitive dysfunction in HIV+ patients as opposed to drug abuse. Either way, both infections must be treated optimally by reducing viral load, especially in cognitively impaired patients.

2.7.2 Drug Abuse

Drugs of abuse contribute to the spread of HIV and are associated with medical noncompliance. There is evidence that alcohol abuse as well as chronic use of cocaine, morphine derivatives, and methamphetamine have direct consequences on the development of neurological complications of HIV infection (Tyor and Middaugh 1999; Nath 2010). Drug abuse likely adversely affects the onset and course of HAND through several mechanisms. Cocaine, methamphetamine, and morphine in combination with Tat cause disruption of the BBB. These drugs also have been shown to directly or indirectly increase HIV replication. So they are likely to contribute to HAND pathogenesis and may have effects on other neurological complications of HIV infection. In addition, drugs of abuse have complex effects on the immune system. In particular, alcohol has many immunosuppressive effects that could adversely predispose to opportunistic infections. Furthermore, drugs of abuse have direct adverse consequences on brain function, often in brain regions that are involved in HAND pathogenesis. They may have synergistic effects with HIV in producing neurotoxicity. These factors emphasize the importance of ensuring that HIV-infected patients who abuse drugs are encouraged and given the opportunity to participate in drug rehabilitation programs.

3 Other Neurological Complications Directly Related to HIV and/or Its Treatment

3.1 HIV-Associated Acute and Chronic Meningitis

Symptomatic acute meningitis with classic symptoms of headache and neck stiffness can follow in a small percentage of individuals with recent HIV exposure (Snider et al. 1983). Preceding or concurrent symptoms such as fever, rash, arthralgias, and lymphadenopathy are nonspecific, but should prompt consideration of HIV infection, especially if the patient is at risk. However, HIV-associated aseptic meningitis can be asymptomatic. HIV infection can also manifest acutely as a meningoencephalitis. Therefore in addition to the potential symptoms associated with acute meningitis, patients may exhibit signs of encephalitis such as behavioral abnormalities, seizures, and alterations in consciousness. Mild CSF abnormalities occur in many HIV-infected patients, including mild elevations in the white count and protein. Chronic meningitis after HIV infection is seen in as many as 40 % of HIV-infected patients and is often asymptomatic (de Almeida et al. 2011). Differentiating HIV meningitis from other important causes of CSF abnormalities may possibly be aided by determination of CSF lactic acid levels, which tend to be higher in patients with TB and cryptococcal meningitis. Otherwise treatment of HIV-associated meningitis is symptomatic, unless viral titers and other potential associated illnesses dictate a need for antiretroviral treatment.

3.2 HIV-Associated Vacuolar Myelopathy

HIV-associated vacuolar myelopathy has been estimated to affect approximately 10 % of AIDS patients. The frequency of this disorder in the cART era, however, is not precisely known. The major clinical manifestations include subacute to chronic slowly progressive gait impairment. Early symptoms include urinary urgency, frequency, and erectile dysfunction. As symptoms progress patients develop painless paraparesis with gait incoordination. Neurological examination reveals spastic paraparesis and impairments in proprioception and vibratory sensation, reflecting corticospinal and/or posterior column dysfunction. A discrete sensory level is absent. Pain and temperature sensation are relatively spared unless there is concomitant polyneuropathy, which is not uncommon (Dal Pan et al. 1994).

Differential diagnosis includes non-HIV-associated myelopathies such as cervical spondylotic myelopathy, disc disease with compressive myelopathy, or B12 deficiency. Multiple coinfections or HIV-associated comorbidities may have spinal cord involvement. Toxoplasmosis and primary CNS lymphoma (PCNSL) rarely present with intramedullary spinal cord disease. PML may rarely have upper cervical cord involvement, though this is almost never in isolation without concomitant brain involvement. Viral myelitis may be related to HSV, VZV, or CMV reactivation. Neurosyphilis may cause a myelitis or, with meningovascular syphilis, an anterior spinal artery syndrome. Tuberculosis may be associated with a meningomyelitis, focal parenchymal disease, or spondylodiscitis (i.e., Pott's disease). Coinfection with HTLV, which causes tropical spastic paraparesis, may increase the risk for developing myelopathy (Harrison et al. 1997).

Diagnostic evaluation should include B12 as well as syphilis and HTLV serology. Additional workup depends on the tempo of onset and progression of symptoms. MRI is the preferred imaging modality, generally of the cervical and thoracic spine. While imaging is commonly normal, the most common abnormalities noted are nonspecific intramedullary T2 hyperintensity and mild thoracic cord atrophy (Chong et al. 1999).

Initiation of cART is a mainstay of management, because case reports have described improvement after antiretroviral therapy; no well designed studies, however, have proven that cART improves outcomes. Spasticity may be treated with baclofen or tizanidine, and urinary symptoms may respond to oxybutynin (Ditropan). Physical and occupational therapy for gait training and fall prevention are important. It is our experience that successful immune reconstitution and viral suppression with effective cART generally arrests myleopathy progression. While some patients endorse clinical improvement, no adequate studies have been performed to date.

The pathogenesis of HIV vacuolar myelopathy is not well known, though it has been postulated that impaired transmethylation may adversely affect myelin formation and repair [reviewed in Tan and Guiloff (1998)]. Pathologically HIV vacuolar myelopathy is characterized by intramyelinolytic and periaxonal vacuoles often with foamy macrophages in the lateral and posterior columns of the spinal cord (Petito et al. 1985). These changes typically involve the middle portion of the thoracic cord. Axonal degeneration and asymmetric distribution are noted with advanced stages.

3.3 *HIV-Associated Neuromuscular Disorders*

3.3.1 Peripheral Neuropathy

HIV-associated distal sensory polyneuropathy (HIV-PN) and antiretroviral toxic neuropathy (ATN) are important primarily due to the need to treat neuropathic pain and impact on quality of life, function, and disability (Pettersen et al. 2006; Ellis et al. 2010). Early in the cART era approximately 35 % of patients with moderate to severe immunodeficiency (CD4 T-cell count <300) had symptomatic

Table 2 Tests to consider in the workup of HIV-PN

• B12 (± methylmalonic acid if low normal)
• Renal, hepatic function
• Inquire about drug and alcohol intake
• Review medication history
• Red blood cell folate
• Fasting blood glucose with 2 h oral glucose tolerance test if normal
• Thyroid stimulating hormone
• Serum, urine immunofixation electrophoresis
• Other tests to consider
• HCV, HBV serology

PN, with HIV-PN and ATN constituting the principal causes (Morgello et al. 2004; Schifitto et al. 2005). Recent data suggest symptomatic PN impacts a smaller percentage (10 %) of the HIV-infected population (Evans et al. 2011). Trends in earlier HIV diagnosis and earlier initiation of effective nonneurotoxic cART likely contribute to these observations. The dideoxynucloside NRTIs didanosine (ddI), zalcitabine (ddC), and stavudine (d4T), the so-called "d-drugs," have all been associated with the development of ATN (Berger et al. 1993; Blum et al. 1996; Abrams et al. 1994). D-drugs are associated with a threefold increased risk for ATN and up to 20 % discontinue them due to this (Schifitto et al. 2002; Moore et al. 2000). While the frequency of ATN has decreased markedly with increased availability of nonneurotoxic antiretroviral options, ATN remains important in resource-limited regions where d4T in particular is a common cART component.

The most common symptom of the HIV-associated PNs is numbness and the most disabling aspect is neuropathic pain. Up to one third of subjects with PN signs suffer from pain (Schifitto et al. 2005; Tagliati et al. 1999). Examination reveals stocking-pattern impairments of pain, temperature, and vibratory sensation, with decreased or absent ankle jerks (Cornblath and McArthur 1988). Weakness may be noted in distal extensor muscles but is not prominent. PN may also be symptom predominant with minimal, if any, abnormal neurological signs, suggesting "small fiber" pathology. ATN shares similar clinical and electrophysiologic features to HIV-PN, the clinical distinction contingent on the timing of symptom onset or worsening related to drug exposure or the patient's response to dosage reduction or drug withdrawal (Berger et al. 1993). At times the neuropathic symptoms of ATN may continue to progress for weeks after NRTI withdrawal, commonly referred to as "coasting."

Important conditions which need to be considered in the differential diagnosis include alcohol or intravenous drug (e.g., heroin) abuse, malnutrition, vitamin (e.g., thiamine, B12) deficiency, renal insufficiency, and disorders of glucose metabolism. Each of these conditions may cause PN or significantly contribute to PN risk in the setting of HIV (Table 2). In considering the approach to workup, determining the prevailing state of immunodeficiency during onset and evolution of PN symptoms is paramount. When PN onset coincides with severe immunodeficiency

Table 3 Nonantiretroviral medications associated with toxic PN

Opportunistic infection prophylaxis
• Trimethoprim–sulfamethoxazole
• Dapsone
• Metronidazole
Other/antimicrobials
• Isoniazid
• Pyridoxine (B6)
• Chloramphenicol
Antineoplastic agents
• Vincristine
• Thalidomide
• Hydoxyurea

in a cART-naïve patient, the high prevalence of HIV-PN in this setting makes alternative causes unlikely. Conversely, if PN symptoms commence with only mild immunodeficiency (e.g., CD4 > 500), HIV-PN is unlikely and other etiologies should be investigated. Worsening neuropathic pain in the setting of stable HIV disease on nonneurotoxic cART with suppressed viral load should also prompt thorough evaluation.

Common tests to consider for laboratory screening, which are consistent with current recommendations for the laboratory evaluation of PN, are detailed in Table 2 (England et al. 2009). Antineuronal antibody panels are not considered generally useful. Syphilis does not cause a generalized PN and is therefore not screened for. It is also important to consider medications other than ARVs which may cause toxic PN (Table 3).

As the combination of neuropathic symptoms and signs is relatively sensitive and specific for a diagnosis of HIV-PN, electrodiagnostic (EDX) testing is not generally required. It may, however, be indicated when atypical features such as severe weakness or sign/symptom asymmetry suggest an alternative diagnosis such as lumbosacral polyradiculopathy, acquired demyelinating PN, or vasculitis. Skin biopsy for epidermal nerve fiber density may be helpful when PN signs do not accompany PN symptoms and diagnosis is uncertain, but is not routinely required. Nerve biopsy may be indicated to support a diagnosis of vasculitis or chronic inflammatory demyelinating PN.

No evidence-based neuroregenerative treatments are available. Recombinant nerve growth factor (rNGF), prosaptide, coenzyme Q10, and acetyl-l-carnitine have not demonstrated efficacy (Schifitto et al. 2001; Valcour et al. 2009; Evans et al. 2007; Youle and Osio 2007; Rabing Christensen et al. 2004). While limited data suggest that cART initiation improves HIV-PN, no large-scale trial documents significant improvement (Martin et al. 2000; Markus and Brew 1998).

Management has therefore focused on neuropathic pain relief. Unfortunately, trials evaluating multiple drugs from different drug classes have failed to show meaningful efficacy: amitriptyline (Kieburtz et al. 1998; Shlay et al. 1998), gabapentin (Hahn et al. 2004), pregabalin (Simpson et al. 2010), mexiletine

(Kemper et al. 1998; Kieburtz et al. 1998), topical lidocaine gel (Estanislao et al. 2004), low-dose topical capsaicin (Paice et al. 2000), Peptide T (an in vivo gp120 inhibitor) (Simpson et al. 1996), and acupuncture (Shlay et al. 1998). Improvement in neuropathic pain has been described with lamotrigine (specifically in ATN), high-dose (8 %) topical capsacin, rNGF, and smoked marijuana (Simpson et al. 2003, 2008; Abrams et al. 2007; Ellis et al. 2009). In select patients opioids may be considered when moderate to severe pain impacts function and persists despite nonnarcotic therapies. No studies have evaluated the short- or long-term analgesic efficacy of opioids for HIV-associated PNs (Eisenberg et al. 2005). The recommendations from the Washington State Agency Medical Directors' Group are particularly helpful in providing guidance for physicians considering opioid therapy (AMDG). It is important to consider not only pain relief but also improvement in pain-related interference (function) when evaluating therapeutic response.

The pathophysiology of HIV-PN is likely an indirect effect of immune activation rather than direct viral infection of peripheral nerves, dorsal root ganglia (DRG), or Schwann cells (Keswani et al. 2003; Rizzuto et al. 1995; Melli et al. 2006). Classic pathological findings of HIV-PN include distal axonal degeneration with reduced unmyelinated fiber density and lesser reductions in small and large myelinated fiber densities (Mah et al. 1988).

D-drug neurotoxicity stems from inhibition of the mitochondrial DNA (mtDNA) gamma polymerase, a key enzyme for mtDNA replication and repair (Dalakas et al. 2001). In ATN, axonal degeneration is most prominent in unmyelinated fibers with abnormal mitochondria in both nerve axons and Schwann cells (Lewis and Dalakas 1995). ATN typically occurs within 1 year of treatment initiation and most commonly within the first 3 months (Lichtenstein et al. 2005; Arenas-Pinto et al. 2008). Genetic makeup may affect risk (Cherry et al. 2008; Kallianpur et al. 2006). Other antiretrovirals such as indinavir, ritonavir, and saquinavir may play a role in neurotoxicity (Pettersen et al. 2006; Ellis et al. 2008; Banerjee et al. 2011). It is not currently known whether PI-related PN risk represents a direct drug effect or is a consequence of PI-induced metabolic derangements (which may include insulin resistance or hyperlipidemia, hypertriglyceridemia).

3.3.2 Autonomic Neuropathy

While autonomic symptoms and asymptomatic signs are frequent in later stages of HIV infection, the prevalence of autonomic neuropathy is not precisely known (Freeman et al. 1990; Gluck et al. 2000; Compostella et al. 2008; Ruttimann et al. 1991). Such is not surprising given that autonomic dysfunction is frequent with PNs affecting small, thinly myelinated or unmyelinated fiber populations. Autonomic neuropathy is clinically important due to cardiac morbidity (Craddock et al. 1987). Clinical features may include resting tachycardia, orthostatic hypotension, impotence or urinary dysfunction, early satiety, constipation and/or diarrhea, and sweating disorders. Study of sympathetic ganglia shows neuronal degeneration and perivascular mononuclear cell infiltration with T-cells and macrophages (Chimelli and Martins 2002). Bedside evaluation may reveal a >20 mmHg drop in systolic or >10 mmHg drop in diastolic blood pressure without an adequate increase in heart rate. In the absence of a dedicated autonomic laboratory or quantitative sensory testing, electrophysiological evaluation of suspected autonomic neuropathy is unfortunately limited. Sympathetic skin response, which evaluates small fiber sudomotor function, may be available on standard EMG equipment but is insensitive. Heart rate variability occurring with Valsalva or deep breathing may also be evaluated with standard electromyography laboratory equipment.

Important considerations when evaluating suspected autonomic dysfunction are anemia and hypovolemia, both of which are common in AIDS. Other important conditions to exclude are medication side effects, cardiomyopathy, and adrenal insufficiency, the latter which is suggested by orthostatic hypotension with generalized fatigue, myalgia, weakness, hyponatremia, and/or hyperkalemia. A random cortisol level and, depending on the clinical suspicion, a cosyntropin stimulation test may be performed to screen for adrenal insufficiency. Treatment of orthostatic hypotension may include NaCl supplementation, thigh-high compression stockings, the mineralocorticoid fludrocortisone, midodrine, or erythropoietin.

3.3.3 Inflammatory Demyelinating Polyneuropathies

Acute and chronic inflammatory demyelinating polyneuropathies (AIDP and CIDP, respectively) are rare in HIV, with descriptions limited to small case series. These PNs are typically observed with mild-to-moderate immunosuppression, although AIDP has been described with both HIV seroconversion syndrome and in severe immunodeficiency (though CD4 counts are rarely <50) (Cornblath et al. 1987; Leger et al. 1989).

Clinical, electrophysiological, and pathological findings do not differ from non-HIV-infected patients. AIDP is characterized by ascending weakness with back pain, distal parasthesias, and areflexia evolving over 1–4 weeks. CIDP is considered when progressive, predominately proximal weakness evolves over at least 2 months with distal sensory complaints and areflexia. An important clinical feature of HIV-associated acquired inflammatory demyelinating PNs is the presence of a mild lymphocytic pleocytosis in CSF, on the order of 20–30 cells/mm^3, contrasting with the traditional profile of acellular albuminocytologic dissociation.

Differential diagnosis of rapidly progressive ascending weakness includes CMV polyradiculopathy or the toxic complication of NRTI therapy known as HIV-associated neuromuscular weakness syndrome (HANWS). While CMV has been associated with AIDS-associated AIDP, it is more commonly associated with ascending weakness due to polyradiculopathy. HANWS is an acute (1–2 weeks) or subacute (>2 weeks) syndrome of rapidly progressive neuromuscular weakness associated with lactic acidosis and possibly hepatic steatosis (Simpson et al. 2004). Patients present with nausea, vomiting, abdominal pain, and a sensorimotor axonal PN, although some have been noted to have mixed features of demyelination and/or concomitant myopathy. While the underlying pathophysiology is not clear, mitochondrial dysfunction is the leading hypothesis. This condition has been linked mostly to d4T and there exists little evidence-based management options outside of NRTI withdrawal and supportive care.

Management of AIDP and CIDP is the same as that for non-HIV-infected patients: plasma exchange (PLEX) or intravenous immunoglobulin (IVIG) for AIDP and oral immunosuppressants with or without IVIG or PLEX for CIDP.

3.3.4 Other Polyneuropathies

Diffuse Infiltrative Lymphomatosis Syndrome

Diffuse infiltrative lymphomatosis syndrome (DILS) is estimated to affect 3–4 % of HIV-infected individuals and is characterized by multisystem CD8 lymphocytic visceral infiltration involving predominately the salivary glands and lungs, but also kidney, gut, and peripheral nerve (Williams et al. 1998). Clinical presentation is of parotid enlargement and sicca symptoms coupled with persistent CD8 hyperlymphocytosis. Diagnosis is based on xerostomia, chronic (>6 months) submandibular enlargement, and lymphocytic infiltration on salivary biopsy or documented by gallium scintigraphy. Patients with peripheral nerve involvement typically present with acute or subacute painful sensorimotor axonal PN (Moulignier et al. 1997). The pathological hallmark consists of angiocentric peripheral nerve infiltration of T lymphocytes, mimicking that seen with T-cell lymphoma (Gherardi et al. 1998). Management includes initiation of cART with or without low to moderate doses of systemic corticosteroids. Approaches to symptomatic pain management depend upon pain severity and its impact on function.

Mononeuritis Multiplex

Mononeuritis multiplex (MM) or PNS vasculitis has a low incidence in HIV (<1 %), but it is important to recognize because it is treatable. Whether isolated or part of a systemic process, PNS vasculitis occurs generally early after initial infection or in late stages with severe immunodeficiency. Whereas an underlying state of immune activation may be at play early in the infection, CMV reactivation or another infectious phenomenon is the major consideration in AIDS.

The classic presentation of AIDS-associated PNS vasculitis is of a painful MM, although overlapping mononeuropathies may present as a prototypical length-dependent painful PN. EDX studies generally document an asymmetric sensorimotor axonal PN. Needle exam may show active denervation coupled with variable degrees of acute, subacute, or chronic reinnervation changes. When rash is present, skin biopsy may be appropriate to look for leukocytoclastic vasculitis. When nerve biopsy is performed (generally sural, superficial peroneal, or radial), concomitant muscle biopsy (gastrocnemius, peroneus brevis, or deltoid) may increase sensitivity.

All recognized vasculitis syndromes—those affecting small, medium, or large vessels—have been reported with HIV, with the pathological hallmark being necrotizing thrombosis or inflammatory infiltration of vessels on nerve biopsy [reviewed in Chetty (2001)]. Early in disease the pathogenic mechanism for HIV-associated MM is not clear, although it is likely a consequence of either CD8 T-cell-mediated or immune complex-mediated inflammation. These findings may reflect a reaction to HIV infection of endothelial cells or, in some cases, neoplasm. Other potential causes for a similar clinical picture include Hepatitis B or C viral coinfection, cryoglobulinemia, immune complex disease, or drug-induced PN vasculitis. Serum autoantibodies are frequently difficult to interpret due to the high frequency of polyclonal B-cell activation in the setting of abnormal T-cell regulation. When MM arises with severe immunodeficiency, CMV infection is the classic pathogen; the utility of plasma CMV PCR in this setting, however, is unclear. Many AIDS patients with MM have a history of prior or current CMV infection (e.g., retinitis, gastroenteritis, or pneumonia). MM has also been reported with mycobacterium tuberculosis and pneumocystic jirovecii.

Management depends mostly on the severity of prevailing immunodeficiency. In early HIV infection, MM may be treated with antiretroviral therapy whereas in later stages treatment is directed towards the underlying coinfection. For those with advanced disease and evidence of CMV, ganciclovir with or without foscarnet is preferred (Anduze-Faris et al. 2000).

3.3.5 HIV-Associated Myopathies

Polymyositis

HIV-associated myopathies are rare, estimated to effect up to 1 % of patients (Vivithanaporn et al. 2010). Most myopathies are likely related to either immune dysfunction associated with chronic infection or drug toxicity. The most common myopathy is HIV-associated polymyositis (HIV-PM) which can present at any stage of infection (Johnson et al. 2003; Heckmann et al. 2010). Other myopathies include toxic myopathy due to drugs, inclusion body myopathy, and nemaline rod myopathy.

Patients with confirmed HIV-PM are essentially indistinguishable from autoimmune non-HIV PM, both clinically and pathologically. Patients present with subacute or insidiously progressive diffuse and symmetric shoulder and hip girdle weakness (Heckmann et al. 2010; Johnson et al. 2003). Myalgia, weight loss, and fatigue are common, although nonspecific in advanced immunodeficiency (Simpson et al. 1993). Esophageal and myocardial involvement, although reported, are rare.

Elevated creatine kinase (CK) levels are the most prominent laboratory abnormality, with values approximately 2,500 IU/L in two of the larger case series; CK levels, however, are not always elevated, and levels do not correlate with weakness severity (Simpson et al. 1993; Kenyon et al. 2010). The presence of autoantibodies is infrequent. Screening for the human T-cell lymphotrophic virus (HTLV 1/2) coinfection, the lentivirus associated with tropical spastic paraparesis, may be considered. Although DILS is a rare HIV-associated complication, series of DILS patients reported a high frequency (9/35, 26 %) of myopathy (Kazi et al. 1996). EDX testing typically demonstrates irritable myopathy although, as in autoimmune PM, severity of EDX abnormalities does not correlate with symptom severity and can be normal. Biopsy shows primary inflammation with CD8 lymphocytes infiltrating nonnecrotic muscle fibers and rare, scattered necrosis (Illa et al. 1991). While HIV-1-RNA has been detected infrequently in affected myocytes (Seidman et al. 1994), the direct role of HIV as the causative agent of HIV-PM is uncertain because productive virus within myocytes has not been documented (Leon-Monzon et al. 1993; Till and MacDonell 1990).

The approach to treatment of HIV-PM is essentially identical to that of non-HIV PM, with prednisone at dose of up to 60 mg/day generally considered first-line therapy. Resolution of symptoms with steroids has been seen in over half of treated patients (Heckmann et al. 2010; Johnson et al. 2003). Second-line therapy, as with autoimmune PM, may include IVIG, methotrexate, and azathioprine, with outcomes not fully described.

Toxic Myopathy

The introduction of AZT was followed by multiple reports of myopathy, which subsequently improved after drug withdrawal (Bessen et al. 1988). The incidence of toxic myopathy has been estimated to be 0.4 % in patients receiving AZT (Simpson et al. 1997). The underlying mechanism is mitochondrial, similar to that described with ATN (Lewis and Dalakas 1995). Although a member of the NRTI drug class, AZT is not associated with PN. Patients present with symmetrical shoulder and hip girdle weakness typically seen after chronic (>6 months) therapy. CK elevations are similar to that observed with HIV-PM. EMG may show myopathic motor units although may be normal (Cupler et al. 1995). Muscle biopsy typically shows ragged red fibers and structurally abnormal mitochondria (Mhiri et al. 1991). AZT discontinuation generally improves weakness and myalgia within weeks, although improvement may be variable (Manji et al. 1993). Symptoms may progress after drug discontinuation for a few weeks to months, a “coasting” phenomenon as seen with ATN.

Several other medications used in HIV may be associated with myotoxicity. Rhabdomyolysis with varied degrees of myalgia and weakness has been reported with didanosine, raltegravir, tenofovir, etravairine, and ritonavir-boosted cART regimens—the latter of which is likely due to ritonavir’s inhibition of the CYP3A4 hepatic enzyme. With concurrent use of protease inhibitors and medications such as simvastatin and lovastatin that rely on CYP3A4 for metabolism, statin levels may increase significantly and increase the potential for rhabdomyolysis. Current recommendations advise against using simvastatin or lovastatin with concomitant use of a protease inhibitor or delaviridine, another CYP3A4 inhibitor (Dube et al. 2003). Pravastatin and fluvastatin have the least potential for drug–drug interactions as they are not metabolized via the CYP3A4 enzyme.

Other Myopathies

Inclusion-body myopathy (IBM) has been rarely described in patients with HIV, with a younger age of onset (average 44 years) than in sporadic IBM (s-IBM). Clinical presentation and EDX are typical of sporadic IBM, with prominent proximal extremity weakness and subsequent progression to distal extremity weakness and dysphagia. Muscle pathology shows viral-specific expanded CD8 clones invading MHC-1-expressing muscle fibers and secondary endomysial inflammation (Dalakas et al. 2007). There exists no proven disease-modifying therapy, and introduction of cART would be indicated (Cupler et al. 1996).

Nemaline rod myopathy has been described in isolated case reports. Symptom onset is often over several months and characterized by progressive, proximal weakness and severe muscle atrophy. Muscle biopsies reveal prominent type 1 fiber atrophy and intracytoplasmic rod bodies with minimal inflammation (Authier et al. 2005). No randomized controlled studies have been conducted of treatments, which have included steroids, IVIG, and PLEX.

4 Opportunistic Infections and Other HIV-Associated Complications

4.1 Toxoplasmosis

4.1.1 Introduction/Epidemiology

Toxoplasmosis gondii is a small, intracellular protozoa, which is acquired through ingesting uncooked meat, contaminated water, or cat feces. Toxoplasmosis is the most common cause of a focal brain mass in HIV patients. It is a late complication of HIV, usually occurring with $CD4^+$ T-lymphocyte counts below 200 cells/μL. The incidence of CNS toxoplasmosis has significantly decreased with the use of cART and the use of sulfamethoxazole–trimethoprim for prophylaxis (Antinori et al. 2004).

4.1.2 Clinical Manifestations

Given that toxoplasmosis causes focal brain lesions, focal neurological deficits can be seen. Other nonspecific symptoms include headache, fever, and altered mental status. Seizures can occur (Antinori et al. 2004).

4.1.3 Differential Diagnosis and Diagnostic Evaluation

Diagnosis can usually be made using serum tests and brain imaging, in the proper clinical setting. A positive serum *T. gondii* IgG antibody indicates exposure, increasing the probability of toxoplasmosis as the cause of a focal brain mass

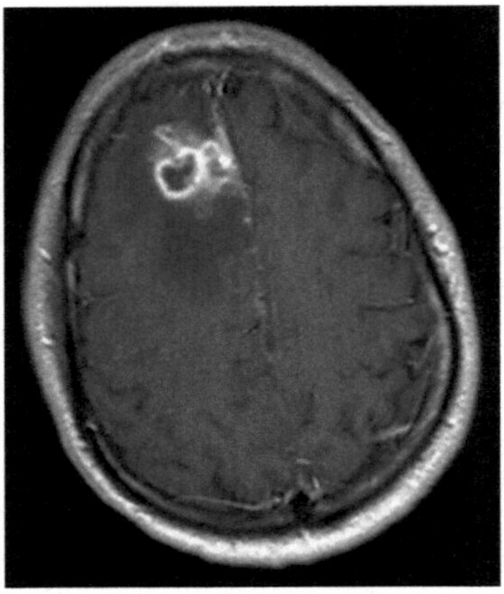

Fig. 2 Gadolinium-enhanced T1 brain MRI showing toxoplasmosis encephalitis

seen on MRI. The antibody test can be a false negative in patients with advanced immunosuppression or in newly acquired infection, before an immune response is mounted (Antinori et al. 2004).

Brain MRI typically shows multifocal ring-enhancing lesions with surrounding vasogenic edema, especially in the basal ganglia or at the gray–white junction (Fig. 2). The main alternative diagnosis for these findings is PCNSL. PCNSL will usually show a solitary CNS lesion, while toxoplasmosis is multifocal. However, this pattern cannot be relied on to distinguish these two conditions. MR perfusion, thallium single photon emission computed tomography (SPECT), and positron emission tomography (PET) show decreased uptake with a toxoplasmosis abscess and increased with PCNSL. MR spectroscopy will show an elevated lactate peak. Such findings on advanced imaging, along with positive toxoplasma IgG, suggest the diagnosis. In actual practice though, patients with such lesions on brain MRI are frequently treated for toxoplasmosis, which should rapidly respond to proper treatment. If there is no clinical and/or radiologic response to toxoplasmosis treatment within 2 weeks, then further consideration is given to PCNSL. Biopsy is usually necessary, which can exclude PCNSL and other causes of brain masses, such as bacterial abscess, tuberculoma, and cryptococcoma (Antinori et al. 2004).

CSF examination typically shows nonspecific findings: a mild-to-moderate pleocytosis, elevated protein concentration, and normal-to-low glucose. CSF *T. gondii* antibodies are not helpful, and PCR is helpful only when positive. Therefore, CSF examination is seldom helpful, and lumbar puncture may be contraindicated due to mass effect (Antinori et al. 2004). An algorithm for diagnostic workup of brain lesions in AIDS is detailed in Fig. 3.

4.1.4 Management

Toxoplasmosis is treated with pyrimethamine, sulfadiazine, and leucovorin. Clindamycin, azithromycin, or atovaquone can be used in sulfa-allergic patients.

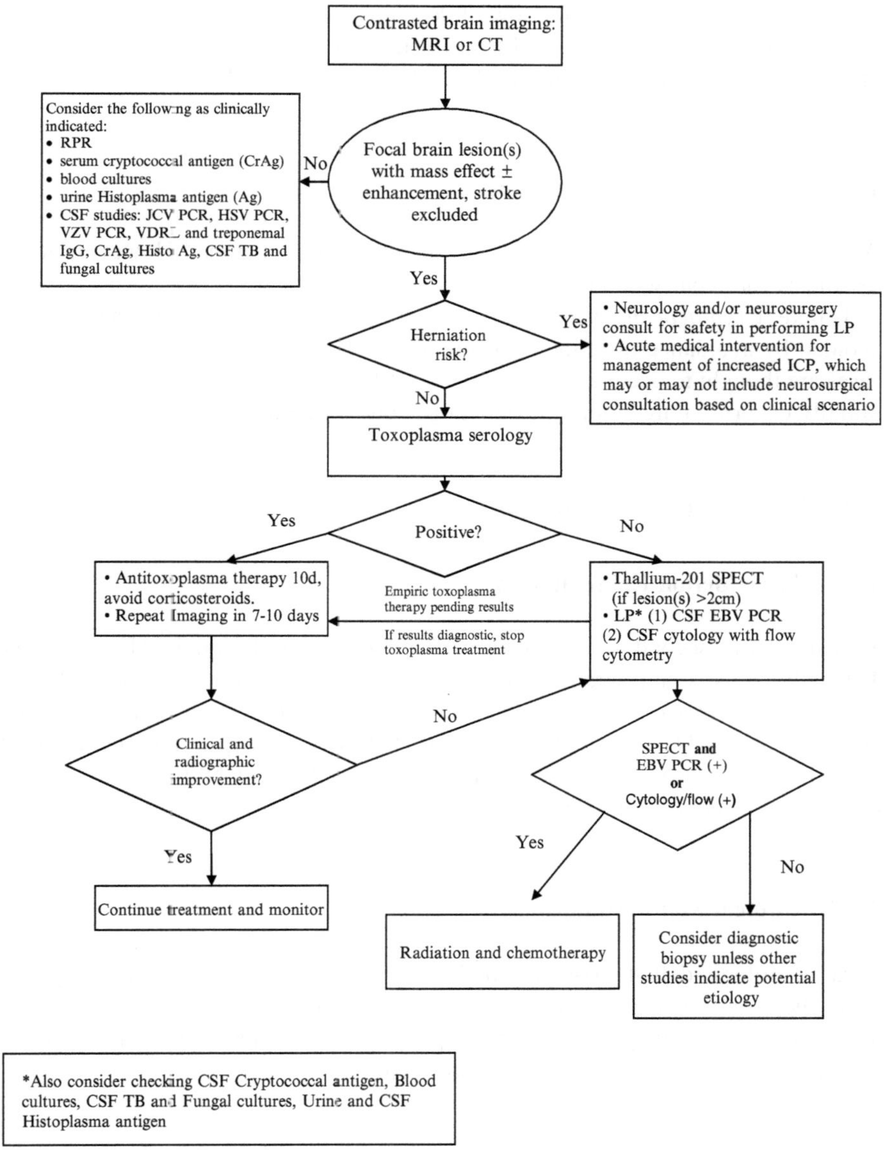

Fig. 3 Diagnostic algorithm for evaluation of brain lesions in HIV/AIDS

Acute therapy is continued for 6 weeks, followed by maintenance therapy with lower doses of the same drugs to prevent recurrence. If the CD4 count rises above 200 cells/mm^3, sulfamethoxazole–trimethoprim prophylaxis may be used.

Steroids are not used unless the patient has life-threatening vasogenic edema, because both PCNSL and toxoplasmosis will respond to their use, and the diagnosis may remain unknown. When glucocorticoid use is unavoidable, they should be used for the least amount of time and lowest dose possible, toxoplasmosis management should continue, and the patient should be monitored for recurrence or worsening off of steroids, which would suggest PCNSL (Dedicoat and Livesley 2006).

4.1.5 Prognosis and Pathogenesis

With proper treatment, prognosis of toxoplasmosis is good. However, toxoplasmosis can recur with immunosuppression.

Unlike *Cryptococcus neoformans*, *T. gondii* is not ubiquitous and maybe avoided by avoidance of feline exposure and undercooked meats. However, most cases are thought to occur as a recrudescence of a latent infection, which may have initially occurred long before the patient became immunosuppressed.

4.2 Primary CNS Lymphoma

4.2.1 Introduction/Epidemiology

PCNSL is the second leading cause of a focal brain lesion in HIV patients. Typically, it is a high-grade, B-cell, non-Hodgkin lymphoma associated with Epstein–Barr virus (EBV) infection and is a late complication of HIV, usually occurring with $CD4^+$ T-lymphocyte counts below 50 cells/μL. The incidence of PCNSL has significantly decreased with the use of cART. PCNSL now affects less than 5 % of patients with HIV.

4.2.2 Clinical Manifestations

Focal neurologic deficits based on the location of the focal mass, encephalopathy, headaches, and seizures are commonly seen. Fever is uncommon, which can help differentiate from toxoplasmosis. PCNSL typically does not spread outside of the CNS and systemic manifestations are not seen.

4.2.3 Differential Diagnosis and Diagnostic Evaluation

Brain MRI usually shows an isolated enhancing lesion with surrounding vasogenic edema, often in a periventricular location. Sometimes there are multiple enhancing lesions. This can be confused with CNS toxoplasmosis (see above). The enhancement tends to be more heterogeneous in PCNSL, and more ring like in toxoplasmosis. MR perfusion, thallium SPECT, and PET show increased uptake with PCNSL and decreased with toxoplasmosis (Fig. 4). MR spectroscopy will show an elevated choline peak.

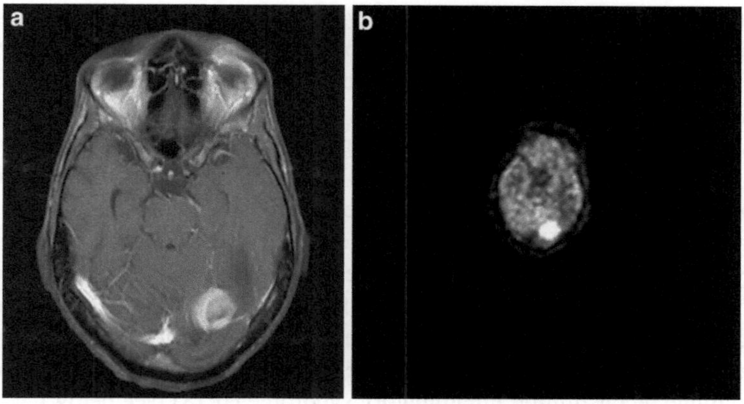

Fig. 4 (**a**) Gadolinium enhanced T1 brain MRI showing ring enhancement and (**b**) SPECT showing increased uptake in primary CNS lymphoma of left occipital lobe

CSF examination typically shows a mild pleocytosis, mildly elevated protein concentration, and normal glucose. These findings are nonspecific. However, a positive EBV PCR done on CSF of an HIV positive patient with a CNS mass lesion is highly predictive of a PCNSL diagnosis, although a negative CSF EBV PCR does not exclude the diagnosis. Cytology and flow cytometry can also be performed on the CSF, although they are often negative because this is a parenchymal tumor.

Brain biopsy is necessary for a definitive diagnosis and usually pursued after toxoplasmosis has been excluded (Hochberg et al. 2007).

4.2.4 Management and Prognosis

Specific treatment protocols are beyond the scope of this chapter. Consultation with a neuro-oncologist is recommended. Frequently, methotrexate and whole brain radiation are used. Steroids are used to control vasogenic edema. cART should be initiated, or changed, to obtain an undetectable plasma HIV RNA level (Skiest and Crosby 2003).

Prognosis for survival is poor, but is best if virological control can be achieved with cART.

4.3 Progressive Multifocal Leukoencephalopathy

Progressive multifocal leukoencephalopathy (PML) is caused by the polyomavirus, JC virus, which primarily infects oligodendrocytes in the CNS. This infection is presented in detail elsewhere (Chap. 4) and so this discussion will focus on aspects unique to HIV infection. PML is typically an opportunistic infection seen in AIDS (Cinque et al. 2009). Briefly, PML is suspected when there is subacute progression of neurological findings including cognitive, motor, and visual deficits. The diagnosis is usually established through a combination of imaging and CSF determination (JC virus DNA) or brain biopsy. The treatment of PML in association with HIV infection is cART. Prior to the introduction of cART PML was almost always fatal, with survival only a few months on average. With cART approximately 50 % of patients will stabilize. In patients already on cART, it is reasonable to consider altering the regimen to obtain undetectable plasma HIV and/or agents that have a higher CPE score (see HAND treatment, section 2.4). Patients are best monitored closely for response to therapy by frequent follow-up and one should consider repeat MRI and repeat lumbar puncture for JC virus DNA if recrudescence is suspected. A number of alternative treatments have been tried, such as interferon-alpha and cytarabine, but none have been clearly effective. PML is also associated with immune reconstitution inflammatory syndrome (IRIS) (see below).

4.4 Cryptococcal Meningitis

4.4.1 Introduction/Epidemiology

C. neoformans is a pathogenic yeast, seen mainly as an opportunistic pathogen of immunocompromised patients. Although usually called meningitis, this infection is more appropriately termed a meningoencephalitis, because the brain parenchyma is almost invariably involved. Up to 15 % of HIV-infected individuals eventually develop cryptoccocal meningitis. In parts of the world with highest HIV prevalence, cryptococcal meningitis is the leading cause of community acquired meningitis, ahead of pneumococcal and meningococcal meningitis (Hakim et al. 2000). The global burden of HIV-related cryptococcosis is estimated to be 1 million cases annually (Park et al. 2009). Even in countries with good healthcare resources and access to ART, there is a persistent burden of cryptococcal disease consisting largely of patients with newly diagnosed HIV infection or patients not on ART. Cryptococcal meningitis is usually a late complication of HIV infection, typically with a $CD4^{+}$ T-lymphocyte counts less than 100 cells/μL. A growing group of non-HIV-infected patients develop cryptococcosis due to immunosuppressive agents being used for an increasingly wide array of conditions.

4.4.2 Clinical Manifestations

Clinical course is usually insidious, rarely presenting acutely, with progression over weeks. The main initial features of cryptococcal meningitis are nonspecific, essentially the same as for subacute or chronic forms of meningitis, and include headache, fever, nausea, vomiting, phonophobia, and photophobia. However, typical signs and symptoms of meningitis may be lacking because of the absence of a vigorous inflammatory response in immunosuppressed patients. As the infection progresses, signs and symptoms associated with elevated intracranial pressure (ICP) develop. These include encephalopathy, diplopia (sixth nerve palsy), other cranial neuropathies, and papilledema.

4.4.3 Differential Diagnosis and Diagnostic Evaluation

Because typical features of meningitis may be lacking in early infection, even atypical headache or subtle signs or symptoms of CNS dysfunction should be evaluated for cryptococcal meningitis in HIV patients, especially with low CD4 counts. CSF is tested for cryptococcal antigen, followed by presence of *Cryptococcus* on CSF culture. India ink staining is not as specific or sensitive (<50 %), but is rapid, and can support a clinical suspicion, while the other tests are pending. Opening pressure is usually high, otherwise, CSF profiles are normal or nonspecifically abnormal with a mild-to-moderate lymphocytic pleocytosis, elevated protein concentration, and normal to moderately low glucose. The CSF profile is likely to be normal or less abnormal in those with more advanced AIDS/immunosuppression.

It is not possible to make a diagnosis of cryptococcal meningitis based on brain imaging. However, brain MRI with contrast is indicated to evaluate for presence of a cryptococcoma and to exclude other alternative or concomitant CNS processes.

4.4.4 Management

Treatment of cryptococcosis is challenging, with little new drug development or recent definitive studies. Amphotericin B is the mainstay of cryptococcal treatment, but is difficult to use, requiring daily intravenous infusion, with the inherent problems of inconvenience and line infections, particularly in immunocompromised patients. Amphotericin also has a narrow therapeutic window because of nephrotoxicity. Kidney function must be carefully monitored. A liposomal formulation is safer, but costs 50–100 times more. Flucytosine is used in combination with amphotericin, but causes myelosuppression and dose adjustment is required for renal insufficiency (Vermes et al. 2000). It also causes peripheral neuropathy, which can confound HIV- and ART PN. Fluconazole is often used in developing countries, but is not as effective without flucytosine (Larsen et al. 1994) and has a slower initial response compared to amphotericin (Saag et al. 1992). Its metabolism also interacts with that of the nucleoside reverse transcriptase inhibitors, complicating management of the patient's underlying HIV infection.

To maintain low CSF pressures and prevent morbidity and mortality associated with high ICP, lumbar drainage or serial lumbar puncture is critical, based on small series and expert opinion, but not adequately designed controlled trials. Brain imaging should be done prior to lumbar puncture to exclude mass effect. Patients who are treated with no more than minor deviations from CSF pressure treatment guidelines have significantly better outcomes (Shoham et al. 2005). Conversely, use of medications to control elevated ICP, such as mannitol, acetazolamide, and steroids, has not been effective. Lumbar (or ventricular) drains can be used for patients in whom serial lumbar punctures are not possible.

4.4.5 Prognosis

Cryptococcal meningitis is a major cause of morbidity and mortality in HIV infection worldwide and raised ICP is a leading cause of morbidity and mortality. Mortality can be as high as 80–100 % in resource-limited settings (Wang and Carm 2001), but still ranges from 20 to 40 % at 10 weeks despite amphotericin B and cART (Lortholary et al. 2006). Morbidities associated with elevated ICP include cognitive impairment, cranial neuropathies, severe headache, progressive loss of vision, hearing impairment, and decreased level of consciousness.

4.4.6 Pathogenesis

Infection takes place through the respiratory route, and then the organism spreads hematogenously to distant sites, including the CNS. It can also remain latent and reactivate years later, usually in the setting of AIDS. The organisms have a polysaccharide capsule, which has profound immune-suppressive actions, and pathogenic cryptococci have neurotropism with a predilection to CNS infection. Raised ICP is common in patients with cryptococcal meningitis: 65–75 % have opening pressure on lumbar puncture of at least 20 cm H_2O, and 20–25 % have opening pressure of at least 35 cm H_2O at time of diagnosis (Bicanic et al. 2009). The mechanism may be blockage of CSF reabsorption at the arachnoid villi because of the presence of organisms that shed polysaccharide. This mechanism is widely hypothesized in the literature, but has not been conclusively demonstrated. A high organism load appears to be necessary but not sufficient for the development of ICP, suggesting that other factors must play a role. There is likely a complex interplay of host and organism factors leading to elevated ICP.

4.5 Syphilis

4.5.1 Introduction/Epidemiology

Syphilis is caused by the spirochetal bacteria, *Treponema pallidum.* Rates of syphilis dropped dramatically in the 1990s but have increased in the 2000s. The presence of syphilis, particularly genital ulcers, increases the risk of HIV infection. It is not clear that the presence of HIV significantly alters the natural course of syphilis. There may be an increased, but still very low risk of neurosyphilis in HIV patients (Zellan and Augenbraun 2004).

4.5.2 Clinical Manifestations

The stages of syphilitic infection are unchanged by HIV. Primary syphilis manifests with painless ulcers at the site of primary infection, usually genital or anal.

Secondary syphilis usually manifests with a rash, which can have a widely variable appearance and can be mild or severe with constitutional symptoms. Syphilis then becomes latent, with serological evidence of infection, but no signs or symptoms. Tertiary syphilis is a late stage and includes cardiovascular and neurological manifestations (e.g., tabes dorsalis). Neurosyphilis can occur either early or late in the course of infection. The spectrum of manifestations from neurosyphilis is very broad and can include altered mental status, signs and symptoms of basilar meningitis, tabes dorsalis, stroke-like symptoms, which are usually more subacute in onset than a typical acute stroke, psychiatric symptoms, and personality change.

4.5.3 Differential Diagnosis and Diagnostic Evaluation

Nontreponemal assays, such as the RPR or VDRL test, measure nonspecific antibodies as a screen for syphilis. Treponemal assays, such as the fluorescent treponemal antibody absorbent (FTA-Abs) test, are used to confirm the results of the screen. However, a negative nontreponemal test does not exclude syphilis, especially in an HIV-positive individual. It is therefore preferable to use treponemal tests in patients with HIV, because treponemal tests have higher sensitivity and specificity. Both nontreponemal and treponemal tests have decreased sensitivity and specificity in patients with HIV infection, so biopsy of suspicious lesions and use of special stains for *T. pallidum* may be helpful (Larsen et al. 1995).

CSF examination is recommended in all patients with syphilis who have neurological signs or symptoms, in any patient found to fail standard treatment, and in syphilis known to be present for more than 1 year or of unknown duration (Sexually transmitted diseases treatment guidelines 2002. Centers for Disease Control and Prevention 2002). Definitive diagnosis of neurosyphilis requires a positive CSF VDRL and pleocytosis. However, there are case reports of negative CSF VDRL in HIV-infected patients who were diagnosed with neurosyphilis. A negative CSF FTA-Abs can help exclude neurosyphilis, but a positive result is nonspecific. Therefore, in a patient with a negative CSF VDRL, a clinician must use judgment to differentiate neurosyphilis from neurological manifestations of HIV itself or other CNS opportunistic infections.

4.5.4 Management

Detailed management recommendations are beyond the scope of this chapter. Treatment of all stages of syphilis is the same with or without HIV infection, and consists of long-acting penicillin G. Oral doxycycline and tetracycline can be used, but are only recommended for patients with definite penicillin allergy. Oral agents are not effective in neurosyphilis. Periodic serological testing after treatment is necessary to ensure adequate treatment and to monitor for re-infection.

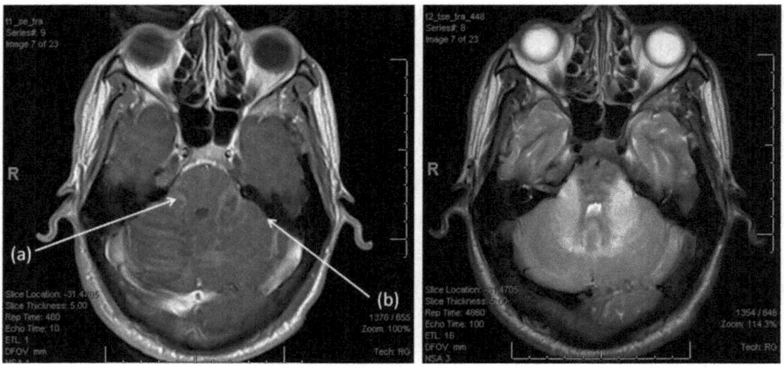

Fig. 5 Brain (*arrow a*) and meningeal (*arrow b*) gadolinium enhancement on T1 brain MRI, (left) and diffuse signal on a T2 image (right) in infratentorial PML IRIS

4.6 Immune Reconstitution Inflammatory Syndrome

4.6.1 Introduction/Epidemiology

Immune restoration through the use of cART controls HIV infection and associated opportunistic infections. However, sometimes the restored immune system responds vigorously to the presence of an opportunistic infection (or HIV itself), producing a paradoxical inflammatory clinical deterioration. This occurs in up to 30 % of patients with cryptococcal meningitis, PML, and pulmonary tuberculosis (Shelburne et al. 2005). Risk factors include greater degree of immunosuppression, higher HIV or opportunistic infection burden, and shorter time between diagnosis of HIV or opportunistic infection and initiation of cART.

4.6.2 Clinical Manifestations

The most common symptoms are altered mental status and fever, sometimes lymphadenopathy, as well as symptoms which could be associated with the underlying opportunistic infection even in the absence of IRIS.

4.6.3 Differential Diagnosis and Diagnostic Evaluation

Other diagnoses to consider with paradoxical deterioration in the setting of cART initiation include medication side effect or toxicity and progression of the underlying disease due to noncompliance with cART, resistance to cART, or inadequate cART levels due to drug–drug interactions or malabsorption (Lipman and Breen 2006).

Brain MRI may show enhancing lesions, indicating presence of inflammation, even in conditions which do not usually enhance, such as PML (Fig. 5). Brain biopsy will often also show a nonspecific inflammatory reaction.

4.6.4 Management

In severe cases with significant brain edema and risk for brain herniation, cART can be stopped and steroids started. Typically, however, IRIS resolves without specific treatment, so cART is continued and the patient is given supportive care. If cART is stopped, IRIS may recur when cART is re-started. IRIS may be preventable by reducing the load of the opportunistic infection with treatments for that infection prior to starting cART. However, initiation of cART is key to controlling most opportunistic infections. The risks versus benefits of delayed cART therapy are unknown and may depend on the specific underlying infection. In the absence of data, most clinicians seem to favor immediate cART initiation, then managing the IRIS if it occurs.

4.6.5 Prognosis and Pathogenesis

IRIS can be severe and morbidity and mortality can be high in the acute setting. If the patient survives and does well through the initial phase of IRIS, prognosis is good. Inflammatory PML-IRIS actually has a better prognosis than typical, noninflammatory PML (Berger et al. 1998), probably because the inflammation, although leading to clinical deterioration and altered mental status, is actually helping to combat the underlying infection.

IRIS may be due to an imbalance in the inflammatory response to the underlying infection, with an excess of cytotoxic CD8 positive T lymphocytes and a lack of regulatory CD4 positive T lymphocytes (Vendrely et al. 2005).

4.7 CMV Polyradiculitis, Encephalitis, and Ventriculitis

4.7.1 CMV Polyradiculopathy

Acute lumbosacral polyradiculopathy in the setting of HIV is important to identify given the characteristically rapid progression and the poor prognosis when left untreated. The most-clearly defined etiology of lumbosacral polyradiculopathy in patients with HIV is CMV, with an estimated incidence of <1 % (Anders and Goebel 1998). CMV polyradiculopathy typically presents with flaccid ascending paraplegia evolving quickly over a period of days to weeks, often accompanied by lumbar pain, lower extremity paresthesias, and bowel and bladder dysfunction. Early urinary retention or bowel obstipation are important clinical signs that help distinguish this syndrome from AIDP. Weakness generally outweighs sensory complaints, is typically mildly asymmetric and involves proximal as well as distal muscles.

EDX typically show evidence for lumbosacral radiculopathy, though features of concomitant HIV-PN are common (Corral et al. 1997; So and Olney 1994). CSF reveals a polymorphonuclear pleocytosis (average cell count of 500–600/mL) with elevated protein and decreased glucose, although may be normal (Kim and Hollander 1993; Corral et al. 1997; Anders and Goebel 1998; So and Olney 1994; Miller et al. 1996). CMV PCR is both sensitive and specific. Lumbosacral MRI is important to rule out a structural or compressive lesion and may reveal nerve root thickening and/or gadolinium enhancement. It is important to consider a dilated eye examination, particularly with complaints of "flashing lights" or "floaters," for CMV retinitis. Treatment for CMV polyradiculopathy with ganciclovir and/or cART shows variable improvement with survival on average of 4 months with death from comorbid conditions (Miller et al. 1996). Without antiviral therapy, survival is on the order of weeks (Kim and Hollander 1993). Other potential causes for acute radiculitis include other herpesviruses (e.g., HSV and VZV). Elsberg syndrome, a combination of bladder dysfunction and polyradiculitis associated with genital HSV, has been described in patients with HIV with significant improvement after intravenous acyclovir (Yoritaka et al. 2005). VZV may cause "segmental zoster paresis," in which intravenous acyclovir and a tapering dose of steroids has been reported to show benefit (Kawajiri et al. 2007). Less commonly described causes of polyradiculopathy include mycobacterium tuberculosis, neoplasm (e.g., leptomeningeal infiltration of systemic lymphoma), syphilis, and toxoplasmosis.

4.7.2 CMV Encephalitis and Ventriculitis

Evidence of CMV infection in postmortem brains of AIDS patients is surprisingly common, affecting over one quarter of severely immunocompromised patients. Clinical infection, however, was rare (less than 5 %) in the pre-cART era and is even more infrequent in the current treatment era. As the vast majority of HIV-infected individuals have had prior CMV infection, it generally represents viral reactivation in the setting of severe immunodeficiency.

CMV infection can result in two distinct CNS syndromes: CMV encephalitis and ventriculitis (Grassi et al. 1998). CMV microglial nodular encephalitis generally presents as a subacute progressive encephalopathy progressing over weeks. A fluctuating, altered sensorium is observed in combination with variable degrees of cognitive, mood, and personality changes. Headache and fever are commonly absent. The presentation can be protean and may also include focal neurological symptoms and/or features of meningitis. CMV ventriculitis may show similar changes in cognition and sensorium, but is also classically characterized by ataxia, nystagmus, as well as cranial neuropathies. These latter symptoms have also been reported with CMV rhomboencephalitis. Patients may have other systemic manifestations of CMV reactivation (e.g., retinitis, esophagitis, colitis), although reactivation may be compartmental and isolated to the CNS.

Major diagnostic considerations of CMV encephalitis include delirium, HAND, and other viral encephalitides, including but not limited to HSV and VZV encephalitis. Diagnostic considerations for CMV ventriculitis primarily include lymphomatous or zoster-related ventriculitis.

CSF abnormalities are variable, ranging from normal cell counts to lymphocytic pleocytosis or predominantly polymorphonuclear pleocytosis. Total protein is typically elevated and hypoglycorrhachia may or may not be present. Diagnosis generally relies upon PCR detection of CMV DNA in CSF. MRI of the brain is commonly normal with diffuse microglial nodular encephalitis. Ventriculitis, however, may show enhancement or signal abnormalities along the ependymal lining of the lateral ventricles. When there is high suspicion for CMV, one should consider workup for extraneurological manifestations, including but not limited to a dilated eye examination to evaluate for retinitis.

Gancyclovir therapy is the mainstay of treatment, and there is some data to suggest that outcomes are improved with combination foscarnet therapy (Anduze-Faris et al. 2000). Prolonged treatment is frequently complicated by toxicities, predominately hematological for gancyclovir and renal for foscarnet. Introduction of cART may prolong survival after CMV disease in patients with AIDS.

4.8 Stroke

Stroke is more common in HIV-infected individuals (Tipping et al. 2007). Most strokes are ischemic and many of these are related to concurrent infection including meningitis and/or vasculitis/vasculopathy. Often these strokes are seen in relatively young people. Therefore in addition to a search for ongoing opportunistic infections, it is important to investigate other causes of stroke in the young such as cardioembolism and coagulopathy.

Ovbiagele and Nath (2011) suggest that a recent increase in the US in hospitalized patients with stroke and HIV infection may be due to the use of cART. cART is often associated with hyperlipidemia and metabolic dysfunction, which could predispose these patients to stroke. However, this issue is complex since other risk factors for stroke such as hypertension are also increasing in an aging population of HIV-positive individuals.

References

Abrams DI, Goldman AI, Launer C, Korvick JA, Neaton JD, Crane LR, Grodesky M, Wakefield S, Muth K, Kornegay S et al (1994) A comparative trial of didanosine or zalcitabine after treatment with zidovudine in patients with human immunodeficiency virus infection. The Terry Beirn Community Programs for Clinical Research on AIDS. N Engl J Med 330 (10):657–662

Abrams DI, Jay CA, Shade SB, Vizoso H, Reda H, Press S, Kelly ME, Rowbotham MC, Petersen KL (2007) Cannabis in painful HIV-associated sensory neuropathy: a randomized placebo-controlled trial. Neurology 68(7):515–521

AMDG Agency Medical Director's Group Interagency Guideline on Opioid Dosing for Chronic Non-cancer Pain 2010 Update. http://www.agencymeddirectors.wa.gov/opioiddosing.asp. Accessed 22 Mar 2011

Anders HJ, Goebel FD (1998) Cytomegalovirus polyradiculopathy in patients with AIDS. Clin Infect Dis 27(2):345–352

Anduze-Faris BM, Fillet AM, Gozlan J, Lancar R, Boukli N, Gasnault J, Caumes E, Livartowsky J, Matheron S, Leport C, Salmon D, Costagliola D, Katlama C (2000) Induction and maintenance therapy of cytomegalovirus central nervous system infection in HIV-infected patients. AIDS 14(5):517–524

Antinori A, Larussa D, Cingolani A, Lorenzini P, Bossolasco S, Finazzi MG, Bongiovanni M, Guaraldi G, Grisetti S, Vigo B, Gigli B, Mariano A, Dalle Nogare ER, De Marco M, Moretti F, Corsi P, Abrescia N, Rellecati P, Castagna A, Mussini C, Ammassari A, Cinque P, D'Arminio Monforte A (2004) Prevalence, associated factors, and prognostic determinants of AIDS-related toxoplasmic encephalitis in the era of advanced highly active antiretroviral therapy. Clin Infect Dis 39(11):1681–1691

Antinori A, Arendt G, Becker JT, Brew BJ, Byrd DA, Cherner M, Clifford DB, Cinque P, Epstein LG, Goodkin K, Gisslen M, Grant I, Heaton RK, Joseph J, Marder K, Marra CM, McArthur JC, Nunn M, Price RW, Pulliam L, Robertson KR, Sacktor N, Valcour V, Wojna VE (2007) Updated research nosology for HIV-associated neurocognitive disorders. Neurology 69 (18):1789–1799

Arenas-Pinto A, Bhaskaran K, Dunn D, Weller IV (2008) The risk of developing peripheral neuropathy induced by nucleoside reverse transcriptase inhibitors decreases over time: evidence from the Delta trial. Antivir Ther 13(2):289–295

Atkinson JH, Grant II (1997) Mood disorder due to human immunodeficiency virus: yes, no, or maybe? Semin Clin Neuropsychiatry 2(4):276–284

Authier FJ, Chariot P, Gherardi RK (2005) Skeletal muscle involvement in human immunodeficiency virus (HIV)-infected patients in the era of highly active antiretroviral therapy (HAART). Muscle Nerve 32(3):247–260

Banerjee S, McCutchan JA, Ances BM, Deutsch R, Riggs PK, Way L, Ellis RJ (2011) Hypertriglyceridemia in combination antiretroviral-treated HIV-positive individuals: potential impact on HIV sensory polyneuropathy. AIDS 25(2):F1–6

Becker JT, Kingsley L, Mullen J, Cohen B, Martin E, Miller EN, Ragin A, Sacktor N, Selnes OA, Visscher BR (2009) Vascular risk factors, HIV serostatus, and cognitive dysfunction in gay and bisexual men. Neurology 73(16):1292–1299

Berger AR, Arezzo JC, Schaumburg HH, Skowron G, Merigan T, Bozzette S, Richman D, Soo W (1993) 2',3'-dideoxycytidine (ddC) toxic neuropathy: a study of 52 patients. Neurology 43 (2):358–362

Berger JR, Levy RM, Flomenhoft D, Dobbs M (1998) Predictive factors for prolonged survival in acquired immunodeficiency syndrome-associated progressive multifocal leukoencephalopathy. Ann Neurol 44(3):341–349

Bessen LJ, Greene JB, Louie E, Seitzman P, Weinberg H (1988) Severe polymyositis-like syndrome associated with zidovudine therapy of AIDS and ARC. N Engl J Med 318(11):708

Bicanic T, Brouwer AE, Meintjes G, Rebe K, Limmathurotsakul D, Chierakul W, Teparrakkul P, Loyse A, White NJ, Wood R, Jaffar S, Harrison T (2009) Relationship of cerebrospinal fluid pressure, fungal burden and outcome in patients with cryptococcal meningitis undergoing serial lumbar punctures. AIDS 23(6):701–706

Blum AS, Dal Pan GJ, Feinberg J, Raines C, Mayjo K, Cornblath DR, McArthur JC (1996) Low-dose zalcitabine-related toxic neuropathy: frequency, natural history, and risk factors. Neurology 46(4):999–1003

Cherry CL, Rosenow A, Affandi JS, McArthur JC, Wesselingh SL, Price P (2008) Cytokine genotype suggests a role for inflammation in nucleoside analog-associated sensory neuropathy (NRTI-SN) and predicts an individual's NRTI-SN risk. AIDS Res Hum Retroviruses 24 (2):117–123

Chetty R (2001) Vasculitides associated with HIV infection. J Clin Pathol 54(4):275–278

Chimelli L, Martins AR (2002) Degenerative and inflammatory lesions in sympathetic ganglia: further morphological evidence for an autonomic neuropathy in AIDS. J NeuroAIDS 2(3):67–82
Chong J, Di Rocco A, Tagliati M, Danisi F, Simpson DM, Atlas SW (1999) MR findings in AIDS-associated myelopathy. AJNR Am J Neuroradiol 20(8):1412–1416
Cinque P, Koralnik IJ, Gerevini S, Miro JM, Price RW (2009) Progressive multifocal leukoencephalopathy in HIV-1 infection. Lancet Infect Dis 9(10):625–636
Clifford DB, Smurzynski M, Park LS, Yeh TM, Zhao Y, Blair L, Arens M, Evans SR (2009) Effects of active HCV replication on neurologic status in HIV RNA virally suppressed patients. Neurology 73(4):309–314
Compostella C, Compostella L, D'Elia R (2008) The symptoms of autonomic dysfunction in HIV-positive Africans. Clin Auton Res 18(1):6–12
Cornblath DR, McArthur JC (1988) Predominantly sensory neuropathy in patients with AIDS and AIDS-related complex. Neurology 38(5):794–796
Cornblath DR, McArthur JC, Kennedy PG, Witte AS, Griffin JW (1987) Inflammatory demyelinating peripheral neuropathies associated with human T-cell lymphotropic virus type III infection. Ann Neurol 21(1):32–40
Corral I, Quereda C, Casado JL, Cobo J, Navas E, Perez-Elias MJ, Pintado V, Fortun J, Guerrero A (1997) Acute polyradiculopathies in HIV-infected patients. J Neurol 244(8):499–504
Craddock C, Pasvol G, Bull R, Protheroe A, Hopkin J (1987) Cardiorespiratory arrest and autonomic neuropathy in AIDS. Lancet 2(8549):16–18. doi:S0140-6736(87)93054-6
Cupler EJ, Danon MJ, Jay C, Hench K, Ropka M, Dalakas MC (1995) Early features of zidovudine-associated myopathy: histopathological findings and clinical correlations. Acta Neuropathol 90(1):1–6
Cupler EJ, Leon-Monzon M, Miller J, Semino-Mora C, Anderson TL, Dalakas MC (1996) Inclusion body myositis in HIV-1 and HTLV-1 infected patients. Brain 119(Pt 6):1887–1893
Dal Pan GJ, Glass JD, McArthur JC (1994) Clinicopathologic correlations of HIV-1-associated vacuolar myelopathy: an autopsy-based case-control study. Neurology 44(11):2159–2164
Dalakas MC, Semino-Mora C, Leon-Monzon M (2001) Mitochondrial alterations with mitochondrial DNA depletion in the nerves of AIDS patients with peripheral neuropathy induced by 2'3'-dideoxycytidine (ddC). Lab Invest 81(11):1537–1544
Dalakas MC, Rakocevic G, Shatunov A, Goldfarb L, Raju R, Salajegheh M (2007) Inclusion body myositis with human immunodeficiency virus infection: four cases with clonal expansion of viral-specific T cells. Ann Neurol 61(5):466–475
de Almeida SM, Boritza K, Cogo LL, Pessa L, Franca J, Rota I, Muro M, Ribeiro C, Raboni SM, Vidal LR, Nogueira MB, Ellis R (2011) Quantification of cerebrospinal fluid lactic acid in the differential diagnosis between HIV chronic meningitis and opportunistic meningitis. Clin Chem Lab Med 49(5):891–896
Dedicoat M, Livesley N (2006) Management of toxoplasmic encephalitis in HIV-infected adults (with an emphasis on resource-poor settings). Cochrane Database Syst Rev 3:CD005420
Dube MP, Stein JH, Aberg JA, Fichtenbaum CJ, Gerber JG, Tashima KT, Henry WK, Currier JS, Sprecher D, Glesby MJ (2003) Guidelines for the evaluation and management of dyslipidemia in human immunodeficiency virus (HIV)-infected adults receiving antiretroviral therapy: recommendations of the HIV Medical Association of the Infectious Disease Society of America and the Adult AIDS Clinical Trials Group. Clin Infect Dis 37(5):613–627
Eisenberg E, McNicol ED, Carr DB (2005) Efficacy and safety of opioid agonists in the treatment of neuropathic pain of nonmalignant origin: systematic review and meta-analysis of randomized controlled trials. JAMA 293(24):3043–3052
Ellis RJ, Marquie-Beck J, Delaney P, Alexander T, Clifford DB, McArthur JC, Simpson DM, Ake C, Collier AC, Gelman BB, McCutchan JA, Morgello S, Grant I (2008) Human immunodeficiency virus protease inhibitors and risk for peripheral neuropathy. Ann Neurol 64(5):566–572

Ellis RJ, Toperoff W, Vaida F, van den Brande G, Gonzales J, Gouaux B, Bentley H, Atkinson JH (2009) Smoked medicinal cannabis for neuropathic pain in HIV: a randomized, crossover clinical trial. Neuropsychopharmacology 34(3):672–680

Ellis RJ, Rosario D, Clifford DB, McArthur JC, Simpson D, Alexander T, Gelman BB, Vaida F, Collier A, Marra CM, Ances B, Atkinson JH, Dworkin RH, Morgello S, Grant I (2010) Continued high prevalence and adverse clinical impact of human immunodeficiency virus-associated sensory neuropathy in the era of combination antiretroviral therapy: the CHARTER Study. Arch Neurol 67(5):552–558

England JD, Gronseth GS, Franklin G, Carter GT, Kinsella LJ, Cohen JA, Asbury AK, Szigeti K, Lupski JR, Latov N, Lewis RA, Low PA, Fisher MA, Herrmann DN, Howard JF, Lauria G, Miller RG, Polydefkis M, Sumner AJ (2009) Practice parameter: the evaluation of distal symmetric polyneuropathy: the role of laboratory and genetic testing (an evidence-based review). Report of the American Academy of Neurology, the American Association of Neuromuscular and Electrodiagnostic Medicine, and the American Academy of Physical Medicine and Rehabilitation. PM R 1(1):5–13

Estanislao L, Carter K, McArthur J, Olney R, Simpson D (2004) A randomized controlled trial of 5% lidocaine gel for HIV-associated distal symmetric polyneuropathy. J Acquir Immune Defic Syndr 37(5):1584–1586

Evans SR, Simpson DM, Kitch DW, King A, Clifford DB, Cohen BA, McArthur JC (2007) A randomized trial evaluating Prosaptide for HIV-associated sensory neuropathies: use of an electronic diary to record neuropathic pain. PLoS One 2(6):e551

Evans SR, Ellis RJ, Chen H, Yeh TM, Lee AJ, Schifitto G, Wu K, Bosch RJ, McArthur JC, Simpson DM, Clifford DB (2011) Peripheral neuropathy in HIV: prevalence and risk factors. AIDS 25(7):919–928

Freeman R, Roberts MS, Friedman LS, Broadbridge C (1990) Autonomic function and human immunodeficiency virus infection. Neurology 40(4):575–580

Gherardi RK, Chretien F, Delfau-Larue MH, Authier FJ, Moulignier A, Roulland-Dussoix D, Belec L (1998) Neuropathy in diffuse infiltrative lymphocytosis syndrome: an HIV neuropathy, not a lymphoma. Neurology 50(4):1041–1044

Gluck T, Degenhardt E, Scholmerich J, Lang B, Grossmann J, Straub RH (2000) Autonomic neuropathy in patients with HIV: course, impact of disease stage, and medication. Clin Auton Res 10(1):17–22

Grassi MP, Clerici F, Perin C, D'Arminio Monforte A, Vago L, Borella M, Boldorini R, Mangoni A (1998) Microglial nodular encephalitis and ventriculoencephalitis due to cytomegalovirus infection in patients with AIDS: two distinct clinical patterns. Clin Infect Dis 27(3):504–508

Hahn K, Arendt G, Braun JS, von Giesen HJ, Husstedt IW, Maschke M, Straube ME, Schielke E (2004) A placebo-controlled trial of gabapentin for painful HIV-associated sensory neuropathies. J Neurol 251(10):1260–1266

Hakim JG, Gangaidzo IT, Heyderman RS, Mielke J, Mushangi E, Taziwa A, Robertson VJ, Musvaire P, Mason PR (2000) Impact of HIV infection on meningitis in Harare, Zimbabwe: a prospective study of 406 predominantly adult patients. AIDS 14(10):1401–1407

Harrison K, Vlahov D, Jones K, Charron K, Clements ML (1995) Medical eligibility, comprehension of the consent process, and retention of injection drug users recruited for an HIV vaccine trial. J Acquir Immune Defic Syndr Hum Retrovirol 10(3):386–390

Harrison LH, Vaz B, Taveira DM, Quinn TC, Gibbs CJ, de Souza SH, McArthur JC, Schechter M (1997) Myelopathy among Brazilians coinfected with human T-cell lymphotropic virus type I and HIV. Neurology 48(1):13–18

Heaton RK, Franklin DR, Ellis RJ, McCutchan JA, Letendre SL, Leblanc S, Corkran SH, Duarte NA, Clifford DB, Woods SP, Collier AC, Marra CM, Morgello S, Mindt MR, Taylor MJ, Marcotte TD, Atkinson JH, Wolfson T, Gelman BB, McArthur JC, Simpson DM, Abramson I, Gamst A, Fennema-Notestine C, Jernigan TL, Wong J, Grant I (2011) HIV-associated neurocognitive disorders before and during the era of combination antiretroviral therapy: differences in rates, nature, and predictors. J Neurovirol 17(1):3–16

Heckmann JM, Pillay K, Hearn AP, Kenyon C (2010) Polymyositis in African HIV-infected subjects. Neuromuscul Disord 20(11):735–739
Hochberg FH, Baehring JM, Hochberg EP (2007) Primary CNS lymphoma. Nat Clin Pract Neurol 3(1):24–35
Illa I, Nath A, Dalakas M (1991) Immunocytochemical and virological characteristics of HIV-associated inflammatory myopathies: similarities with seronegative polymyositis. Ann Neurol 29(5):474–481
Janssen ES (1991) Nomenclature and research case definitions for neurologic manifestations of the human immunodeficiency virus-type 1 infection. Neurology 41:778–785
Johnson RW, Williams FM, Kazi S, Dimachkie MM, Reveille JD (2003) Human immunodeficiency virus-associated polymyositis: a longitudinal study of outcome. Arthritis Rheum 49 (2):172–178
Kallianpur AR, Hulgan T, Canter JA, Ritchie MD, Haines JL, Robbins GK, Shafer RW, Clifford DB, Haas DW (2006) Hemochromatosis (HFE) gene mutations and peripheral neuropathy during antiretroviral therapy. AIDS 20(11):1503–1513
Kawajiri S, Tani M, Noda K, Fujishima K, Hattori N, Okuma Y (2007) Segmental zoster paresis of limbs: report of three cases and review of literature. Neurologist 13(5):313–317
Kazi S, Cohen PR, Williams F, Schempp R, Reveille JD (1996) The diffuse infiltrative lymphocytosis syndrome. Clinical and immunogenetic features in 35 patients. AIDS 10(4):385–391
Kemper CA, Kent G, Burton S, Deresinski SC (1998) Mexiletine for HIV-infected patients with painful peripheral neuropathy: a double-blind, placebo-controlled, crossover treatment trial. J Acquir Immune Defic Syndr Hum Retrovirol 19(4):367–372
Kenyon C, Pillay K, Heckmann JM (2010) Beware of 'normal' creatine kinase levels in HIV-associated polymyositis. S Afr Med J 100(3):156–157
Keswani SC, Polley M, Pardo CA, Griffin JW, McArthur JC, Hoke A (2003) Schwann cell chemokine receptors mediate HIV-1 gp120 toxicity to sensory neurons. Ann Neurol 54 (3):287–296
Kieburtz K, Simpson D, Yiannoutsos C, Max MB, Hall CD, Ellis RJ, Marra CM, McKendall R, Singer E, Dal Pan GJ, Clifford DB, Tucker T, Cohen B (1998) A randomized trial of amitriptyline and mexiletine for painful neuropathy in HIV infection. AIDS Clinical Trial Group 242 Protocol Team. Neurology 51(6):1682–1688
Kim YS, Hollander H (1993) Polyradiculopathy due to cytomegalovirus: report of two cases in which improvement occurred after prolonged therapy and review of the literature. Clin Infect Dis 17(1):32–37
Larsen RA, Bozzette SA, Jones BE, Haghighat D, Leal MA, Forthal D, Bauer M, Tilles JG, McCutchan JA, Leedom JM (1994) Fluconazole combined with flucytosine for treatment of cryptococcal meningitis in patients with AIDS. Clin Infect Dis 19(4):741–745
Larsen SA, Steiner BM, Rudolph AH (1995) Laboratory diagnosis and interpretation of tests for syphilis. Clin Microbiol Rev 8(1):1–21
Leger JM, Bouche P, Bolgert F, Chaunu MP, Rosenheim M, Cathala HP, Gentilini M, Hauw JJ, Brunet P (1989) The spectrum of polyneuropathies in patients infected with HIV. J Neurol Neurosurg Psychiatry 52(12):1369–1374
Leon-Monzon M, Lamperth L, Dalakas MC (1993) Search for HIV proviral DNA and amplified sequences in the muscle biopsies of patients with HIV polymyositis. Muscle Nerve 16 (4):408–413
Letendre S, Paulino AD, Rockenstein E, Adame A, Crews L, Cherner M, Heaton R, Ellis R, Everall IP, Grant I, Masliah E (2007) Pathogenesis of hepatitis C virus coinfection in the brains of patients infected with HIV. J Infect Dis 196(3):361–370
Letendre S, Marquie-Beck J, Capparelli E, Best B, Clifford D, Collier AC, Gelman BB, McArthur JC, McCutchan JA, Morgello S, Simpson D, Grant I, Ellis RJ (2008) Validation of the CNS penetration-effectiveness rank for quantifying antiretroviral penetration into the central nervous system. Arch Neurol 65(1):65–70
Lewis W, Dalakas MC (1995) Mitochondrial toxicity of antiviral drugs. Nat Med 1(5):417–422

Lichtenstein KA, Armon C, Baron A, Moorman AC, Wood KC, Holmberg SD (2005) Modification of the incidence of drug-associated symmetrical peripheral neuropathy by host and disease factors in the HIV outpatient study cohort. Clin Infect Dis 40(1):148–157

Lipman M, Breen R (2006) Immune reconstitution inflammatory syndrome in HIV. Curr Opin Infect Dis 19(1):20–25

Lortholary O, Poizat G, Zeller V, Neuville S, Boibieux A, Alvarez M, Dellamonica P, Botterel F, Dromer F, Chene G (2006) Long-term outcome of AIDS-associated cryptococcosis in the era of combination antiretroviral therapy. AIDS 20(17):2183–2191

Mah V, Vartavarian LM, Akers MA, Vinters HV (1988) Abnormalities of peripheral nerve in patients with human immunodeficiency virus infection. Ann Neurol 24(6):713–717

Manji H, Harrison MJ, Round JM, Jones DA, Connolly S, Fowler CJ, Williams I, Weller IV (1993) Muscle disease, HIV and zidovudine: the spectrum of muscle disease in HIV-infected individuals treated with zidovudine. J Neurol 240(8):479–488

Markus R, Brew BJ (1998) HIV-1 peripheral neuropathy and combination antiretroviral therapy. Lancet 352(9144):1906–1907

Martin C, Solders G, Sonnerborg A, Hansson P (2000) Antiretroviral therapy may improve sensory function in HIV-infected patients: a pilot study. Neurology 54(11):2120–2127

McArthur JC, Brew BJ (2010) HIV-associated neurocognitive disorders: is there a hidden epidemic? AIDS 24(9):1367–1370

McArthur JC, Hoover DR, Bacellar H, Miller EN, Cohen BA, Becker JT, Graham NM, McArthur JH, Selnes OA, Jacobson LP et al (1993) Dementia in AIDS patients: incidence and risk factors. Multicenter AIDS Cohort Study. Neurology 43(11):2245–2252

Melli G, Keswani SC, Fischer A, Chen W, Hoke A (2006) Spatially distinct and functionally independent mechanisms of axonal degeneration in a model of HIV-associated sensory neuropathy. Brain 129(Pt 5):1330–1338

Mhiri C, Baudrimont M, Bonne G, Geny C, Degoul F, Marsac C, Roullet E, Gherardi R (1991) Zidovudine myopathy: a distinctive disorder associated with mitochondrial dysfunction. Ann Neurol 29(6):606–614

Miller RF, Fox JD, Thomas P, Waite JC, Sharvell Y, Gazzard BG, Harrison MJ, Brink NS (1996) Acute lumbosacral polyradiculopathy due to cytomegalovirus in advanced HIV disease: CSF findings in 17 patients. J Neurol Neurosurg Psychiatry 61(5):456–460

Mintz M (2005) HIV encephalopathy: unique aspects in children and adolescents. Clinical features. The Neurology of AIDS. Oxford University Press, London

Moore RD, Wong WM, Keruly JC, McArthur JC (2000) Incidence of neuropathy in HIV-infected patients on monotherapy versus those on combination therapy with didanosine, stavudine and hydroxyurea. AIDS 14(3):273–278

Morgello S, Estanislao L, Simpson D, Geraci A, DiRocco A, Gerits P, Ryan E, Yakoushina T, Khan S, Mahboob R, Naseer M, Dorfman D, Sharp V (2004) HIV-associated distal sensory polyneuropathy in the era of highly active antiretroviral therapy: the Manhattan HIV Brain Bank. Arch Neurol 61(4):546–551

Moulignier A, Authier FJ, Baudrimont M, Pialoux G, Belec L, Polivka M, Clair B, Gray F, Mikol J, Gherardi RK (1997) Peripheral neuropathy in human immunodeficiency virus-infected patients with the diffuse infiltrative lymphocytosis syndrome. Ann Neurol 41(4):438–445

Nath A (2010) Human immunodeficiency virus-associated neurocognitive disorder: pathophysiology in relation to drug addiction. Ann N Y Acad Sci 1187:122–128

Nath A, Schiess N, Venkatesan A, Rumbaugh J, Sacktor N, McArthur J (2008) Evolution of HIV dementia with HIV infection. Int Rev Psychiatry 20(1):25–31

Navia BA, Jordan BD, Price RW (1986) The AIDS dementia complex: I Clinical features. Ann Neurol 19:517–524

Ovbiagele B, Nath A (2011) Increasing incidence of ischemic stroke in patients with HIV infection. Neurology 76:444–450

Paice JA, Ferrans CE, Lashley FR, Shott S, Vizgirda V, Pitrak D (2000) Topical capsaicin in the management of HIV-associated peripheral neuropathy. J Pain Symptom Manage 19(1):45–52

Park BJ, Wannemuehler KA, Marston BJ, Govender N, Pappas PG, Chiller TM (2009) Estimation of the current global burden of cryptococcal meningitis among persons living with HIV/AIDS. AIDS 23(4):525–530

Petito CK, Navia BA, Cho ES, Jordan BD, George DC, Price RW (1985) Vacuolar myelopathy pathologically resembling subacute combined degeneration in patients with the acquired immunodeficiency syndrome. N Engl J Med 312(14):874–879

Pettersen JA, Jones G, Worthington C, Krentz HB, Keppler OT, Hoke A, Gill MJ, Power C (2006) Sensory neuropathy in human immunodeficiency virus/acquired immunodeficiency syndrome patients: protease inhibitor-mediated neurotoxicity. Ann Neurol 59(5):816–824

Price RW, Brew BJ (1988) The AIDS dementia complex. J Infect Dis 158(5):1079–1083

Rabing Christensen E, Stegger M, Jensen-Fangel S, Laursen AL, Ostergaard L (2004) Mitochondrial DNA levels in fat and blood cells from patients with lipodystrophy or peripheral neuropathy and the effect of 90 days of high-dose coenzyme Q treatment: a randomized, double-blind, placebo-controlled pilot study. Clin Infect Dis 39(9):1371–1379

Rizzuto N, Cavallaro T, Monaco S, Morbin M, Bonetti B, Ferrari S, Galiazzo-Rizzuto S, Zanette G, Bertolasi L (1995) Role of HIV in the pathogenesis of distal symmetrical peripheral neuropathy. Acta Neuropathol (Berl) 90(3):244–250

Robertson KR, Su Z, Margolis DM, Krambrink A, Havlir DV, Evans S, Skiest DJ (2010) Neurocognitive effects of treatment interruption in stable HIV-positive patients in an observational cohort. Neurology 74(16):1260–1266

Ruttimann S, Hilti P, Spinas GA, Dubach UC (1991) High frequency of human immunodeficiency virus-associated autonomic neuropathy and more severe involvement in advanced stages of human immunodeficiency virus disease. Arch Intern Med 151(12):2441–2443

Ryan EL, Morgello S, Isaacs K, Naseer M, Gerits P (2004) Neuropsychiatric impact of hepatitis C on advanced HIV. Neurology 62(6):957–962

Saag MS, Powderly WG, Cloud GA, Robinson P, Grieco MH, Sharkey PK, Thompson SE, Sugar AM, Tuazon CU, Fisher JF et al (1992) Comparison of amphotericin B with fluconazole in the treatment of acute AIDS-associated cryptococcal meningitis. The NIAID Mycoses Study Group and the AIDS Clinical Trials Group. N Engl J Med 326(2):83–89

Schifitto G, Yiannoutsos C, Simpson DM, Adornato BT, Singer EJ, Hollander H, Marra CM, Rubin M, Cohen BA, Tucker T, Koralnik IJ, Katzenstein D, Haidich B, Smith ME, Shriver S, Millar L, Clifford DB, McArthur JC (2001) Long-term treatment with recombinant nerve growth factor for HIV-associated sensory neuropathy. Neurology 57(7):1313–1316

Schifitto G, McDermott MP, McArthur JC, Marder K, Sacktor N, Epstein L, Kieburtz K (2002) Incidence of and risk factors for HIV-associated distal sensory polyneuropathy. Neurology 58 (12):1764–1768

Schifitto G, McDermott MP, McArthur JC, Marder K, Sacktor N, McClernon DR, Conant K, Cohen B, Epstein LG, Kieburtz K (2005) Markers of immune activation and viral load in HIV-associated sensory neuropathy. Neurology 64(5):842–848

Schouten J, Cinque P, Gisslen M, Reiss P, Portegies P (2011) HIV-1 infection and cognitive impairment in the cART era: a review. AIDS 25(5):561–575

Seidman R, Peress NS, Nuovo GJ (1994) In situ detection of polymerase chain reaction-amplified HIV-1 nucleic acids in skeletal muscle in patients with myopathy. Mod Pathol 7(3):369–375

Seifert F, Struffert T, Hildebrandt M, Blumcke I, Bruck W, Staykov D, Huttner HB, Hilz MJ, Schwab S, Bardutzky J (2008) In vivo detection of hepatitis C virus (HCV) RNA in the brain in a case of encephalitis: evidence for HCV neuroinvasion. Eur J Neurol 15(3):214–218

Sexually transmitted diseases treatment guidelines 2002. Centers for Disease Control and Prevention (2002). MMWR Recomm Rep 51 (RR-6):1–78

Shelburne SA, Visnegarwala F, Darcourt J, Graviss EA, Giordano TP, White AC Jr, Hamill RJ (2005) Incidence and risk factors for immune reconstitution inflammatory syndrome during highly active antiretroviral therapy. AIDS 19(4):399–406

Shlay JC, Chaloner K, Max MB, Flaws B, Reichelderfer P, Wentworth D, Hillman S, Brizz B, Cohn DL (1998) Acupuncture and amitriptyline for pain due to HIV-related peripheral neuropathy: a randomized controlled trial. Terry Beirn Community Programs for Clinical Research on AIDS. JAMA 280(18):1590–1595

Shoham S, Cover C, Donegan N, Fulnecky E, Kumar P (2005) Cryptococcus neoformans meningitis at 2 hospitals in Washington, D.C.: adherence of health care providers to published practice guidelines for the management of cryptococcal disease. Clin Infect Dis 40(3):477–479

Simpson DM, Citak KA, Godfrey E, Godbold J, Wolfe DE (1993) Myopathies associated with human immunodeficiency virus and zidovudine: can their effects be distinguished? Neurology 43(5):971–976

Simpson DM, Dorfman D, Olney RK, McKinley G, Dobkin J, So Y, Berger J, Ferdon MB, Friedman B (1996) Peptide T in the treatment of painful distal neuropathy associated with AIDS: results of a placebo-controlled trial. The Peptide T Neuropathy Study Group. Neurology 47(5):1254–1259

Simpson DM, Slasor P, Dafni U, Berger J, Fischl MA, Hall C (1997) Analysis of myopathy in a placebo-controlled zidovudine trial. Muscle Nerve 20(3):382–385

Simpson DM, McArthur JC, Olney R, Clifford D, So Y, Ross D, Baird BJ, Barrett P, Hammer AE (2003) Lamotrigine for HIV-associated painful sensory neuropathies: a placebo-controlled trial. Neurology 60(9):1508–1514

Simpson DM, Estanislao L, Evans S, McArthur J, Marcus K, Truffa M, Lucey B, Naismith R, Longergan JL, Clifford D (2004) HIV-associated neuromuscular weakness syndrome. AIDS 18(10):1403–1412

Simpson DM, Brown S, Tobias J (2008) Controlled trial of high-concentration capsaicin patch for treatment of painful HIV neuropathy. Neurology 70(24):2305–2313

Simpson DM, Schifitto G, Clifford DB, Murphy TK, Durso-De Cruz E, Glue P, Whalen E, Emir B, Scott GN, Freeman R (2010) Pregabalin for painful HIV neuropathy: a randomized, double-blind, placebo-controlled trial. Neurology 74(5):413–420

Skiest DJ, Crosby C (2003) Survival is prolonged by highly active antiretroviral therapy in AIDS patients with primary central nervous system lymphoma. AIDS 17(12):1787–1793

Snider WD, Simpson DM, Nielsen S, Gold JW, Metroka CE, Posner JB (1983) Neurological complications of acquired immune deficiency syndrome: analysis of 50 patients. Ann Neurol 14(4):403–418

So YT, Olney RK (1994) Acute lumbosacral polyradiculopathy in acquired immunodeficiency syndrome: experience in 23 patients. Ann Neurol 35(1):53–58

Tagliati M, Grinnell J, Godbold J, Simpson DM (1999) Peripheral nerve function in HIV infection: clinical, electrophysiologic, and laboratory findings. Arch Neurol 56(1):84–89

Tan SV, Guiloff RJ (1998) Hypothesis on the pathogenesis of vacuolar myelopathy, dementia, and peripheral neuropathy in AIDS. J Neurol Neurosurg Psychiatry 65(1):23–28

Till M, MacDonell KB (1990) Myopathy with human immunodeficiency virus type 1 (HIV-1) infection: HIV-1 or zidovudine? Ann Intern Med 113(7):492–494

Tipping B, de Villiers L, Wainwright H, Candy S, Bryer A (2007) Stroke in patients with human immunodeficiency virus infection. J Neurol Neurosurg Psychiatry 78(12):1320–1324

Tozzi VBP, Libertone R, Antinori A (2010) Cognitive function in treated HIV patients. Neurobehav HIV Med 2:95–113. doi:10.2147/NBHIV.S13936

Tyor W (2009) Pathogenesis and treatment of HIV-associated dementia: recent studies in a SCID mouse model, Handbook of neurochemistry and molecular neurobiology. Springer, Berlin

Tyor WR, Middaugh LD (1999) Do alcohol and cocaine abuse alter the course of HIV-associated dementia complex? J Leukoc Biol 65(4):475–481

UNAIDS UNAIDS Report on the Global AIDS Epidemic 2010. http://www.unaids.org/en/dataanalysis/epidemiology/. Accessed 22 Mar 2011

Valcour V, Shikuma C, Shiramizu B, Watters M, Poff P, Selnes O, Holck P, Grove J, Sacktor N (2004) Higher frequency of dementia in older HIV-1 individuals: the Hawaii Aging with HIV-1 Cohort. Neurology 63(5):822–827

Valcour V, Yeh TM, Bartt R, Clifford D, Gerschenson M, Evans SR, Cohen BA, Ebenezer GJ, Hauer P, Millar L, Gould M, Tran P, Shikuma C, Souza S, McArthur JC (2009) Acetyl-l-carnitine and nucleoside reverse transcriptase inhibitor-associated neuropathy in HIV infection. HIV Med 10(2):103–110

Van Rie A, Harrington PR, Dow A, Robertson K (2007) Neurologic and neurodevelopmental manifestations of pediatric HIV/AIDS: a global perspective. Eur J Paediatr Neurol 11(1):1–9

Vendrely A, Bienvenu B, Gasnault J, Thiebault JB, Salmon D, Gray F (2005) Fulminant inflammatory leukoencephalopathy associated with HAART-induced immune restoration in AIDS-related progressive multifocal leukoencephalopathy. Acta Neuropathol (Berl) 109(4):449–455

Vermes A, Guchelaar HJ, Dankert J (2000) Flucytosine: a review of its pharmacology, clinical indications, pharmacokinetics, toxicity and drug interactions. J Antimicrob Chemother 46 (2):171–179

Vivithanaporn P, Heo G, Gamble J, Krentz HB, Hoke A, Gill MJ, Power C (2010) Neurologic disease burden in treated HIV/AIDS predicts survival: a population-based study. Neurology 75 (13):1150–1158

Vivithanaporn P, Nelles K, Deblock L, Newman SC, Gill MJ, Power C (2012) Hepatitis C virus co-infection increases neurocognitive impairment severity and risk of death in treated HIV/AIDS. J Neurol Sci 312(1–2):45–51

Wang W, Carm AR (2001) Clinical manifestations of AIDS with cryptococcal meningitis. Chin Med J (Engl) 114(8):841–843

Williams FM, Cohen PR, Jumshyd J, Reveille JD (1998) Prevalence of the diffuse infiltrative lymphocytosis syndrome among human immunodeficiency virus type 1-positive outpatients. Arthritis Rheum 41(5):863–868

Yoritaka A, Ohta K, Kishida S (2005) Herpetic lumbosacral radiculoneuropathy in patients with human immunodeficiency virus infection. Eur Neurol 53(4):179–181

Youle M, Osio M (2007) A double-blind, parallel-group, placebo-controlled, multicentre study of acetyl L-carnitine in the symptomatic treatment of antiretroviral toxic neuropathy in patients with HIV-1 infection. HIV Med 8(4):241–250

Zellan J, Augenbraun M (2004) Syphilis in the HIV-infected patient: an update on epidemiology, diagnosis, and management. Curr HIV/AIDS Rep 1(3):142–147

Human T-Cell Lymphotropic Virus Type 1 Infection

Steven Jacobson and Raya Massoud

Abstract Human T-cell lymphotropic virus type 1 (HTLV-1) was the first retrovirus to be identified as pathogenic in humans. Since its discovery in 1980 and its association with a progressive myelopathy termed HTLV-1 associated myelopathy/tropical spastic paraparesis (HAM/TSP) in 1985, there has been accumulating evidence that HTLV-1 can cause multiple neurologic conditions, suggesting a wider impact of this infection on the human nervous system. In this review we describe the clinical presentation of the various HTLV-1-associated neurologic conditions with a focus on HAM/TSP. We also discuss the current status of knowledge regarding the immunopathogenesis, pathology, and radiologic findings in HAM/TSP. Finally, we highlight the therapeutic strategies that have been taken to date and discuss the clinical trials in HAM/TSP.

Keywords HTLV-1 • retrovirus • HTLV-1 associated myelopathy (HAM/TSP) • taxonomy • cell-to-cell transmission • immunopathogenesis • Tax • HBZ • pathology • spinal cord atrophy • myopathy • neuropathy • cognitive deficits • imunomodulation • antiviral

1 The Virus

1.1 Discovery

Human T-cell lymphotropic virus-1 (HTLV-1) was the first human oncoretrovirus to be discovered. It was first isolated in 1979 by Poiesz et al. ((Rho et al. 1981) from

S. Jacobson (✉) • R. Massoud
Viral Immunology Section, Neuroimmunology Branch, National Institute of Neurological Disorders and Stroke, National Institutes of Health, Bethesda, MD 20892, USA
e-mail: jacobsons@ninds.nih.gov; massoudr@ninds.nih.gov

A.C. Jackson (ed.), *Viral Infections of the Human Nervous System*, Birkhäuser Advances in Infectious Diseases,
DOI 10.1007/978-3-0348-0425-7_8,

peripheral blood lymphocytes (PBLs) of a patient thought to have cutaneous T-cell lymphoma but subsequently diagnosed with adult T-cell leukemia/lymphoma (ATLL). Shortly thereafter, Yoshida and Yamamoto (1986) isolated a retrovirus from a T-cell line of a patient with ATLL in Japan and named it ATLV. The similarity of the two viruses was quickly confirmed by comparing their genomes and the name HTLV-1 was agreed on. Thus a human equivalent of the animal oncogenic retroviruses was finally discovered, validating decades of research and clarifying the cause of ATLL, a leukemia with epidemiological characteristics that have long suggested a strong environmental factor.

Gessain et al. (1985) first suggested an association between HTLV-1 and tropical spastic paraparesis (TSP), a myelopathy endemic to the tropics and recognized since 1969. Independently, Osame and collegues in Japan recognized a similar syndrome in HTLV-1 seropositive subjects and named it HTLV-1-associated myelopathy (HAM) (Kitajima et al. 1988). In 1988 a consensus conference by the World Health Organization recommended the hybrid name HAM/TSP and set extensive diagnostic criteria (World Health Organization 1988).

Since then, HTLV-1 infection has been associated with a wide spectrum of inflammatory conditions such as dermatitis, arthritis, interstitial pneumonitis, uveitis, sicca syndrome, as well as an increasing number of neurologic manifestations and a higher prevalence of infections (tuberculosis, strongyloidosis, scabies, infectious dermatitis) in seropositive subjects (Marinho et al. 2005).

The discovery of HTLV-2, a closely related retrovirus, came 2 years later when Kalyanaraman et al. (1982) isolated it from splenic cells of an individual with hairy T-cell leukemia. Subsequently, it was found to be endemic in native Amerindian populations, Central/West Africa and epidemic among intravenous drug users (IVDUs) in the United States and Europe (Roucoux and Murphy 2004). However, its association with human neurologic disease remains controversial and will not be discussed here.

1.2 Structure, Tropism, and Replication

1.2.1 Taxonomy

HTLV-1 belongs to the Deltaretrovirus genus of the Retroviridae family (Fig. 1). This genus includes the genomically similar HTLV-2, bovine leukemia virus (BLV), simian T-cell leukemia viruses (STLV-1, -2, -3), HTLV-3, and HTLV-4, all of which are characterized by an extra genomic region pX between the envelope gene (env) and the 3′ long terminal repeat (3′ LTR).

Unlike other retroviruses, HTLV-1 is very stable genetically. This is thought to result from its mode of replication as the virus uses the cellular DNA polymerase, which is more reliable than viral reverse transcriptase (RT), to duplicate its genome. Still, mild variations in the restriction maps of the LTR led to the identification of four slightly different genotypes reflecting the geographic origin of the strain. These are the cosmopolitan (C), Japanese (J), African (A), and Melanesian (M) subtypes. Notably,

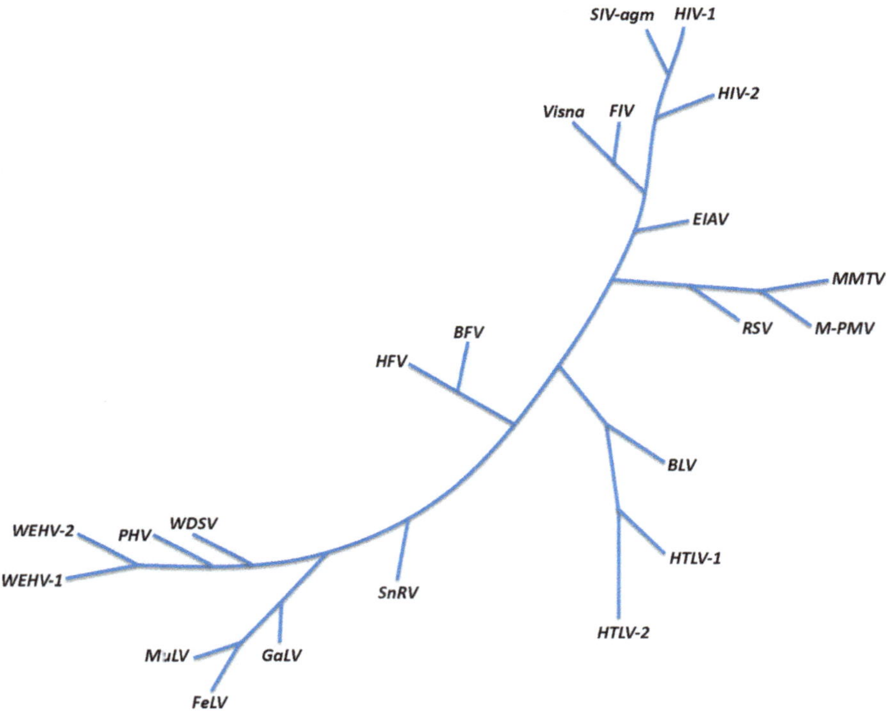

Fig. 1 Retroviral phylogenetic tree based on analysis of the reverse transcriptase gene. Bovine leukemia virus (BLV), Human T-cell lymphotropic virus-1 and -2 (HTV-1 and HTLV-2), Human immunodeficiency virus-1 and -2 (HIV-1 and HIV-2), Simian immunodeficiency virus of African green monkey (SIV-agm), Equine infectious anemia virus (EIAV), Feline immunodeficiency virus (FIV), Mouse mammary tumor virus (MMTV), Mason-Pfizer monkey virus (M-MPV), Rous sarcoma virus (RSV), Bovine foamy virus (BFV), human foamy retrovirus (HFV), snakehead retrovirus (snRV), Walleye dermal sarcoma virus (WDSV), Walleye epidermal hyperplasia virus-1 and 2 (WEHV-1,-2), Murine leukemia virus (MuLV), Feline leukemia virus (FeLV), Gibbon ape leukema virus (GaLV)

these isolates cluster closely with simian isolates in the same geographic area, suggesting simian-to-human transmission and an African origin, as Africa is the only continent where all the HTLVs and STLVs have been found (Van Dooren et al. 2001).

1.2.2 Microscopic Structure

The virion is an enveloped, spherical, 80–110 nm diameter particle containing two molecules of positive (sense) single-stranded RNA (ssRNA). The envelope is a lipid bilayer derived from cellular membrane and embedded with viral glycoproteins (gp46 and gp21). It is internally lined by the viral matrix protein (p24), which encases the nucleocapsid (p15) that carries the RNA and a functional reverse transcriptase (RT), protease and integrase (IN).

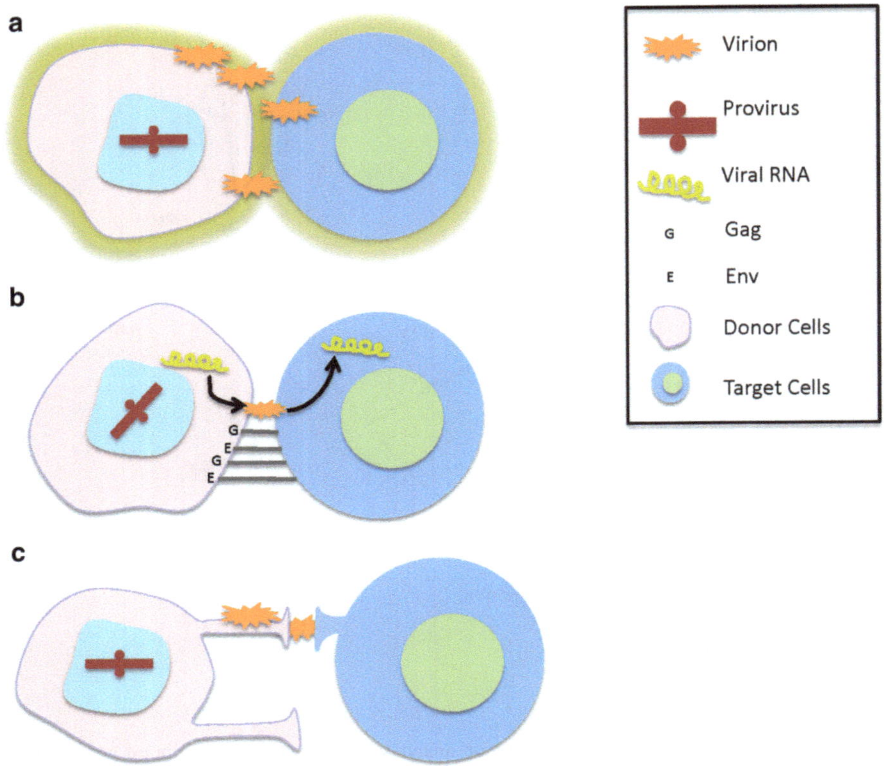

Fig. 2 Cell-to-cell transmission of HTLV-1. (**a**) Biofilm like extracellular assemblies trap the virion at the surface of an infected cell and rapidly transfer it to a target cell after conjugation. (**b**) Virological synapse formation at the point of contact. (**c**) Intercellular conduit

1.2.3 Receptor and Infectivity

Unlike HIV, the HTLV-1 free virion is poorly infectious and transmission requires live infected cells. Several distinct mechanisms of cell-to-cell transmission have been described and in vivo the virus probably combines multiple strategies (Fig. 2):

1. Virological synapses (VS): These are virus induced, specialized cell–cell contacts composed of cellular and viral molecules; they optimize transmission by protecting the virus from the surrounding environment. Tax plays an important role in their formation by causing polarization of the cytoskeleton (Igakura et al. 2003; Pais-Correia et al. 2010).
2. Extracellular viral assemblies (biofilm): After viral budding the infected cell will retain viral particles at its surface, trapped in complexes of collagen, agrin, and linker proteins. These can be rapidly transferred to the surface of an uninfected target cell upon contact. They are the viral equivalent of bacterial and fungal biofilms (Pais-Correia et al. 2010).

3. Role of Protein 8 (p8): p8 is a viral protein encoded in the pX region and generated by processing of p12; it was recently proven to enhance T-cell conjugation by interacting with lymphocyte function antigen-1 (LFA-1) and ICAM-1 and can also lead to the formation of intercellular conduits that transmit the virus. (Van Prooyen et al. 2010).

Following the establishment of cell-to-cell contact by one of the previous mechanisms, HTLV Env will fuse with the target cell's membrane and viral material is transferred. Since in vitro HTLV-1 can infect multiple cell types, it has long been suggested that its receptor is a ubiquitous molecule (Trejo and Ratner 2000). Currently there is evidence for a glucose transporter (GLUT-1), heparan sulfate protegoglycan (HSPG) and neuropilin-1 as important mediators of this virus–cell membrane interaction (Jones et al. 2005).

Despite a broad range of target cells in vitro (human T- and B-lymphocytes, fibroblasts, and cells from monkeys, rats, rabbits, and hamsters, but not from mice), in vivo HTLV-1 mainly infects T-cells ($CD4^+$ and $CD8^+$) and its reservoir is in the memory/effector $CD4^+$ T-cells and T-regulatory cells (Richardson et al. 1990). Dentritic cells (DC) can also be infected and were found to have viral assemblies on their surface suggesting that they can transmit the virus to the T-lymphocyte during antigen-presenting cell (APC)–lymphocyte interaction (Jones et al. 2008).

1.2.4 Molecular Structure and Replication

After entry into the target cell, the viral capsid dissociates releasing its contents into the cytoplasm. Viral RNA is then retro transcribed into dsDNA by the RT and transported into the nucleus where it is integrated into the host DNA via IN, thus becoming a provirus.

The proviral genome is 9 kb and encodes structural proteins (Gag: capsid, nucleocapsid, and matrix—Env), regulators of viral expression (Tax and Rex), enzymes [IN, RT, protease (pro)], and accessory proteins (HBZ, p12, p30, p13) (Fig. 3). The accessory and regulatory proteins are encoded from the pX region by alternative splicing and internal initiation codons. pX contains four open reading frames (ORFs) coding for p12 (ORF-1), p13 (ORF-2), p30 (ORF-3), and Tax (ORF-4). Enhancer elements for transcription initiation are contained in the LTR and the polyadenylation signal for positive strand transcription is in the 3′ LTR. The structural proteins and enzymes are encoded from unspliced or singly spliced mRNAs whereas the regulatory and accessory proteins are transcribed from alternatively spliced transcripts.

Tax and Rex regulate positive strand transcription: Tax controls transcription and Rex regulates nucleo-cytoplasmic transport of the unspliced and partially spliced mRNA. It is thought that early transcription is Rex independent and involves multiple spliced mRNA, whereas late phase transcription is Rex dependent and consists of unspliced mRNA. In the initial stages of the infection, translation of Tax is favored over that of Rex resulting in the export of spliced

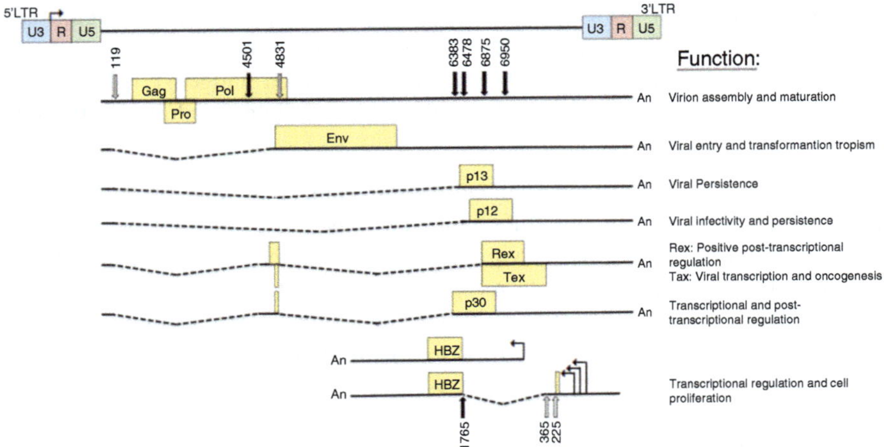

Fig. 3 The HTLV-1 proviral genome and main products. *Solid lines* indicate exons; *dotted lines* indicate introns. The name of the gene product is depicted inside each *box*. The key functions for each protein are listed to the *right*

viral mRNA (Green and Chen 1990). Eventually there is accumulation of sufficient levels of Rex resulting in cytoplasmic transport of incompletely spliced mRNA that encodes structural and enzymatic proteins and leads to viral assembly. Thus Rex might be a critical regulator of the switch between latent and productive infection.

Tax is a pleiotropic transcription factor that drives the transcription of all HTLV-1 genes from the LTR by recruiting members of the CRE-binding/activating transcription factors (CREB/ATF) family of viral promoter. It can also activate many cellular transcription factors such as the nuclear factor-κB (NF-κB) and activator protein-1 (AP-1) promoting cellular proliferation and survival. These functions, in addition to the fact that Tax contains immunodominant epitopes for MHC class I restricted cytotoxic T-cell responses, suggest that Tax is implicated in HAM/TSP pathogenesis.

P30 promotes early viral spread and T-cell survival as its expression in Jurkat cells results in the activation of the G2-M cell cycle checkpoint (Datta et al. 2007). Additionally, p30 can act as a negative posttranscriptional regulator of HTLV-1 expression by binding and retaining the doubly spliced Tax/rex mRNA in the nucleus (Datta et al. 2007). This effect is in opposition to the positive posttranscriptional regulation mediated by Rex. The two proteins are maintained in equilibrium by a feedback loop allowing viral persistence and evasion from the immune system (Baydoun et al. 2008).

P13 localizes in the mitochondria and interferes with Ras and Myc oncogenes to suppress tumor growth (Hiraragi et al. 2005). It sensitizes the infected cells to FasL and C-2 ceramide-induced apoptosis. These effects on the apoptosis are opposite to the effects of Tax.

P12 is required for efficient infection of primary T-cells and syncytial formation; it interacts with interleukin-2 receptor beta and gamma chains leading to activation of the Janus kinase/signal transducer and activator of transcription 5 (JAK/STAT5) (Albrecht et al. 2000). It also promotes immune evasion by inducing proteasomal degradation of the newly synthesized MHC class I molecules (Johnson et al. 2001). By inducing LFA-1 clustering on the T-cells, p12 is implicated in the cell-to-cell transmission of the virus. Finally, in vivo animal models have demonstrated that it is essential for viral persistence (Kim et al. 2006).

The HBZ gene is the only viral gene that is consistently expressed in all HTLV-1 infected cells, suggesting that it might be critical for viral persistence and pathogenesis. It is transcribed from the negative strand of the proviral genome into two different forms: spliced and unspliced HBZ (sHBZ and usHBZ). The sHBZ expression predominates and sHBZ RNA rather than usHBZ RNA promotes the growth of T-cells. The regulation of its transcription remains unclear; notably 90 % of HBZ mRNA is retained in the nucleus.

Like Tax, HBZ can affect the transcription of many cellular and viral genes. It interacts with cAMP response element binding (CREB) and CREB binding protein (CBP)/p300 to suppress Tax-mediated viral gene transcription. It can also interact and modulate the activity of many AP-1 transcriptional family members (c-Jun, JunB, JunD) via its bZIP domain. In particular, HBZ decreases c-Jun and Jun-B mediated transcription by decreasing their DNA binding and by enhancing degradation of c-Jun and its sequestration into nuclear bodies. In contrast, HBZ activates JunD-mediated transcription by forming heterodimers with JunD via its bZIP domain (Matsuoka 2010).

Notably, HBZ can inhibit the classical NF-κB pathway by inhibiting DNA binding of p65 and increasing the expression of PDLIM2, the E3 ubiquitin ligase of p65, leading to increased degradation of p65. It was also noted that HBZ expression is associated with a phenotype of regulatory T-cells. In one study by Saito et al. (2009), sHBZ RNA levels correlated with the PVL and disease severity in HAM/TSP patients; however, its role in that context remains to be further studied.

1.3 Epidemiology

Worldwide it is estimated that 15–20 million persons are infected with HTLV-1 (de The and Kazanji 1996; Proietti et al. 2005). Geographically there are clusters of highly endemic regions contiguous to areas that are almost completely free of the virus (Fig. 4). This distribution argues for a founder effect followed by transmission through close contact such as sexual intercourse and breastfeeding. The highly endemic areas are southwestern Japan (5–15 % seropositive), the Caribbean islands, foci in South America (Colombia, French Guyana, parts of Brazil), equatorial Africa, Papua New Guinea, and the Solomon Islands, some areas of the Middle East (Mashhad region in Iran) and isolated clusters in Melanesia. In North America and Australia the highest prevalence is among the indigenous populations; in native

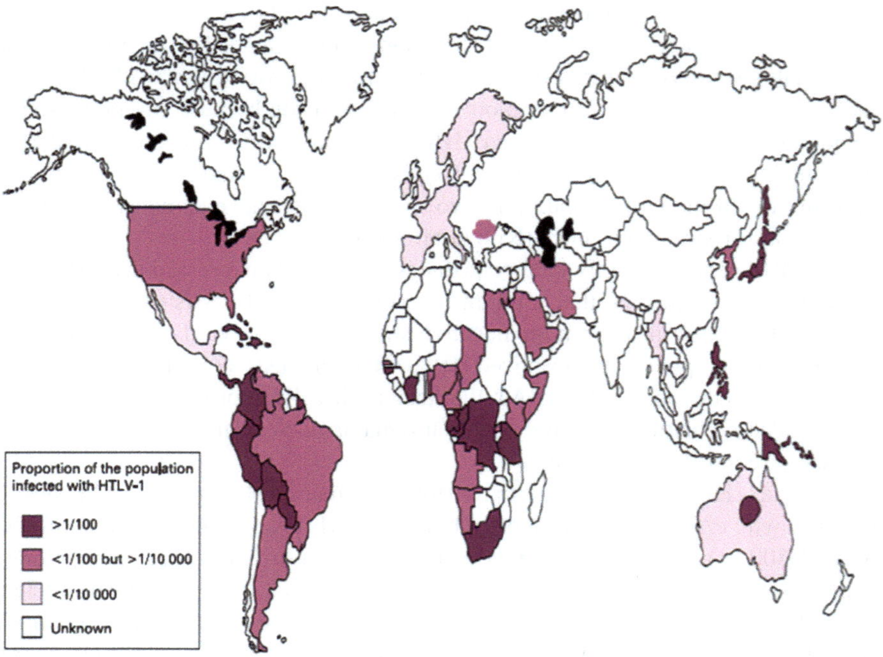

Fig. 4 Endemic distribution of HTLV-1. Reproduced from Cooper et al (2009) with permission

Europeans the highest prevalence is in Romania. Notably, seroprevalence increases with age, peaking between 40 and 60 years, and is slightly higher in women.

Transmission of HTLV-1 requires transfer of infected T-cells and occurs via one of three routes: mother to child by breastfeeding, sexual contact, and intravenous exposure to infected cells.

1. Mother to child by breastfeeding: the major mode of transmission in endemic areas with up to 30 % rate of transmission if breastfeeding is prolonged for greater than 6 months. Less than 5 % of vertical transmission occurs in utero or at birth (Furnia et al. 1999).
2. Sexual contact: in contrast to HIV, the transmission is mainly from male to female and is thought to be mediated by infected T-cells in semen.
3. Intravenous exposure to infected lymphoid cells: blood transfusion will transmit the virus very efficiently with reported rates up to 70 %.

Epidemiologically, the route of infection has been linked to disease outcome: vertical transmission is a risk factor for developing ATLL later in life and sexual or blood mediated infection predisposes for HAM/TSP. Immunologically this has been explained by the observation that oral inoculation of the virus leads to T-cell tolerance and a higher proviral load (PVL), typical findings in ATLL patients (Uchiyama et al. 1977; Fujino and Nagata 2000).

2 HTLV-1 Associated Neurologic Disorders

2.1 Introduction

The clinical outcome of HTLV-1 infection is highly variable and depends mainly on host immunologic and genetic factors rather than on viral variants. It is currently accepted that the majority of infected subjects remain lifetime asymptomatic carriers (AC), 3–5 % develop ATLL, and 2–3 % develop HAM/TSP. However with the recognition of a wider spectrum of systemic and neurologic manifestations the real proportion of AC is probably closer to 70 %.

Neurologically the first recognized outcome was a progressive myelopathy (HAM/TSP). Multiple prospective natural history studies have shown that HTLV-1 can cause more widespread neurologic damage resulting in a variety of clinical syndromes such as a myopathy, peripheral neuropathy (PN), autonomic dysfunction, motor-neuron disease (ALS-like syndrome), and cognitive dysfunction. Some experts refer to this constellation as the HTLV-1 "neurologic complex," but it remains to be seen whether isolated syndromes represent a new neurologic outcome or an early/subclinical presentation of HAM/TSP (Araujo and Silva 2006).

2.2 HTLV-1 Associated Myelopathy/Tropical Spastic Paraparesis

2.2.1 Clinical Presentation and Diagnosis

HAM/TSP is a progressive myelopathy occurring in up to 3 % of HTLV-1 infected subjects. The diagnostic guidelines were first set in 1988 by the World Health Organization (1988) that outlined a diffuse neurologic dysfunction rather than a pure myelopathy. In 2006, a panel of experts refined the criteria by adding levels of ascertainment to include patients who might be presenting early in the course or with an incomplete evaluation and not fulfilling all the criteria (Table 1) (De Castro-Costa et al. 2006).

Clinically, the most common initial symptom in HAM/TSP (in 60 %) is weakness of the lower extremities. The onset can be asymmetric, but the weakness usually progresses to a symmetrical spastic gait. Other common symptoms include back pain, paresthesias of the lower extremities, constipation, and sexual dysfunction. Bladder dysfunction is also very common and can predate the weakness by many years (Araujo et al. 1998).

On neurologic examination the prominent findings are pyramidal in type: proximal weakness of the lower extremities with hyperreflexia, Babinski sign, and a spastic tone. In the upper extremities strength is often normal, but hyperreflexia and Hoffmann's sign are commonly found. In contrast to the motor findings, the sensory signs are usually subtle with mild loss of vibration and preserved proprioception. A sensory level is rare, but can be detected in patients with a rapid course.

Table 1 Levels of ascertainment for the diagnosis of HAM/TSP

Definite	Probable	Possible
1. A nonremitting progressive spastic paraparesis with sufficiently impaired gait to be perceived by the patient. Sensory symptoms or signs may or may not be present. When present, they remain subtle and without a clear-cut sensory level. Urinary and anal sphincter signs or symptoms may or may not be present	1. Monosymptomatic presentation: spasticity or hyperreflexia in the lower limbs or isolated Babinski sign with or without subtle sensory signs or symptoms, or neurogenic bladder only confirmed by urodynamic tests	1. Complete or incomplete clinical presentation
2. Presence of HTLV-I antibodies in serum and CSF confirmed by Western blot and/or a positive PCR for HTLV-I in blood and/or CSF	2. Presence of HTLV-I antibodies in serum and/or CSF confirmed by Western blot and/or a positive PCR for HTLV-I in blood and/or CSF	2. Presence of HTLV-I antibodies in serum and/or CSF confirmed by Western blot and/or a positive PCR for HTLV-I in blood and/or CSF
3. Exclusion of other disorders that can resemble TSP/HAM	3. Exclusion of other disorders that can resemble TSP/HAM	3. Disorders that can resemble TSP/HAM have not been excluded

Reproduced from De Castro-Costa et al. (2006) with permission

Disability ultimately results from advancing age, weakness of the lower extremities, and back pain with gait and sphincter dysfunction being of major impact (Franzoi and Araujo 2007).

Typically, the onset is insidious with relentless progression leading to disability within 2 years of onset. The rate of progression is variable and seems to be faster in women with premenopausal onset of disease, in subjects with higher proviral loads and in immunosuppressed individuals contracting the infection by transfusion. A rapidly progressive form with acute to subacute onset and complete paraplegia in less than 2 years can also occur (Yamashita, Ueda et al. 2011).

Many conditions can mimic HAM/TSP and should be excluded by appropriate laboratory, radiological, and clinical evaluations (Table 2). It is particularly difficult to distinguish HAM/TSP from the myelopathy of primary progressive multiple sclerosis (PPMS); in this context, a ratio of HTLV-1 PVL in the CSF to PVL in the blood >2 usually suggests the diagnosis of HAM/TSP rather than multiple sclerosis.

2.2.2 Pathology of HAM/TSP

Histopathologic analysis has been an invaluable tool in understanding the pathogenesis, extent, and dynamics of HAM/TSP. Overall, the findings point to an early inflammatory stage followed by late degeneration that evolve simultaneously in the brain and spinal cord and predominate in watershed areas (spine: mid-lower

Table 2 Differential diagnosis of HAM/TSP

Structural lesions	Inflammatory	Nutritional	Infectious	Uncertain etiology	Genetic
• Spinal cord compression	• Multiple sclerosis	• Vitamin B12 deficiency	• Neuroschistosomiasis	• HTLV-1 negative TSP (in endemic areas)	• Familial spastic paraparesis
• Syringomyelia	• Idiopathic transverse myelitis (TM)	• Lathyrism	• Neurocysticercosis	• Primary lateral sclerosis	
• Spinal arteriovenous fistula	• TM associated with collagen vascular diseases	• Konzo	• Lyme disease	• Vacuolar myelopathy of AIDS	
	• Sjogren's syndrome		• Neurosyphilis		
	• Behcet disease		• Neurotuberculosis		
	• Paraneoplastic myelopathy		• Fungal myelopathy		

thoracic region, brain: deep white matter and junction of the cortex-white matter). This nonrandom distribution probably reflects slow blood flow in affected areas favoring transmigration of inflammatory cells that are thought to be pathogenic. The pathology predominates in the lateral columns of the thoracic cord (corticospinal, spinocerebellar, and spinothalamic tracts), whereas the anterior and posterior columns are less severely affected, a finding consistent with the semiology of the disease.

HAM/TSP lesions consist of axonal loss, demyelination, reactive astrocytosis, and fibrillary gliosis in close association with a perivascular/parenchymal inflammatory process. The inflammatory component evolves during the course of the disease allowing a classification of the lesions as active-chronic or inactive-chronic lesions (Umehara et al. 1993; Aye et al. 2000).

Early in the course (less than 5 years) the infiltrates consist of an equal number of $CD4^+$, $CD8^+$ T-cells, and foamy macrophages and show increased expression of MHC class I molecules and β2 microglobulin on the infiltrating cells and endothelial cells. Also there is increased expression of MHC class II molecules on endothelial cells, microglia, and mononuclear cells. Immunohistochemistry shows increased inflammatory cytokines (TNF-α, IL-1β, and INF-γ) on the infiltrating macrophages, astrocytes, and microglia.

Later in the course (more than 8 years) the lesions become less inflammatory with predominance of $CD8^+$ T-cells, upregulation of HLA class I, and downregulation of all the proinflammatory cytokines except for INF-γ.

Using the monoclonal antibody TIA1 [which recognizes cytolytic-granule associated protein, a marker of cytolytic T-cells (CTL) and NK cells], Umehara et al. (1994) demonstrated that many CTLs exist in the active-chronic lesions, but decrease with disease duration. Notably, their number correlated with HTLV-1 PVL in the lesion suggesting that they were HTLV-1-specific CTLs. Furthermore, a high rate of apoptosis, especially affecting the T-helper inducer cells ($CD3^+$, $OPD4^+$, $CD45RO^+$ T-cells) was found in the early lesions and decreased with chronicity. Together these findings suggest that early in the disease, HTLV-1-specific CTLs are recruited into the lesions and eliminate HTLV-1-infected Th cells by apoptosis. At a later stage the inflammation becomes more sterile.

Additionally, the infiltrating mononuclear cells have increased expression of very late antigen four (VLA-4) and the endothelial cells show increased expression of vascular adhesion molecule type 1 (VCAM-1) (Umehara et al. 1996). Given that in vitro VCAM-1-mediated matrix metalloproteinase type 2 (MMP-2) induction was upregulated in HAM/TSP T-lymphocytes, it is possible that VCAM-1/VLA-4 interaction, followed by MMP secretion is an important mechanism of lesion formation in HAM/TSP.

Finally, axonal damage is an early phenomenon in HAM/TSP and is closely associated with the inflammatory process. Also MMP 2 and 9, which are important mediators of blood–brain barrier breakdown in transmigration of lymphocytes, are expressed on infiltrating cells and HAM/TSP patients have higher CSF levels of MMP9 and tissue inhibitors of MMP (TIMP) type 3 than AC (Kambara et al. 1999).

2.2.3 Radiologic Abnormalities in HAM/TSP

MRI studies in patients with HAM/TSP show a high prevalence of brain and spinal cord abnormalities (Godoy et al. 1995; Howard et al. 2003; Bagnato et al. 2005). In the brain, both HTLV-1AC and HAM/TSP patients have a high prevalence of white matter lesions (33–100 %); these are thought to represent chronic perivascular inflammation and are of unclear clinical significance. In particular, they cannot distinguish between AC and HAM/TSP patients and in the latter group they did not correlate with the degree of disability (Morgan et al. 2007).

In the spinal cord, thoracic cord atrophy without an abnormal signal is accepted to be the characteristic finding in HAM/TSP (Alcindor et al. 1992; Bagnato et al. 2005). However, its incidence has varied from 20 to 74 % and unlike in multiple sclerosis in which brain and spinal cord atrophy are associated with disability and a poor response to immunomodulatory treatment, in HAM/TSP cord atrophy did not correlate with disease duration, severity, or response to INF-α (Yukitake et al. 2008).

T-2 hyperintensities and contrast enhancing lesions have also been shown to occur in a subset of HAM/TSP patients (<10 %), which may indicate a more malignant form of the disease characterized by rapid progression and severe motor impairment (Yukitake et al. 2008).

2.2.4 Risk Factors for HAM/TSP

Three fundamental questions concerning HTLV-1-associated disease remain unanswered: (1) Why only a small proportion of infected subjects will ever develop disease? (2) How can the outcome of HTLV-1 infection be so different (HAM/TSP vs. ATLL)? (3) What accounts for the long latency period?

Answers to these questions will further clarify the pathogenesis of HTLV-1-associated diseases; allow risk stratification and, hopefully, the development of prophylactic and therapeutic strategies.

Viral variants: It is still unclear if specific variants of HTLV-1 are preferentially associated with HAM/TSP. This hypothesis is supported by a twin study demonstrating different HTLV-1 variants in monozygotic twins with discordant clinical outcomes (Nakane et al. 2000). Also in a study by Furukawa et al. (2000) a Tax variant of the cosmopolitan A subgroup was associated with HAM/TSP. Since Tax is a strong transactivator of many host genes, including inflammatory cytokines and a dominant epitope recognized by specific CTLs, it is plausible that a particular variant can affect disease outcome.

High PVL: HTLV-1 PVL is a reflection of the total number of infected cells in the peripheral blood occurring both through de novo infection and proliferation of infected cells. Many studies suggest that a high PVL is critical in the pathogenesis of HAM/TSP; the median PVL is 16-fold higher in HAM/TSP patients than in AC (but the ranges overlap) and the risk of disease rises exponentially once the PVL exceeds 1 % (Nagai et al. 1998). The PVL, however, might reflect the interplay of multiple risk factors as it is the end result of complex host–virus interactions influenced by the host genetic background, route of viral inoculation, and possibly viral variants (Nagai et al. 1998).

Host genetic background: The host genetic background is linked to disease outcome by influencing the PVL, degree of viral expression, and the resulting specific immune response. Its role in regulating the PVL is supported by a study showing that AC related to HAM/TSP patients have a higher PVL than unrelated AC (Nagai et al. 1998). Also certain HLA alleles have been reported to modulate the risk of HAM/TSP. HLA-A*02 and Cw*08 have been associated with lower risk whereas HLA-B*5401 was associated with a higher risk through their effects on PVL (Usuku et al. 1988; Jeffery et al. 1999). Some Class II alleles such as HLA-DRB1*0101 are also associated with increased risk of HAM/TSP and polymorphisms in the TNF-alpha and IL-15 genes and SDF-1 have been implicated in susceptibility.

Higher levels of HTLV-1 viral expression: HTLV-1 infected subjects develop a strong virus-specific immune response suggesting that viral expression is occurring in vivo. However, even in patients with HAM/TSP who characteristically have a stronger virus-specific immune response, viral protein expression cannot be detected in freshly isolated PBLs. HTLV-1 mRNA, on the other hand, can be detected in freshly isolated PBLs and is higher in HAM/TSP patients than in AC and even higher in HAM/TSP CSF than in PBLs.

This suppression of viral expression in vivo is known to be transient and at least partially mediated by the innate immune system, in particular type-I IFNs. It is possible that patients with HAM/TSP have higher levels of expression based on genetic factors related to the IFN system and that the level of expression in lymphocytes may differ between tissues depending on the strength of the IFN response, explaining preferential sites of involvement in HTLV-1 related inflammation.

Spontaneous lymphoproliferation (SP): SP is the ability of HTLV-1-infected peripheral blood mononuclear cells (PBMC) to spontaneously proliferate in vitro. This immunologic alteration occurs in all HTLV-1-infected cells but at a higher magnitude in patients with HAM/TSP compared to AC.

Both $CD4^+$ and $CD8^+$ T-cells are involved in the SP but the $CD8^+$ T-cells, in particular HTLV-1-specific CTLs, predominate (Sakai et al. 2001). While the cause of this phenomenon remains unclear, it is possible that it is mediated by certain viral products. Viral gene expression is suppressed in vivo, but will occur after several hours in culture. In particular Tax can transactivate the expression of IL-2/IL-2 R, which will lead to the proliferation of $CD4^+$ T-cells in an autocrine fashion. Also, Tax can induce IL-15 expression, which is known to be important for the maintenance of $CD8^+$ T-cells.

SP might be an important occurrence in situ in HAM/TSP patients: It is proposed that HTLV-1-infected $CD4^+$ T-cells and HTLV-1-specific $CD8^+$ CTLs enter the central nervous system (CNS) and proliferate similar to the SP in vitro. Bystander damage of neural tissues may be a consequence of their interactions.

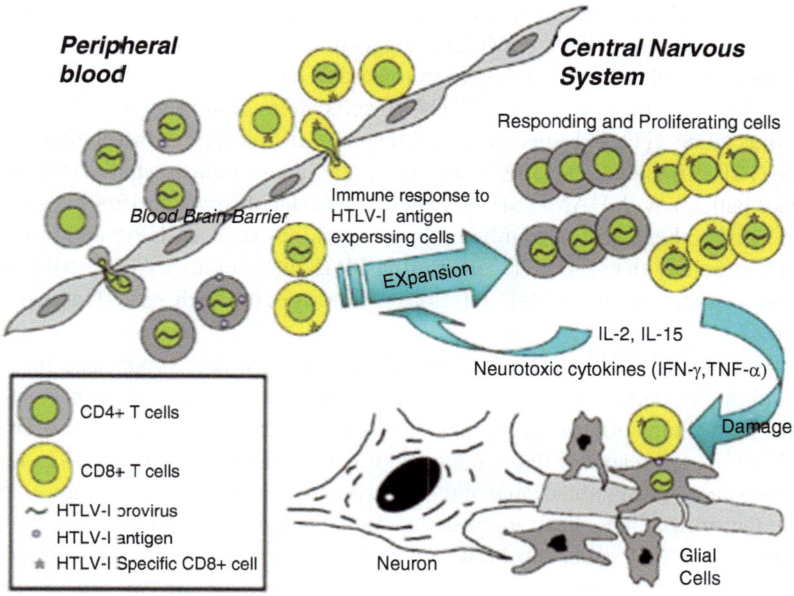

Fig. 5 Bystander damage in HAM/TSP

2.2.5 Immunopathogenesis of HAM/TSP

The precise mechanisms of HAM/TSP pathogenesis remain unclear; however, many findings suggest that a high PVL in the PBLs can directly lead to immunologic abnormalities that result in the pathologic process.

The pathogenic hypothesis with the most supportive evidence both in vivo and in vitro is that of bystander damage to the CNS (Fig. 5) (Ijichi et al. 1993). Infected $CD4^+$ T-cells become activated in the periphery with increased adherence to human endothelial cells that allow for enhanced transmigration into the CNS and a proinflammatory Th1 profile. HTLV-1-specific CTLs are also recruited into the lesions in an attempt to eliminate the infected $CD4^+$ T-cells and this interaction may lead to bystander damage of neural tissues. As the process evolves, the frequency of $CD4^+$ T-cells in the lesions decreases (possibly through an apoptotic mechanism triggered by the $CD8^+$ T-cells) as does the PVL and frequency of HTLV-1-specific CTLS. However, $CD8^+$ T-cells persist and become the dominant cell type at the chronic stage of the lesion. This persistence may result from their expansion or resistance to apoptosis.

Thus, infected $CD4^+$ T-cells seem to be a major early player in the pathogenesis by invading the CNS and triggering bystander damage. This hypothesis is supported by many findings including that $CD4^+$ T-cells in PBL of HAM/TSP patients have increased expression of the adhesion molecule LFA-1 and increased adherence to human endothelial cells (Ichinose et al. 1992). Also, the infected $CD4^+$ T-cells demonstrate accelerated transmigration through basement membrane. $CD4^+$

T-cells from HAM PBL have increased spontaneous production of inflammatory cytokines (INF-G, TNF-A, GMCSF) suggesting that there is a disruption of the immunologic balance with a deviation towards Th1 (Nishiura et al. 1996).

In addition to the major role of $CD4^+$ T-cells, there is substantial evidence that a vigorous $CD8^+$ HTLV-1-specific T-cell response is involved in the pathogenesis of HAM/TSP. Virus-specific $CD8^+$ CTLs occur at higher frequency in the PBL and the CSF of patients with HAM/TSP compared to AC (Jacobson et al. 1990). These cells have increased expression of inflammatory cytokines such as INF-γ and proliferate more vigorously than $CD4^+$ T-cells during SP (Kubota et al. 1998). Finally, $CD8^+$ T-cells are detected in HAM/TSP lesions at all stages of the disease (Umehara et al. 1993).

Given that $CD8^+$ T-cells are activated in HAM/TSP but suppressed in ATLL, they might be an important determinant of disease outcome. They also influence the PVL given their antiviral properties. The paradoxical dissociation between PVL and $CD8^+$ T-cell response in the majority of HAM/TSP patients (high PVL, high T-cell responses) is an indication that the PVL is affected by many other factors. Some of the factors that have been suggested include the number of $CD4^+$ FoxpP3+ cells, frequency of iNKT cells, and MHC-I favorable for HBZ-specific T-cell responses.

In addition to the hypothesis of bystander damage, there are two other hypothetical mechanisms of HTLV-1-mediated neuronal injury: (1) direct toxicity, whereby HTLV-1-infected glial cells express viral antigens on their surface and are then directly damaged by cytotoxic T-cells and (2) molecular mimicry, in which anti-HTLV-1 antibodies cross react with self-antigens on neural tissues. Notably some have suggested that anti-Tax antibodies can cross react with the heterogeneous nuclear ribonucleoprotein A1 (hnRNP-A1) leading to neuronal dysfunction (Fujinami and Oldstone 1985).

2.2.6 Therapeutic Strategy for HAM/TSP

Advances in understanding the pathogenesis of HAM/TSP have highlighted many therapeutic targets and biomarkers of disease activity. Namely, the proviral load, spontaneous lymphoproliferation, viral expression (particularly levels of Tax mRNA level), and HTLV-1-specific T-cell responses can be used to gauge the effect of therapeutic agents. A therapeutic strategy in HAM/TSP can also be directed at immunomodulation, antiviral effects, or a combination of both.

Many clinical trials have been conducted to date. However, most of them are small, observational, and have included patients at different stages of the disease.

Corticosteroids remain the most widely prescribed therapy for HAM/TSP despite the lack of evidence for long-term benefit. The rationale for their use is based on their immunomodulatory action, in particular, the ability to increase interferon-β and foxp3 expression. The latter mechanism is very relevant to HAM/TSP, in which there is evidence for Tax-dependent reduction of foxp3+ T-regulatory cells (Yamano et al. 2005). Recently Alberti et al. reported clinical improvement in a cohort of Chilean patients treated with a single dose of betamethasone; notably, the clinical improvement correlated with an increase in

Foxp3 and a decrease in Tax expression at the mRNA and protein levels for both (Alberti et al. 2011). However, a large double-blinded placebo-controlled trial is still needed to definitely prove or refute their usefulness.

Danazol, an attenuated anabolic steroid, has been tested in a total of 15 patients with HAM/TSP in two different studies. There was improved mobility and bladder function; however, liver dysfunction was a frequent complication. Its action probably results from a combination of immunomodulatory and anabolic effects on the muscle (Harrington et al. 1991).

Small observational studies of heparin to decrease the migration of lymphocytes into the CNS and of pentoxyfilline to downregulate the Th1 phenotype showed some benefit.

Plasmapheresis and intravenous immunoglobulins have been tried as a therapeutic option for HAM/TSP given their ability to modulate the humoral response to HTLV-1 (Matsuo et al. 1988; Kuroda et al. 1991a, b). Both seem to lead to transient benefit and it remains to be seen if periodic treatment could lead to sustained improvement.

Class I interferons (INF-α and -β) have cytostatic and antiviral properties and in HAM/TSP, they seem to have profound effects on viral transcriptional activity and the cellular immune response. In a blinded clinical trial, INF-α was proven to be beneficial in HAM/TSP in a dose-dependent manner (Izumo et al. 1996). However there was no placebo arm to the trial and despite good evidence for short-term benefit (66 % clinical response at 4 weeks) long-term effects have not been well studied. Notably, INF-α treatment was associated with a decrease in the PVL, central memory $CD8^+$ T-cells, and $CD8^+$ T-cell perforin expression.

Similarly INF-β was shown to have profound effects on the virus-specific cellular immune response: In an open label, single center trial in 12 HAM/TSP patients, no clinical changes were observed over the trial duration (28 weeks), but there was a significant decrease in spontaneous lymphoproliferation, the frequency of Tax-specific $CD8^+$ T-cells, and Tax mRNA load (Oh et al. 2005).

The nucleoside reverse transcriptase inhibitors (NRTIs) zidovudine and dideoxycytidine are effective against HTLV-1 in vitro (Derse et al. 2001). However in a randomized double-blind placebo-controlled trial of a combination regimen (zidovudine 250 mg and lamivudine 150 mg twice daily) in 16 patients there was no clinical viral or immunological response (Taylor et al. 2006). This questions the importance of HTLV-1 RT in vivo.

Cyclosporine was studied in a small group of HAM/TSP patients with early progressive disease and showed clinical improvement in 5/7, correlating with a reduction of the PVL in CSF and other inflammatory markers.

Many inflammatory pathways are deregulated in HAM/TSP and one therapeutic approach is to block them with monoclonal antibodies. Particularly relevant to HAM/TSP are the interleukins 2 and 15 (IL-2 and IL-15). HTLV-1Tax can upregulate the expression of IL-2 and CD25 (the alpha subunit of the IL-2 receptor in T-cells) and that of IL-15 and its specific receptor (IL-15 Rα). These signaling pathways are key to the spontaneous lymphoproliferation observed in vitro and which might be a surrogate marker for important in vivo pathogenic events (see above).

A clinical trial of humanized anti-CD25 monoclonal antibody (Daclizumab, antitac) was conducted in 1998 in which nine patients received the drug, resulting in significant reduction in $CD4^+$ $CD25^+$ T-cells, proviral load, and spontaneous lymphoproliferation. Notably, 3/9 patients had clinical improvement and none showed clinical deterioration. However, the follow-up was only for 14 weeks (Lehky et al. 1998). A monocloncal antibody targeting CD122 (IL-15 receptor β chain) is currently being tested in a phase I/II study.

Valproic acid (VPA), via its histone deacetylase activity, can activate HTLV-1 expression in the infected cells. This mechanism might reduce the PVL by exposing the infected cells to the immune system; it might also sensitize the virus to RT inhibitors. Thus, VPA alone or in combination with RT inhibitors could be beneficial. However, in a recent single center trial, Olindo et al. (2011) treated 19 HAM/TSP patients with VPA for 2 years, there was no significant effect on the PVL, CD38/HLA expression, or the efficiency of $CD8^+$ lysis. However, VPA was found to be safe.

In an animal study, 20 baboons naturally infected with STLV-1 (similar homology to HTLV-1) had significant reduction in their PVL when treated with a combination of zidovudine and sodium valproate in human-equivalent doses (Afonso et al. 2010).

2.3 *Other Neurologic Manifestations of HTLV-1 Infection*

2.3.1 HTLV-1-Associated Myopathy

HTLV-1 can also cause polymyositis (PM) and inclusion body myositis, and in endemic areas may explain cases of isolated muscle cramps or otherwise unexplained elevations of creatine kinase (CK) (Gabbai et al. 1994; Gilbert et al. 2001).

HTLV-1 PM occurs mostly in the context of HAM/TSP and should be suspected if the patients develop a more proximal and symmetric weakness with myalgias and elevated CK. Bulbar muscle involvement occurs in up to 30 % and the course is more protracted than that of idiopathic PM with a poor response to steroids. Pathologically, the muscle is infiltrated by mononuclear cells and shows fiber size variability and regeneration. Although the pathogenesis is not well defined, a direct myotoxic effect of the viral proteins, particularly tax, and host cytokines is suspected. The virus does not infect the muscle directly, but can be detected in the infiltrating $CD4^+$ lymphocytes.

2.3.2 HTLV-1-Associated Polyneuropathy

HTLV-1 infection has been associated with a peripheral neuropathy (PN) in the context of HAM/TSP (up to 30 %) and as an isolated neurologic outcome of the infection (up to 6.3 % of HTLV-1 seropositive subjects without HAM/TSP) (Kiwaki et al. 2003; Leite et al. 2004). This diagnosis, however, requires a high index of suspicion as a majority of the affected subjects are asymptomatic and even their neurologic examination and electrophysiologic studies can be unrevealing. When symptomatic the patients report paresthesias, burning sensations, and distal hypoesthesias. The neurologic examination can reveal stock-and-glove sensory loss with decreased/absent ankle jerks. These symptoms and signs, however, are not specific to the PN as they can be the result of spinal roots or posterior column involvement in the absence of PN.

HTLV-1-associated PN can be a mixed sensory-motor neuropathy (mostly in the context of HAM/TSP), a pure sensory PN, or rarely a mononeuritis multiplex (Kiwaki et al. 2003). Its pathogenesis remains unknown as there is no evidence for direct infection of the nerves and the pathologic findings have been variable. In the context of HAM/TSP, some have shown that the spinal roots can be affected by the spinal cord inflammation, postulating that the distal changes are secondary (Kiwaki et al. 2003). Others have reported axonal atrophy, demyelination, remyelination, and fibrosis in sural nerve biopsies of patients with HAM/TSP (Bhigjee et al. 1993). Interestingly, some have reported perivascular inflammatory infiltrates in the vasa nervorum though this has not been consistent. Similar pathologic findings have been reported in HTLV-1-associated PN without HAM/TSP.

2.3.3 Amyotrophic Lateral Sclerosis

HTLV-1 can cause an Amyotrophic Lateral Sclerosis (ALS)-like syndrome in association with lymphocytic infiltrates throughout the CNS (Kuroda and Sugihara 1991). Features that can differentiate it from seronegative-ALS are: a slower course resembling that of HAM/TSP with progression over a median of 9.4 years, early bladder dysfunction, and sensory signs. Some patients have responded at least partially to steroids. The pathology in a Brazilian patient with this syndrome revealed discrete diffuse lymphocytic infiltrates in the CNS, cell loss and gliosis of the hypoglossal nuclei, anterior horn cell loss with gliosis, and axonal loss in the corticospinal tract (Silva et al. 2005). These findings along with hand atrophy and bulbar symptoms suggest that the process is not restricted to the thoracic spinal cord as is the case for HAM/TSP.

2.3.4 HTLV-1-Associated Cognitive Deficits

The description of brain white matter lesions in association with HTLV-1 infection, similar to those seen in the context of multiple sclerosis and HIV infection, prompted investigations of a resulting cognitive disturbance. Most of the studies were uncontrolled and reported a subcortical type of dysfunction with psychomotor slowing and impaired attention, visual and working memory (Cartier et al. 1997). In one controlled study from Brazil (Silva et al. 2003), the investigators used extensive neuropsychological testing to compare HAM/TSP patients and AC to matched healthy controls. They found HTLV-1 infection to be associated with mild deficits in psychomotor speed, verbal fluency, verbal and visual memory, selective and alternate attention, and visuoconstructive abilities. There was no difference between HAM/TSP and AC and no association with the degree of motor impairment (Silva et al. 2003).

2.3.5 Dysautonomia

Dysautonomia is often an underestimated finding in HAM/TSP patients. It predominantly affects the sympathetic nervous system, impairing cardiovascular control and sweating. It can also cause neurogenic bladder in HTLV-1-infected patients without HAM/TSP (Alamy et al. 2001).

2.3.6 Other Neurologic Associations

In highly endemic areas, HTLV-1 infection has been associated with various neurologic syndromes, but whether or not this is a true association remains to be seen. This is probably the case for the association with acute disseminated encephalomyelitis (ADEM), cranial neuropathies, various movement disorders, and cerebellar ataxia (Araujo and Silva 2006). Meningeal signs in the context of HTLV-1 infection may signal leukemic infiltrates or another infection. Chronic hypertrophic pachymeningitis with cranial neuropathies has been reported as a possible association with HTLV-1 infection (Kitajima et al. 2002).

3 Conclusions

The role of HTLV-I in chronic, neurologic disease will undoubtedly lead to a better understanding of infectious triggers, immunological mediated neuropathological responses and neurodegenerative mechanisms that may also play a role in other neurologic diseases. Importantly, clinical interventional trials can target many of these pathways that can be experimentally monitored in a disease such as HAM/TSP that may have applications to other chronic, inflammatory diseases of the nervous system, such as multiple sclerosis, in which viruses have been suggested to play a role.

References

Afonso PV, Mekaouche M, Mortreux F, Toulza F, Moriceau A, Wattel E, Gessain A, Bangham CR, Dubreuil G, Plumelle Y, Hermine O, Estaquier J, Mahieux R (2010) Highly active antiretroviral treatment against STLV-1 infection combining reverse transcriptase and HDAC inhibitors. Blood 116(19):3802–3808

Alamy AH, Menezes FB, Leite ACB, Nascimento OM, Araujo AQC (2001) Dysautonomia in human T-cell lymphotrophic virus type I-associated myelopathy/tropical spastic paraparesis. Ann Neurol 50(5):681–685

Alberti C, Cartier L, Valenzuela MA, Puente J, Tanaka Y, Ramirez E (2011) Molecular and clinical effects of betamethasone in human T-cell lymphotropic virus type-I-associated myelopathy/tropical spastic paraparesis patients. J Med Virol 83(9):1641–1649

Albrecht B, Collins ND, Burniston MT, Nisbet JW, Ratner L, Green PL, Lairmore MD (2000) Human T-lymphotropic virus type 1 open reading frame I p12(I) is required for efficient viral infectivity in primary lymphocytes. J Virol 74(21):9828–9835

Alcindor F, Valderrama R, Canavaggio M, Lee H, Katz A, Montesinos C, Madrid RE, Merino RR, Pipia PA (1992) Imaging of human T-lymphotropic virus type I-associated chronic progressive myeloneuropathies. Neuroradiology 35(1):69–74

Araujo AQ, Andrade-Filho AS, Castro-Costa CM, Menna-Barreto M, Almeida SM (1998) HTLV-I-associated myelopathy/tropical spastic paraparesis in Brazil: a nationwide survey. HAM/TSP Brazilian Study Group. J Acquir Immune Defic Syndr Hum Retrovirol 19(5):536–541

Araujo AQ, Silva MT (2006) The HTLV-1 neurological complex. Lancet Neurol 5 (12):1068–1076

Aye MM, Matsuoka E, Moritoyo T, Umehara F, Suehara M, Hokezu Y, Yamanaka H, Isashiki Y, Osame M, Izumo S (2000) Histopathological analysis of four autopsy cases of HTLV-I-associated myelopathy/tropical spastic paraparesis: inflammatory changes occur simultaneously in the entire central nervous system. Acta Neuropathol 100(3):245–252

Bagnato F, Butman JA, Mora CA, Gupta S, Yamano Y, Tasciyan TA, Solomon JM, Santos WJ, Stone RD, McFarland HF, Jacobson S (2005) Conventional magnetic resonance imaging features in patients with tropical spastic paraparesis. J Neurovirol 11(6):525–534

Baydoun HH, Bellon M, Nicot C (2008) HTLV-1 Yin and Yang: Rex and p30 master regulators of viral mRNA trafficking. AIDS Rev 10(4):195–204

Bhigjee AI, Bill PL, Wiley CA, Windsor IM, Matthias DA, Amenomori T, Wachsman W, Moorhouse D (1993) Peripheral nerve lesions in HTLV-I associated myelopathy (HAM/TSP). Muscle Nerve 16(1):21–26

Cartier L, Gormaz A, Kleinsteuber K, Ramirez E, Galeno H (1997) New form of subcortical dementia: encephalopathy due to infection with human lymphotropic T virus (HTLV-1). Clinical case. Rev Med Chil 125(2):209–213

Cooper SA, van der Loeff MS, Taylor GP (2009) The neurology of HTLV-1 infection. Pract Neurol 9(1):16–26

Datta A, Silverman L, Phipps AJ, Hiraragi H, Ratner L, Lairmore MD (2007) Human T-lymphotropic virus type-1 p30 alters cell cycle G2 regulation of T lymphocytes to enhance cell survival. Retrovirology 4:49

De Castro-Costa CM, Araujo AQ, Barreto MM, Takayanagui OM, Sohler MP, da Silva EL, de Paula SM, Ishak R, Ribas JG, Rovirosa LC, Carton H, Gotuzzo E, Hall WW, Montano S, Murphy EL, Oger J, Remondegui C, Taylor GP (2006) Proposal for diagnostic criteria of tropical spastic paraparesis/HTLV-I-associated myelopathy (TSP/HAM). AIDS Res Hum Retroviruses 22(10):931–935

de The G, Kazanji M (1996) An HTLV-I/II vaccine: from animal models to clinical trials? J Acquir Immune Defic Syndr Hum Retrovirol 13(Suppl 1):S191–S198

Derse D, Hill SA, Lloyd PA, Chung H, Morse BA (2001) Examining human T-lymphotropic virus type 1 infection and replication by cell-free infection with recombinant virus vectors. J Virol 75 (18):8461–8468

Franzoi AC, Araujo AQ (2007) Disability and determinants of gait performance in tropical spastic paraparesis/HTLV-I associated myelopathy (HAM/TSP). Spinal Cord 45(1): 64-68

Fujinami RS, Oldstone MB (1985) Amino acid homology between the encephalitogenic site of myelin basic protein and virus: mechanism for autoimmunity. Science 230(4729):1043–1045

Fujino T, Nagata Y (2000) HTLV-I transmission from mother to child. J Reprod Immunol 47 (2):197–206

Furnia A, Lal R, Maloney E, Wiktor S, Pate E, Rudolph D, Waters D, Blattner W, Manns A (1999) Estimating the time of HTLV-I infection following mother-to-child transmission in a breast-feeding population in Jamaica. J Med Virol 59(4):541–546

Furukawa Y, Yamashita M, Usuku K, Izumo S, Nakagawa M, Osame M (2000) Phylogenetic subgroups of human T cell lymphotropic virus (HTLV) type I in the tax gene and their association with different risks for HTLV-I-associated myelopathy/tropical spastic paraparesis. J Infect Dis 182(5):1343–1349

Gabbai AA, Wiley CA, Oliveira AS, Smith R, Schmidt B, Nobrega JA, Bordin JO, Roman GC (1994) Skeletal muscle involvement in tropical spastic paraparesis/HTLV-1-associated myelopathy. Muscle Nerve 17(8):923–930

Gessain A, Barin F, Vernant JC, Gout O, Maurs L, Calender A, de The G (1985) Antibodies to human T-lymphotropic virus type-I in patients with tropical spastic paraparesis. Lancet 2 (8452):407–410

Gilbert DT, Morgan O, Smikle MF, Simeon D, Barton EN (2001) HTLV-1 associated polymyositis in Jamaica. Acta Neurol Scand 104(2):101–104

Godoy AJ, Kira J, Hasuo K, Goto I (1995) Characterization of cerebral white matter lesions of HTLV-I-associated myelopathy/tropical spastic paraparesis in comparison with multiple sclerosis and collagen-vasculitis: a semiquantitative MRI study. J Neurol Sci 133(1–2):102–111

Green PL, Chen IS (1990) Regulation of human T cell leukemia virus expression. FASEB J 4 (2):169–175

Harrington WJ Jr, Sheremata WA, Snodgrass SR, Emerson S, Phillips S, Berger JR (1991) Tropical spastic paraparesis/HTLV-1-associated myelopathy (TSP/HAM): treatment with an anabolic steroid danazol. AIDS Res Hum Retroviruses 7(12):1031–1034

Hiraragi H, Michael B, Nair A, Silic-Benussi M, Ciminale V, Lairmore M (2005) Human T-lymphotropic virus type 1 mitochondrion-localizing protein p13II sensitizes Jurkat T cells to Ras-mediated apoptosis. J Virol 79(15):9449–9457

Howard AK, Li DK, Oger J (2003) MRI contributes to the differentiation between MS and HTLV-I associated myelopathy in British Columbian coastal natives. Can J Neurol Sci 30 (1):41–48

Ichinose K, Nakamura T, Kawakami A, Eguchi K, Nagasato K, Shibayama K, Tsujihata M, Nagataki S (1992) Increased adherence of T cells to human endothelial cells in patients with human T-cell lymphotropic virus type I-associated myelopathy. Arch Neurol 49(1):74–76

Igakura T, Stinchcombe JC, Goon PK, Taylor GP, Weber JN, Griffiths GM, Tanaka Y, Osame M, Bangham CR (2003) Spread of HTLV-I between lymphocytes by virus-induced polarization of the cytoskeleton. Science 299(5613):1713–1716

Ijichi S, Izumo S, Eiraku N, Machigashira K, Kubota R, Nagai M, Ikegami N, Kashio N, Umehara F, Maruyama I et al (1993) An autoaggressive process against bystander tissues in HTLV-I-infected individuals: a possible pathomechanism of HAM/TSP. Med Hypotheses 41 (6):542–547

Izumo S, Goto I, Itoyama Y, Okajima T, Watanabe S, Kuroda Y, Araki S, Mori M, Nagataki S, Matsukura S, Akamine T, Nakagawa M, Yamamoto I, Osame M (1996) Interferon-alpha is effective in HTLV-I-associated myelopathy: a multicenter, randomized, double-blind, controlled trial. Neurology 46(4):1016–1021

Jacobson S, Shida H, McFarlin DE, Fauci AS, Koenig S (1990) Circulating CD8+ cytotoxic T lymphocytes specific for HTLV-I pX in patients with HTLV-I associated neurological disease. Nature 348(6298):245–248

Jeffery KJ, Usuku K, Hall SE, Matsumoto W, Taylor GP, Procter J, Bunce M, Ogg GS, Welsh KI, Weber JN, Lloyd AL, Nowak MA, Nagai M, Kodama D, Izumo S, Osame M, Bangham CR (1999) HLA alleles determine human T-lymphotropic virus-I (HTLV-I) proviral load and the risk of HTLV-I-associated myelopathy. Proc Natl Acad Sci USA 96(7):3848–3853

Johnson JM, Nicot C, Fullen J, Ciminale V, Casareto L, Mulloy JC, Jacobson S, Franchini G (2001) Free major histocompatibility complex class I heavy chain is preferentially targeted for degradation by human T-cell leukemia/lymphotropic virus type 1 p12(I) protein. J Virol 75 (13):6086–6094

Jones KS, Petrow-Sadowski C, Bertolette DC, Huang Y, Ruscetti FW (2005) Heparan sulfate proteoglycans mediate attachment and entry of human T-cell leukemia virus type 1 virions into CD4+ T cells. J Virol 79(20):12692–12702

Jones KS, Petrow-Sadowski C, Huang YK, Bertolette DC, Ruscetti FW (2008) Cell-free HTLV-1 infects dendritic cells leading to transmission and transformation of CD4(+) T cells. Nat Med 14(4):429–436

Kalyanaraman VS, Sarngadharan MG, Robert-Guroff M, Miyoshi I, Golde D, Gallo RC (1982) A new subtype of human T-cell leukemia virus (HTLV-II) associated with a T-cell variant of hairy cell leukemia. Science 218(4572):571–573

Kambara C, Nakamura T, Furuya T, Migita K, Ida H, Kawakami A, Shirabe S, Nakane S, Kinoshita I, Eguchi K (1999) Vascular cell adhesion molecule-1-mediated matrix metalloproteinase-2 induction in peripheral blood T cells is up-regulated in patients with HTLV-I-associated myelopathy. J Neuroimmunol 99(2):242–247

Kim SJ, Nair AM, Fernandez S, Mathes L, Lairmore MD (2006) Enhancement of LFA-1-mediated T cell adhesion by human T lymphotropic virus type 1 p12I1. J Immunol 176(9):5463–5470

Kitajima I, Osame M, Izumo S, Igata A (1988) Immunological studies of HTLV-I associated myelopathy. Autoimmunity 1(2):125–131

Kitajima M, Korogi Y, Shigematsu Y, Liang L, Matsuoka M, Yamamoto T, Jhono M, Eto K, Takahashi M (2002) Central nervous system lesions in adult T-cell leukaemia: MRI and pathology. Neuroradiology 44(7):559–567

Kiwaki T, Umehara F, Arimura Y, Izumo S, Arimura K, Itoh K, Osame M (2003) The clinical and pathological features of peripheral neuropathy accompanied with HTLV-I associated myelopathy. J Neurol Sci 206(1):17–21

Kubota R, Kawanishi T, Matsubara H, Manns A, Jacobson S (1998) Demonstration of human T lymphotropic virus type I (HTLV-I) tax-specific CD8+ lymphocytes directly in peripheral blood of HTLV-I-associated myelopathy/tropical spastic paraparesis patients by intracellular cytokine detection. J Immunol 161(1):482–488

Kuroda R, Nakatani J, Yamada Y, Yorimae A, Kitano M (1991a) Location of a DBS-electrode in lateral thalamus for deafferentation pain. An autopsy case report. Acta Neurochir Suppl (Wien) 52:140–142

Kuroda Y, Sugihara H (1991) Autopsy report of Htlv-I-associated myelopathy presenting with Als-like manifestations. J Neurol Sci 106(2):199–205

Kuroda Y, Takashima H, Ikeda A, Endo C, Neshige R, Kakigi R, Shibasaki H (1991b) Treatment of HTLV-I-associated myelopathy with high-dose intravenous gammaglobulin. J Neurol 238 (6):309–314

Lehky TJ, Levin MC, Kubota R, Bamford RN, Flerlage AN, Soldan SS, Leist TP, Xavier A, White JD, Brown M, Fleisher TA, Top LE, Light S, McFarland HF, Waldmann TA, Jacobson S (1998) Reduction in HTLV-I proviral load and spontaneous lymphoproliferation in HTLV-I-associated myelopathy/tropical spastic paraparesis patients treated with humanized anti-Tac. Ann Neurol 44(6):942–947

Leite AC, Silva MT, Alamy AH, Afonso CR, Lima MA, Andrada-Serpa MJ, Nascimento OJ, Araujo AQ (2004) Peripheral neuropathy in HTLV-I infected individuals without tropical spastic paraparesis/HTLV-I-associated myelopathy. J Neurol 251(7):877–881

Marinho J, Galvao-Castro B, Rodrigues LC, Barreto ML (2005) Increased risk of tuberculosis with human T-lymphotropic virus-1 infection: a case-control study. J Acquir Immune Defic Syndr 40(5):625–628

Matsuo H, Nakamura T, Tsujihata M, Kinoshita I, Satoh A, Tomita I, Shirabe S, Shibayama K, Nagataki S (1988) Plasmapheresis in treatment of human T-lymphotropic virus type-I associated myelopathy. Lancet 2(8620):1109–1113

Matsuoka M (2010) HTLV-1 bZIP factor gene: Its roles in HTLV-1 pathogenesis. Mol Aspects Med 31(5):359–366

Morgan DJ, Caskey MF, Abbehusen C, Oliveira-Filho J, Araujo C, Porto AF, Santos SB, Orge GO, Joia MJ, Muniz AL, Siqueira I, Glesby MJ, Carvalho E (2007) Brain magnetic resonance imaging white matter lesions are frequent in HTLV-I carriers and do not discriminate from HAM/TSP. AIDS Res Hum Retroviruses 23(12):1499–1504

Nagai M, Usuku K, Matsumoto W, Kodama D, Takenouchi N, Moritoyo T, Hashiguchi S, Ichinose M, Bangham CR, Izumo S, Osame M (1998) Analysis of HTLV-I proviral load in 202 HAM/TSP patients and 243 asymptomatic HTLV-I carriers: high proviral load strongly predisposes to HAM/TSP. J Neurovirol 4(6):586–593

Nakane S, Shirabe S, Moriuchi R, Mizokami A, Furuya T, Nishiura Y, Okazaki S, Yoshizuka N, Suzuki Y, Nakamura T, Katamine S, Gojobori T, Eguchi K (2000) Comparative molecular analysis of HTLV-I proviral DNA in HTLV-I infected members of a family with a discordant HTLV-I-associated myelopathy in monozygotic twins. J Neurovirol 6(4):275–283

Nishiura Y, Nakamura T, Ichinose K, Shirabe S, Tsujino A, Goto H, Furuya T, Nagataki S (1996) Increased production of inflammatory cytokines in cultured CD4+ cells from patients with HTLV-I-associated myelopathy. Tohoku J Exp Med 179(4):227–233

Oh U, Yamano Y, Mora CA, Ohayon J, Bagnato F, Butman JA, Dambrosia J, Leist TP, McFarland H, Jacobson S (2005) Interferon-beta1a therapy in human T-lymphotropic virus type I-associated neurologic disease. Ann Neurol 57(4):526–534

Olindo S, Belrose G, Gillet N, Rodriguez S, Boxus M, Verlaeten O, Asquith B, Bangham C, Signate A, Smadja D, Lezin A, Cesaire R, Willems L (2011) Safety of long-term treatment of HAM/TSP patients with valproic acid. Blood 118(24):6306–6309

Pais-Correia AM, Sachse M, Guadagnini S, Robbiati V, Lasserre R, Gessain A, Gout O, Alcover A, Thoulouze MI (2010) Biofilm-like extracellular viral assemblies mediate HTLV-1 cell-to-cell transmission at virological synapses. Nat Med 16(1):83–89

Proietti FA, Carneiro-Proietti AB, Catalan-Soares BC, Murphy EL (2005) Global epidemiology of HTLV-I infection and associated diseases. Oncogene 24(39):6058–6068

Rho HM, Poiesz B, Ruscetti FW, Gallo RC (1981) Characterization of the reverse transcriptase from a new retrovirus (HTLV) produced by a human cutaneous T-cell lymphoma cell line. Virology 112(1):355–360

Richardson JH, Edwards AJ, Cruickshank JK, Rudge P, Dalgleish AG (1990) In vivo cellular tropism of human T-cell leukemia virus type 1. J Virol 64(11):5682–5687

Roucoux DF, Murphy EL (2004) The epidemiology and disease outcomes of human T-lymphotropic virus type II. AIDS Rev 6(3):144–154

Saito M, Matsuzaki T, Satou Y, Yasunaga J, Saito K, Arimura K, Matsuoka M, Ohara Y (2009) In vivo expression of the HBZ gene of HTLV-1 correlates with proviral load, inflammatory markers and disease severity in HTLV-1 associated myelopathy/tropical spastic paraparesis (HAM/TSP). Retrovirology 6:19

Sakai JA, Nagai M, Brennan MB, Mora CA, Jacobson S (2001) In vitro spontaneous lymphoproliferation in patients with human T-cell lymphotropic virus type I-associated neurologic disease: predominant expansion of CD8+ T cells. Blood 98(5):1506–1511

Silva MT, Leite AC, Alamy AH, Chimelli L, Andrada-Serpa MJ, Araujo AQ (2005) ALS syndrome in HTLV-I infection. Neurology 65(8):1332–1333

Silva MTT, Mattos P, Alfano A, Araujo AQC (2003) Neuropsychological assessment in HTLV-1 infection: a comparative study among TSP/HAM, asymptomatic carriers, and healthy controls. J Neurol Neurosurg Psychiatry 74(8):1085–1089

Taylor GP, Goon P, Furukawa Y, Green H, Barfield A, Mosley A, Nose H, Babiker A, Rudge P, Usuku K, Osame M, Bangham CR, Weber JN (2006) Zidovudine plus lamivudine in Human T-Lymphotropic Virus type-I-associated myelopathy: a randomised trial. Retrovirology 3:63

Trejo SR, Ratner L (2000) The HTLV receptor is a widely expressed protein. Virology 268(1):41–48

Uchiyama T, Yodoi J, Sagawa K, Takatsuki K, Uchino H (1977) Adult T-cell leukemia: clinical and hematologic features of 16 cases. Blood 50(3):481–492

Umehara F, Izumo S, Nakagawa M, Ronquillo AT, Takahashi K, Matsumuro K, Sato E, Osame M (1993) Immunocytochemical analysis of the cellular infiltrate in the spinal-cord lesions in Htlv-I-associated myelopathy. J Neuropathol Exp Neurol 52(4):424–430

Umehara F, Izumo S, Takeya M, Takahashi K, Sato E, Osame M (1996) Expression of adhesion molecules and monocyte chemoattractant protein-1 (MCP-1) in the spinal cord lesions in HTLV-I-associated myelopathy. Acta Neuropathol 91(4):343–350

Umehara F, Nakamura A, Izumo S, Kubota R, Ijichi S, Kashio N, Hashimoto K, Usuku K, Sato E, Osame M (1994) Apoptosis of T-lymphocytes in the spinal-cord lesions in Htlv-I-associated myelopathy—a possible mechanism to control viral-infection in the central-nervous-system. J Neuropathol Exp Neurol 53(6):617–624

Usuku K, Sonoda S, Osame M, Yashiki S, Takahashi K, Matsumoto M, Sawada T, Tsuji K, Tara M, Igata A (1988) HLA haplotype-linked high immune responsiveness against HTLV-I in HTLV-I-associated myelopathy: comparison with adult T-cell leukemia/lymphoma. Ann Neurol 23(Suppl): S143–150. World Health Organization (1988) Report of the Scientific Group on HTLV-I Infections and Associated Diseases. World Health Organization Report (WP) CDS (S)/ICP/PHC/ 14. 1988.

Van Dooren S, Salemi M, Vandamme AM (2001) Dating the origin of the African human T-cell lymphotropic virus type-i (HTLV-I) subtypes. Mol Biol Evol 18(4):661–671

Van Prooyen N, Gold H, Andresen V, Schwartz O, Jones K, Ruscetti F, Lockett S, Gudla P, Venzon D, Franchini G (2010) Human T-cell leukemia virus type 1 p8 protein increases cellular conduits and virus transmission. Proc Natl Acad Sci USA 107(48):20738–20743

World Health Organization (1988) Report of the Scientific Group on HTLV-I Infections and Associated Diseases. World Health Organization Report (WP) CDS (S)/ICP/PHC/014. 1988.

Yamano Y, Takenouchi N, Li HC, Tomaru U, Yao K, Grant CW, Maric DA, Jacobson S (2005) Virus-induced dysfunction of CD4+ CD25+ T cells in patients with HTLV-I-associated neuroimmunological disease. J Clin Invest 115(5):1361–1368

Yamashita S, Ueda A, Hirahara T, Kimura E, Hirano T, Uchino M (2011) Fulminant myelopathy following neurogenic proximal weakness associated with human T-cell lymphotropic virus type I infection. Intern Med 50(8):919–924

Yoshida T, Yamamoto N (1986) Detection of adult T-cell leukemia virus and ATLV antibodies. Nihon Rinsho 44(10):2256–2262

Yukitake M, Takase Y, Nanri Y, Kosugi M, Eriguchi M, Yakushiji Y, Okada R, Mizuta H, Kuroda Y (2008) Incidence and clinical significances of human T-cell lymphotropic virus type I-associated myelopathy with T2 hyperintensity on spinal magnetic resonance images. Intern Med 47(21):1881–1886

Part IV
Viral Zoonoses

Rabies

Alan C. Jackson

Abstract Rabies is an ancient disease, but it remains an important problem in countries with endemic dog rabies, especially in Asia and Africa. Rabies in wildlife, particularly from bats, is the main threat to humans in North America. Experimental studies in animals have given us detailed knowledge about the neural pathways of viral spread through the host. Human rabies often has distinctive clinical features reflecting the early brainstem involvement, including hydrophobia, but physicians in North America and Europe may not consider a diagnosis of rabies because the disease is rare and their lack of familiarity with the clinical manifestations. There is a progressive clinical course to coma and the disease is virtually always fatal. When rabies is treated aggressively there are often many medical complications, including multiple organ failure. Therapeutic attempts have been disappointing, but new approaches need to be taken in the future. An improved understanding of rabies pathogenesis might lead to important insights into the development of new therapeutic approaches.

Keywords Encephalitis • Rabies • Virus

1 Introduction

Although rabies is a very old disease it remains an important public health problem in humans in many resource-poor and resource-limited countries, particularly in Asia and Africa, because of the ongoing threat related to the presence of endemic dog rabies. Rabies is often not recognized by North American physicians, even if typical clinical features are present, due to lack of familiarity with the disease.

A.C. Jackson (✉)
Departments of Internal Medicine (Neurology) and of Medical Microbiology, University of Manitoba, Winnipeg, MB, Canada
e-mail: ajackson2@hsc.mb.ca

A.C. Jackson (ed.), *Viral Infections of the Human Nervous System*, Birkhäuser Advances in Infectious Diseases,
DOI 10.1007/978-3-0348-0425-7_9,

The clinical features of rabies encephalitis are unusual because consciousness is relatively unaffected at a stage in which there are symptoms and signs due to brainstem involvement. Although rabies can be effectively prevented after recognized exposures, the therapy of human rabies remains an important challenge for the future. Our understanding of the pathogenesis of rabies has been gained through a variety of experimental studies performed in animal models. An improved understanding of rabies pathogenesis may be necessary for the development of novel therapies of rabies in the future.

2 Rabies Virus

Rabies virus is a RNA virus in the Rhabdoviridae family (genus *Lyssavirus*), and the genome is single-stranded, negative-sense (antisense), and nonsegmented. The genome consists of 11,932 nucleotides and encodes five proteins: nucleocapsid protein (N), matrix protein (M), phosphoprotein (P), glycoprotein (G), and a RNA-dependent RNA polymerase or large polymerase protein (L) (Wunner 2007). Virus particles are bullet shaped and at the center of the viral particle is encapsidated RNA forming a ribonucleoprotein (RNP) core consisting of helical genomic RNA associated with the N, P, and L proteins. The RNP is a functional template for transcription and replication (Schnell et al. 2010). The G and M proteins are associated with the lipid-bilayer envelope that surrounds the RNP core. The M protein lines the viral envelope and forms an inner leaflet between the envelope and RNP core and the G protein produces spike-like projections on the surface of the viral envelope (Wunner 2007). The glycoprotein is the major surface antigen of the virus and induces and binds virus-neutralizing antibodies and is important for immunity. Rabies virus belongs to Genotype 1 of the lyssaviruses and there are ten other genotypes of lyssaviruses and five of these ten genotypes have been recognized to very rarely cause human disease indistinguishable from rabies (Table 1).

3 Pathogenesis

3.1 Routes of Transmission

Rabies virus is almost always transmitted by an animal bite. Exposure of mucosal membranes or contamination of skin lesions with infectious rabies virus may much less frequently lead to transmission. Rarely, transmission has been documented by aerosol transmission in a laboratory (Tillotson et al. 1977b; Winkler et al. 1973) or in caves containing millions of bats (Constantine 1962). In these cases viral entry into the host is thought to occur by an olfactory pathway into the brain. Iatrogenic transmission either by transplantation of tissues (cornea or vascular conduit) or

Table 1 Reported human rabies cases due to other *Lyssavirus* genotypes (Adapted from Jackson AC: Human disease, in Rabies, Second Edition, edited by AC Jackson and WH Wunner, 2007, Elsevier Academic, London, pp 309–340; Copyright Elsevier)

Virus (genotype)	Year	Location	Age of patient	References
Mokola (3)[a]	1968	Nigeria	3.5	Familusi and Moore (1972)
Mokola (3)	1971	Nigeria	6	Familusi et al. (1972)
Duvenhage (4)	1970	South Africa	31	Meredith et al. (1971)
Duvenhage (4)	2006	South Africa	77	Paweska et al. (2006)
Duvenhage (4)	2007	Kenya	34	van Thiel et al. (2009)
European Bat Lyssavirus 1 (5)	1985	Russia	11	Selimov et al. (1989)
European Bat Lyssavirus 1 (5)	2002	Ukraine	34	Botvinkin et al. (2005)
European Bat Lyssavirus 2 (6)	1985	Finland	30	Roine et al. (1988)
European Bat Lyssavirus 2 (6)	2002	Scotland	55	Johnson et al. (2012), Nathwani et al. (2003)
Australian Bat Lyssavirus (7)	1996	Australia	39	Samaratunga et al. (1998)
Australian Bat Lyssavirus (7)	1998	Australia	37	Hanna et al. (2000)
Irkut (pending)	2007	Russia	20	Leonova et al. (2009)

[a]It is doubtful that this patient's clinical picture was actually caused by Mokola virus infection

Table 2 Cases of human rabies associated with organ transplantation in the USA (Burton et al. 2005; Srinivasan et al. 2005) and Germany (Maier et al. 2010) (Adapted with permission from Jackson AC: Human disease, in Rabies, Second Edition, edited by AC Jackson and WH Wunner, 2007, Elsevier Academic, London, pp 309-340; Copyright Elsevier)

	Sex/age	Organ transplanted	Onset of clinical rabies posttransplantation
Donor in USA	Male/20	–	–
Recipient 1	Male/53	Liver	21 days
Recipient 2	Female/50	Kidney	27 days
Recipient 3	Male/18	Kidney	27 days
Recipient 4	Female/55	Iliac artery segment (for a liver)	27 days
Donor in Germany	Female/26	–	–
Recipient 1	Female/46	Lung	6 weeks
Recipient 2	Male/72	Kidney	5 weeks
Recipient 3	Male/47	Kidney and pancreas	5 weeks

solid organs (e.g., kidney and liver) (Jackson 2007a; Maier et al. 2010; Srinivasan et al. 2005) (Table 2) has been well documented. There have also been three reported cases with corneal transplantation from donors with rabies in which the recipients did not develop rabies (Sureau et al. 1981; Vetter et al. 2011). Oral transmission has been reported experimentally in a variety of animal species (Charlton and Casey 1979a, c; Fischman and Ward 1968) and may occur naturally with consumption of carcasses of rabies animals by wildlife and also when humans eat raw dog meat (Wallerstein 1999). Viral entry likely occurs due to breaks in the integrity of the gastrointestinal mucosa. Human cases of rabies that occur without a history of an exposure or contact with animals are not unusual, and are most commonly due to infection with bat rabies virus variants. These cases are thought to be due to unrecognized or forgotten bites.

3.2 Events at Site of Viral Entry/Exposure and Viral Spread to the CNS

The sequential pathogenetic steps that typically occur after an animal bite are outlined in Fig. 1. Studies in experimental animals have indicated that rabies virus remains close to the site of entry for the vast majority of the incubation period in rabies (Charlton et al. 1997), which typically lasts for 20–90 days. With bites involving muscles, the virus binds to the nicotinic acetylcholine receptor, which is present on the postsynaptic side of the neuromuscular junction and functions to localize and concentrate the virus and facilitates its subsequent uptake and transfer to peripheral motor neurons (Lentz et al. 1982). The virus spreads within axons of peripheral nerves by retrograde fast axonal transport. This was experimentally demonstrated using colchicine, a microtubule-disrupting agent that is an inhibitor of fast axonal transport, which was applied locally to the sciatic nerve of rats using elastomer cuffs and inhibited viral spread (Tsiang 1979). The earliest neurologic symptoms in rabies are likely related to infection and inflammation in local dorsal root ganglia, which may cause paresthesias, pain, and pruritus.

Similar experimental studies have not been reported in a model using bat rabies virus in which the exposures are more superficial than due to wounds associated with dog or wild carnivore exposures and typically involve the epidermis, dermis, and/or subcutaneous tissues. The skin is richly innervated and likely rabies virus gains access to small sensory or autonomic nerves, but no information is available about the sequential steps involved in this process during the incubation period. It is very doubtful that hematogenous spread plays any significant role in natural rabies (see Sect. 3.5).

3.3 Viral Spread Within the CNS

Rabies virus also disseminates throughout the central nervous system in axons by fast axonal transport along neuroanatomical connections. This was determined using stereotaxic brain inoculation in rats (Gillet et al. 1986) with the administration of colchicine, which inhibited viral transport in the CNS (Ceccaldi et al. 1989, 1990). Rabies virus is now a very important tool used for transneuronal tracing studies that provide an understanding of functional neuronal networks (Ugolini 2010, 2011). Ultrastructural studies in a skunk model have shown that viral budding occurs at synapses or in adjacent plasma membranes of dendrites (Charlton and Casey 1979b).

3.4 Viral Spread from the CNS

There is centrifugal spread of the virus to multiple organs in the body along autonomic and/or sensory nerves. In rabies vectors there is spread of the virus

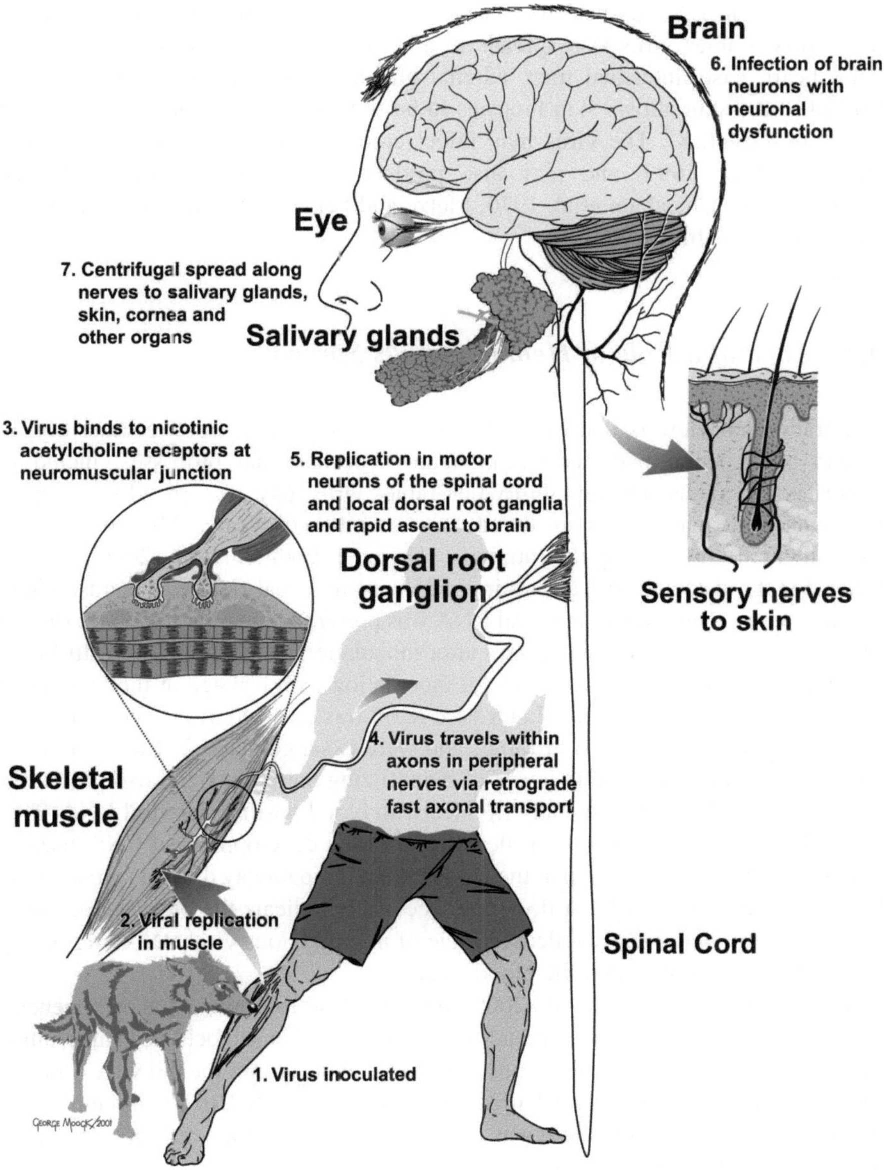

Fig. 1 Schematic diagram showing the sequential steps in the pathogenesis of rabies after an animal bite/peripheral inoculation of rabies virus (From Jackson AC: Pathogenesis, in Rabies, Second Edition, edited by AC Jackson and WH Wunner, 2007, Elsevier Academic, London, pp 341–381; Copyright Elsevier)

to the salivary glands along multiple terminal axons (Charlton et al. 1983) with viral budding on the apical plasma membrane into the acinar lumen (Balachandran and Charlton 1994), resulting in the secretion of high-titer virus in the saliva (higher than in CNS tissues) that is necessary for transmission in oral fluids to new hosts. Rabies virus also spreads to a variety of other organs, including the eyes, heart, gastrointestinal tract, adrenal gland, and skin (Jackson et al. 1999). Spread to the skin allows skin biopsies to be used as a diagnostic test for human rabies (see Sect. 8). The virus may infect cardiac ganglia and the myocardium (Jackson et al. 1999) and may also produce a myocarditis causing cardiac complications (Cheetham et al. 1970; Metze and Feiden 1991; Raman et al. 1988; Ross and Armentrout 1962).

3.5 Lack of a Role of Hematogenous Spread

The traditional view is that rabies virus spreads within its hosts exclusively by spread by a neural pathway within neurons or neuronal processes, including dendrites and axons. The possibility that rabies virus may also spread by a hematogenous route has been considered for a number of years. After mice were experimentally infected by the intramuscular route, Lodmell et al. (2006) assayed blood for the detection of rabies virus RNA using a real-time polymerase chain reaction assay. They found that viral RNA was present in the blood of 30/32 (94 %) of mice from between 1 h and 2 days after inoculation. This finding was attributed to trauma to blood vessels at the site of inoculation with leakage of the inoculated virus into the blood circulation. Later viral RNA was detected in 21/25 (84 %) mice that developed clinical signs of rabies and were exsanguinated 2–4 days after the onset of paralysis. Detectable levels of neutralizing antibody were also present in the sera of 11/21 (52 %) clinically ill mice with blood positive for viral RNA. This study showed that viral RNA was detected after the development of CNS disease, but did not provide evidence that the bloodstream is a pathway of viral spread to the CNS. Detection of viral RNA does not necessarily indicate the presence of infectious virus and may merely reflect leakage of noninfectious viral RNA originating in infected CNS tissues into the bloodstream.

In another study, Preuss et al. (2009) made an effort to evaluate the pathogenetic importance of hematogenous spread of rabies virus. They inoculated mice intramuscularly in the gastrocnemius muscle or intravenously into the tail vein with 10^6 focus forming units of silver-haired bat rabies virus (SHBRV) variant. After intramusclular inoculation there was initial infection of spinal cord motor neurons and brainstem neurons and subsequent spread to involve forebrain neurons. After intravenous inoculation SHBRV initially infected hypothalamic nuclei that are connected by neurosecretory fibers to the neurohypophysis (posterior pituitary) and median eminence, which are circumventricular organs with an incomplete blood–brain barrier. These findings indicate that hematogenous spread of SHBRV leads to retrograde invasion of the CNS from the neurovascular interface of the hypothalamus–pituitary system. It is not unexpected that massive intravenous viral doses would initiate brain infection at sites in which the blood–brain barrier is relatively poorly developed (e.g., circumventricular organs). However, this does not shed any light on the sequence of events that likely occur during natural infection. Hence, there is no strong evidence that hematogenous spread plays a significant role in the spread of rabies virus to the CNS under natural conditions.

4 Pathology

Although there is severe clinical disease in rabies with a poor outcome, the histopathological changes in the brain and spinal cord are typically relatively mild. In rabies there are variable amounts of inflammatory changes with mononuclear cell, inflammatory cell infiltration of the leptomeninges, and perivascular mononuclear inflammatory infiltrates in the parenchyma. There are also microglial nodules called Babes' nodules in the parenchyma consisting of activated microglia and monocytes, which were described by Babes (1892). There may also be neuronophagia, which is a microscopic pattern characterized by accumulations of activated microglia/macrophages in the process of phagocytosing degenerating and/or dying neurons (Rossiter and Jackson 2007). In a series of 49 autopsies on human rabies cases, neuronophagia was observed in 57 % of the cases (Dupont and Earle 1965).

Neurons are the neural cell type predominantly infected by rabies virus, although other cell types are observed to be infected relatively infrequently (e.g., astrocytes) (Jackson et al. 2000). Overall, there are few degenerative changes in neurons. Infected neurons may contain eosinophilic inclusions in the cytoplasm called Negri bodies (Fig. 2), which were described by Adelchi Negri in the early 1900s (Negri 1903a, b) and are considered a pathological hallmark of rabies. Negri bodies are most prominent in large neurons (e.g., Purkinje cells) and, ultrastructurally, they are composed of large aggregates of granulofilamentous matrix material and variable numbers of viral particles (Rossiter and Jackson 2007).

5 Bases for Neuronal Dysfunction

Because the histopathological findings in human rabies, including the lack of neuronal death or apoptosis (Jackson et al. 2008), do not adequately explain the severe clinical disease in rabies, it has been understood that the CNS disease must be due to neuronal dysfunction (Fu and Jackson 2005). A number of hypotheses have been explored to determine the bases for the neuronal dysfunction in rabies, including abnormalities of a variety of neurotransmitters, nitric oxide neurotoxicity, and excitotoxicity (Jackson 2007b), but none of these mechanisms have proved to be satisfactory. Recent work argues against the mechanism of excitotoxicity in rabies (Scott et al. 2008; Weli et al. 2006). Previously unrecognized structural abnormalities affecting neuronal processes, including dendrites and axons, have recently been recognized in rabies virus-infected transgenic mice expressing the

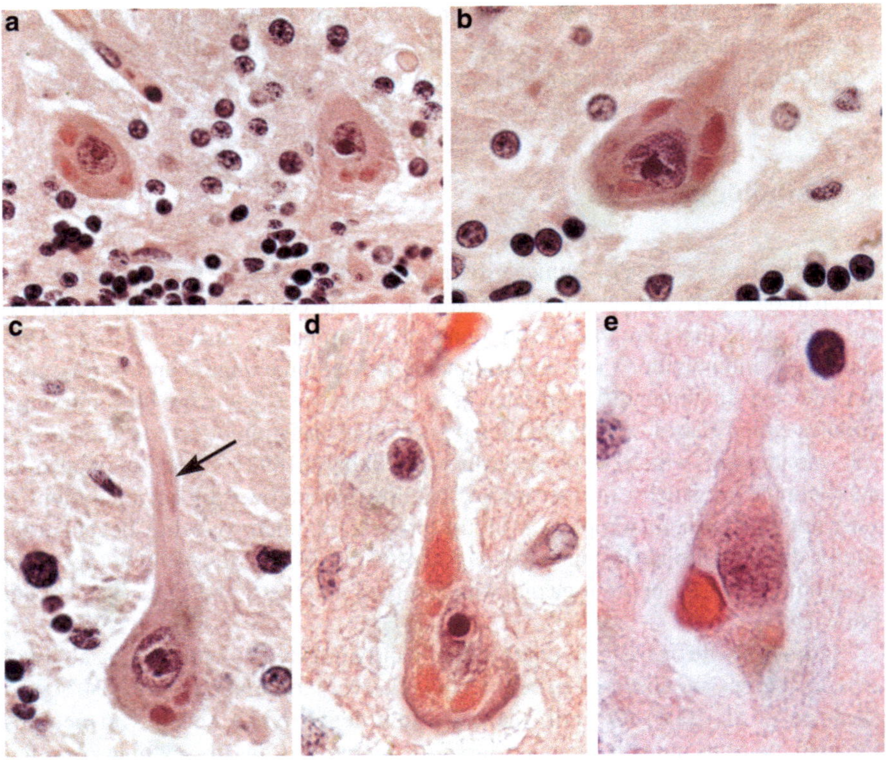

Fig. 2 Hematoxylin- and eosin-stained sections showing Negri bodies in the perikarya of (**a–c**) cerebellar Purkinje cells and (**d**, **e**) pyramidal neurons in the cerebral cortex of human rabies cases. The *arrow* in (**c**) indicates a Negri body in an apical dendrite (Magnifications: A, ×390, B, ×550, C, ×690, D, ×920, E, ×1085). (Adapted from Rossiter JP and Jackson AC: Pathology, in Rabies, Second Edition, edited by AC Jackson and WH Wunner, 2007, Elsevier Academic, London, pp 383–409; Copyright Elsevier)

yellow fluorescent protein using hindlimb footpad inoculation (Scott et al. 2008). These changes may explain the severe clinical disease in rabies. Further recent studies in primary neuronal cultures using dorsal root ganglion neurons indicate that oxidative stress likely plays an important role in the degeneration of neuronal process (Jackson et al. 2010).

6 Epidemiology

Worldwide, dogs are the most important vector of rabies and there is dog-to-dog transmission of rabies virus in resource-poor and resource-limited countries (Knobel et al. 2007). There are probably more than 75,000 human deaths per year worldwide, but the data are not very accurate and the number of cases is frequently underestimated. Political, economic, cultural, and religious barriers have all contributed to the persisting problem of human rabies. Unfortunately, at least half of the victims of rabies are children.

In North America the reservoir for rabies virus infection is in wildlife, including bats, raccoons, skunks, and foxes (Blanton et al. 2011). Rabies virus variants can be characterized by using reverse transcription polymerase chain reaction (RT-PCR) amplification and sequencing or with the use of monoclonal antibodies, which allows identification of the species of the vector that the variant is associated and can also be very useful in identifying the source of human cases in situations in which there is no history of an animal exposure. In the USA and Canada most human cases are due to transmission of rabies virus from bats, and the majority (60 %) of human cases of rabies due to bat variants do not give a history of a bat bite or scratch and 33 % have no history of contact with bats at all (Jackson 2011). In North America the most common bat rabies virus variant associated with human cases is found in silver-hair bats (*Lasionycteris noctivagans*) and eastern pipistrelle bats. In the USA the second most common variant associated with human cases is found in Brazilian (Mexican) free-tail bats (*Tadarida braziliensis*). Laboratory testing indicates that bats commonly found in houses, including big brown bats (*Eptesicus fuscus*) and little brown bats (*Myotis lucifugus*), are not infrequently found to be rabid, but the variants are much less frequently associated with human cases of rabies. Vampire bats also transmit rabies virus to humans and cattle in Latin America, and rabies in cattle is an important economic problem.

Terrestrial rabies in the United States is associated with rabies virus variants that cause rabies in a variety of carnivore species in specific geographical regions (Fig. 3). Raccoons have endemic rabies throughout the Eastern United States with over 2,200 laboratory diagnosed cases per year determined by using passive surveillance (Blanton et al. 2011). Raccoon rabies was present in Florida in the 1940s and spread northward and arrived in Canada in 1999 (Wandeler et al. 2000). Raccoon rabies has been well controlled in Canadian provinces with active rabies control programs (Rosatte et al. 2009). Transmission of the raccoon rabies virus variant resulting in human rabies is very rare and only a single case has been reported to date (Silverstein et al. 2003). Unlike bat exposures, the vast majority of raccoon exposures are recognized, which allows for the opportunity for the initiation of rabies postexposure prophylaxis.

Endemic skunk rabies is present in the Midwestern United States, California, and in the prairie provinces of Canada. Rabies in red foxes has been well controlled with the use of oral immunization programs in Canada (Ontario) and Europe (Rosatte et al. 2007). Worldwide, a variety of other animals are vectors of rabies, including mongooses, jackals, coyotes, and wolves. However, most human rabies exposures are actually associated with domestic animals, particularly companion animals such as dogs and cats. Rodents are not reservoirs of rabies virus. Small rodents, including mice, rats, chipmunks, squirrels, gerbils, hamsters, and guinea pigs, and lagomorphs (e.g., rabbits and hares) are rarely infected with rabies virus and have not been known to transmit rabies to humans (Manning et al. 2008). Woodchucks account for the majority of rabies in rodents that is reported to the United States Centers for Disease Control and Prevention (Childs et al. 1997).

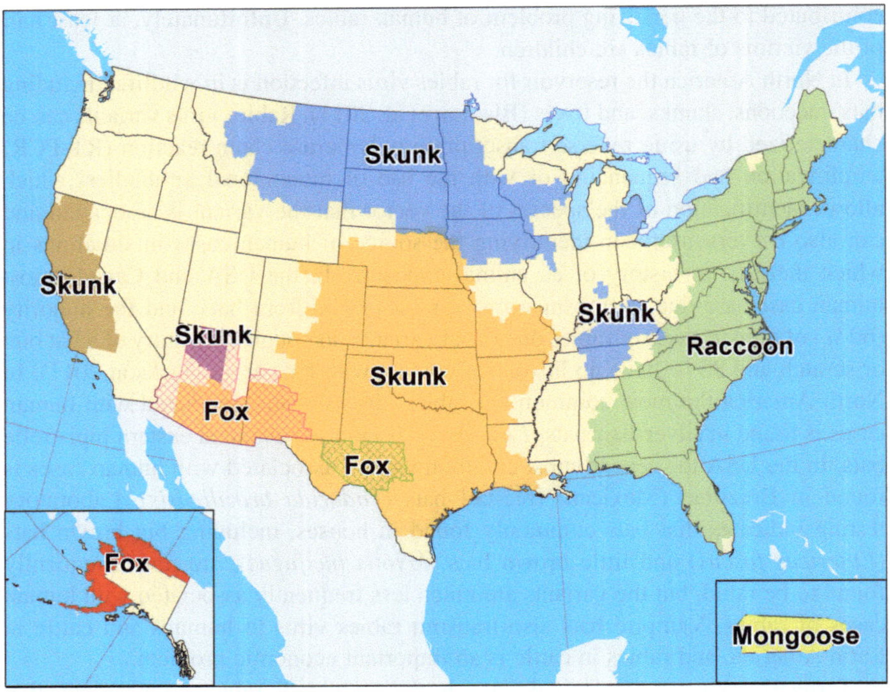

Fig. 3 Distribution of the major rabies virus variants among wild terrestrial reservoirs in the United States and Puerto Rico, 2010 (From JD Blanton et al., J Am Vet Med Assoc 239:773–783, 2011, Centers for Disease Control and Prevention)

All mammals are considered potentially susceptible to rabies. However, opossums are considered relatively resistant to rabies (Baer et al. 1990).

7 Clinical Features of Human Disease

The incubation period of rabies is usually 20–90 days, but there are cases with an incubation period of only a few days and others with incubation periods in excess of a year. In general, severe multiple bites and facial bites are associated with shorter incubation periods (Warrell and Warrell 1991). The best documented long incubation periods were cases who had immigrated to the United States that had incubation periods of up to 6 years with identification of a rabies virus variant from the country of origin (Smith et al. 1991). A 10-year-old Vietnamese girl in Australia developed rabies and transmission likely occurred at least 5 years earlier (Bek et al. 1992; McColl et al. 1993). The first prodromal symptoms of rabies are usually nonspecific and include fever, chills, malaise, fatigue, insomnia, anorexia, headache, anxiety, and irritability. These symptoms may last for up to 10 days before the onset of neurological symptoms. The first specific symptoms, which are neurological and highly suggestive of rabies, are paresthesias, pain, and pruritus at the site of the exposure; any associated wound may have completely healed at this time.

There are two clinical forms of rabies: encephalitic (furious) in 80 % of patients and paralytic (dumb) in 20 %. In encephalitic rabies it is thought that the burden of the involvement is in the brain, whereas in paralytic rabies the burden likely involves the spinal cord, spinal nerve roots, and nerve plexuses. In encephalitic rabies fever is common and patients have episodes of generalized arousal or hyperexcitability, which are separated by lucid periods (Warrell 1976). They may have aggressive behavior, confusion, and hallucinations. Signs of autonomic dysfunction include hypersalivation, piloerection (gooseflesh), sweating, and priapism. About half of patients develop hydrophobia, a characteristic manifestation of rabies. Patients may initially experience pain in the throat or have difficulty swallowing. On attempts to swallow, they experience contractions of the diaphragm and other inspiratory muscles, which last for about 5–15 s. Subsequently, the sight, sound, or even mention of water (or of any liquids) may trigger the spasms. A draft of air on the skin may have the same effect (aerophobia). The disease usually progresses through paralysis and coma, resulting in a fatal outcome.

In paralytic rabies flaccid muscle weakness develops early in the course of the disease and often begins in the bitten extremity and then spreads to involve other extremities and the facial muscles. Sphincter involvement, pain, and sensory disturbances also occur. Hydrophobia is unusual, although bulbar and respiratory muscles eventually become involved. Patients with paralytic rabies also progress to coma and death, although they may survive longer than patients with encephalitic rabies.

There are many potential medical complications in rabies patients who are treated aggressively in a critical care unit. Cardiac disorders are common complications, including sinus tachycardia, heart failure, hypotension, a variety of cardiac arrhythmias, and cardiac arrest (Hattwick 1974; Warrell et al. 1976). The cardiac manifestations probably reflect infection involving the autonomic nervous system (e.g., cardiac ganglia) or myocardium (Jackson et al. 1999) and there may be associated myocarditis (Cheetham et al. 1970; Raman et al. 1988; Ross and Armentrout 1962). Respiratory complications include hyperventilation, hypoxemia, respiratory depression with apnea, atelectasis, and aspiration pneumonia (Hattwick 1974). Either hyperthermia or hypothermia may be present and reflect hypothalamic involvement in the infection. Endocrine complications include inappropriate secretion of antidiuretic hormone and diabetes insipidus (Bhatt et al. 1974; Hattwick 1974). If patients are treated aggressively in critical care units, then multiple organ failure commonly develops.

8 Investigations

Routine laboratory investigations are usually normal in rabies. Magnetic resonance imaging may show signal abnormalities in brain, spinal cord, nerve roots, and/or plexuses, but these are not specific for the disease (Laothamatas et al. 2011). Cerebrospinal fluid (CSF) typically shows a mononuclear cell pleocytosis, usually with a cell count less than 100 cells/μL. Neutralizing antirabies virus antibodies may be detected in serum in a previously unvaccinated patient, but at this time clinical illness may be present for days to weeks. The presence of neutralizing antirabies virus antibodies in CSF is thought to be diagnostic of rabies, whereas there is an absence of CSF antibodies in patients who receive rabies vaccine.

A laboratory diagnosis of rabies can be made with the detection of rabies virus antigen or RNA in a body fluid or a tissue and rabies virus can be isolated (cultured) from CNS tissues. Rabies virus antigen may be detected using the direct fluorescent antibody technique in tissue sections from a full thickness skin biopsy (usually 5–6 mm in diameter) containing hair follicles (minimum of 10), which can be obtained from the posterior region of the neck at the hairline (Warrell et al. 1988). Corneal impression smears have also been used in the past, but they have relatively low sensitivity for rabies diagnosis (Warrell et al. 1988). The development of an assay for the presence of rabies virus RNA in saliva detected by using reverse transcription polymerase chain reaction (RT-PCR) amplification has been an important advance for the confirmation of rabies diagnosis. RT-PCR can also be used on skin biopsy specimens (Dacheux et al. 2008) and in CSF, although the latter is a much less sensitive test. A negative test for the detection of rabies virus antigen or RNA (except on brain tissues) does not exclude rabies, and repeat testing may need to be performed for diagnostic confirmation. Brain tissues may be obtained by brain biopsy, which is only very rarely performed, or postmortem with detection of rabies virus antigen and also by rabies virus isolation using culture techniques.

9 Human Rabies Due to Other Lyssaviruses

Rabies virus (Genotype 1) is in the genus *Lyssavirus* that includes ten other genotypes. Five of these ten genotypes have been recognized to very rarely cause human disease in which the clinical and laboratory features are indistinguishable from rabies (Table 1), excluding the case of a 3.5-year-old girl who had a convulsion associated with fever, which was reported to be due to Mokola virus infection (Familusi and Moore 1972). Mokola virus infection was a highly unlikely cause of the convulsion, and cross-contamination in the laboratory is the probable explanation for the viral isolation (Jackson 2007a). Cases due to other lyssaviruses have occurred in Africa (5) and Europe (4) and less commonly in Australia (2) and Asia (1) and no cases have occurred in the Americas.

10 Prognosis in Human Rabies

Rabies is virtually always fatal. Most surviving cases have received rabies vaccine prior to the onset of their illness (Table 3). Another case did not receive rabies vaccine, but had neutralizing antirabies virus antibodies at the time of clinical presentation (Willoughby et al. (2005)).

Table 3 Cases of Human Rabies with Recovery (Adapted from Jackson AC: Human disease, in Rabies, Second Edition, edited by AC Jackson and WH Wunner, 2007, Elsevier Academic, London, pp 309-340; Copyright Elsevier)

Location	Year	Age of patient	Transmission	Immunization prior to onset	Outcome	References
United States	1970	6	Bat bite	Duck embryo vaccine	Complete recovery	Hattwick et al. (1972)
Argentina	1972	45	Dog bites	Suckling mouse brain vaccine	Mild sequelae	Porras et al. (1976)
United States	1977	32	Laboratory (vaccine strain)	Preexposure vaccination	Sequelae	Tillotson et al. (1977a, b)
Mexico	1992	9	Dog bites	Postexposure vaccination (combination)	Severe sequelae[a]	Alvarez et al. (1994)
India	2000	6	Dog bites	Postexposure vaccination (combination)	Severe sequelae[b]	Madhusudana et al. (2002)
United States	2004	15	Bat bite	No postexposure therapy	Mild sequelae	Hu et al. (2007), Willoughby et al. (2005)
Brazil	2008	15	Vampire bat bite	Postexposure vaccination	Severe sequelae	Ministerio da Saude in Brazil (2008)

Recovery of cases with atypical features of rabies without the development of rabies virus-neutralizing antibodies has not been not included because they are likely not cases of rabies (Blanton et al. 2011; Holzmann-Pazgal et al. 2010)

[a]Patient died less than 4 years after developing rabies with marked neurological sequelae (L. Alvarez, personal communication)

[b]Patient died about 2 years after developing rabies with marked neurological sequelae (S. Mahusudana, personal communication)

A 17-year-old patient recently survived rabies virus infection, but she did not have typical clinical features of rabies and did not require intensive care (Holzmann-Pazgal et al. 2010). She had fever, headache, nuchal rigidity, disorientation, and limb weakness, and had a CSF pleocytosis and enlarged lateral ventricles on MR imaging. She developed only a low titer of rabies virus-neutralizing antibodies in sera (up to 1:14) after receiving human rabies immune globulin and one dose of rabies vaccine, and she had no detectable neutralizing antirabies virus antibodies in CSF (Holzmann-Pazgal et al. 2010). Other diagnostic tests for rabies, which are based on detection of rabies virus antigen and RNA, were negative. A similar case of an 8-year-old female from California (Wiedeman et al. 2012) experienced sore throat and vomiting and later over a few days she developed swallowing difficulties. A few days later she developed abdominal pain and neck and back pain, and then on the next day she had sore throat and abdominal pain and was noted to be confused. She deteriorated rapidly and required endotracheal intubation. CSF showed 6 leukocytes/μL with a protein of 62 mg/dL. Over the next few days she developed fever, decreased level of consciousness, and ascending flaccid paralysis. MR brain imaging showed multiple T2 and FLAIR signal abnormalities in cortical and subcortical regions and in the periventricular white matter. Electrophysiological studies were consistent with a demyelinating and predominantly motor polyneuropathy. She had rabies virus-specific IgG and IgM in her serum and CSF, but she did not develop rabies virus-neutralizing antibodies. All other diagnostic tests for rabies were also negative. She showed progressive improvement after just over 2 weeks in hospital and she was discharged home after another 5 weeks in rehabilitation. Both of these cases had atypical clinical features for rabies, and the lack or minimal development of rabies virus-neutralizing antibodies indicates that it is unlikely that either of these patients actually developed rabies and recovered. However, the exact etiology and pathogenetic mechanisms involved in their illnesses remain uncertain.

11 Precautions to Prevent Exposures from Rabies Patients

Excluding transmission from transplantation of tissues or organs, there is only a single report from Ethiopia indicating probable human-to-human transmission of rabies virus in two cases (Fekadu et al. 1996). In this report, a 41-year-old female died of rabies 33 days after her 5-year-old son died of rabies. He had bitten his mother on her little finger. Additionally, a 5-year-old boy presented with rabies 36 days after his mother died of rabies. He had repeatedly received kisses on his mouth from his mother during her illness. There are no reports of transmission to health care workers, although this remains an important theoretical concern. As a result of the possibility of transmission, health care workers should initiate body substance precautions as soon as diagnosis of rabies is seriously considered and wear gowns, gloves, masks, and eye protection. They may also require postexposure prophylaxis after high-risk contact with a patient with rabies. Oral secretions are of particular concern because of the possibility that they may contain infectious rabies virus. Failure to consider a diagnosis of rabies and initiate appropriate precautions may lead to subsequent recommendations for postexposure rabies prophylaxis to be given to a large number of health care workers at very high cost. For example, 440 individuals were administered postexposure rabies prophylaxis after postmortem diagnosis of a case of rabies in British Columbia (Parker et al. 2003).

12 Therapy of Human Rabies

Although rabies may be effectively prevented, medical management after clinical disease develops almost universally results in a fatal outcome. In 2003 a group of physicians with expertise in rabies and rabies researchers made recommendations on therapies that could be considered for an aggressive approach for a patient with rabies (Jackson et al. 2003). Young and previously healthy patients with an early clinical diagnosis of rabies were felt to be the best potential candidates for aggressive therapy (Jackson et al. 2003). Therapies suggested for consideration include rabies vaccine, human rabies immune globulin, monoclonal antibodies (for the future), ribavirin, interferon-α, and ketamine. The recommendation for therapy with ketamine was based on experimental animal work that was performed at the Institut Pasteur in Paris in the early 1990s (Lockhart et al. 1991). Like current therapies for a variety of infectious and other noninfectious diseases, it was felt that a combination of therapies might improve efficacy in situations in which specific therapies used individually had failed in the past.

In 2004 a patient survived from rabies who had not received rabies vaccine prior to the onset of clinical disease (Willoughby et al. (2005)). This 15-year-old female was bitten by a bat on a finger and did not receive postexposure prophylaxis therapy for rabies. About a month after the bite she developed typical clinical features of rabies encephalitis and had a CSF pleocytosis. Five days after the onset of neurological symptoms she was transferred to a tertiary care hospital in Milwaukee, Wisconsin. Neutralizing antirabies virus antibodies were detected in sera and CSF (initially at titers of 1:102 and 1:47, respectively). Nuchal skin biopsies were negative for rabies virus antigen and rabies virus RNA was not detected in the skin biopsies or in saliva by RT-PCR, and viral isolation on saliva was negative. The patient was intubated, and put into a drug-induced coma, which included the noncompetitive *N*-methyl-D-aspartate (NMDA) antagonist ketamine at 48 mg/kg/day as a continuous infusion and intravenous midazolam for 7 days. A burst-suppression pattern on her electroencephalogram was maintained and supplemental phenobarbital was given as needed. She also received antiviral therapy, including intravenous ribavirin and amantadine 200 mg/day administered enterally. She improved and was discharged from hospital with neurologic deficits and she subsequently demonstrated further progressive neurologic improvement (Hu et al. 2007).

Table 4 Cases of human rabies with treatment failures that used the main components of the "Milwaukee Protocol." (Updated from Jackson AC: Therapy in human rabies, in Research Advances in Rabies, Alan C. Jackson (ed), Advances in Virus Research 79:365–375, 2011; Copyright Elsevier)

Case No.	Year of death	Age and sex of patient	Virus source	Country	References
1	2005	47 male	Kidney and pancreas transplant (dog)	Germany	Maier et al. (2010)
2	2005	46 female	Lung transplant (dog)	Germany	Maier et al. (2010)
3	2005	72 male	Kidney transplant (dog)	Germany	Maier et al. (2010)
4	2005	Unknown	Dog	India	Bagchi (2005)
5	2005	7 male	Vampire bat	Brazil	[a]
6	2005	20-30 female	Vampire bat	Brazil	[a]
7	2006	33 male	Dog	Thailand	Hemachudha et al. (2006)
8	2006	16 male	Bat	USA (Texas)	Houston Chronicle (2006)
9	2006	10 female	Bat	USA (Indiana)	Christenson et al. (2007)
10	2006	11 male	Dog (Philippines)	USA (California)	Aramburo et al. (2011), Christenson et al. (2007)
11	2007	73 male	Bat	Canada (Alberta)	McDermid et al. (2008)
12	2007	55 male	Dog (Morocco)	Germany	Drosten (2007)
13	2007	34 female	Bat (Kenya)	The Netherlands	van Thiel et al. (2009)
14	2008	5 male	Dog	Equatorial Guinea	Rubin et al. (2009)
15	2008	55 male	Bat	USA (Missouri)	Pue et al. (2009), Turabelidze et al. (2009)
16	2008	8 female	Cat	Colombia	Juncosa (2008)
17	2008	15 male	Vampire bat	Colombia	Badillo et al. (2009)
18	2009	37 female	Dog (South Africa)	Northern Ireland	Hunter et al. (2010)
19	2009	42 male	Dog (India)	USA (Virginia)	Blanton et al. (2010)
20	2010	11 female	Cat	Romania	Luminos et al. (2011)
21	2012	63 male	Brown bat	USA (Massachusetts)	Mccormick (2012)

[a]Personal communication from Dr. Rita Medeiros, University of Para, Belem, Brazil

This patient is the first documented survivor who did not receive any rabies vaccine prior to the onset of clinical rabies. It is uncertain if therapy with one or more specific agents played a significant role in her favorable outcome (Jackson 2005). Since that time there have been over 20 cases in which the main components of this approach (the "Milwaukee Protocol") have been used and the therapy failed with fatal outcomes (Table 4). The induction of coma per se has no established benefit for the management of infectious diseases of the nervous system, and to date there is no evidence supporting this approach in rabies or other viral encephalitides. Hence, there is little justification for therapeutic coma becoming a routine therapy for the management of rabies. Experimental evidence does not support excitotoxicity in rabies; recent evidence in an animal model argues against this hypothesis and there was lack of efficacy of ketamine therapy both in vitro and in vivo (Weli et al. 2006). In situations in which there is strong experimental evidence of excitotoxicity in animal models, multiple clinical trials in humans have shown a lack of efficacy of neuroprotective agents in stroke (Ginsberg 2009). Hence, a neuroprotective effect of a therapy given to a single patient without a credible scientific rationale is very doubtful. It is highly probable that this patient would also have recovered with only supportive therapy.

Neutralizing antirabies virus antibodies are an important marker of an adaptive immune response that is essential for viral clearance (Lafon 2007) and recovery. The presence of serum-neutralizing antirabies virus antibodies early in a patient's clinical course, probably occurring in less than 20 % of all patients with rabies, is likely an important factor contributing to a favorable outcome. There have been six survivors of rabies who received rabies vaccine prior to the onset of their disease and only one without vaccine (Table 3). This information suggests that an early immune response is associated with a positive outcome. Recovery of cases with atypical clinical features of rabies without the development of rabies virus-neutralizing antibodies (Blanton et al. 2011; Holzmann-Pazgal et al. 2010) is likely not due to rabies and these patients should not be considered survivors. It should be noted that therapy of encephalitic (furious) rabies with massive doses of human rabies immune globulin resulted in the rapid development of quadriplegia and bilateral facial paralysis (Hemachudha et al. 2003), suggesting an immunopathological mechanism of neuronal injury. Bat rabies virus variants may be less neurovirulent than canine virus variants or other variants that are responsible for most human cases of rabies (Lafon 2005), and rabies due to canine rabies virus variants likely has a less favorable outcome than cases caused by bat rabies virus variants. One previous survivor of rabies, who was also infected with a bat rabies virus variant received rabies vaccine prior to the onset of disease and made a good neurological recovery (Hattwick et al. 1972). It is unknown if the causative bat rabies virus variant in the Milwaukee case was attenuated and had different biological properties than other isolated variants because there was no viral isolation with this case. Finally, most survivors of rabies to date have shown neutralizing antirabies virus antibodies in sera and CSF. However, other diagnostic laboratory tests are usually negative for rabies virus antigen and RNA in fluids and tissues, and brain tissues have not been tested. This may be because viral clearance was so effective that centrifugal spread of the infection to peripheral organ sites was reduced or very rapid clearance occurred through immune-mediated mechanisms.

A Canadian case of rabies was treated with the Milwaukee Protocol and after therapy with therapeutic coma remained in a brain death-like state for about 4 weeks. At autopsy there was complete loss of neurons in the cerebral cortex and positive staining for rabies virus antigen was observed in both brainstem and cerebellar neurons, indicating a failure of clearance of the viral infection from the brain and also failure of protection against neuronal injury and loss (McDermid et al. 2008). In Germany lung and kidney/pancreas recipients from a rabies virus-infected donor developed rabies and were treated with major components of the Milwaukee Protocol, including intravenous midazolam, ketamine, and phenobarbital (in one) (Maier et al. 2010). One patient died within 2 days whereas the other survived 64 days after the onset of clinical rabies. At autopsy the two patients had $1.2–2.3 \times 10^9$ RNA copies/mg of CNS tissue, which indicates ineffective viral clearance with the therapy. The long surviving patient did show viral clearance from systemic organs and peripheral nerve. Hence, Milwaukee Protocol therapy has proved ineffective in promoting viral clearance from the CNS in rabies. It remains highly doubtful that the Milwaukee Protocol will prove to be useful in the management of human rabies. Unfortunately, promotion and repetition of this flawed therapy may impede progress in developing new effective therapies for rabies. We need a better understanding of basic mechanisms underlying rabies pathogenesis in humans and animals, which may be helpful in the development of novel therapeutic approaches for the management of this disease.

13 Prevention of Human Rabies

The most efficient method of reducing human rabies is to eliminate rabies in animal vectors. Worldwide, endemic dog rabies in many countries remains an important public health problem and the elimination of canine rabies needs greater attention. Oral vaccination of wildlife vectors such as foxes, raccoons, and coyotes has been very useful in controlling wildlife rabies in a variety of geographical regions (Rosatte et al. 2007). There is no effective method of controlling rabies in insectivorous bats.

Rabies can be very effectively prevented after recognized rabies exposures if current guidelines are observed from the Centers for Disease Control and Prevention (Manning et al. 2008) or World Health Organization (World Health Organization 2005), which are available on the Morbidity and Mortality Weekly Report (http://www.cdc.gov/mmwr/) and World Health Organization (http://www.who.int/en/) websites, respectively. Algorithms can be helpful in making management decisions after an exposure (Fig. 4). Assessment of the risk of an exposure is based on the nature of the exposure, the species, and clinical status of an animal and whether a dog, cat, or ferret is available for an observation period of 10 days. Laboratory testing on brain tissues is needed for a definitive diagnosis of rabies in animals. If a dog, cat, or ferret remains healthy for a period of 10 days after an exposure, then one can confidently conclude that transmission of rabies virus did not occur during that exposure. Effective rabies prophylaxis in humans includes wound cleansing, rabies immunization with a modern cell culture rabies vaccine (recently reduced from five to four doses at appropriate intervals), and injection of human rabies immune globulin (dose depends on body weight) into

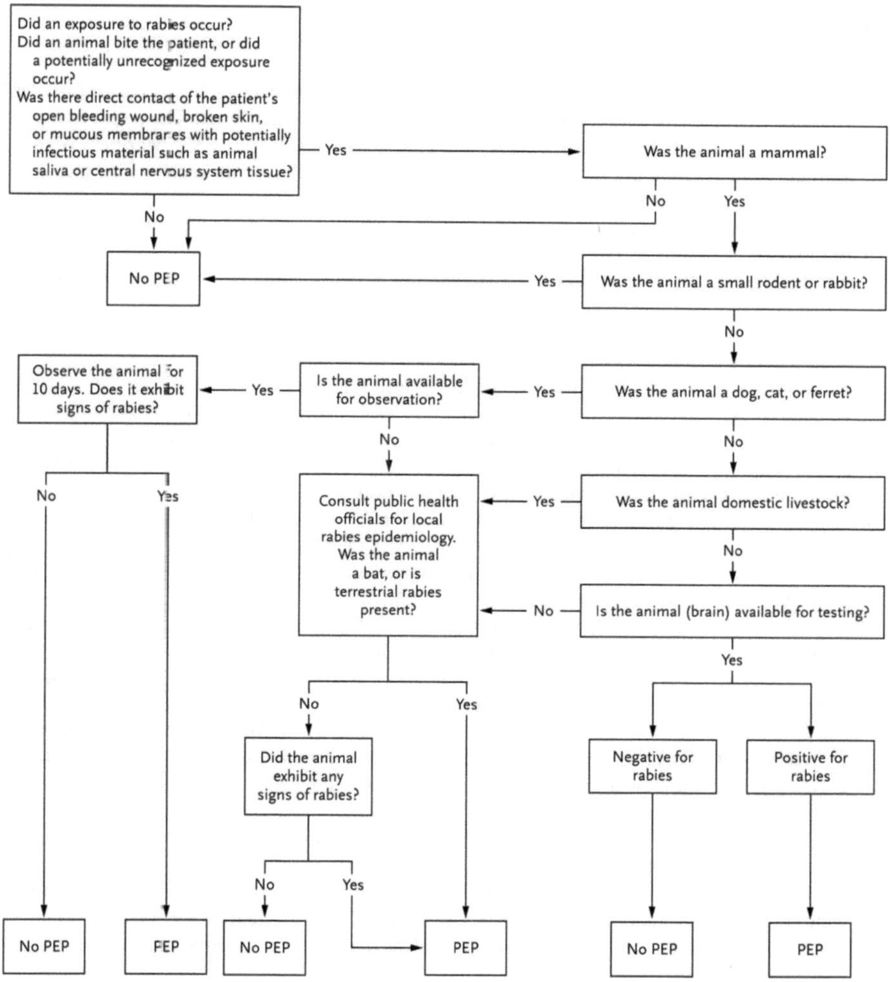

Fig. 4 Algorithm for rabies postexposure prophylaxis (PEP) in the United States. Reproduced from Rupprecht CE and Gibbons RV: Prophylaxis against rabies. New England Journal of Medicine 351:2626–2635, 2004

and around the wound with the residual quantity given intramuscularly at a distant site from where the vaccine was administered (Manning et al. 2008; Rupprecht et al. 2010).

Preexposure rabies vaccination is available to persons at risk of having rabies exposures, which simplifies postexposure therapy. Booster doses of rabies vaccine can be given at periodic intervals based on serum antirabies virus-neutralizing antibody titers. Normally, three doses of rabies vaccine (intramuscularly or intradermally) are given at intervals (on days 0, 7, and 21) for preexposure vaccination. After an exposure in preimmunized individuals, two doses of rabies vaccine are given (on days 0 and 3) and no human rabies immune globulin should be given.

References

Alvarez L, Fajardo R, Lopez E, Pedroza R, Hemachudha T, Kamolvarin N, Cortes G, Baer GM (1994) Partial recovery from rabies in a nine-year-old boy. Pediatr Infect Dis J 13:1154–1155

Aramburo A, Willoughby RE, Bollen AW, Glaser CA, Hsieh CJ, Davis SL, Martin KW, Roy-Burman A (2011) Failure of the Milwaukee Protocol in a child with rabies. Clin Infect Dis 53:572–574. doi:10.1093/cid/cir483

Babes VM (1892) Sur certains caracteres des lesions histologiques de la rage. Ann Inst Pasteur 6:209–223

Badillo R, Mantilla JC, Pradilla G (2009) Human rabies encephalitis by a vampire bat bite in an urban area of Colombia (Spanish). Biomedica 29:191–203

Baer GM, Shaddock JH, Quirion R, Dam TV, Lentz TL (1990) Rabies susceptibility and acetylcholine receptor (Letter). Lancet 335:664–665

Bagchi S (2005) Coma therapy The Telegraph Calcutta July 4:2005

Balachandran A, Charlton K (1994) Experimental rabies infection of non-nervous tissues in skunks (*Mephitis mephitis*) and foxes (*Vulpes vulpes*). Vet Pathol 31:93–102

Bek MD, Smith WT, Levy MH, Sullivan E, Rubin GL (1992) Rabies case in New South Wales, 1990: public health aspects. Med J Aust 156:596–600

Bhatt DR, Hattwick MAW, Gerdsen R, Emmons RW, Johnson HN (1974) Human rabies: diagnosis, complications, and management. Am J Dis Child 127:862–869

Blanton JD, Palmer D, Rupprecht CE (2010) Rabies surveillance in the United States during 2009. J Am Vet Med Assoc 237:646–657. doi:10.2460/javma.237.6.646

Blanton JD, Palmer D, Dyer J, Rupprecht CE (2011) Rabies surveillance in the United States during 2010. J Am Vet Med Assoc 239:773–783. doi:10.2460/javma.239.6.773

Botvinkin AD, Selnikova OP, Antonova LA, Moiseeva AB, Nesterenko EY, Gromashevsky LV (2005) Human rabies case caused from a bat bite in Ukraine. Rabies Bull Europe 29:5–7

Burton EC, Burns DK, Opatowsky MJ, El-Feky WH, Fischbach B, Melton L, Sanchez E, Randall H, Watkins DL, Chang J, Klintmalm G (2005) Rabies encephalomyelitis: clinical, neuroradiological, and pathological findings in 4 transplant recipients. Arch Neurol 62:873–882. doi:10.1001/archneur.62.6.873

Ceccaldi PE, Gillet JP, Tsiang H (1989) Inhibition of the transport of rabies virus in the central nervous system. J Neuropathol Exp Neurol 48:620–630

Ceccaldi P-E, Ermine A, Tsiang H (1990) Continuous delivery of colchicine in the rat brain with osmotic pumps for inhibition of rabies virus transport. J Virol Meth 28:79–84

Charlton KM, Casey GA (1979a) Experimental oral and nasal transmission of rabies virus in mice. Can J Comp Med 43:10–15

Charlton KM, Casey GA (1979b) Experimental rabies in skunks: immunofluorescence light and electron microscopic studies. Lab Invest 41:36–44

Charlton KM, Casey GA (1979c) Experimental rabies in skunks: oral, nasal, tracheal and intestinal exposure. Can J Comp Med 43:168–172

Charlton KM, Casey GA, Campbell JB (1983) Experimental rabies in skunks: mechanisms of infection of the salivary glands. Can J Comp Med 47:363–369

Charlton KM, Nadin-Davis S, Casey GA, Wandeler AI (1997) The long incubation period in rabies: delayed progression of infection in muscle at the site of exposure. Acta Neuropathol 94:73–77

Cheetham HD, Hart J, Coghill NF, Fox B (1970) Rabies with myocarditis: two cases in England. Lancet 1:921–922

Childs JE, Colby L, Krebs JW, Strine T, Feller M, Noah D, Drenzek C, Smith JS, Rupprecht CE (1997) Surveillance and spatiotemporal associations of rabies in rodents and lagomorphs in the United States, 1985-1994. J Wildl Dis 33:20–27

Christenson JC, Holm BM, Lechlitner S, Howell JF, Wenger M, Roy-Burman A, Hsieh CJ, LaBar L, Petru A, Davis S, Simons N, Apolinario P, Furst K, Benjamin B, Christie LJ, Schnurr D,

Glaser CA, Rosenberg J, Sun B, Willoughby RE, Rupprecht CE (2007) Human rabies—Indiana and California, 2006. MMWR Morb Mortal Wkly Rep 56:361–365
Constantine DG (1962) Rabies transmission by nonbite route. Public Health Rep 77:287–289
Dacheux L, Reynes JM, Buchy P, Sivuth O, Diop BM, Rousset D, Rathat C, Jolly N, Dufourcq JB, Nareth C, Diop S, Iehle C, Rajerison R, Sadorge C, Bourhy H (2008) A reliable diagnosis of human rabies based on analysis of skin biopsy specimens. Clin Infect Dis 47:1410–1417. doi:10.1086/592969
Drosten C (2007) Rabies—Germany (Hamburg) ex Morocco. ProMED-mail 20070419.1287. Available http://www.promedmail.org/. Accessed 20 Jan 2012
Dupont JR, Earle KM (1965) Human rabies encephalitis. A study of forty-nine fatal cases with a review of the literature. Neurology 15:1023–1034
Familusi JB, Moore DL (1972) Isolation of a rabies related virus from the cerebrospinal fluid of a child with "aseptic meningitis". Afr J Med Sci 3:93–96
Familusi JB, Osunkoya BO, Moore DL, Kemp GE, Fabiyi A (1972) A fatal human infection with Mokola virus. Am J Trop Med Hyg 21:959–963
Fekadu M, Endeshaw T, Alemu W, Bogale Y, Teshager T, Olson JG (1996) Possible human-to-human transmission of rabies in Ethiopia. Ethiop Med J 34:123–127
Fischman HR, Ward FE (1968) Oral transmission of rabies virus in experimental animals. Am J Epidemiol 88:132–138
Fu ZF, Jackson AC (2005) Neuronal dysfunction and death in rabies virus infection. J Neurovirol 11:101–106. doi:10.1080/13550280590900445
Gillet JP, Derer P, Tsiang H (1986) Axonal transport of rabies virus in the central nervous system of the rat. J Neuropathol Exp Neurol 45:619–634
Ginsberg MD (2009) Current status of neuroprotection for cerebral ischemia: synoptic overview. Stroke 40(Suppl 3):S111–S114. doi:10.1161/STROKEAHA.108.528877
Hanna JN, Carney IK, Smith GA, Tannenberg AEG, Deverill JE, Botha JA, Serafin IL, Harrower BJ, Fitzpatrick PF, Searle JW (2000) Australian bat lyssavius infection: a second human case, with a long incubation period. Med J Aust 172:597–599
Hattwick MAW (1974) Human rabies. Public Health Rev 3:229–274
Hattwick MAW, Weis TT, Stechschulte CJ, Baer GM, Gregg MB (1972) Recovery from rabies: a case report. Ann Intern Med 76:931–942
Hemachudha T, Sunsaneewitayakul B, Desudchit T, Suankratay C, Sittipunt C, Wacharapluesadee S, Khawplod P, Wilde H, Jackson AC (2006) Failure of therapeutic coma and ketamine for therapy of human rabies. J Neurovirol 12:407–409. doi:10.1080/13550280600902295
Holzmann-Pazgal G, Wanger A, Degaffe G, Rose C, Heresi G, Amaya R, Eshofonie A, Lee-Han H, Awosika-Olumo A, Kuzmin I, Rupprecht CE (2010) Presumptive abortive human rabies—Texas, 2009. MMWR 59:185–190
Houston Chronicle. (2006). Rabies, human—USA (Texas). ProMED-mail 20060513.1360. Available http://www.promedmail.org/. Accessed 20 Jan 2012
Hu WT, Willoughby RE Jr, Dhonau H, Mack KJ (2007) Long-term follow-up after treatment of rabies by induction of coma (Letter). N Engl J Med 357:945–946
Hunter M, Johnson N, Hedderwick S, McCaughey C, Lowry K, McConville J, Herron B, McQuaid S, Marston D, Goddard T, Harkess G, Goharriz H, Voller K, Solomon T, Willoughby RE, Fooks AR (2010) Immunovirological correlates in human rabies treated with therapeutic coma. J Med Virol 82:1255–1265. doi:10.1002/jmv.21785
Jackson AC (2005) Recovery from rabies (Editorial). N Engl J Med 352:2549–2550. doi:10.1056/NEJMe058092
Jackson AC (2007a) Human disease. In: Jackson AC, Wunner WH (eds) Rabies, 2nd edn. Elsevier Academic, London, pp 309–340
Jackson AC (2007b) Pathogenesis. In: Jackson AC, Wunner WH (eds) Rabies, 2nd edn. Elsevier Academic, London, pp 341–381
Jackson AC (2011) Update on rabies. Research and Reports in Tropical Medicine 2:31–43. doi:10.2147/RRTM.S16013

Jackson AC, Ye H, Phelan CC, Ridaura-Sanz C, Zheng Q, Li Z, Wan X, Lopez-Corella E (1999) Extraneural organ involvement in human rabies. Lab Invest 79:945–951
Jackson AC, Phelan CC, Rossiter JP (2000) Infection of Bergmann glia in the cerebellum of a skunk experimentally infected with street rabies virus. Can J Vet Res 64:226–228
Jackson AC, Warrell MJ, Rupprecht CE, Ertl HCJ, Dietzschold B, O'Reilly M, Leach RP, Fu ZF, Wunner WH, Bleck TP, Wilde H (2003) Management of rabies in humans. Clin Infect Dis 36:60–63. doi:10.1086/344905
Jackson AC, Randle E, Lawrance G, Rossiter JP (2008) Neuronal apoptosis does not play an important role in human rabies encephalitis. J Neurovirol 14:368–375. doi:10.1080/13550280802216502
Jackson AC, Kammouni W, Zherebitskaya E, Fernyhough P (2010) Role of oxidative stress in rabies virus infection of adult mouse dorsal root ganglion neurons. J Virol 84:4697–4705. doi:10.1128/JVI.02654-09
Johnson N, Brookes SM, Healy DM, Spencer Y, Hicks D, Nunez A, Wells G, Fooks AR (2012) Pathology associated with a human case of rabies in the United Kingdom caused by European bat lyssavirus type-2 (In press). Intervirology. doi: 10.1159/000333019
Juncosa B (2008) Hope for rabies victims: unorthodox coma therapy shows promise. First a U.S. girl—and now two South American kids survive onset of the deadly virus. Available http://www.scientificamerican.com/. Accessed 20 Jan 2012
Knobel DL, Kaare M, Fèvre E, Cleaveland S (2007) Dog rabies and its control. In: Jackson AC, Wunner WH (eds) Rabies, 2nd edn. Elsevier Academic, London, pp 573–594
Lafon M (2005) Bat rabies–the Achilles heel of a viral killer? Lancet 366:876–877
Lafon M (2007) Immunology. In: Jackson AC, Wunner WH (eds) Rabies, 2nd edn. Elsevier Academic, London, pp 489–504
Laothamatas J, Sungkarat W, Hemachudha T (2011) Neuroimaging in rabies. Adv Virus Res 79:309–327. doi:10.1016/B978-0-12-387040-7.00014-7
Lentz TL, Burrage TG, Smith AL, Crick J, Tignor GH (1982) Is the acetylcholine receptor a rabies virus receptor? Science 215:182–184
Leonova GN, Belikov SI, Kondratov IG, Krylova NV, Pavlenko EV, Romanova EV, Chentsova IV, Petukhova SA (2009) A fatal case of bat lyssavirus infection in Primorye Territory of the Russian Far East. Rabies Bull Europe 33:5–8
Lockhart BP, Tsiang H, Ceccaldi PE, Guillemer S (1991) Ketamine-mediated inhibition of rabies virus infection *in vitro* and in rat brain. Antivir Chem Chemother 2:9–15
Lodmell DL, Dimcheff DE, Ewalt LC (2006) Viral RNA in the bloodstream suggests viremia occurs in clinically ill rabies-infected mice. Virus Res 116:114–118
Luminos M, Barboi G, Draganescu A, Streinu Cercel A, Staniceanu F, Jugulete G, Visan A, Negulescu C, Turcitu MA (2011) Human rabies in a Romanian boy—an ante mortem case study. Rabies Bull Europe 35:5–10
Madhusudana SN, Nagaraj D, Uday M, Ratnavalli E, Kumar MV (2002) Partial recovery from rabies in a six-year-old girl (Letter). Int J Infect Dis 6:85–86
Maier T, Schwarting A, Mauer D, Ross RS, Martens A, Kliem V, Wahl J, Panning M, Baumgarte S, Muller T, Pfefferle S, Ebel H, Schmidt J, Tenner-Racz K, Racz P, Schmid M, Struber M, Wolters B, Gotthardt D, Bitz F, Frisch L, Pfeiffer N, Fickenscher H, Sauer P, Rupprecht CE, Roggendorf M, Haverich A, Galle P, Hoyer J, Drosten C (2010) Management and outcomes after multiple corneal and solid organ transplantations from a donor infected with rabies virus. Clin Infect Dis 50:1112–1119. doi:10.1086/651267
Manning SE, Rupprecht CE, Fishbein D, Hanlon CA, Lumlertdacha B, Guerra M, Meltzer MI, Dhankhar P, Vaidya SA, Jenkins SR, Sun B, Hull HF (2008) Human rabies prevention—United States, 2008: recommendations of the Advisory Committee on Immunization Practices. MMWR Morb Mortal Wkly Rep 57(No. RR-3):1–28
McColl KA, Gould AR, Selleck PW, Hooper PT, Westbury HA, Smith JS (1993) Polymerase chain reaction and other laboratory techniques in the diagnosis of long incubation rabies in Australia. Aust Vet J 70:84–89

Mccormick C (2012) Rabies killed Marston Mills man. Cape Cod Times. Cape Cod Media Group. January 28, 2012

McDermid RC, Saxinger L, Lee B, Johnstone J, Noel Gibney RT, Johnson M, Bagshaw SM (2008) Human rabies encephalitis following bat exposure: failure of therapeutic coma. Can Med Assoc J 178:557–561. doi:10.1503/cmaj.071326

Meredith CD, Rossouw AP, HvP K (1971) An unusual case of human rabies thought to be of chiropteran origin. S Afr Med J 45:767–769

Metze K, Feiden W (1991) Rabies virus ribonucleoprotein in the heart (Letter). N Engl J Med 324:1814–1815

Ministerio da Saude in Brazil (2008) Rabies, human survival, bat-Brazil: (Pernambuco). ProMED-mail 20081114.3599. Available http://www.promedmail.org/. Accessed 20 Jan 2012

Nathwani D, McIntyre PG, White K, Shearer AJ, Reynolds N, Walker D, Orange GV, Fooks AR (2003) Fatal human rabies caused by European bat lyssavirus type 2a infection in Scotland. Clin Infect Dis 37:598–601

Negri A (1903a) Beitrag zum Studium der Aetiologie der Tollwuth. Z Hyg Infektionskr 43:507–528

Negri A (1903b) Zur Aetiologie der Tollwuth. Die diagnose der Tollwuth auf Grund der Neuen Befunde. Z Hyg Infektionskr 44:519

Parker R, McKay D, Hawes C, Daly P, Bryce E, Doyle P, Moore W, McKenzie I, Roscoe D, Weatherill S, Skowronski DM, Petric M, Pielak K, Naus M (2003) Human rabies, British Columbia—January 2003. Can Commun Dis Rep 29:137–138

Paweska JT, Blumberg LH, Liebenberg C, Hewlett RH, Grobbelaar AA, Leman PA, Croft JE, Nel LH, Nutt L, Swanepoel R (2006) Fatal human infection with rabies-related Duvenhage virus, South Africa. Emerg Infect Dis 12:1965–1967

Porras C, Barboza JJ, Fuenzalida E, Adaros HL, Oviedo AM, Furst J (1976) Recovery from rabies in man. Ann Intern Med 85:44–48

Preuss MA, Faber ML, Tan GS, Bette M, Dietzschold B, Weihe E, Schnell MJ (2009) Intravenous inoculation of a bat-associated rabies virus causes lethal encephalopathy in mice through invasion of the brain via neurosecretory hypothalamic fibers. PLoS Pathog 5:e1000485. doi:10.1371/journal.ppat.1000485

Pue HL, Turabelidze G, Patrick S, Grim A, Bell C, Reese V, Basilan R, Rupprecht C, Robertson K (2009) Human rabies—Missouri, 2008. MMWR Morb Mortal Wkly Rep 58:1207–1209

Raman GV, Prosser A, Spreadbury PL, Cockcroft PM, Okubadejo OA (1988) Rabies presenting with myocarditis and encephalitis. J Infect 17:155–158

Roine RO, Hillbom M, Valle M, Haltia M, Ketonen L, Neuvonen E, Lumio J, Lahdevirta J (1988) Fatal encephalitis caused by a bat-borne rabies-related virus: clinical finding. Brain 111:1505–1516

Rosatte RC, Tinline RR, Johnston DH (2007) Rabies control in wild carnivores. In: Jackson AC, Wunner WH (eds) Rabies, 2nd edn. Elsevier Academic, London, pp 595–634

Rosatte RC, Donovan D, Allan M, Bruce L, Buchanan T, Sobey K, Stevenson B, Gibson M, Macdonald T, Whalen M, Davies JC, Muldoon F, Wandeler A (2009) The control of raccoon rabies in Ontario Canada: proactive and reactive tactics, 1994-2007. J Wildl Dis 45:772–784

Ross E, Armentrout SA (1962) Myocarditis associated with rabies: report of a case. N Engl J Med 266:1087–1089

Rossiter JP, Jackson AC (2007) Pathology. In: Jackson AC, Wunner WH (eds) Rabies, 2nd edn. Elsevier Academic, London, pp 383–409

Rubin J, David D, Willoughby RE Jr, Rupprecht CE, Garcia C, Guarda DC, Zohar Z, Stamler A (2009) Applying the Milwaukee Protocol to treat canine rabies in Equatorial Guinea. Scand J Infect Dis 41:372–375. doi:10.1080/00365540902798333

Rupprecht CE, Briggs D, Brown CM, Franka R, Katz SL, Kerr HD, Lett SM, Levin R, Meltzer MI, Schaffner W, Cieslak PR (2010) Use of a reduced (4-dose) vaccine schedule for postexposure prophylaxis to prevent human rabies: recommendations of the advisory committee on immunization practices. MMWR Recomm Rep 59:1–9

Samaratunga H, Searle JW, Hudson N (1998) Non-rabies lyssavirus human encephalitis from fruit bats: Australian bat lyssavirus (pteropid lyssavirus) infection. Neuropathol Appl Neurobiol 24:331–335

Schnell MJ, McGettigan JP, Wirblich C, Papaneri A (2010) The cell biology of rabies virus: using stealth to reach the brain. Nat Rev Microbiol 8:51–61. doi:10.1038/nrmicro2260

Scott CA, Rossiter JP, Andrew RD, Jackson AC (2008) Structural abnormalities in neurons are sufficient to explain the clinical disease and fatal outcome in experimental rabies in yellow fluorescent protein-expressing transgenic mice. J Virol 82:513–521. doi:10.1128/JVI.01677-07

Selimov MA, Tatarov AG, Botvinkin AD, Klueva EV, Kulikova LG, Khismatullina NA (1989) Rabies-related Yuli virus; Identification with a panel of monoclonal antibodies. Acta Virol 33:542–546

Silverstein MA, Salgado CD, Bassin S, Bleck TP, Lopes MB, Farr BM, Jenkins SR, Sockwell DC, Marr JS, Miller GB (2003) First human death associated with raccoon rabies—Virginia, 2003. MMWR Morb Mortal Wkly Rep 52:1102–1103

Smith JS, Fishbein DB, Rupprecht CE, Clark K (1991) Unexplained rabies in three immigrants in the United States: a virologic investigation. N Engl J Med 324:205–211

Srinivasan A, Burton EC, Kuehnert MJ, Rupprecht C, Sutker WL, Ksiazek TG, Paddock CD, Guarner J, Shieh WJ, Goldsmith C, Hanlon CA, Zoretic J, Fischbach B, Niezgoda M, El-Feky WH, Orciari L, Sanchez EQ, Likos A, Klintmalm GB, Cardo D, LeDuc J, Chamberland ME, Jernigan DB, Zaki SR (2005) Transmission of rabies virus from an organ donor to four transplant recipients. N Engl J Med 352:1103–1111. doi:10.1056/NEJMoa043018

Sureau P, Portnoi D, Rollin P, Lapresle C, Chaouni-Berbich A (1981) Prevention of inter-human rabies transmission after corneal graft (French). Comptes Rendus de l'Académie des Sciences III, Sciences de la Vie 293:689–692

Tillotson JR, Axelrod D, Lyman DO (1977a) Follow-up on rabies—New York. MMWR 26:249–250

Tillotson JR, Axelrod D, Lyman DO (1977b) Rabies in a laboratory worker—New York. MMWR 26:183–184

Tsiang H (1979) Evidence for an intraaxonal transport of fixed and street rabies virus. J Neuropathol Exp Neurol 38:286–296

Turabelidze G, Pue H, Grim A, Patrick S (2009) First human rabies case in Missouri in 50 years causes death in outdoorsman. Mo Med 106:417–419

Ugolini G (2010) Advances in viral transneuronal tracing. J Neurosci Methods 194:2–20. doi:10.1016/j.jneumeth.2009.12.001

Ugolini G (2011) Rabies virus as a transneuronal tracer of neuronal connections. Adv Virus Res 79:165–202. doi:10.1016/B978-0-12-387040-7.00010-X

van Thiel PP, de Bie RM, Eftimov F, Tepaske R, Zaaijer HL, van Doornum GJ, Schutten M, Osterhaus AD, Majoie CB, Aronica E, Fehlner-Gardiner C, Wandeler AI, Kager PA (2009) Fatal human rabies due to Duvenhage virus from a bat in Kenya: failure of treatment with coma-induction, ketamine, and antiviral drugs. PLoS Negl Trop Dis 3:e428. doi:10.1371/journal.pntd.0000428

Vetter JM, Frisch L, Drosten C, Ross RS, Roggendorf M, Wolters B, Muller T, Dick HB, Pfeiffer N (2011) Survival after transplantation of corneas from a rabies-infected donor. Cornea 30:241–244. doi:10.1097/ICO.0b013e3181e4572a

Wallerstein C (1999) Rabies cases increase in the Philippines. Br Med J 318:1306

Wandeler AI, Rosatte RC, Williams D, Lee TK, Gensheimer KF, Montero JT, Trimarchi CV, Morse DL, Eidson M, Smith PF, Hunter JL, Smith KA, Johnson RH, Jenkins SR, Berryman C (2000) Update: raccoon rabies epizootic—United States and Canada, 1999. MMWR Morb Mortal Wkly Rep 49:31–35

Warrell DA (1976) The clinical picture of rabies in man. Trans R Soc Trop Med Hyg 70:188–195

Warrell DA, Warrell MJ (1991) Rabies. In: Lambert HP (ed) Infections of the central nervous system. B.C. Decker, Philadelphia, PA, pp 317–328

Warrell DA, Davidson NM, Pope HM, Bailie WE, Lawrie JH, Ormerod LD, Kertesz A, Lewis P (1976) Pathophysiologic studies in human rabies. Am J Med 60:180–190

Warrell MJ, Looareesuwan S, Manatsathit S, White NJ, Phuapradit P, Vejjajiva A, Hoke CH, Burke DS, Warrell DA (1988) Rapid diagnosis of rabies and post-vaccinal encephalitides. Clin Exp Immunol 71:229–234

Weli SC, Scott CA, Ward CA, Jackson AC (2006) Rabies virus infection of primary neuronal cultures and adult mice: failure to demonstrate evidence of excitotoxicity. J Virol 80:10270–10273. doi:10.1128/JVI.01272-06

Wiedeman J, Plant J, Glaser C, Messenger S, Wadford D, Sheriff H, Fritz C, Lindsay A, McKenzie M, Hammond C, Gordon E, Rupprecht CE, Petersen BW (2012) Recovery of a patient from clinical rabies—California, 2011. MMWR Morb Mortal Wkly Rep 61:61–65

Willoughby RE Jr, Tieves KS, Hoffman GM, Ghanayem NS, Amlie-Lefond CM, Schwabe MJ, Chusid MJ, Rupprecht CE (2005) Survival after treatment of rabies with induction of coma. N Engl J Med 352:2508–2514

Winkler WG, Fashinell TR, Leffingwell L, Howard P, Conomy JP (1973) Airborne rabies transmission in a laboratory worker. JAMA 226:1219–1221

World Health Organization (2005) WHO expert consultation on rabies: first report. World Health Organization, Geneva

Wunner WH (2007) Rabies virus. In: Jackson AC, Wunner WH (eds) Rabies, 2nd edn. Elsevier Academic, London, pp 23–68

West Nile Virus Infection

James J. Sejvar and Marc Fischer

Abstract Although long recognized as a human pathogen, West Nile virus (WN virus) emerged as a significant public health problem following its introduction and spread across North America. Subsequent years have seen a greater understanding of all aspects of the viral infection. The North American epidemic witnessed a further understanding of the virology, pathogenesis, clinical features, and epidemiology of WN virus infection. Approximately 80 % of human WN virus infections are asymptomatic. Most symptomatic persons experience an acute systemic febrile illness; less than 1 % of infected persons develop neuroinvasive disease, which typically manifests as meningitis, encephalitis, or anterior myelitis resulting in acute flaccid paralysis. Older age is associated with more severe illness and higher mortality; other risk factors for poor outcome have been challenging to identify. In addition to natural infection through mosquito bites, transfusion- and organ transplant-associated infections have occurred. Since there is no definitive treatment for WN virus infection, protection from mosquito bites and other preventative measures are critical. WN virus has reached an endemic pattern in North America, but the future epidemiologic pattern is uncertain.

Keywords Acute flaccid paralysis • Arbovirus • Encephalitis • West Nile virus

---

J.J. Sejvar (✉)
Division of High-Consequence Pathogens and Pathology, National Center for Emerging and Zoonotic Infectious Diseases (NCEZID), Centers for Disease Control and Prevention (CDC), 1600 Clifton Road, Mailstop A-30, Atlanta, GA 30333, USA
e-mail: zea3@cdc.gov

M. Fischer
Division of Vector-Borne Diseases, National Center for Emerging and Zoonotic Infectious Diseases (NCEZID), Centers for Disease Control and Prevention (CDC), 3150 Rampart Road, Fort Collins, CO 80521, USA

A.C. Jackson (ed.), *Viral Infections of the Human Nervous System*, Birkhäuser Advances in Infectious Diseases,
DOI 10.1007/978-3-0348-0425-7_10,

1 Virology

West Nile (WN) virus [family, *Flaviviridae*] is a member of the Japanese encephalitis serologic complex, which includes Japanese encephalitis, St. Louis encephalitis, Murray Valley encephalitis, and Kunjin viruses (Mackenzie et al. 2002). The virus was first isolated in 1937 from a febrile patient in the West Nile district of Uganda, making it one of the first arthropod-borne viruses (arboviruses) to be identified. Structurally, it is an enveloped, spherical virus of approximately 40–50 nm in diameter with a lipid bilayer membrane surrounding a nucleocapsid core (Chambers et al. 1998).

The WN virus genome is a single stranded, positive-sense RNA ~11,000 nucleotides in length. The genome codes for seven nonstructural proteins, and three structural proteins—the envelope, membrane, and capsid proteins (Chambers et al. 1998; Monath and Heinz 1996; Lanciotti et al. 1999). The envelope protein is a class II fusion protein that interacts with cellular receptors and is important in eliciting humoral immunity (Monath and Heinz 1996). The nonstructural proteins have varying functions in viral replication and immune response (Falkler et al. 1973; Murthy et al. 1999; Clum et al. 1997).

Serologic and molecular studies in the 1960s suggested significant strain variation among WN viruses (Parks et al. 1958; Nir et al. 1968; Savage et al. 1999). More recent genomic sequencing of the envelope protein has demonstrated two main lineages of WN virus (lineage 1 and lineage 2 viruses) (Lanciotti et al. 2002). Lineage 1 viruses are widespread and have caused recent large epizootics with high equine mortality in Europe (Castillo-Olivares and Wood 2004; Murgue et al. 2001b) and epidemics of human encephalitis in several African countries, the Middle East, eastern Europe, and North America (L'Vov et al. 2004; Zeller and Schuffenecker 2004; Weinberger et al. 2001; Tsai et al. 1998). Lineage 2 viruses have been found primarily in southern and central Africa, where they are usually associated with systemic febrile illness without involvement of the central nervous system (CNS). Recent geographic spread of lineage 2 viruses has been observed in the Mediterranean region and Europe and a recent outbreak due to a lineage 2 virus resulted in cases of severe human neurologic illness in Greece in 2010 (Anastasiadou et al. 2011; Danis et al. 2011a, b; Papa et al. 2011).

The lineage 1 virus which began circulating in the USA in 1999 has an over 99.8 % nucleotide homology with viruses isolated in Israel in 1998 and 1999 (Nir et al. 1968; Lanciotti et al. 2002). In addition, this North American strain was closely related to WN virus isolates from Romania, Italy, and Russia, which had also experienced outbreaks of WN virus with substantial neurologic disease (Nir et al. 1968; Savage et al. 1999; Lanciotti et al. 2002).

Analysis of WN virus isolates in the USA has indicated a slowly evolving genetic divergence in different geographic areas; these variations usually represent less than 0.5 % of the genomic sequence and result in only a few amino acid changes in any given isolate (Blitvich et al. 2004; Beasley et al. 2003; Davis et al. 2003, 2004; Ebel et al. 2004; Granwehr et al. 2004a). In 2001, a distinct genetic variant emerged (WN02 strain), which displaced the original New York 1999 strain of the virus as the predominant circulating virus by 2004 (Herring et al. 2007; Grinev et al. 2008), and all currently circulating viruses in North America are derived from the WN02 strain. The virus subsequently appears to have reached relative genetic stasis within North America (Davis et al. 2007).

2 Ecology

WN virus is an arbovirus that is maintained in an enzootic cycle between mosquitoes and vertebrate hosts, primarily birds (Hubalek and Halouzka 1999; Hayes 2001). In North America, although the WN virus genome has been found in more than 58 mosquito species, *Culex pipiens* (the northern house mosquito), *Cx. quinquefasciatus* (the southern house mosquito), and *Cx. tarsalis* are the most important vectors (Centers for Disease Control and Prevention 2002b, e) (Brault 2009; Campbell et al. 2002) WN virus has also been isolated from ticks in the eastern hemisphere (Hayes 1989) but their role in the enzootic cycle is unclear (Lawrie et al. 2004).

In temperate regions, the virus is thought to overwinter in adult mosquitoes (Hayes 1989; Nasci et al. 2001). The enzootic cycle begins when infected mosquitoes feed on nonimmune birds. High-titer viremia develops in some birds, allowing for transmission of the virus to uninfected mosquitoes and continuation of the cycle. Although WN virus infection has been identified in more than 150 North American bird species, not all develop high-titer viremia. Some avian species develop mild or no illness but other species, particularly birds of the *Corvidae* family in North America, often succumb to infection, resulting in large epizootics with high mortality (Komar et al. 2001; Eidson et al. 2001). These bird die-offs have served as important sentinel surveillance indicators of subsequent human infections (Eidson et al. 2001).

Once a sufficient proportion of mosquitoes and amplifying vertebrate hosts are infected, "bridging" mosquito species that feed more frequently on mammals and other nonavian vertebrates become infected. Naturally acquired WN virus infection has been seen in numerous avian, reptile, and mammalian species, (Marfin et al. 2001; Komar 2000; Heinz-Taheny et al. 2004; Miller et al. 2003; Bunning et al. 2002). Humans and horses are most commonly recognized to develop symptomatic infections, however, neither species is thought to significantly contribute to the viral amplification cycle because viremia is brief and of low titer (Hayes 1989).

2.1 Nonmosquito Transmission Routes

Nearly all human infections of WN virus are due to bites from infected mosquitoes; however, other modes of transmission in humans have been noted. Transfusion-associated

WN virus transmission was first identified in 2002 when 23 people in the USA were infected after receiving platelets, red blood cells, or plasma from 16 viremic blood donors (Pealer et al. 2003; Harrington et al. 2003). Routine screening of blood donations was initiated in the USA in 2003 and more than 3,000 infected blood products have been removed from the blood supply (Centers for Disease Control and Prevention 2011). However, rare cases of transfusion-associated transmission continue to occur due to blood donations that have virus levels below the limit of detection (Petersen and Epstein 2005; Centers for Disease Control and Prevention 2007, 2009).

Also in 2002, transmission via donated organs was first documented when WN virus infection was identified in four recipients of organs from a common donor (Iwamoto et al. 2003). A second transmission occurred in 2005 in which WN virus infection occurred in three of four organ recipients from a common WN virus-infected donor, who was seropositive for WN virus but negative for WN virus nucleic acid, suggesting that transmission may be possible in the absence of detectable serum viremia. Since that time, several other cases or clusters of WN virus transmitted through solid organ transplants have been reported in the USA and Europe (Rhee et al. 2011; Nett et al. in press; Costa et al. 2011).

Other rare transmission circumstances have been identified. Intrauterine transmission has been documented in one case of a mother who was infected with WN virus at approximately 27 weeks gestation and later delivered an infant with severe chorioretinitis and lissencephaly (Alpert et al. 2003; Centers for Disease Control and Prevention 2002a, d). Cord blood and heel-stick blood from the infant were positive for WN virus-specific IgM and neutralizing antibodies. In 2002, possible transmission via human breast milk was identified (Centers for Disease Control and Prevention 2002b, e). Human laboratory-acquired WN virus infections have been acquired by the percutaneous route (Centers for Disease Control and Prevention 2002a, d) and transmission through conjunctival exposure in an occupational setting has been suspected (Fonseca et al. 2005). Dialysis-associated WN virus transmission has also been suspected (Centers for Disease Control and Prevention 2004).

3 Geographic Distribution

WN virus is one of the most widely distributed of all arboviruses, with an extensive distribution throughout Africa, the Middle East, parts of Europe and the former Soviet Union, South Asia, Australia (Petersen and Roehrig 2001; Campbell et al. 2002), and more recently, North, Central, and South America and the Caribbean (Gubler 2001; Estrada-Franco et al. 2003; Cruz et al. 2005; Health Canada 2005).

The first recorded instance of severe neurological disease during an outbreak occurred in Israeli nursing homes in 1957 (Spigland et al. 1958). The virus was then isolated from the brains of three children who died of encephalitis in India in 1980 and 1981 (George et al. 1984). Nevertheless, severe neurological disease during

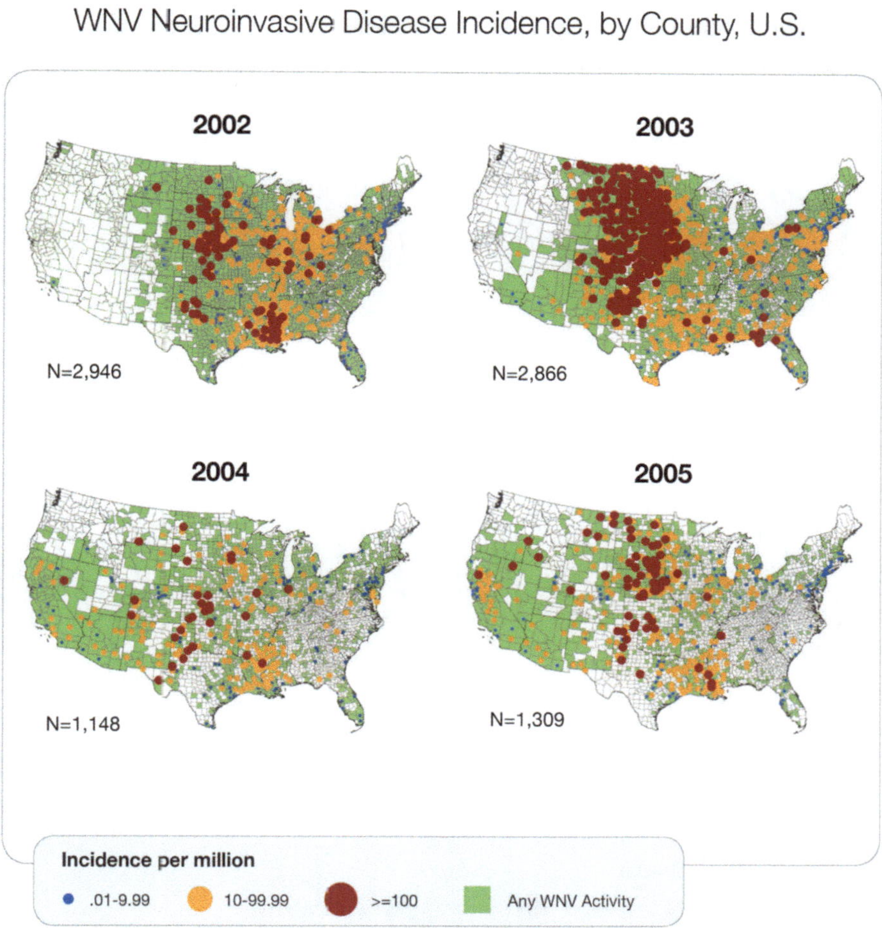

Fig. 1 Geographic distribution and spread of West Nile virus, United States, 2002–2005

outbreaks remained uncommon until recent outbreaks in Algeria, 1994 (Le Guenno et al. 1996); Romania, 1996 (Tsai et al. 1998); Tunisia, 1997 (Murgue et al. 2001a); Russia, 1999 (Platonov 2001); USA and Canada, 1999–2004 (Petersen and Hayes 2004); Israel, 2000 (Chowers et al. 2001); and Sudan, 2002 (Depoortere et al. 2004).

The most notable epidemiologic development associated with WN virus infection has been the unprecedented outbreak in North America. The virus was first identified in the western hemisphere following an outbreak in the New York City area in the summer of 1999. Subsequently the virus spread rapidly across the USA (Figs. 1 and 2) (Petersen and Hayes 2004); the 2002 and 2003 seasons were the largest outbreaks of neurologic illness with WN virus recorded to date. The incidence of reported WN virus illnesses has subsequently decreased, and since approximately 2004, the number of reported cases in the USA has remained

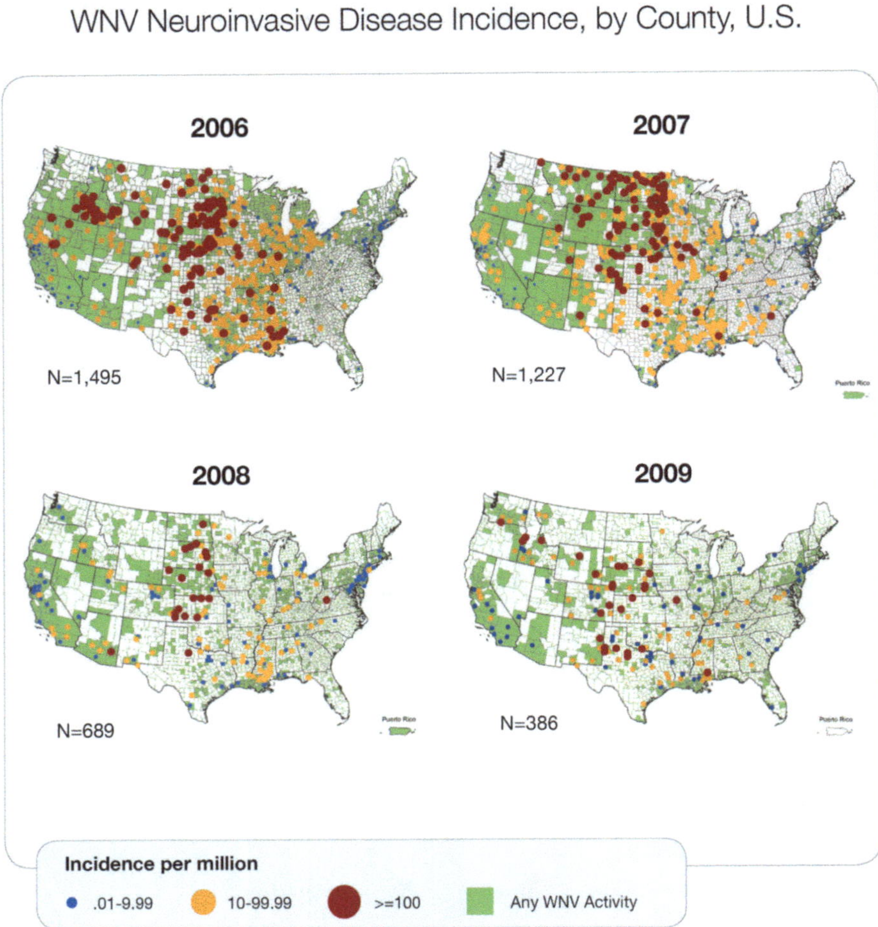

Fig. 2 Geographic distribution and spread of West Nile virus, United States, 2006–2009

relatively stable or decreased, suggesting a more endemic pattern (Table 1). As of 2011, over 30,000 human cases of WNV illness and 1,220 deaths had been cumulatively reported in the USA with the highest incidence of disease in the Western and Mountain regions (Centers for Disease Control and Prevention 2011; Lindsey et al. 2010). In Canada, WNV was first detected in southern Ontario in 2001 and, by 2002, had spread to Manitoba, Quebec, Nova Scotia, and Saskatchewan (Drebot et al. 2003). By 2003, the virus' distribution extended into New Brunswick and Alberta. Since that time, cases of human illness have continued to be reported from British Columbia, Quebec, Alberta, Manitoba, Saskatchewan, and Ontario (Public Health Agency of Canada 2011).

WN virus activity in the remainder of the world has been less dramatic. WN virus was first detected south of the US border in 2001, when a resident of the Cayman Islands developed WN virus encephalitis (Centers for Disease Control and

Table 1 Reported West Nile virus disease cases in humans, United States, 1999–2010

Number of cases					
Year(s)	Total	WNND	WNF/other	Deaths	# US states
1999–2002	4,305	3,088	1,217	303	39[a]
2003	9,862	2,866	6,996	264	45[a]
2004	2,539	1,148	1,391	100	40[a]
2005	3,000	1,309	1,691	119	43[a]
2006	4,269	1,495	2,774	177	43[a]
2007	3,630	1,227	2,403	124	43
2008	1,356	689	667	44	45[a]
2009	720	386	334	32	37[a]
2010	1,021	629	392	57	40[a]

[a]Plus DC

Prevention 2002c). In 2002, WN virus was isolated from a dead common raven in Tabasco State, Mexico (Estrada-Franco et al. 2003) and in 2003, WN virus RNA was detected in brain tissue of a dead horse from Nuevo Leon State, Mexico (Blitvich et al. 2004). Since that time, serologic evidence in birds and horses suggests that WN virus is present in multiple Caribbean, Central, and South American countries, with evidence of infection as far south as Argentina (Adrian Diaz et al. 2008; Morales et al. 2006; Petersen and Hayes 2008; Pauvolid-Correa et al. 2011). Despite the apparent widespread distribution of WN virus in Latin America and the Caribbean, as determined by serological studies in birds and horses, few cases of human illness have been reported. The reason for this is unclear, but may be due to the presence of other endemic flaviviruses (such as dengue virus) which may provide some cross-protection against severe WN virus disease, a decrease in virulence of the strains in Latin America due to selection of attenuated strains in migrating birds or ecological conditions, less robust surveillance, or other geographic or epidemiologic factors (Petersen and Hayes 2008).

In the Mediterranean and Europe, WN virus circulation has been recognized since the 1960s. Large outbreaks of human illness have occurred in Romania in 1996 and Russia in 1999 caused by a lineage 1 virus and resulting in large numbers of cases of encephalitis. Sporadic human and equine WN virus outbreaks have been described in France, Spain, the Czech Republic, and Italy. In 2004, a lineage 2 virus of central African origin was detected in birds in Hungary, the first time a lineage 2 virus was identified in Europe (Calistri et al. 2010); a lineage 2 virus was also responsible for recent outbreaks of human illness in Greece and Italy (Anastasiadou et al. 2011; Danis et al. 2011a, b; Papa et al. 2011; Barzon et al. 2011).

Several large outbreaks of human encephalitic illness have occurred in Israel since the 1950s, and the WN virus strain that emerged in North America is thought to have derived from a virus causing epizootic illness among farmed geese in Israel in 1998. Serologic evidence of WN virus infection in humans has also been identified in a number of other Middle Eastern and South Asian countries.

WN viruses in sub-Saharan Africa have predominantly been lineage 2 viruses, historically associated with sporadic cases of zoonotic and human illness. However, a large outbreak of human WN virus illness was described in South Africa in 1974, with over 18,000 cases of febrile illness reported (McIntosh et al. 1976). Lineage 1 viruses have also been identified in several African countries (Weinberger et al. 2001). In Australia, Kunjin virus, a lineage 1 variant of WNV, is endemic and has been recognized to cause sporadic cases and small outbreaks of human and equine illness. Human illness with Kunjin virus is generally mild and nonencephalitic (Hall et al. 2001; Gray et al. 2011).

4 Pathogenesis of Human Infection

Following natural infection through a mosquito bite, WN virus is thought to replicate in regional lymphoid tissue and the spleen, after which a viremia develops. In otherwise healthy people, viremia usually peaks between 2 and 4 days after infection and prior to illness onset (Hayes and O'Leary 2004; Southam and Moore 1952, 1954); viremia begins to decline once clinical illness begins. In immunodeficient persons, persistence of virus may be more prolonged (Iwamoto et al. 2003; Southam and Moore 1952). In less than 1 % of infected people, viremia results in dissemination to the CNS. The exact mechanism by which WN virus is able to invade the CNS is unknown, but data from animal models suggest that peripheral production of TNF-α leads to increased blood–brain barrier permeability, perhaps facilitating virus entry (Diamond and Klein 2004). Other hypothesized entry routes include infection of cerebral endothelial cells and migration across the cell to brain parenchyma; migration of WN virus-infected leukocytes through tight junctions; and direct viral shedding through choroid plexus. Data also suggest that WN virus may be transported through peripheral nerve axons in a retrograde fashion, leading to CNS invasion (Samuel et al. 2007). Following neuroinvasion, the virus directly infects neurons, and less frequently, astrocytes, leading to neuronophagia and cell death. Histopathologic changes in human WN virus infection are characterized by the presence of microglial nodules composed of lymphocytes and histiocytes; leptomeningeal mononuclear inflammatory infiltrates are present in cases of meningitis. CD8 T-lymphocytes represent the predominant inflammatory cell type in the nodules and infiltrates (Sampson et al. 2000; Doron et al. 2003). While nearly all brain regions may be affected, WN virus appears to have a specific neurotropism for neurons in the basal ganglia, thalamus, brainstem (particularly the medulla and pons). This correlates well with the predominant clinical symptomatology demonstrated by persons with WN encephalitis. Spinal cord pathology is significant for involvement of ventral and dorsal gray and white matter and nerve roots, with a particular predilection for spinal cord anterior horn cells (Doron et al. 2003; Agamanolis et al. 2003). This results in multisegmental, patchy involvement similar to that seen with poliovirus infection (Fratkin et al. 2004). This involvement correlates clinically with the multifocal and often segmental distribution of weakness observed in cases of WN poliomyelitis. Inflammation of spinal and cranial nerve roots, resulting in radiculitis, may also be seen (Park et al. 2003).

5 Epidemiology of Human Infection and Illness

Epidemiologic studies of human WN virus infection in the USA suggest that the majority of infections are clinically silent (Tsai et al. 1998; Mostashari et al. 2001). Following large epidemics in North America and Europe, an estimated 2–3 % of the population were infected in a given year (Tsai et al. 1998; Mostashari et al. 2001; Centers for Disease Control and Prevention 2001; McCarthy et al. 2001). Serological surveys indicate that even in areas experiencing outbreaks, less than 5 % of the population may have been exposed to the virus (Petersen and Hayes 2004; Mostashari et al. 2001; Tsai et al. 1998; Centers for Disease Control and Prevention 2001).

In temperate regions, human WN virus infection incidence increases in early summer and peaks in August or early September. In milder climates, transmission may be seen year round. In the USA, within large regional WN virus epidemics, the incidence of human disease varies markedly from county to county, suggesting the importance of local ecological conditions (Lindsey et al. 2010).

6 Risk Factors for Severe Disease and Death

Of all infected persons, less than 1 % develops West Nile virus neuroinvasive disease (WNND), which is manifested as aseptic meningitis, encephalitis, or anterior (polio)myelitis. Although WNND has been reported among all ages, the proportion of persons who progress to WNND is greater among older than younger persons (O'Leary et al. 2004). Serologic surveys in Romania and New York City indicate that WN virus infection incidence is constant across all age groups during outbreaks (Tsai et al. 1998; Mostashari et al. 2001). However, surveillance data from the USA indicate that age is the most important host risk factor for development of neuroinvasive disease after infection. The incidence of neuroinvasive disease increases approximately 1.5-fold for each decade of life, resulting in a risk approximately 30 times greater for persons 80–90 years old compared to children younger than 10 years old (O'Leary et al. 2004). During outbreaks, hospitalized persons older than 70 years of age had case fatality rates of 15 % in Romania (Tsai et al. 1998) and 29 % in Israel (Chowers et al. 2001). Encephalitis with severe muscle weakness and change in the level of consciousness were also prominent clinical risk factors predicting fatal outcome (Chowers et al. 2001; Nash et al. 2001).

Based upon a limited number of cases, patients who acquire WN virus from infected donor organs are likely at higher risk for severe neurologic disease and death compared with patients infected through mosquito bites (Rhee et al. 2011; Nett et al. in press). The risk of severe neurologic disease among other organ transplant recipients is not well-defined and may be related to the interval between infection and transplantation or type of post-transplant immunosuppressive therapy. A seroprevalence study carried out in a Canadian outpatient transplant clinic following a WN virus epidemic in 2002 indicated that the risk of neuroinvasive disease following infection was 40 % (95 % confidence interval 16–80 %) (Kumar et al. 2004a). During that epidemic, transplant patients were approximately 40 times more likely than the population-at-large to develop WN neuroinvasive disease (Kumar et al. 2004b). However, another study assessed seropositivity and incidence of WNND among 194 solid organ transplant recipients and 195 controls, and found no significant difference in seropositivity for WNV IgG between the groups, and determined that incidence of WNND among the transplant recipients was low among the seropositive transplant patients (Freifeld et al. 2010).

In addition to increased age and organ transplantation, hypertension, cerebrovascular disease, renal disease, and diabetes have also been identified as possible risk factors for WNND, and prior immunosuppression has been associated with a fatal outcome (Nash et al. 2001; Murray et al. 2006, 2009; Patnaik et al. 2006; Jean et al. 2007; Chowers et al. 2001; Bode et al. 2006). The incidence of neuroinvasive disease and the probability of death after acquiring neuroinvasive disease are slightly higher in men than women (O'Leary et al. 2004).

7 Clinical Manifestations of Human West Nile Virus Infection

The understanding of the spectrum of illness due to WN virus infection in humans has expanded during the North American outbreak, and a number of previously under-recognized syndromes have been characterized (Sejvar et al. 2003a; Emig and Apple 2004; Pepperell et al. 2003; Sayao et al. 2004). It is estimated that more than 80 % of infected persons remain asymptomatic. Of those who develop symptoms, the vast majority develop an acute, systemic febrile illness ("West Nile fever"; WNF); less than 1 % of infected persons develop neuroinvasive disease including aseptic meningitis ("West Nile meningitis"; WNM), encephalitis ("West Nile encephalitis" WNE), or an acute poliomyelitis-like syndrome ("West Nile poliomyelitis" WNP) (Campbell et al. 2002; Granwehr et al. 2004b). WN virus meningitis makes up the largest percentage of neuroinvasive disease cases in younger age groups, while the proportion of encephalitis increases in older age groups. The clinical picture among persons with WN virus infection may be not always be clear-cut, and sometimes the clinical features of WNF, WNM, and WNE may overlap. For example, patients may develop altered mental status due to severe systemic illness, without true histopathologic or radiologic evidence of cerebral inflammation or "encephalitis" in the pathophysiologic sense. Similarly, patients presenting with fever, headache, and "neck stiffness" may not undergo lumbar puncture to demonstrate pleocytosis and, consequently, a diagnosis of WN "meningitis" may not be reported. Despite these limitations, the clinical syndrome in most persons with WN virus illness may be discerned on clinical grounds.

7.1 West Nile Fever

WNF is the predominant clinical syndrome seen in most infected persons. All ages may be affected, but data suggest that the incidence of WNF may be higher among younger individuals (Brown 1995; Pepperell et al. 2003; O'Leary et al. 2004; Hayes and Gubler 2006). Following an incubation period of approximately 2–14 days, infected persons experience the abrupt onset of fever, headache, fatigue, and myalgias. Gastrointestinal complaints, including nausea and vomiting, may be predominant, leading to dehydration.

Occasionally a rash may be noted; this rash tends to be morbilliform, maculopapular, and nonpruritic, and predominates over the torso and extremities, sparing the palms and soles (Ferguson et al. 2005; Del Giudice et al. 2005; Gorsche and Tilley 2005; Anderson et al. 2004). The rash may be transient, lasting less than 24 h in some persons. Rash appears to be more frequently seen in WNF than in more severe illness manifestations (WNM or WNE) (Ferguson et al. 2005). In addition, rash is more frequently observed among younger persons than among older persons (Ferguson et al. 2005). Whether the presence of a rash correlates with host immune or cytokine response to infection remains unknown.

Although elderly persons with symptomatic disease have a higher mortality rate than younger symptomatic persons (O'Leary et al. 2004), most patients experience complete recovery. Some otherwise healthy persons, however, may continue to experience persistent fatigue, headaches, and difficulties concentrating for days or weeks following infection (Watson et al. 2004). In particular, profound fatigue, sometimes interfering with work or school activities, may last for months among persons recovering from WNF (Sejvar et al. 2008). Deaths among persons with WNF occur primarily among older persons and the immunocompromised population, and are frequently attributable to cardiopulmonary complications (Sejvar et al. 2011).

7.2 *Neuroinvasive Disease*

7.2.1 Meningitis

WNM is clinically similar to that of other viral meningitides. Affected persons experience the abrupt onset of fever and headache, and demonstrate meningeal signs, including nuchal rigidity, Kernig's and/or Brudzinski's signs, and photophobia or phonophobia. The associated headache may be severe, requiring hospitalization for pain control; associated gastrointestinal disturbance may result in dehydration, exacerbating head pain, and systemic symptoms (Sejvar et al. 2008). WNM is generally associated with a favorable outcome, though, similar to WNF, some patients experience persistent headache, fatigue, and myalgias (Sejvar et al. 2003a, 2008).

Cerebrospinal fluid (CSF) examination is characterized by a modest pleocytosis, generally less than 500 cells/mm^3. While this pleocytosis is usually lymphocytic, CSF obtained soon after the onset of symptoms may show a neutrophilic predominance (Crichlow et al. 2004; Sejvar et al. 2005). The presence of plasma cells has been suggested to be indicative of WN virus infection (Carson et al. 2003), a finding requiring further substantiation.

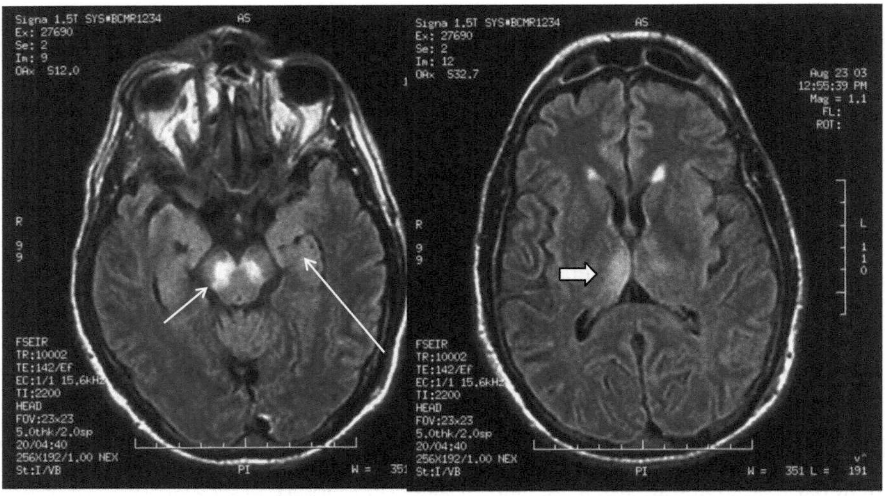

Fig. 3 Fluid-attenuated inversion recovery magnetic resonance imaging sequence of the brain in a patient with West Nile virus encephalitis with associated parkinsonism and tremor, displaying signal abnormality in the substantia nigra (*short arrow*), the mesial temporal lobe (*long arrow*), and right posterior thalamus (*thick arrow*)

7.2.2 Encephalitis

WNE may range in severity from a mild, self-limited confusional state to severe encephalopathy, coma, and death. Several neurological syndromes, primarily extra-pyramidal disorders, have been observed in patients with WN virus encephalitis (Sejvar et al. 2003a; Pepperell et al. 2003; Burton et al. 2004; Sayao et al. 2004). Increased intracranial pressure and cerebral edema are infrequently associated with WNE.

An estimated 20–70 % of patients with WNE have abnormal findings on brain magnetic resonance imaging (MRI); however, even in cases of severe WNE, the MRI may be normal, or abnormal findings may not be apparent until several weeks after onset of illness, or may be apparent only on diffusion-weighted imaging (Brilla et al. 2004; Ali et al. 2005; Petropoulou et al. 2005). The most characteristic MRI findings in patients with WNE are bilateral signal abnormalities in the basal ganglia and thalami on T2-, FLAIR, and diffusion-weighted image sequences, indicating the viral neurotropism for these deep gray structures (Fig. 3). These MRI findings, which may be seen in other flaviviral encephalitides, may be indicative but not diagnostic for WNE. Electroencephalographic (EEG) abnormalities may be present in the form of generalized slowing, frequently anteriorly or temporally predominant, and triphasic sharp waves (Rodriguez and Westmoreland 2007; Gandelman-Marton et al. 2003). These EEG abnormalities, however, are also nonspecific. Overt seizures appear to be relatively uncommon with WNE and are estimated to occur in 3–6 % of patients (Doron et al. 2003). CSF abnormalities in patients with WNE are the same as those seen in WNM, characterized by moderate lymphocytic pleocytosis, elevated protein, and normal glucose. One large study suggested that the mean CSF white blood cell count in patients with WNE was 227 cells/mm^3 (median, 90 cells/mm^3) (Tyler et al. 2006).

Patients with WN virus encephalitis frequently develop a coarse tremor, particularly in the upper extremities. The tremor tends to be postural, and may have a kinetic component (Burton et al. 2004; Sejvar et al. 2003a; Emig and Apple 2004; Sayao et al. 2004). Myoclonus, predominantly of the upper extremities and facial muscles, may occur, and may be present during sleep. Features of parkinsonism, including hypomimia, bradykinesia, and postural instability, may be seen and can be associated with falls and functional difficulties (Sejvar et al. 2003a; Robinson et al. 2003). Cerebellar ataxia, with associated truncal instability and gait disturbance leading to falls, has been described (Kanagarajan et al. 2003; Burton et al. 2004; Sayao et al. 2004). These abnormal movements usually follow the onset of mental status changes and typically resolve over time; however, tremor and parkinsonism may persist in patients recovering from severe encephalitis (Sejvar et al. 2003a; Pepperell et al. 2003).

The development of these movement disorders in WNE is due to specific neurotropism of WN virus for extrapyramidal structures; there is frequent involvement of the brainstem (particularly the medulla and pons), the deep gray matter nuclei, particularly the substantia nigra of the basal ganglia and the thalami, and the cerebellum (Kelley et al. 2003; Guarner et al. 2004; Bosanko et al. 2003). This clinico-pathologic correlation may be extended to the neuroimaging findings in WNE, as noted above.

7.2.3 Weakness and Paralysis

Acute weakness is associated with WN virus infection; most cases of paralysis are due to viral involvement of the lower motor neurons of the spinal cord (anterior horn cells), resulting in poliomyelitis (Sejvar et al. 2003b, 2005; Jeha et al. 2003; Leis et al. 2002; Glass et al. 2002; Li et al. 2003), a syndrome more typically associated with poliovirus infection. The clinical features of WNP are characteristic and dramatic and should be easily differentiated from the characteristic diffuse "muscle weakness" described by many persons with severe fatigue associated with WN virus infection (Table 2). WNP generally develops soon after illness onset, usually within the first 24–48 h. Limb paralysis generally develops rapidly and may be abrupt, occasionally raising clinical concern about stroke (Kulstad and Wichter 2003; Berner et al. 2002). The weakness is usually asymmetric, and often results in monoplegia. Patients with severe and extensive spinal cord involvement develop a more symmetric dense quadriplegia. Central facial weakness, frequently bilateral, can also be seen (Jeha et al. 2003). Sensory loss or numbness is generally absent though some patients experience intense pain in the affected limbs just before or during the onset of weakness, and this limb pain may be persistent (Sejvar et al. 2005).

Table 2 Clinical and electrodiagnostic features of West Nile virus-associated acute flaccid paralysis

Characteristic	West Nile poliomyelitis	Guillain–Barré syndromes	Fatigue
Timing of onset	Acute phase of infection	One to eight weeks following acute infection	Acute infection
Fever and leukocytosis	Present	Absent	Present
Weakness distribution	Asymmetric; occasional monoplegia	Generally symmetric; proximal and distal muscles	Generalized, subjective, but neurologic examination normal
Sensory symptoms	Absence of numbness, paresthesias, or sensory loss; pain often present	Painful distal paresthesias and sensory loss	Generally absent
Bowel/bladder involvement	Often present	Rare	Not present
Concurrent encephalopathy	Often present	Generally absent	May be seen with fever, meningitis, or encephalitis
CSF profile	Pleocytosis and elevated protein	No pleocytosis; elevated protein (albuminocytologic dissociation)	Pleocytosis and elevated protein in the setting of meningitis/encephalitis

In some persons, involvement of respiratory muscle innervation leading to diaphragmatic and intercostal muscle paralysis may result in respiratory failure requiring emergent endotracheal intubation (Sejvar et al. 2005; Fan et al. 2004). Involvement of the lower brainstem, including the motor nuclei of the vagus and glossopharyngeal nerves, is similar to that seen in poliovirus infection, and appears to be the genesis of this syndrome (Doron et al. 2003; Agamanolis et al. 2003). Respiratory involvement in WNP is associated with high morbidity and mortality, and among survivors, prolonged ventilatory support lasting months may be required (Sejvar et al. 2005). Patients who develop bulbar findings, such as dysarthria, dysphagia or loss of gag reflex, are at greater risk for respiratory failure and should be monitored closely.

Electrodiagnostic studies (electromyography/nerve conduction studies) will display findings consistent with a motor axonopathy with little or no demyelinating changes and preservation of sensory nerve potentials (Leis et al. 2003; Al-Shekhlee and Katirji 2004). Spinal MRI may show signal abnormalities in the anterior spinal cord, consistent with anterior horn cell damage; ventral nerve root enhancement may be seen as well (Fig. 4).

Other forms of acute flaccid paralysis, including radiculopathy and the acute demyelinating polyradiculoneuropathy form of Guillain–Barré syndrome, have

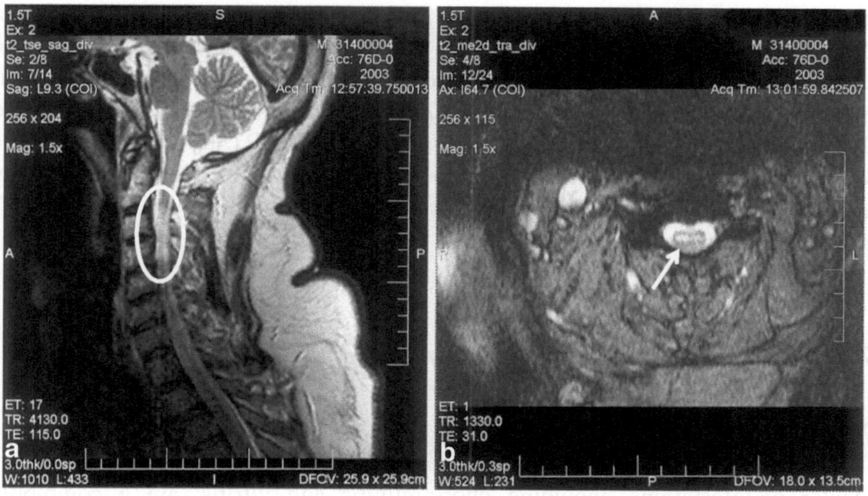

Fig. 4 Sagittal (**a**) and axial (**b**) T2-weighted magnetic resonance imaging of the cervical spinal cord of a patient with bilateral upper extremity paralysis and respiratory failure from West Nile poliomyelitis, displaying increased signal in the anterior spinal cord (*circle* and *arrow*)

also been associated with WN virus infection (Park et al. 2003; Ahmed et al. 2000). However, these syndromes appear to be far less common than poliomyelitis, and may be differentiated on the basis of clinical and electrophysiologic features. (Table 2) The weakness associated with the Guillain–Barré syndrome is usually symmetric and ascending, and is associated with sensory and autonomic dysfunction. Additionally, CSF examination will generally show elevated protein in the absence of pleocytosis (cytoalbuminologic dissociation), and electodiagnostic studies will be consistent with a predominantly demyelinating polyneuropathy.

Recovery of limb strength in persons with WNP is variable (Sejvar et al. 2005; Cao et al. 2005). However, persistent weakness and associated functional disability appears to be the rule, at least in the short term, and prolonged physical and occupational therapy may be required. Most limb strength recovery occurs within the first 6–8 months after acute illness, following which improvement appears to plateau (Sejvar et al. 2006; Cao et al. 2005). In particular, quadriplegia and respiratory failure are associated with high morbidity and mortality, and recovery is slow and incomplete (Sejvar et al. 2005). More than 50 % of the mortality associated with WNP occurs in patients with acute neuromuscular respiratory failure; of patients who survive respiratory failure due to WNP, a substantial number require prolonged tracheostomy or long-term supplemental oxygen (Sejvar et al. 2005, 2006). In general, less profound initial weakness may be associated with more rapid and more complete strength recovery (Sejvar et al. 2005). However, even patients with initially severe and profound paralysis may experience substantial recovery (Sejvar et al. 2005; Cao et al. 2005); thus, the initial severity of paralysis should not be used as a prognosticator of eventual outcome. This recovery phenomenon is thought to be due to the involvement of a large number of motor neurons which may initially be reversibly damaged, but are able to recover (Li et al. 2003). Electrodiagnostic studies may be useful in predicting recovery of muscle strength, with subsequent improvement correlating with motor unit number estimate (MUNE) values (Li et al. 2003) (97). Although case reports have suggested the occurrence of relapsing or delayed-onset cases of WNP (Sejvar et al. 2010), the long-term clinical and functional outcomes in patients with WNP is still emerging, and whether there may be the subsequent development of a delayed, "post-polio"-like syndrome years after acute illness is unknown.

7.3 Other Clinical Manifestations

7.3.1 Ocular Manifestations

Ocular manifestations, including chorioretinitis and vitritis, are a commonly reported clinical manifestation (Hershberger et al. 2003; Adelman et al. 2003; Bains et al. 2003; Kuchtey et al. 2003; Shaikh and Trese 2004; Vandenbelt et al. 2003). The chorioretinal lesions have been described as multifocal and with a "target-like" appearance (Adelman et al. 2003); retinal hemorrhages have also been noted. Lesions tend to be clustered primarily in the temporal and nasal regions of the periphery of the fundus. This distribution and appearance of the chorioretinal lesions have been suggested to be distinctive for WN virus infection (Hershberger et al. 2003). One study in Tunisia identified chorioretinitis in 20 (69 %) of 29 patients with laboratory-confirmed, symptomatic WN virus infection. (Khairallah et al. 2004); the authors concluded that ophthlamoscopic examination should be performed on all patients with suspected WN virus disease.

An inflammatory vitritis has occurred concomitantly with the chorioretinitis, and may be significant enough to obscure the optic disc. Symptomatic persons describe gradual visual blurring and loss, floaters and flashes. Although experience with management is limited, improvement both in symptoms and in underlying chorioretinal lesions has been observed following treatment with intraocular corticosteroids (Adelman et al. 2003). To date, WN virus has not been isolated intra-orbitally.

7.3.2 Miscellaneous Manifestations

Several other clinical manifestations have been described in association with WN virus infection, but have been described mostly in case reports or small case series, and a definitive causal association with WN virus infection is less substantiated. Rhabdomyolysis has been reported in the temporal setting of WN virus infection (Jeha et al. 2003; Kulstad and Wichter 2003), suggesting a viral myositis, but the presence of virus in muscle tissue has not been observed. Hepatitis and pancreatitis have been reported in cases of severe WN virus infection (Sampson et al. 2000;

Perelman and Stern 1974), and WN virus has been identified in hepatic and pancreatic specimens at pathology, suggesting that viscerotropic WN virus disease may be an infrequent manifestation of infection. Myocarditis has been seen pathologically in WN virus infection and cardiac arrhythmias have occurred in persons with WNP, suspected to be due to autonomic dysfunction (Fratkin et al. 2004).

A study published in 2010 identified WN virus RNA in the urine of 5 (20 %) of 25 patients at up to 7 years following acute WN virus illness (Murray et al. 2010). Four of the patients with WN virus RNA in their urine reported persistent subjective symptoms and one patient had developed renal failure after their acute illness. However, one subsequent study found no WN virus RNA in urine samples collected from 40 patients at 6.5 years after acute WN virus disease and another study detected WN virus RNA in the urine of only one (1.6 %) of 63 persons tested <5 months after initial acute WN virus infection (Gibney et al. 2011; Baty et al. in press). The frequency and clinical implications of persistent WN virus RNA in urine are unknown, and additional data are needed.

8 WN Virus Long-Term Outcomes

Few data exist regarding the long-term neurologic and functional outcomes of WN virus encephalitis. As with many other viral encephalitides, initial severe neurologic illness does not necessarily correlate with eventual outcome, and some patients with initial severe encephalopathy with associated coma may experience dramatic recovery and minimal sequelae (Sejvar et al. 2003a). However, others experience persistent neurologic dysfunction, including movement disorders, headaches, fatigue, and cognitive complaints. Large hospital-based series suggest that patients with severe WN virus encephalitis frequently require assistance with daily activities following acute care discharge (Emig and Apple 2004; Pepperell et al. 2003). Patients often report substantial functional and cognitive difficulties for up to a year following acute infection, and only 37 % of patients in the 1999 New York City outbreak achieved full recovery at 1 year (Klee et al. 2004). Of 265 persons developing symptomatic WN virus infection in Idaho between 2006 and 2008, 53 % reported one or more persistent symptom 6 months or more following acute illness; most frequent complaints were fatigue, muscle aches, and difficulties with memory and concentration (Cook et al. 2010). Cognitive complaints including difficulties with attention and concentration have been described among patients recovering from WNE, and suggest a predominantly subcortical type of cognitive dysfunction based on prominent thalamic and basal ganglia involvement (Sejvar et al. 2003a). However, limited formal neuropsychometric assessments have been performed. A few studies have shown that persons recovering from WN virus illness demonstrate measurable neurocognitive deficits on standardized testing as long as 1 year after acute illness (Haaland et al. 2006; Sadek et al. 2010). Other studies have shown that persons recovering from WN virus illness do not perform significantly differently on standardized neurocognitive assessments based upon the nature of clinical illness or from unaffected persons (Sejvar et al. 2008). However, self-reported fatigue, somatic, and cognitive complaints are common among persons recovering from WN virus illness, and subjective complaints and poorer performance on self-reported functionality indices have been seen in patients months or years following acute illness (Klee et al. 2004; Carson et al. 2006). One study has suggested normalization of self-reported symptoms within 1 year of acute illness (Loeb et al. 2008). Neuropsychiatric symptoms, including depression, anxiety, and apathy, have been reported by patients recovering from WNE (Sejvar et al. 2003a; Berg et al. 2010; Murray et al. 2007).

Fatality rates range from 10 % to 20 % among patients with severe neuroinvasive disease (O'Leary et al. 2004; Pepperell et al. 2003), and mortality is higher among the elderly and those with immunocompromising conditions.

9 Pediatric WN Virus Infection

Most children with symptomatic WN virus infection present with WNF; neuroinvasive disease, when it occurs, is frequently meningitis (Lindsey et al. 2009; Civen et al. 2006; Hayes and O'Leary 2004). However, poliomyelitis (Heresi et al. 2004; Vidwan et al. 2003), fatal encephalitis (Carey et al. 1968), rhomboencephalitis (Nichter et al. 2000), and hepatitis (Yim et al. 2004) have been described in children with WN virus infection. Similar to adults, immunocompromised children may be more susceptible to more severe illness (Ravindra et al. 2004).

10 Diagnosis

The diagnosis of WN virus infection may be suspected clinically by the onset of fever, aseptic meningitis, encephalitis, or paralysis in the setting of known WN virus transmission or mosquito activity. The clinical features of tremor, myoclonus, and parkinsonism may be helpful in diagnosing WNE in the setting of a WN virus epidemic. WNP often has a dramatic clinical presentation, and should be suspected in any person who develops acute flaccid paralysis during periods of WN virus transmission. Other etiologies of flaccid paralysis, most importantly Guillain–Barré syndrome, must also be considered, however, and can be distinguished from WNP on the basis of clinical and laboratory findings (Table 2). In general, routine laboratory, radiologic, and neuroimaging testing are unhelpful in the confirmation of WN virus infection, but are useful in excluding other etiologies of meningitis, encephalitis, and flaccid paralysis.

The diagnosis of WN virus infection is typically made by demonstrating WN virus-specific immunoglobulin (Ig)M antibodies in serum or CSF. The presence of anti-WN virus IgM is usually good evidence of recent WN virus infection, but may indicate infection with another closely related flavivirus (e.g., St. Louis encephalitis

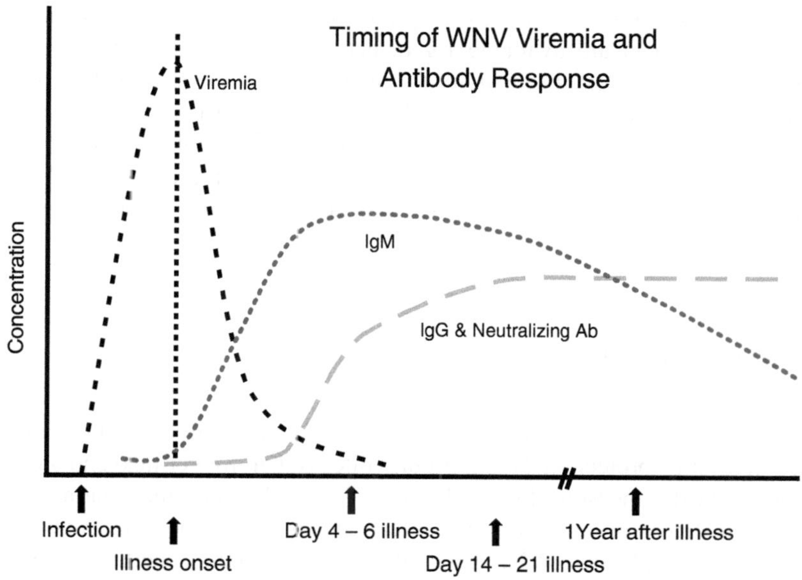

Fig. 5 Kinetics and temporal profile of viremia and humoral IgM and IgG responses to acute West Nile virus infection in humans

or dengue virus) (Martin et al. 2002). Because anti-WN virus IgM can persist in some patients for over a year, a positive test result occasionally may reflect past infection (Fig. 5) (Roehrig et al. 2003). The detection of WN virus-specific IgM in the CSF of a patient with a clinically compatible illness is generally considered evidence of recent infection, although IgM antibodies have been found in three patients at 3–6 months after their acute illness (Kapoor et al. 2004). Serum collected within 8 days of illness onset may lack detectable IgM, and the test may need to be repeated on a convalescent-phase sample. IgG antibody generally is detectable shortly after IgM and persists for years. Plaque-reduction neutralization tests (PRNT) can be performed to measure virus-specific neutralizing antibodies (Martin et al. 2000, 2002; Johnson et al. 2000). A fourfold or greater rise in virus-specific neutralizing antibodies between acute- and convalescent-phase serum specimens collected 2–3 weeks apart may be used to confirm recent WN virus infection and to discriminate between cross-reacting antibodies from closely related flaviviruses. While commercial assays for detection of WN virus-specific antibodies are widely available, their sensitivities and specificities may vary. In the USA, confirmatory testing is available at the Centers for Disease Control and Prevention Division of Vector-Borne Diseases and many local and state health department laboratories.

Although the gold standard for diagnosis of WN virus infection is isolation of virus from biological specimens (e.g., blood, serum, CSF, or tissues, particularly CNS), this is infrequently achieved because the virus is often cleared from the blood by the time of illness onset. Virus isolation by culture is difficult to perform, is of low yield, and can be performed only in laboratories with proper biosafety containment facilities. Nucleic acid amplification (NAA) tests (polymerase chain reaction techniques) are available for detection of WN virus RNA in clinical samples, but are of limited sensitivity for many of the same reasons that isolation is not ideal for diagnosis. Results of viral culture and NAA tests are more likely to be positive with samples from elderly or immunocompromised hosts (Koepsell et al. 2010; Penn et al. 2006). Immunohistochemical staining can detect WN virus antigen in formalin-fixed tissue (Guarner et al. 2004). Negative results of these tests do not exclude WN virus infection.

11 Management

There is currently no definitive treatment for WN virus infection. Prevention of infection through protection from mosquito bites is therefore critical and an important public health measure. In the absence of definitive antiviral treatment, management of illness due to WN virus infection remains supportive. Patients with otherwise uncomplicated WNF generally do not require specific intervention, though control of headache and rehydration may sometimes be needed. However, persons with documented WN viremia and patients with WNF in which other risk factors, including older age and underlying immunosuppression, are present should be observed for progression to more severe neuroinvasive disease. Patients with severe WNM may also require pain control for severe headache, and dehydration due to associated nausea and vomiting may require hospitalization for rehydration. In patients with WNE, attention to level of alertness and airway protection is important. While seizures and increased intracranial pressure have been infrequently reported with WNE, if present, they should be managed appropriately.

Patients with WNP may not have concurrent meningitis or encephalitis, and thus WN virus infection may not initially be suspected. This may result in the implementation of inappropriate diagnostic procedures or treatment modalities, including anticoagulation for suspected acute stroke or muscle biopsy for suspected myopathy. WN virus infection should be suspected in persons developing acute asymmetric paralysis, particularly if accompanied by other signs of infection. Patients developing early dysarthria and dysphagia are at higher risk for subsequent acute respiratory failure (Sejvar et al. 2005); for this reason, hospitalization and observation of patients with poliomyelitis is advised, and the development of dysarthria and dysphagia should be viewed with concern. Management of poliomyelitis due to poliovirus suggests that initiation of aggressive physical activity during the acute febrile period of illness is associated with more profound and persistent weakness (Guyton 1949); in the absence of additional data, avoidance of aggressive physical activity during the acute febrile illness or during the initial 48–72 h of weakness in WNP would be reasonable.

The fact that WN viremia in humans is short-lived and is usually cleared by the time of clinical presentation, presents a substantial theoretical obstacle for specific antiviral therapies. Any therapeutic agent would have to be effective at reducing intracellular virus load and/or prevent viral spread in the CNS and, possibly reduce the inflammatory response to infection in order to be effective. The recent use of several therapeutic modalities, including antiviral agents, nucleic acid analogues and missense sequences, immunomodulating agents, and angiotensin-receptor blockers, has been outside the setting of carefully controlled, randomized, blinded, placebo-controlled trials; thus, anecdotal reports of effectiveness of these agents are unsubstantiated. However, such anecdotal reports of effectiveness and the clinical desire to provide an intervention have led to empiric use.

The antiviral agent ribavirin has demonstrated in vitro activity against WN virus infection, but efficacy has not yet been demonstrated in animal models or humans (Jordan et al. 2000; Ferrara et al. 1981). The agent has been used in an uncontrolled, nonblinded fashion in a group of patients with WN virus neuroinvasive disease in Israel in 2000, where it was found to be ineffective and potentially harmful (Chowers et al. 2001). The immunomodulating agent interferon-α, while again showing in vitro inhibition of cytotoxicity due to WN virus (Morrey et al. 2004), has not been fully evaluated in animal models, and data from an open-label, nonblinded trial in the USA have not suggested clear benefit. The use of interferon-α in the treatment of the closely related Japanese encephalitis virus suggested no benefit (Solomon et al. 2003).

Animal models and anecdotal reports have suggested the efficacy of high-titer WN virus-specific intravenous immune globulin (IVIG) from pooled donors (Omr-IgG-am®) (Diamond et al. 2003; Shimoni et al. 2001; Agrawal and Petersen 2003), and humanized monoclonal WN virus antibodies targeting the envelope protein of the virus (MGAWN1) (Oliphant et al. 2005; Beigel et al. 2010; Morrey et al. 2006; Smeraski et al. 2011). However, animal models suggest that efficacy is greatest if these therapeutics are given prior to or very shortly after onset of clinical illness, and attempts at human randomized clinical trials to assess the efficacy of these therapeutic agents have been unsuccessful, largely due to the challenge of enrolling a sufficient number of subjects within a likely therapeutic window. Neither of these products is licensed or available for use in the USA.

12 Prevention

Given the absence of definitive treatment of WN virus infection, prevention remains the cornerstone of management of human WN virus from a public health standpoint. Prevention may take the form of community-based programs and personal protection. Reduction of vector populations by public mosquito control programs are employed to different degrees in various communities in North America, and may involve removal of mosquito breeding sites, larviciding, and spraying for adult mosquitoes. Personal protective measures include limiting outdoor activities at dawn and dusk when mosquito activity is high, covering exposed skin with long sleeves and pants, and using insect repellent. The most effective repellents for use on skin are products that contain either diethyltoluamide (DEET), picaridin (KBR 3023), IR3535, or oil of lemon eucalyptus (Barnard and Xue 2004; Frances et al. 2002, 2004; Costantini et al. 2004; Zielinski-Gutierrez et al. 2012). In general, higher concentrations of active ingredient provide longer duration of protection, regardless of the active ingredient. DEET efficacy tends to plateau at a concentration of approximately 50 % (Buescher et al. 1983). Permethrin is an effective insecticide and repellent approved for use on clothing or fabrics, but not on skin.

Although several candidate WN virus vaccines are being evaluated, none are licensed or available for use in humans (Hall and Khromykh 2004; Chang et al. 2004; Monath et al. 2006; Pletnev et al. 2003). Four WN virus vaccines are licensed in the USA for use in horses (Chang et al. 2004; Siger et al. 2006; Ng et al. 2003). It is unclear if vaccination with related flavivirus vaccines (e.g., Japanese encephalitis or yellow fever) provide significant protection against WN virus disease (Monath 2002; Mansfield et al. 2011; Johnson et al. 2005; Tang et al. 2008; Yamshchikov et al. 2005; Kanesa-Thasan et al. 2002; Lobigs and Diamond 2012). Though it is likely that an effective WN virus vaccine for humans can be developed, the cost-effectiveness and commercial viability of such a vaccine remains uncertain.

13 Conclusions

The arrival and subsequent spread of WN virus infection throughout North America has served as a reminder of the capacity for emerging and re-emerging pathogens to move into and thrive in new settings. The future epidemiologic pattern of WN virus infection in North America, and indeed worldwide, remains unclear. Whereas WN virus infection in the USA appears to have reached an endemic pattern, the possibility of large future outbreaks remains, and re-emergence of WN virus in European countries raises the possibility of significant WN virus activity globally. The long-term functional and cognitive outcomes of persons with WN virus illness continues to be elucidated, and will have significant implications on the assessment of overall disease burden. The challenges of definitive treatment and prevention of WN virus infection will require further research.

Disclosure The authors have no financial relationships relevant to this article to disclose. The findings and conclusions in this report are those of the authors and do not necessarily represent the official position of the Centers for Disease Control and Prevention.

References

Adelman RA, Membreno JH, Afshari NA, Stoessel KM (2003) West Nile virus chorioretinitis. Retina 23(1):100–101

Adrian Diaz L, Komar N, Visintin A, Dantur Juri MJ, Stein M, Lobo Allende R, Spinsanti L, Konigheim B, Aguilar J, Laurito M, Almiron W, Contigiani M (2008) West Nile virus in birds, Argentina. Emerg Infect Dis 14(4):689–691

Agamanolis DP, Leslie MJ, Caveny EA, Guarner J, Shieh WJ, Zaki SR (2003) Neuropathological findings in West Nile virus encephalitis: a case report. Ann Neurol 54(4):547–551
Agrawal AG, Petersen LR (2003) Human immunoglobulin as a treatment for West Nile virus infection. J Infect Dis 188(1):1–4
Ahmed S, Libman R, Wesson K, Ahmed F, Einberg K (2000) Guillain-Barre syndrome: an unusual presentation of West Nile virus infection. Neurology 55(1):144–146
Ali M, Safriel Y, Sohi J, Llave A, Weathers S (2005) West Nile virus infection: MR imaging findings in the nervous system. AJNR Am J Neuroradiol 26(2):289–297
Alpert SG, Fergerson J, Noel LP (2003) Intrauterine West Nile virus: ocular and systemic findings. Am J Ophthalmol 136(4):733–735
Al-Shekhlee A, Katirji B (2004) Electrodiagnostic features of acute paralytic poliomyelitis associated with West Nile virus infection. Muscle Nerve 29(3):376–380
Anastasiadou A, Economopoulou A, Kakoulidis I, Zilidou R, Butel D, Zorpidou D, Ferentinos G, Markou P, Kougas E, Papa A (2011) Non-neuroinvasive West Nile virus infections during the outbreak in Greece. Clin Microbiol Infect 17(11):1681–1683. doi:10.1111/j.1469-0691.2011.03642.x
Anderson RC, Horn KB, Hoang MP, Gottlieb E, Bennin B (2004) Punctate exanthem of West Nile Virus infection: report of 3 cases. J Am Acad Dermatol 51(5):820–823
Bains HS, Jampol LM, Caughron MC, Parnell JR (2003) Vitritis and chorioretinitis in a patient with West Nile virus infection. Arch Ophthalmol 121(2):205–207
Barnard DR, Xue RD (2004) Laboratory evaluation of mosquito repellents against *Aedes albopictus*, *Culex nigripalpus*, and *Ochierotatus triseriatus* (Diptera: Culicidae). J Med Entomol 41(4):726–730
Barzon L, Pacenti M, Cusinato R, Cattai M, Franchin E, Pagni S, Martello T, Bressan S, Squarzon L, Cattelan A, Pellizzer G, Scotton P, Beltrame A, Gobbi F, Bisoffi Z, Russo F, Palu G (2011) Human cases of West Nile Virus infection in north-eastern Italy, 15 June to 15 November 2010. Euro Surveill 16(33)
Baty SA, Gibney KB, Staples JE, Patterson AB, Levy C, Lehman J, Wadleigh T, Feld J, Lanciotti R, Nugent CT, Fischer M (in press) Evaluation for West Nile Virus (WNV) RNA in urine of patients within 5 months of WNV infection. J Infect Dis
Beasley DW, Davis CT, Guzman H, Vanlandingham DL, Travassos da Rosa AP, Parsons RE, Higgs S, Tesh RB, Barrett AD (2003) Limited evolution of West Nile virus has occurred during its southwesterly spread in the United States. Virology 309(2):190–195
Beigel JH, Nordstrom JL, Pillemer SR, Roncal C, Goldwater DR, Li H, Holland PC, Johnson S, Stein K, Koenig S (2010) Safety and pharmacokinetics of single intravenous dose of MGAWN1, a novel monoclonal antibody to West Nile virus. Antimicrob Agents Chemother 54(6):2431–2436. doi:10.1128/aac.01178-09
Berg PJ, Smallfield S, Svien L (2010) An investigation of depression and fatigue post West Nile virus infection. S D Med 63(4):127–129, 131–133
Berner YN, Lang R, Chowers MY (2002) Outcome of West Nile fever in older adults. J Am Geriatr Soc 50(11):1844–1846
Blitvich BJ, Fernandez-Salas I, Contreras-Cordero JF, Lorono-Pino MA, Marlenee NL, Diaz FJ, Gonzalez-Rojas JI, Obregon-Martinez N, Chiu-Garcia JA, Black WC 4th, Beaty BJ (2004) Phylogenetic analysis of West Nile virus, Nuevo Leon State, Mexico. Emerg Infect Dis 10(7):1314–1317
Bode AV, Sejvar JJ, Pape WJ, Campbell GL, Marfin AA (2006) West Nile virus disease: a descriptive study of 228 patients hospitalized in a 4-county region of Colorado in 2003. Clin Infect Dis 42(9):1234–1240
Bosanko CM, Gilroy J, Wang AM, Sanders W, Dulai M, Wilson J, Blum K (2003) West nile virus encephalitis involving the substantia nigra: neuroimaging and pathologic findings with literature review. Arch Neurol 60(10):1448–1452. doi:10.1001/archneur.60.10.1448
Brault AC (2009) Changing patterns of West Nile virus transmission: altered vector competence and host susceptibility. Vet Res 40(2):43. doi:10.1051/vetres/2009026

Brilla R, Block M, Geremia G, Wichter M (2004) Clinical and neuroradiologic features of 39 consecutive cases of West Nile Virus meningoencephalitis. J Neurol Sci 220(1–2):37–40
Brown J (1995) West Nile virus in blood donors: Colorado cohort study, 2003. In: Fifth national conference on West Nile virus in the United States, Denver, CO, 3–5 Feb 2004
Buescher MD, Rutledge LC, Wirtz RA, Nelson JH (1983) The dose-persistence relationship of DEET against *Aedes aegypti*. Mosq News 43:364–366
Bunning ML, Bowen RA, Cropp CB, Sullivan KG, Davis BS, Komar N, Godsey MS, Baker D, Hettler DL, Holmes DA, Biggerstaff BJ, Mitchell CJ (2002) Experimental infection of horses with West Nile virus. Emerg Infect Dis 8(4):380–386
Burton JM, Kern RZ, Halliday W, Mikulis D, Brunton J, Fearon M, Pepperell C, Jaigobin C (2004) Neurological manifestations of West Nile virus infection. Can J Neurol Sci 31(2):185–193
Calistri P, Giovannini A, Hubalek Z, Ionescu A, Monaco F, Savini G, Lelli R (2010) Epidemiology of West Nile in Europe and in the Mediterranean basin. Open Virol J 4:29–37. doi:10.2174/1874357901004020029
Campbell GL, Marfin AA, Lanciotti RS, Gubler DJ (2002) West Nile virus. Lancet Infect Dis 2 (9):519–529
Cao NJ, Ranganathan C, Kupsky WJ, Li J (2005) Recovery and prognosticators of paralysis in West Nile virus infection. J Neurol Sci 236(1–2):73–80
Carey DE, Rodrigues FM, Myers RM, Webb JK (1968) Arthropod-borne viral infections in children in Vellore, South India, with particular reference to dengue and West Nile viruses. Indian Pediatr 5(7):285–296
Carson PJ, Steidler T, Patron R, Tate JM, Tight R, Smego RA Jr (2003) Plasma cell pleocytosis in cerebrospinal fluid in patients with West Nile virus encephalitis. Clin Infect Dis 37(1):e12–e15
Carson PJ, Konewko P, Wold KS, Mariani P, Goli S, Bergloff P, Crosby RD (2006) Long-term clinical and neuropsychological outcomes of West Nile virus infection. Clin Infect Dis 43 (6):723–730
Castillo-Olivares J, Wood J (2004) West Nile virus infection of horses. Vet Res 35(4):467–483
Centers for Disease Control and Prevention (2001) Serosurveys for West Nile virus infection – New York and Connecticut counties, 2000. MMWR Morb Mortal Wkly Rep 50(3):37–39
Centers for Disease Control and Prevention (2002a) Laboratory-acquired West Nile virus infections – United States, 2002. MMWR Morb Mortal Wkly Rep 51(50):1133–1135
Centers for Disease Control and Prevention (2002b) Provisional surveillance summary of the West Nile virus epidemic – United States, January-November 2002. MMWR Morb Mortal Wkly Rep 51(50):1129–1133
Centers for Disease Control and Prevention (2002c) West Nile Virus activity – United States, 2001. MMWR Morb Mortal Wkly Rep 51(23):497–501
Centers for Disease Control and Prevention (2002d) Intrauterine West Nile virus infection – New York, 2002. MMWR Morb Mortal Wkly Rep 51(50):1135–1136
Centers for Disease Control and Prevention (2002e) Possible West Nile virus transmission to an infant through breast-feeding – Michigan, 2002. MMWR Morb Mortal Wkly Rep 51 (39):877–878
Centers for Disease Control and Prevention (2004) Possible dialysis-related West Nile virus transmission – Georgia, 2003. MMWR Morb Mortal Wkly Rep 53(32):738–739
Centers for Disease Control and Prevention (2007) West Nile virus transmission through blood transfusion – South Dakota, 2006. MMWR Morb Mortal Wkly Rep 56(4):76–79
Centers for Disease Control and Prevention (2009) West Nile virus transmission via organ transplantation and blood transfusion - Louisiana, 2008. MMWR Morb Mortal Wkly Rep 58 (45):1263–1267
Centers for Disease Control and Prevention (2011) West Nile virus disease and other arboviral diseases – United States, 2010. MMWR Morb Mortal Wkly Rep 60(30):1009–1013
Chambers TJ, Halevy M, Nestorowicz A, Rice CM, Lustig S (1998) West Nile virus envelope proteins: nucleotide sequence analysis of strains differing in mouse neuroinvasiveness. J Gen Virol 79(Pt 10):2375–2380

Chang GJ, Kuno G, Purdy DE, Davis BS (2004) Recent advancement in flavivirus vaccine development. Expert Rev Vaccines 3(2):199–220

Chowers MY, Lang R, Nassar F, Ben-David D, Giladi M, Rubinshtein E, Itzhaki A, Mishal J, Siegman-Igra Y, Kitzes R, Pick N, Landau Z, Wolf D, Bin H, Mendelson E, Pitlik SD, Weinberger M (2001) Clinical characteristics of the West Nile fever outbreak, Israel, 2000. Emerg Infect Dis 7(4):675–678

Civen R, Villacorte F, Robles DT, Dassey DE, Croker C, Borenstein L, Harvey SM, Mascola L (2006) West Nile virus infection in the pediatric population. Pediatr Infect Dis J 25(1):75–78

Clum S, Ebner KE, Padmanabhan R (1997) Cotranslational membrane insertion of the serine proteinase precursor NS2B-NS3(Pro) of dengue virus type 2 is required for efficient in vitro processing and is mediated through the hydrophobic regions of NS2B. J Biol Chem 272(49):30715–30723

Cook RL, Xu X, Yablonsky EJ, Sakata N, Tripp JH, Hess R, Piazza P, Rinaldo CR (2010) Demographic and clinical factors associated with persistent symptoms after West Nile virus infection. Am J Trop Med Hyg 83(5):1133–1136. doi:10.4269/ajtmh.2010.09-0717

Costa AN, Capobianchi MR, Ippolito G, Palu G, Barzon L, Piccolo G, Andreetta B, Filippetti M, Fehily D, Lombardini L, Grossi P (2011) West Nile virus: the Italian national transplant network reaction to an alert in the north-eastern region, Italy 2011. Euro Surveill 16(41)

Costantini C, Badolo A, Ilboudo-Sanogo E (2004) Field evaluation of the efficacy and persistence of insect repellents DEET, IR3535, and KBR 3023 against *Anopheles gambiae* complex and other Afrotropical vector mosquitoes. Trans R Soc Trop Med Hyg 98(11):644–652. doi:10.1016/j.trstmh.2003.12.015

Crichlow R, Bailey J, Gardner C (2004) Cerebrospinal fluid neutrophilic pleocytosis in hospitalized West Nile virus patients. J Am Board Fam Pract 17(6):470–472

Cruz L, Cardenas VM, Abarca M, Rodriguez T, Reyna RF, Serpas MV, Fontaine RE, Beasley DW, Da Rosa AP, Weaver SC, Tesh RB, Powers AM, Suarez-Rangel G (2005) Short report: serological evidence of West Nile virus activity in El Salvador. Am J Trop Med Hyg 72(5):612–615

Danis K, Papa A, Papanikolaou E, Dougas G, Terzaki I, Baka A, Vrioni G, Kapsimali V, Tsakris A, Kansouzidou A, Tsiodras S, Vakalis N, Bonovas S, Kremastinou J (2011a) Ongoing outbreak of West Nile virus infection in humans, Greece, July to August 2011. Euro Surveill 16(34)

Danis K, Papa A, Theocharopoulos G, Dougas G, Athanasiou M, Detsis M, Baka A, Lytras T, Mellou K, Bonovas S, Panagiotopoulos T (2011b) Outbreak of West Nile virus infection in Greece, 2010. Emerg Infect Dis 17(10):1868–1872. doi:10.3201/eid1710.110525

Davis CT, Beasley DW, Guzman H, Raj R, D'Anton M, Novak RJ, Unnasch TR, Tesh RB, Barrett AD (2003) Genetic variation among temporally and geographically distinct West Nile virus isolates, United States, 2001, 2002. Emerg Infect Dis 9(11):1423–1429

Davis CT, Beasley DW, Guzman H, Siirin M, Parsons RE, Tesh RB, Barrett AD (2004) Emergence of attenuated West Nile virus variants in Texas, 2003. Virology 330(1):342–350. doi:10.1016/j.virol.2004.09.016

Davis CT, Li L, May FJ, Bueno R Jr, Dennett JA, Bala AA, Guzman H, Quiroga-Elizondo D, Tesh RB, Barrett AD (2007) Genetic stasis of dominant West Nile virus genotype, Houston, Texas. Emerg Infect Dis 13(4):601–604

Del Giudice P, Schuffenecker I, Zeller H, Grelier M, Vandenbos F, Dellamonica P, Counillon E (2005) Skin manifestations of West Nile virus infection. Dermatology 211(4):348–350

Depoortere E, Kavle J, Keus K, Zeller H, Murri S, Legros D (2004) Outbreak of West Nile virus causing severe neurological involvement in children, Nuba Mountains, Sudan, 2002. Trop Med Int Health 9(6):730–736

Diamond MS, Klein RS (2004) West Nile virus: crossing the blood-brain barrier. Nat Med 10(12):1294–1295. doi:10.1038/nm1204-1294

Diamond MS, Shrestha B, Marri A, Mahan D, Engle M (2003) B cells and antibody play critical roles in the immediate defense of disseminated infection by West Nile encephalitis virus. J Virol 77(4):2578–2586

Doron SI, Dashe JF, Adelman LS, Brown WF, Werner BG, Hadley S (2003) Histopathologically proven poliomyelitis with quadriplegia and loss of brainstem function due to West Nile virus infection. Clin Infect Dis 37(5):e74–e77
Drebot MA, Lindsay R, Barker IK, Buck PA, Fearon M, Hunter F, Sockett P, Artsob H (2003) West Nile virus surveillance and diagnostics: a Canadian perspective. Can J Infect Dis 14(2):105–114
Ebel GD, Carricaburu J, Young D, Bernard KA, Kramer LD (2004) Genetic and phenotypic variation of West Nile virus in New York, 2000–2003. Am J Trop Med Hyg 71(4):493–500
Eidson M, Komar N, Sorhage F, Nelson R, Talbot T, Mostashari F, McLean R (2001) Crow deaths as a sentinel surveillance system for West Nile virus in the northeastern United States, 1999. Emerg Infect Dis 7(4):615–620
Emig M, Apple DJ (2004) Severe West Nile virus disease in healthy adults. Clin Infect Dis 38 (2):289–292
Estrada-Franco JG, Navarro-Lopez R, Beasley DW, Coffey L, Carrara AS, Travassos da Rosa A, Clements T, Wang E, Ludwig GV, Cortes AC, Ramirez PP, Tesh RB, Barrett AD, Weaver SC (2003) West Nile virus in Mexico: evidence of widespread circulation since July 2002. Emerg Infect Dis 9(12):1604–1607
Falkler WA Jr, Diwan AR, Halstead SB (1973) Human antibody to dengue soluble complement-fixing (SCF) antigens. J Immunol 111(6):1804–1809
Fan E, Needham DM, Brunton J, Kern RZ, Stewart TE (2004) West Nile virus infection in the intensive care unit: a case series and literature review. Can Respir J 11(5):354–358
Ferguson DD, Gershman K, LeBailly A, Petersen LR (2005) Characteristics of the rash associated with West Nile virus fever. Clin Infect Dis 41(8):1204–1207
Ferrara EA, Oishi JS, Wannemacher RW Jr, Stephen EL (1981) Plasma disappearance, urine excretion, and tissue distribution of ribavirin in rats and rhesus monkeys. Antimicrob Agents Chemother 19(6):1042–1049
Fonseca K, Prince GD, Bratvold J, Fox JD, Pybus M, Preksaitis JK, Tilley P (2005) West Nile virus infection and conjunctival exposure. Emerg Infect Dis 11(10):1648–1649
Frances SP, Van Dung N, Beebe NW, Debboun M (2002) Field evaluation of repellent formulations against daytime and nighttime biting mosquitoes in a tropical rainforest in northern Australia. J Med Entomol 39(3):541–544
Frances SP, Waterson DG, Beebe NW, Cooper RD (2004) Field evaluation of repellent formulations containing deet and picaridin against mosquitoes in Northern Territory, Australia. J Med Entomol 41(3):414–417
Fratkin JD, Leis AA, Stokic DS, Slavinski SA, Geiss RW (2004) Spinal cord neuropathology in human West Nile virus infection. Arch Pathol Lab Med 128(5):533–537
Freifeld AG, Meza J, Schweitzer B, Shafer L, Kalil AC, Sambol AR (2010) Seroprevalence of West Nile virus infection in solid organ transplant recipients. Transpl Infect Dis 12(2):120–126. doi:10.1111/j.1399-3062.2009.00464.x
Gandelman-Marton R, Kimiagar I, Itzhaki A, Klein C, Theitler J, Rabey JM (2003) Electroencephalography findings in adult patients with West Nile virus-associated meningitis and meningoencephalitis. Clin Infect Dis 37(11):1573–1578
George S, Gourie-Devi M, Rao JA, Prasad SR, Pavri KM (1984) Isolation of West Nile virus from the brains of children who had died of encephalitis. Bull World Health Organ 62(6):879–882
Gibney KB, Lanciotti RS, Sejvar JJ, Nugent CT, Linnen JM, Delorey MJ, Lehman JA, Boswell EN, Staples JE, Fischer M (2011) West nile virus RNA not detected in urine of 40 people tested 6 years after acute West Nile virus disease. J Infect Dis 203(3):344–347. doi:10.1093/infdis/jiq057
Glass JD, Samuels O, Rich MM (2002) Poliomyelitis due to West Nile virus. N Engl J Med 347(16):1280–1281
Gorsche R, Tilley P (2005) The rash of West Nile virus infection. CMAJ 172(11):1440

Granwehr BP, Li L, Davis CT, Beasley DW, Barrett AD (2004a) Characterization of a West Nile virus isolate from a human on the Gulf Coast of Texas. J Clin Microbiol 42(11):5375–5377. doi:10.1128/jcm.42.11.5375-5377.2004

Granwehr BP, Lillibridge KM, Higgs S, Mason PW, Aronson JF, Campbell GA, Barrett AD (2004b) West Nile virus: where are we now? Lancet Infect Dis 4(9):547–556

Gray TJ, Burrow JN, Markey PG, Whelan PI, Jackson J, Smith DW, Currie BJ (2011) West nile virus (Kunjin subtype) disease in the northern territory of Australia – a case of encephalitis and review of all reported cases. Am J Trop Med Hyg 85(5):952–956. doi:10.4269/ajtmh.2011.11-0165

Grinev A, Daniel S, Stramer S, Rossmann S, Caglioti S, Rios M (2008) Genetic variability of West Nile virus in US blood donors, 2002–2005. Emerg Infect Dis 14(3):436–444

Guarner J, Shieh WJ, Hunter S, Paddock CD, Morken T, Campbell GL, Marfin AA, Zaki SR (2004) Clinicopathologic study and laboratory diagnosis of 23 cases with West Nile virus encephalomyelitis. Hum Pathol 35(8):983–990

Gubler DJ (2001) Human arbovirus infections worldwide. Ann N Y Acad Sci 951:13–24

Guyton A (1949) Reaction of the body to poliomyelitis and the recovery process. Arch Int Med 83:27

Haaland KYSJ, Pergam S, Echevarria LA, Davis LE, Goade D, Harnar J, Nfchissey RA, Sewel CM, Ettestad P (2006) Mental status after West Nile virus infection. Emerg Infect Dis 12(8):1260–1262

Hall RA, Khromykh AA (2004) West Nile virus vaccines. Expert Opin Biol Ther 4(8):1295–1305. doi:10.1517/14712598.4.8.1295

Hall RA, Scherret JH, Mackenzie JS (2001) Kunjin virus: an Australian variant of West Nile? Ann N Y Acad Sci 951:153–160

Harrington T, Kuehnert MJ, Kamel H, Lanciotti RS, Hand S, Currier M, Chamberland ME, Petersen LR, Marfin AA (2003) West Nile virus infection transmitted by blood transfusion. Transfusion 43(8):1018–1022

Hayes C (1989) West Nile fever. In: Monath T (ed) The arboviruses: epidemiology and ecology, vol V. CRC, Boca Raton, FL, pp 59–88

Hayes CG (2001) West Nile virus: Uganda, 1937, to New York City, 1999. Ann N Y Acad Sci 951:25–37

Hayes EB, Gubler DJ (2006) West Nile virus: epidemiology and clinical features of an emerging epidemic in the United States. Annu Rev Med 57:181–194

Hayes EB, O'Leary DR (2004) West Nile virus infection: a pediatric perspective. Pediatrics 113 (5):1375–1381

Health Canada (2005) West Nile virus weekly surveillance report. Centre for Infectious Disease Prevention & Control, and the National Microbiology Laboratory of the Public Health Agency of Canada

Heinz-Taheny KM, Andrews JJ, Kinsel MJ, Pessier AP, Pinkerton ME, Lemberger KY, Novak RJ, Dizikes GJ, Edwards E, Komar N (2004) West Nile virus infection in free-ranging squirrels in Illinois. J Vet Diagn Invest 16(3):186–190

Heresi GP, Mancias P, Mazur LJ, Butler IJ, Murphy JR, Cleary TG (2004) Poliomyelitis-like syndrome in a child with West Nile virus infection. Pediatr Infect Dis J 23(8):788–789

Herring BL, Bernardin F, Caglioti S, Stramer S, Tobler L, Andrews W, Cheng L, Rampersad S, Cameron C, Saldanha J, Busch MP, Delwart E (2007) Phylogenetic analysis of WNV in North American blood donors during the 2003–2004 epidemic seasons. Virology 363(1):220–228. doi:10.1016/j.virol.2007.01.019

Hershberger VS, Augsburger JJ, Hutchins RK, Miller SA, Horwitz JA, Bergmann M (2003) Chorioretinal lesions in nonfatal cases of West Nile virus infection. Ophthalmology 110(9):1732–1736

http://www.phac-aspc.gc.ca/wnv-vwn/mon-hmnsurv-2005-eng.php. Accessed 27 Jan 2012

Hubalek Z, Halouzka J (1999) West Nile fever – a reemerging mosquito-borne viral disease in Europe. Emerg Infect Dis 5(5):643–650

Iwamoto M, Jernigan DB, Guasch A, Trepka MJ, Blackmore CG, Hellinger WC, Pham SM, Zaki S, Lanciotti RS, Lance-Parker SE, DiazGranados CA, Winquist AG, Perlino CA, Wiersma S, Hillyer KL, Goodman JL, Marfin AA, Chamberland ME, Petersen LR (2003) Transmission of West Nile virus from an organ donor to four transplant recipients. N Engl J Med 348 (22):2196–2203

Jean CM, Honarmand S, Louie JK, Glaser CA (2007) Risk factors for West Nile virus neuroinvasive disease, California, 2005. Emerg Infect Dis 13(12):1918–1920

Jeha LE, Sila CA, Lederman RJ, Prayson RA, Isada CM, Gordon SM (2003) West Nile virus infection: a new acute paralytic illness. Neurology 61(1):55–59

Johnson AJ, Martin DA, Karabatsos N, Roehrig JT (2000) Detection of anti-arboviral immunoglobulin G by using a monoclonal antibody-based capture enzyme-linked immunosorbent assay. J Clin Microbiol 38(5):1827–1831

Johnson BW, Kosoy O, Martin DA, Noga AJ, Russell BJ, Johnson AA, Petersen LR (2005) West nile virus infection and serologic response among persons previously vaccinated against yellow Fever and Japanese encephalitis viruses. Vector Borne Zoonotic Dis 5(2):137–145

Jordan I, Briese T, Fischer N, Lau JY, Lipkin WI (2000) Ribavirin inhibits West Nile virus replication and cytopathic effect in neural cells. J Infect Dis 182(4):1214–1217

Kanagarajan K, Ganesh S, Alakhras M, Go ES, Recco RA, Zaman MM (2003) West Nile virus infection presenting as cerebellar ataxia and fever: case report. South Med J 96(6):600–601

Kanesa-Thasan N, Putnak JR, Mangiafico JA, Saluzzo JE, Ludwig GV (2002) Short report: absence of protective neutralizng antibodies to West Nile virus in subjects following vaccination with Japanese encephalitis or dengue vaccines. Am J Trop Med Hyg 66(2):115–116

Kapoor H, Signs K, Somsel P, Downes FP, Clark PA, Massey JP (2004) Persistence of West Nile Virus (WNV) IgM antibodies in cerebrospinal fluid from patients with CNS disease. J Clin Virol 31(4):289–291

Kelley TW, Prayson RA, Ruiz AI, Isada CM, Gordon SM (2003) The neuropathology of West Nile virus meningoencephalitis. A report of two cases and review of the literature. Am J Clin Pathol 119(5):749–753

Khairallah M, Ben Yahia S, Ladjimi A, Zeghidi H, Ben Romdhane F, Besbes L, Zaouali S, Messaoud R (2004) Chorioretinal involvement in patients with West Nile virus infection. Ophthalmology 111(11):2065–2070

Klee AL, Maidin B, Edwin B, Poshni I, Mostashari F, Fine A, Layton M, Nash D (2004) Long-term prognosis for clinical West Nile virus infection. Emerg Infect Dis 10(8):1405–1411

Koepsell SA, Freifeld AG, Sambol AR, McComb RD, Kazmi SA (2010) Seronegative naturally acquired West Nile virus encephalitis in a renal and pancreas transplant recipient. Transpl Infect Dis 12(5):459–464. doi:10.1111/j.1399-3062.2010.00530.x

Komar N (2000) West Nile viral encephalitis. Rev Sci Tech 19(1):166–176

Komar N, Panella NA, Burns JE, Dusza SW, Mascarenhas TM, Talbot TO (2001) Serologic evidence for West Nile virus infection in birds in the New York City vicinity during an outbreak in 1999. Emerg Infect Dis 7(4):621–625

Kuchtey RW, Kosmorsky GS, Martin D, Lee MS (2003) Uveitis associated with West Nile virus infection. Arch Ophthalmol 121(11):1648–1649

Kulstad EB, Wichter MD (2003) West Nile encephalitis presenting as a stroke. Ann Emerg Med 41 (2):283

Kumar D, Drebot MA, Wong SJ, Lim G, Artsob H, Buck P, Humar A (2004a) A seroprevalence study of west nile virus infection in solid organ transplant recipients. Am J Transplant 4(11):1883–1888. doi:10.1111/j.1600-6143.2004.00592.x

Kumar D, Prasad GV, Zaltzman J, Levy GA, Humar A (2004b) Community-acquired West Nile virus infection in solid-organ transplant recipients. Transplantation 77(3):399–402. doi:10.1097/01.tp. 0000101435.91619.31

L'Vov DK, Kovtunov AI, Iashkulov KB, Gromashevskii VL, Dzharkenov AF, Shchelkanov M, Kulikova LN, Savage HM, Chimidova NM, Mikhaliaeva LB, Vasil'ev AV, Galkina IV, Prilipov AG, Kinney RM, Samokhvalov EI, Bushkieva B, Gubler DJ, Al'khovskii SK,

Aristova VA, Deriabin PG, Butenko AM, Moskvina TM, L'Vov DN, Zlobina LV, Liapina OV, Sadykova GK, Shatalov AG, Usachev VE, Voronina AG, Luneva LI (2004) [Circulation of West Nile virus (Flaviviridae, Flavivirus) and some other arboviruses in the ecosystems of Volga delta, Volga-Akhtuba flood-lands and adjoining arid regions (2000–2002)]. Vopr Virusol 49(3):45–51

Lanciotti RS, Roehrig JT, Deubel V, Smith J, Parker M, Steele K, Crise B, Volpe KE, Crabtree MB, Scherret JH, Hall RA, MacKenzie JS, Cropp CB, Panigrahy B, Ostlund E, Schmitt B, Malkinson M, Banet C, Weissman J, Komar N, Savage HM, Stone W, McNamara T, Gubler DJ (1999) Origin of the West Nile virus responsible for an outbreak of encephalitis in the northeastern United States. Science 286(5448):2333–2337

Lanciotti RS, Ebel GD, Deubel V, Kerst AJ, Murri S, Meyer R, Bowen M, McKinney N, Morrill WE, Crabtree MB, Kramer LD, Roehrig JT (2002) Complete genome sequences and phylogenetic analysis of West Nile virus strains isolated from the United States, Europe, and the Middle East. Virology 298(1):96–105

Lawrie CH, Uzcategui NY, Gould EA, Nuttall PA (2004) Ixodid and argasid tick species and West Nile virus. Emerg Infect Dis 10(4):653–657

Le Guenno B, Bougermouh A, Azzam T, Bouakaz R (1996) West Nile: a deadly virus? Lancet 348(9037):1315

Leis AA, Stokic DS, Polk JL, Dostrow V, Winkelmann M (2002) A poliomyelitis-like syndrome from West Nile virus infection. N Engl J Med 347(16):1279–1280

Leis AA, Stokic DS, Webb RM, Slavinski SA, Fratkin J (2003) Clinical spectrum of muscle weakness in human West Nile virus infection. Muscle Nerve 28(3):302–308

Li J, Loeb JA, Shy ME, Shah AK, Tselis AC, Kupski WJ, Lewis RA (2003) Asymmetric flaccid paralysis: a neuromuscular presentation of West Nile virus infection. Ann Neurol 53(6):703–710

Lindsey NP, Hayes EB, Staples JE, Fischer M (2009) West Nile virus disease in children, United States, 1999–2007. Pediatrics 123(6):e1084–e1089. doi:10.1542/peds.2008-3278

Lindsey NP, Staples JE, Lehman JA, Fischer M, Centers for Disease Control and Prevention (2010) Surveillance for human West Nile virus disease – United States, 1999–2008. MMWR Surveill Summ 59(2):1–17

Lobigs M, Diamond MS (2012) Feasibility of cross-protective vaccination against flaviviruses of the Japanese encephalitis serocomplex. Expert Rev Vaccines 11(2):177–187. doi:10.1586/erv.11.180

Loeb M, Hanna S, Nicolle L, Eyles J, Elliott S, Rathbone M, Drebot M, Neupane B, Fearon M, Mahony J (2008) Prognosis after West Nile virus infection. Ann Intern Med 149(4):232–241

Mackenzie JS, Barrett AD, Deubel V (2002) The Japanese encephalitis serological group of flaviviruses: a brief introduction to the group. Curr Top Microbiol Immunol 267:1–10

Mansfield KL, Horton DL, Johnson N, Li L, Barrett AD, Smith DJ, Galbraith SE, Solomon T, Fooks AR (2011) Flavivirus-induced antibody cross-reactivity. J Gen Virol 92(Pt 12):2821–2829. doi:10.1099/vir.0.031641-0

Marfin AA, Petersen LR, Eidson M, Miller J, Hadler J, Farello C, Werner B, Campbell GL, Layton M, Smith P, Bresnitz E, Cartter M, Scaletta J, Obiri G, Bunning M, Craven RC, Roehrig JT, Julian KG, Hinten SR, Gubler DJ (2001) Widespread West Nile virus activity, eastern United States, 2000. Emerg Infect Dis 7(4):730–735

Martin DA, Muth DA, Brown T, Johnson AJ, Karabatsos N, Roehrig JT (2000) Standardization of immunoglobulin M capture enzyme-linked immunosorbent assays for routine diagnosis of arboviral infections. J Clin Microbiol 38(5):1823–1826

Martin DA, Biggerstaff BJ, Allen B, Johnson AJ, Lanciotti RS, Roehrig JT (2002) Use of immunoglobulin m cross-reactions in differential diagnosis of human flaviviral encephalitis infections in the United States. Clin Diagn Lab Immunol 9(3):544–549

McCarthy TA, Hadler JL, Julian K, Walsh SJ, Biggerstaff BJ, Hinten SR, Baisley C, Iton A, Brennan T, Nelson RS, Achambault G, Marfin AA, Petersen LR (2001) West Nile virus serosurvey and assessment of personal prevention efforts in an area with intense epizootic activity: Connecticut, 2000. Ann N Y Acad Sci 951:307–316

McIntosh BM, Jupp PG, dos Santos I, Meenehan GM (1976) Epidemics of West Nile and Sindbis viruses in South Africa with Culex (Culex) univittatus Theobald as vector. S Afr J Sci 72:295–300

Miller DL, Mauel MJ, Baldwin C, Burtle G, Ingram D, Hines ME 2nd, Frazier KS (2003) West Nile virus in farmed alligators. Emerg Infect Dis 9(7):794–799

Monath TP (2002) Editorial: jennerian vaccination against West Nile virus. Am J Trop Med Hyg 66(2):113–114

Monath T, Heinz F (1996) Flaviviruses. In: Fields B, Knipe D, Howley P et al (eds) Field's virology, vol I, 3rd edn. Lippincott-Raven, Philadelphia, PA, pp 961–1034

Monath TP, Liu J, Kanesa-Thasan N, Myers GA, Nichols R, Deary A, McCarthy K, Johnson C, Ermak T, Shin S, Arroyo J, Guirakhoo F, Kennedy JS, Ennis FA, Green S, Bedford P (2006) A live, attenuated recombinant West Nile virus vaccine. Proc Natl Acad Sci USA 103(17):6694–6699. doi:10.1073/pnas.0601932103

Morales MA, Barrandeguy M, Fabbri C, Garcia JB, Vissani A, Trono K, Gutierrez G, Pigretti S, Menchaca H, Garrido N, Taylor N, Fernandez F, Levis S, Enria D (2006) West Nile virus isolation from equines in Argentina, 2006. Emerg Infect Dis 12(10):1559–1561

Morrey JD, Day CW, Julander JG, Blatt LM, Smee DF, Sidwell RW (2004) Effect of interferon-alpha and interferon-inducers on West Nile virus in mouse and hamster animal models. Antivir Chem Chemother 15(2):101–109

Morrey JD, Siddharthan V, Olsen AL, Roper GY, Wang H, Baldwin TJ, Koenig S, Johnson S, Nordstrom JL, Diamond MS (2006) Humanized monoclonal antibody against West Nile virus envelope protein administered after neuronal infection protects against lethal encephalitis in hamsters. J Infect Dis 194(9):1300–1308. doi:10.1086/508293

Mostashari F, Bunning ML, Kitsutani PT, Singer DA, Nash D, Cooper MJ, Katz N, Liljebjelke KA, Biggerstaff BJ, Fine AD, Layton MC, Mullin SM, Johnson AJ, Martin DA, Hayes EB, Campbell GL (2001) Epidemic West Nile encephalitis, New York, 1999: results of a household-based seroepidemiological survey. Lancet 358(9278):261–264

Murgue B, Murri S, Triki H, Deubel V, Zeller HG (2001a) West Nile in the Mediterranean basin: 1950–2000. Ann N Y Acad Sci 951:117–126

Murgue B, Murri S, Zientara S, Durand B, Durand JP, Zeller H (2001b) West Nile outbreak in horses in southern France, 2000: the return after 35 years. Emerg Infect Dis 7(4):692–696

Murray K, Baraniuk S, Resnick M, Arafat R, Kilborn C, Cain K, Shallenberger R, York TL, Martinez D, Hellums JS, Hellums D, Malkoff M, Elgawley N, McNeely W, Khuwaja SA, Tesh RB (2006) Risk factors for encephalitis and death from West Nile virus infection. Epidemiol Infect 134(6):1325–1332

Murray KO, Resnick M, Miller V (2007) Depression after infection with West Nile virus. Emerg Infect Dis 13(3):479–481

Murray KO, Koers E, Baraniuk S, Herrington E, Carter H, Sierra M, Kilborn C, Arafat R (2009) Risk factors for encephalitis from West Nile Virus: a matched case-control study using hospitalized controls. Zoonoses Public Health 56(6–7):370–375. doi:10.1111/j.1863-2378.2008.01197.x

Murray K, Walker C, Herrington E, Lewis JA, McCormick J, Beasley DW, Tesh RB, Fisher-Hoch S (2010) Persistent infection with West Nile virus years after initial infection. J Infect Dis 201 (1):2–4. doi:10.1086/648731

Murthy HM, Clum S, Padmanabhan R (1999) Dengue virus NS3 serine protease. Crystal structure and insights into interaction of the active site with substrates by molecular modeling and structural analysis of mutational effects. J Biol Chem 274(9):5573–5580

Nasci RS, Savage HM, White DJ, Miller JR, Cropp BC, Godsey MS, Kerst AJ, Bennett P, Gottfried K, Lanciotti RS (2001) West Nile virus in overwintering Culex mosquitoes, New York City, 2000. Emerg Infect Dis 7(4):742–744

Nash D, Mostashari F, Fine A, Miller J, O'Leary D, Murray K, Huang A, Rosenberg A, Greenberg A, Sherman M, Wong S, Layton M (2001) The outbreak of West Nile virus infection in the New York City area in 1999. N Engl J Med 344(24):1807–1814

Nett RJ, Kuehnert MJ, Ison MG, Orlowski JP, Fischer M, Staples JE (in press) Current practices and evaluation of screening solid organ donors for West Nile virus. Transpl Infect Dis

Ng T, Hathaway D, Jennings N, Champ D, Chiang YW, Chu HJ (2003) Equine vaccine for West Nile virus. Dev Biol 114:221–227

Nichter CA, Pavlakis SG, Shaikh U, Cherian KA, Dobrosyzcki J, Porricolo ME, Chatturvedi I (2000) Rhombencephalitis caused by West Nile fever virus. Neurology 55(1):153

Nir Y, Goldwasser R, Lasowski Y, Margalit J (1968) Isolation of West Nile virus strains from mosquitoes in Israel. Am J Epidemiol 87(2):496–501

O'Leary DR, Marfin AA, Montgomery SP, Kipp AM, Lehman JA, Biggerstaff BJ, Elko VL, Collins PD, Jones JE, Campbell GL (2004) The epidemic of West Nile virus in the United States, 2002. Vector Borne Zoonotic Dis 4(1):61–70

Oliphant T, Engle M, Nybakken GE, Doane C, Johnson S, Huang L, Gorlatov S, Mehlhop E, Marri A, Chung KM, Ebel GD, Kramer LD, Fremont DH, Diamond MS (2005) Development of a humanized monoclonal antibody with therapeutic potential against West Nile virus. Nat Med 11(5):522–530. doi:10.1038/nm1240

Papa A, Xanthopoulou K, Gewehr S, Mourelatos S (2011) Detection of West Nile virus lineage 2 in mosquitoes during a human outbreak in Greece. Clin Microbiol Infect 17(8):1176–1180. doi:10.1111/j.1469-0691.2010.03438.x

Park M, Hui JS, Bartt RE (2003) Acute anterior radiculitis associated with West Nile virus infection. J Neurol Neurosurg Psychiatry 74(6):823–825

Parks JJ, Ganaway JR, Price WH (1958) Studies on immunologic overlap among certain arthropod-borne viruses. III. A laboratory analysis of three strains of West Nile virus which have been studied in human cancer patients. Am J Hyg 68(2):106–119

Patnaik JL, Harmon H, Vogt RL (2006) Follow-up of 2003 human West Nile virus infections, Denver, Colorado. Emerg Infect Dis 12(7):1129–1131

Pauvolid-Correa A, Morales MA, Levis S, Figueiredo LT, Couto-Lima D, Campos Z, Nogueira MF, da Silva EE, Nogueira RM, Schatzmayr HG (2011) Neutralising antibodies for West Nile virus in horses from Brazilian Pantanal. Mem Inst Oswaldo Cruz 106(4):467–474

Pealer LN, Marfin AA, Petersen LR, Lanciotti RS, Page PL, Stramer SL, Stobierski MG, Signs K, Newman B, Kapoor H, Goodman JL, Chamberland ME (2003) Transmission of West Nile virus through blood transfusion in the United States in 2002. N Engl J Med 349(13):1236–1245

Penn RG, Guarner J, Sejvar JJ, Hartman H, McComb RD, Nevins DL, Bhatnagar J, Zaki SR (2006) Persistent neuroinvasive West Nile virus infection in an immunocompromised patient. Clin Infect Dis 42(5):680–683

Pepperell C, Rau N, Krajden S, Kern R, Humar A, Mederski B, Simor A, Low DE, McGeer A, Mazzulli T, Burton J, Jaigobin C, Fearon M, Artsob H, Drebot MA, Halliday W, Brunton J (2003) West Nile virus infection in 2002: morbidity and mortality among patients admitted to hospital in southcentral Ontario. CMAJ 168(11):1399–1405

Perelman A, Stern J (1974) Acute pancreatitis in West Nile Fever. Am J Trop Med Hyg 23 (6):1150–1152

Petersen LR, Epstein JS (2005) Problem solved? West Nile virus and transfusion safety. N Engl J Med 353(5):516–517

Petersen LR, Hayes EB (2004) Westward ho? – the spread of West Nile virus. N Engl J Med 351(22):2257–2259. doi:10.1056/NEJMp048261

Petersen LR, Hayes EB (2008) West Nile virus in the Americas. Med Clin North Am 92(6):1307–1322, ix

Petersen LR, Roehrig JT (2001) West Nile virus: a reemerging global pathogen. Emerg Infect Dis 7(4):611–614

Petropoulou KA, Gordon SM, Prayson RA, Ruggierri PM (2005) West Nile virus meningoencephalitis: MR imaging findings. AJNR Am J Neuroradiol 26(8):1986–1995

Platonov AE (2001) West Nile encephalitis in Russia 1999–2001: were we ready? Are we ready? Ann N Y Acad Sci 951:102–116

Pletnev AG, Claire MS, Elkins R, Speicher J, Murphy BR, Chanock RM (2003) Molecularly engineered live-attenuated chimeric West Nile/dengue virus vaccines protect rhesus monkeys from West Nile virus. Virology 314(1):190–195

Public Health Agency of Canada (2011) West Nile Virus *MONITOR*. http://www.phac-aspc.gc.ca/wnv-vwn/index-eng.php. Accessed 2 Feb 2012

Ravindra KV, Freifeld AG, Kalil AC et al (2004) West Nile virus-associated encephalitis in recipients of renal and pancreas transplants: case series and literature review. Clin Infect Dis 38:1257–1260

Rhee C, Eaton EF, Concepcion W, Blackburn BG (2011) West Nile virus encephalitis acquired via liver transplantation and clinical response to intravenous immunoglobulin: case report and review of the literature. Transpl Infect Dis 13(3):312–317. doi:10.1111/j.1399-3062.2010.00595.x

Robinson RL, Shahida S, Madan N, Rao S, Khardori N (2003) Transient parkinsonism in West Nile virus encephalitis. Am J Med 115(3):252–253

Rodriguez AJ, Westmoreland BF (2007) Electroencephalographic characteristics of patients infected with west nile virus. J Clin Neurophysiol 24(5):386–389

Roehrig JT, Nash D, Maldin B, Labowitz A, Martin DA, Lanciotti RS, Campbell GL (2003) Persistence of virus-reactive serum immunoglobulin m antibody in confirmed west nile virus encephalitis cases. Emerg Infect Dis 9(3):376–379

Sadek JR, Pergam SA, Harrington JA, Echevarria LA, Davis LE, Goade D, Harnar J, Nofchissey RA, Sewell CM, Ettestad P, Haaland KY (2010) Persistent neuropsychological impairment associated with West Nile virus infection. J Clin Exp Neuropsychol 32(1):81–87. doi:10.1080/13803390902881918

Sampson BA, Ambrosi C, Charlot A, Reiber K, Veress JF, Armbrustmacher V (2000) The pathology of human West Nile Virus infection. Hum Pathol 31(5):527–531

Samuel MA, Wang H, Siddharthan V, Morrey JD, Diamond MS (2007) Axonal transport mediates West Nile virus entry into the central nervous system and induces acute flaccid paralysis. Proc Natl Acad Sci USA 104(43):17140–17145

Savage HM, Ceianu C, Nicolescu G, Karabatsos N, Lanciotti R, Vladimirescu A, Laiv L, Ungureanu A, Romanca C, Tsai TF (1999) Entomologic and avian investigations of an epidemic of West Nile fever in Romania in 1996, with serologic and molecular characterization of a virus isolate from mosquitoes. Am J Trop Med Hyg 61(4):600–611

Sayao AL, Suchowersky O, Al-Khathaami A, Klassen B, Katz NR, Sevick R, Tilley P, Fox J, Patry D (2004) Calgary experience with West Nile virus neurological syndrome during the late summer of 2003. Can J Neurol Sci 31(2):194–203

Sejvar JJ, Haddad MB, Tierney BC, Campbell GL, Marfin AA, Van Gerpen JA, Fleischauer A, Leis AA, Stokic DS, Petersen LR (2003a) Neurologic manifestations and outcome of West Nile virus infection. JAMA 290(4):511–515

Sejvar JJ, Leis AA, Stokic DS, Van Gerpen JA, Marfin AA, Webb R, Haddad MB, Tierney BC, Slavinski SA, Polk JL, Dostrow V, Winkelmann M, Petersen LR (2003b) Acute flaccid paralysis and West Nile virus infection. Emerg Infect Dis 9(7):788–793

Sejvar JJ, Bode AV, Marfin AA, Campbell GL, Ewing D, Mazowiecki M, Pavot PV, Schmitt J, Pape J, Biggerstaff BJ, Petersen LR (2005) West Nile virus-associated flaccid paralysis. Emerg Infect Dis 11(7):1021–1027

Sejvar JJ, Bode AV, Marfin AA, Campbell GL, Pape J, Biggerstaff BJ, Petersen LR (2006) West Nile virus-associated flaccid paralysis outcome. Emerg Infect Dis 12(3):514–516

Sejvar JJ, Curns AT, Welburg L, Jones JF, Lundgren LM, Capuron L, Pape J, Reeves WC, Campbel GL (2008) Neurocognitive and functional outcomes in persons recovering from West Nile virus illness. J Neuropsychol 2(Pt 2):477–499

Sejvar JJ, Davis LE, Szabados E, Jackson AC (2010) Delayed-onset and recurrent limb weakness associated with West Nile virus infection. J Neurovirol 16(1):93–100. doi:10.3109/13550280903586378

Sejvar JJ, Lindsey NP, Campbell GL (2011) Primary causes of death in reported cases of fatal West Nile Fever, United States, 2002–2006. Vector Borne Zoonotic Dis 11(2):161–164. doi:10.1089/vbz.2009.0086

Shaikh S, Trese MT (2004) West Nile virus chorioretinitis. Br J Ophthalmol 88(12):1599–1660. doi:10.1136/bjo.2004.049460

Shimoni Z, Niven MJ, Pitlick S, Bulvik S (2001) Treatment of West Nile virus encephalitis with intravenous immunoglobulin. Emerg Infect Dis 7(4):759

Siger L, Bowen R, Karaca K, Murray M, Jagannatha S, Echols B, Nordgren R, Minke JM (2006) Evaluation of the efficacy provided by a Recombinant Canarypox-Vectored Equine West Nile Virus vaccine against an experimental West Nile Virus intrathecal challenge in horses. Vet Ther 7(3):249–256

Smeraski CA, Siddharthan V, Morrey JD (2011) Treatment of spatial memory impairment in hamsters infected with West Nile virus using a humanized monoclonal antibody MGAWN1. Antivir Res 91(1):43–49. doi:10.1016/j.antiviral.2011.04.011

Solomon T, Dung NM, Wills B, Kneen R, Gainsborough M, Diet TV, Thuy TT, Loan HT, Khanh VC, Vaughn DW, White NJ, Farrar JJ (2003) Interferon alfa-2a in Japanese encephalitis: a randomised double-blind placebo-controlled trial. Lancet 361(9360):821–826

Southam CM, Moore AE (1952) Clinical studies of viruses as antineoplastic agents with particular reference to Egypt 101 virus. Cancer 5(5):1025–1034

Southam CM, Moore AE (1954) Induced virus infections in man by the Egypt isolates of West Nile virus. Am J Trop Med Hyg 3(1):19–50

Spigland I, Jasinska-Klingberg W, Hofshi E, Goldblum N (1958) Clinical and laboratory observations in an outbreak of West Nile fever in Israel in 1957. Harefuah 54(11):275–280, English & French abstracts 280–281

Tang F, Zhang JS, Liu W, Zhao QM, Zhang F, Wu XM, Yang H, Ly H, Cao WC (2008) Failure of Japanese encephalitis vaccine and infection in inducing neutralizing antibodies against West Nile virus, People's Republic of China. Am J Trop Med Hyg 78(6):999–1001

Tsai TF, Popovici F, Cernescu C, Campbell GL, Nedelcu NI (1998) West Nile encephalitis epidemic in southeastern Romania. Lancet 352(9130):767–771

Tyler KL, Pape J, Goody RJ, Corkill M, Kleinschmidt-DeMasters BK (2006) CSF findings in 250 patients with serologically confirmed West Nile virus meningitis and encephalitis. Neurology 66(3):361–365

Vandenbelt S, Shaikh S, Capone A Jr, Williams GA (2003) Multifocal choroiditis associated with West Nile virus encephalitis. Retina 23(1):97–99

Vidwan G, Bryant KK, Puri V, Stover BH, Rabalais GP (2003) West Nile virus encephalitis in a child with left-side weakness. Clin Infect Dis 37(6):e91–e94. doi:10.1086/377263

Watson JT, Pertel PE, Jones RC, Siston AM, Paul WS, Austin CC, Gerber SI (2004) Clinical characteristics and functional outcomes of West Nile Fever. Ann Intern Med 141(5):360–365

Weinberger M, Pitlik SD, Gandacu D, Lang R, Nassar F, Ben David D, Rubinstein E, Izthaki A, Mishal J, Kitzes R, Siegman-Igra Y, Giladi M, Pick N, Mendelson E, Bin H, Shohat T (2001) West Nile fever outbreak, Israel, 2000: epidemiologic aspects. Emerg Infect Dis 7(4):686–691

Yamshchikov G, Borisevich V, Kwok CW, Nistler R, Kohlmeier J, Seregin A, Chaporgina E, Benedict S, Yamshchikov V (2005) The suitability of yellow fever and Japanese encephalitis vaccines for immunization against West Nile virus. Vaccine 23(39):4785–4792. doi:10.1016/j.vaccine.2005.04.036

Yim R, Posfay-Barbe KM, Nolt D, Fatula G, Wald ER (2004) Spectrum of clinical manifestations of West Nile virus infection in children. Pediatrics 114(6):1673–1675

Zeller HG, Schuffenecker I (2004) West Nile virus: an overview of its spread in Europe and the Mediterranean basin in contrast to its spread in the Americas. Eur J Clin Microbiol Infect Dis 23(3):147–156

Zielinski-Gutierrez E, Wirtz RA, Nasci RS, Brogdon WG (2012) Chapter 2: the pre-travel consultation counseling & advice for travelers; protection against mosquitoes, ticks, & other insects & arthropods. http://wwwnc.cdc.gov/travel/page/yellowbook-2012-home.htm. Accessed 9 Feb

Japanese Encephalitis Virus Infection

Lance Turtle and Tom Solomon

Abstract Japanese encephalitis (JE) is a viral encephalitis caused by Japanese encephalitis virus (JEV), a single stranded positive sense RNA virus of the genus *Flavivirus*. The reservoir of JEV is wild birds and pigs; it is spread to humans by mosquitoes. There are approximately 68,000 cases and 15,000 deaths per year. JE is endemic to South and Southeast Asia where 2.5 billion people live. Although most of this population are exposed to JEV, only a minority develop JE. JE is a disease of children and the adult population is immune as a result of childhood exposure. The factors that determine whether or not disease develops relate largely to the host immune response and are still being actively investigated. Patients with JE develop a febrile illness followed by the onset of headache, vomiting, clouding of consciousness and possibly seizures. A wide range of neurological abnormalities are possible; for example focal upper motor neuron lesions, acute flaccid paralysis and extrapyramidal features. There are no proven effective therapies and treatment is supportive; however, JE can be prevented by vaccination. A number of vaccines are available including both live attenuated and inactivated cell culture derived vaccines.

Keywords Encephalitis • Encephalitis in Asia • Flaviviridae • Flavivirus • Japanese encephalitis • JE vaccine

L. Turtle
Institute of Infection and Global Health, University of Liverpool, 8 West Derby Street, Liverpool L69 7BE, UK
e-mail: lturtle@liverpool.ac.uk

T. Solomon (✉)
Walton Centre NHS Foundation Trust and Institute of Infection and Global Health, University of Liverpool, 8 West Derby Street, Liverpool L69 7BE, UK
e-mail: tsolomon@liverpool.ac.uk

A.C. Jackson (ed.), *Viral Infections of the Human Nervous System*,
Birkhäuser Advances in Infectious Diseases,
DOI 10.1007/978-3-0348-0425-7_11,

1 Introduction

Japanese encephalitis (JE) is a viral encephalitis caused by Japanese encephalitis virus (JEV), an arthopod-borne member of the genus *Flavivirus*, family *Flaviviridae*. JE is the most important epidemic encephalitis in the world, causing a greater loss of disability adjusted life years (DALYs) than any other arthropod-borne virus (arbovirus) (WHO 2004). So named because it was first described in Japan, the disease is endemic to much of South and Southeast Asia with over two billion people living in JE endemic areas.

2 History

Encephalitis was first described in Japan in 1871, followed by outbreaks every few years. In 1924 the disease was named Japanese B encephalitis to distinguish it from von Economo's encephalitis, or type A encephalitis (encephalitis lethargica) (Solomon 2006); this was probably a form of autoimmune or post-infectious encephalitis, which caused an outbreak from 1916 to 1927. However, since then, the term "type A encephalitis" has not been used, and the term Japanese B encephalitis has also fallen out of use.

A filterable agent was found to reproduce encephalitis in a monkey in 1933, and the causative agent, JEV, was isolated in 1935. JEV was later classified as a *Flavivirus*, a genus named after the prototype species, yellow fever virus (flavi is Latin for yellow). After the Second World War JE was recognised to occur elsewhere is Asia. JEV was isolated in India in the 1950s, and was soon observed to cause outbreaks of encephalitis throughout much of Asia. Despite the existence of vaccines since the 1950s, the disease continues to be a major public health problem in Asia to the present day.

3 Virology

JEV is, like the other members of the genus *Flavivirus*, a single stranded, positive strand (sense) RNA virus. The virion is approximately 50 nm across, enveloped and contains three structural viral proteins. A heterodimer of envelope (E) glycoprotein and membrane (M) protein embedded in the viral envelope surrounds a nucleocapsid (C protein) containing the viral RNA. The viral genome is 11 kilobases (kb) in size and contains a single open reading frame (ORF) slightly over 10 kb in size (Sumiyoshi et al. 1987) (Fig. 1). The ORF is translated as a single large polyprotein, which is then cleaved by a mixture of host and viral proteases to produce mature viral proteins. The ORF is surrounded by 3′ and 5′ untranslated regions (UTR). The 5′ UTR is 95–96 nucleotides in length and the 3′ UTR is more variable in both size and sequence, usually several hundred nucleotides in length (Chambers et al. 1990;

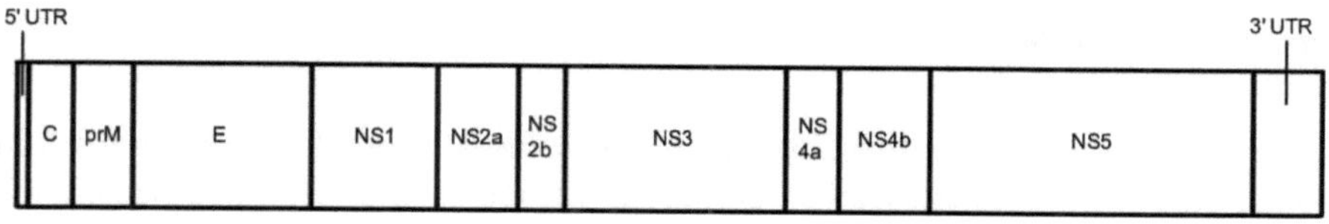

Fig. 1 JEV genome organisation. The layout of the JEV genome is shown. The first three genes at the 5′ (*left*) end encode the structural proteins core (C), pre-membrane (prM) and envelope (E). The seven remaining genes are non-structural (NS) proteins and are involved in all the other functions of the viral replication. At the far 5′ and 3′ ends there are untranslated regions (UTR); these regions do not encode proteins but instead have regulatory function relating to viral replication. The 5′ UTR of JEV is 95–96 nucleotides in length; the 3′ UTR is more variable in both length and sequence and is typically around 400–600 nucleotides in length

Lindenbach et al. 2007). There are ten viral proteins in all. As well as the three structural proteins (C, M and E) mentioned above, there are seven non-structural (NS) proteins, denoted NS1, NS2a, NS2b, NS3, NS4a, NS4b and NS5. M protein is encoded as a precursor "pre-membrane" (prM) and is cleaved during replication. In addition, there is an alternate form of NS1 called NS1' which is produced by a −1 frameshift towards the end of the NS1 gene (Melian et al. 2010). This results in an elongated gene product that ends at the site of a premature stop codon found in the −1 reading frame.

Infection with JEV is cytolytic in mammalian cells. Although no single flavivirus entry receptor has been determined, several putative receptors have been identified in different cell lines; these include the C-type lectin DC-SIGN on dendritic cells for dengue virus entry, the αvβ3 integrin, GRP78 (BiP) and CD14 or a related molecule (Mukhopadhyay et al. 2005). Heparan sulphates, and other highly sulphated glycosaminoglycans are also important in initiating flavivirus attachment to target cells (Chen et al. 1997). Many of these experiments have been performed on flaviviruses other than JEV; whether the lessons learned apply to JEV is not fully understood. Viral replication takes place in the cytoplasm (as is common with many RNA viruses) in specialised lipid structures that are probably derived from the endoplasmic reticulum and are surrounded by a double layer of lipid bilayer making the replicating virus relatively inaccessible to the host cell milieu (Uchil and Satchidanandam 2003). A full-length negative sense RNA strand is generated to act as a template for making new copies of the full-length positive sense viral genome. The viral RNA in encapsidated then the nascent virion is exported via the endoplasmic reticulum. On the way, immature prM is cleaved to form the M protein in the mature virion (Lindenbach et al. 2007).

4 Pathogenesis

As discussed earlier, the majority of JEV exposed individuals will not develop disease, but a minority will develop a severe illness. The factors that determine this are numerous, involving complex interactions between virus and host, and are still being elucidated.

4.1 Virulence Factors

The degree of sequence diversity seen in flaviviruses is less than that observed in other RNA viruses that establish chronic infection, such as HIV and hepatitis C virus. Although viral variation does play an important role in the pathogenesis of some flaviviruses, it has not been shown to be a major factor in JEV pathogenesis. It should be noted, however, that due to the difficulty in isolating JEV from clinical specimens and finding the genome by reverse transcriptase polymerase chain reaction, there is a relative lack of JEV sequence information compared with some other flaviviruses. In JE the host immune response is probably a more critical determinant of clinical outcome, at least in humans.

An overview of the pathogenesis of flavivirus encephalitis is presented in Fig. 2. Following inoculation from the bite of an infected mosquito, JEV is thought to undergo replication in the local tissues mainly in Langerhans cells. Following this it disseminates to the local lymph nodes where further replication in monocyte lineage cells takes place, causing a viraemia, which is followed later by spread to the brain. In mice, high rates of JEV replication within dendritic cells are linked to higher overall mortality (Wang and Deubel 2011). Autophagy (internal phagocytosis for recycling cell components), which is known to be important in dengue virus and HCV replication, is also involved in early JEV replication. Enhancement of autophagy increases JEV replication and inhibition reduces JEV replication (Li et al. 2012). How JEV gains access to the central nervous system (CNS) remains incompletely understood, but interactions at the blood brain barrier (BBB) are probably critical. Proposed mechanisms include active replication within endothelial cells, passive transfer across the BBB or within leukocytes that migrate across the barrier.

4.2 Innate Responses

JEV interacts with host cytosolic pathogen recognition receptors that detect RNA viruses, such as the retinoic acid inducible gene (RIG)-I (Kato et al. 2006) and probably Toll-like receptor (TLR)3 and TLR7. In the case of the related flavivirus West Nile virus (WNV) interactions with TLR3 may promote inflammation that contributes to BBB breakdown (Wang et al. 2004), but whether the same is true in JE is not known. Clinical and pathological studies of JE have suggested that co-infection with neurocysticercosis (which compromises the BBB) increases the risk of developing encephalitis following infection with JEV, consistent with the hypothesis that BBB breakdown contributes to viral entry into the CNS (Desai et al. 1997). However, the possibility that an association between neurocysticercosis and JE merely reflects exposure to pigs had not been fully excluded. Engagement of these pattern recognition receptors triggers the innate immune response with production of type I interferon (interferon-α and -β) and induction of an antiviral state. However, as is the case with many other flaviviruses that have been studied, JEV is able to interfere with the innate immune response and antagonise interferon signalling (Lin et al. 2004, 2006).

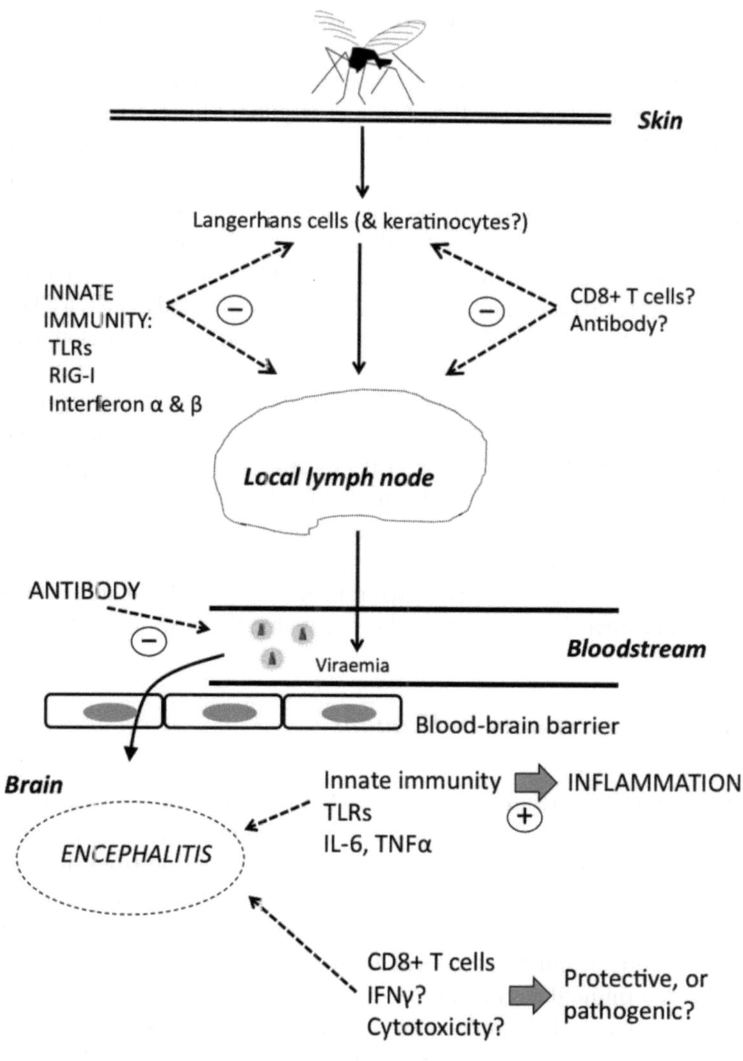

Fig. 2 Overview of the pathogenesis of Japanese encephalitis. Key mechanisms known to be important in immune control or immunopathology of Japanese encephalitis are shown in capitals. *Question mark* indicates areas of uncertainty. *TLR* toll like receptor, *RIG* retinoic acid inducible gene, *IL-6* interleukin 6, *TNF* tumour necrosis factor, *IFN* interferon (Adapted from Turtle et al. 2012 with permission)

4.3 Adaptive Immunity in Animal Models

Experiments in animal models of JE have highlighted some of the key components of adaptive immunity that result in virus clearance. It has been known for many years from passive immunisation experiments that antibody protects from JE in mice (Kimura-Kuroda and Yasui 1988; Lubiniecki et al. 1973). Not surprisingly, when such experiments were repeated using B cell deficient mice, the animals were highly susceptible to JEV infection (Larena et al. 2011). Similarly, transfer of immune B cells was protective.

The role of T cells in JE is less clear. Adoptive transfer of cells from immune mice is protective in animal models of JE (Mathur et al. 1983; Murali-Krishna et al. 1996), although for optimal protection all cellular components of the immune system should be transferred (Larena et al. 2011). Further evidence for cooperation between adaptive immune components was shown by the role of Th2 type CD4+ T cells in protection from JEV through participation in antibody responses (Biswas et al. 2009). However, some observations suggest an important role for CD8+ T cells as well. The dominant type of lymphocyte that accumulates in the CNS of JEV infected mice is a CD8+ T cell (Fujii et al. 2008; Larena et al. 2011). T cell receptor usage is highly clonal, a finding which suggests that this occurs in an antigen driven manner (Fujii et al. 2008). Moreover, JEV is able to inhibit CD8+ T cell responses (to other antigens as well) by interfering with antigen presentation, an effect reversible by stimulating the innate immune system with TLR3, TLR4 or TLR9 ligands (Aleyas et al. 2010). There is a modest degree of protection seen when immune T cells are transferred alone into naïve mice, but when CD8+ T cells were depleted in this model the viral load in nervous tissue increased markedly, but survival did not change (Larena et al. 2011). Together, these data suggest that there may be a dual role of CD8+ T cells in both clearing viral infection but also mediating immunopathology within the CNS in animal models.

4.4 Adaptive Immunity in Humans

Antibody responses are readily detectable in humans forming the basis for diagnostic tests (IgM) and vaccine trial endpoints (virus neutralisation). The development of neutralising antibody after vaccination parallels protection from disease (Hoke et al. 1988) and lower antibody levels are associated with poor outcome (Leake et al. 1986; Libraty et al. 2002; Winter et al. 2004). However, many patients still have high neutralising antibody titres suggesting that once virus is in the CNS other factors may determine the outcome.

T cell responses are detectable in humans, though such responses are often small (Konishi et al. 1995; Kumar et al. 2003, 2004a, c). A mixture of CD4+ and CD8+ T cell responses is seen in humans who have been naturally exposed in JEV endemic areas, whereas generally only CD4 responses are seen in people who have been given inactivated JE vaccine. In a cohort of children who had recovered from JE, there was a correlation between interferon-γ (IFNγ) production and better clinical outcome, whereas proliferative responses were preserved reflecting a relatively selective immune deficit of the antiviral cytokine IFNγ (Kumar et al. 2004b).

4.5 Histopathology

Once in the CNS, there are many mechanisms by which JEV induces neuronal damage and cell death leading to the clinical manifestations of disease. Early

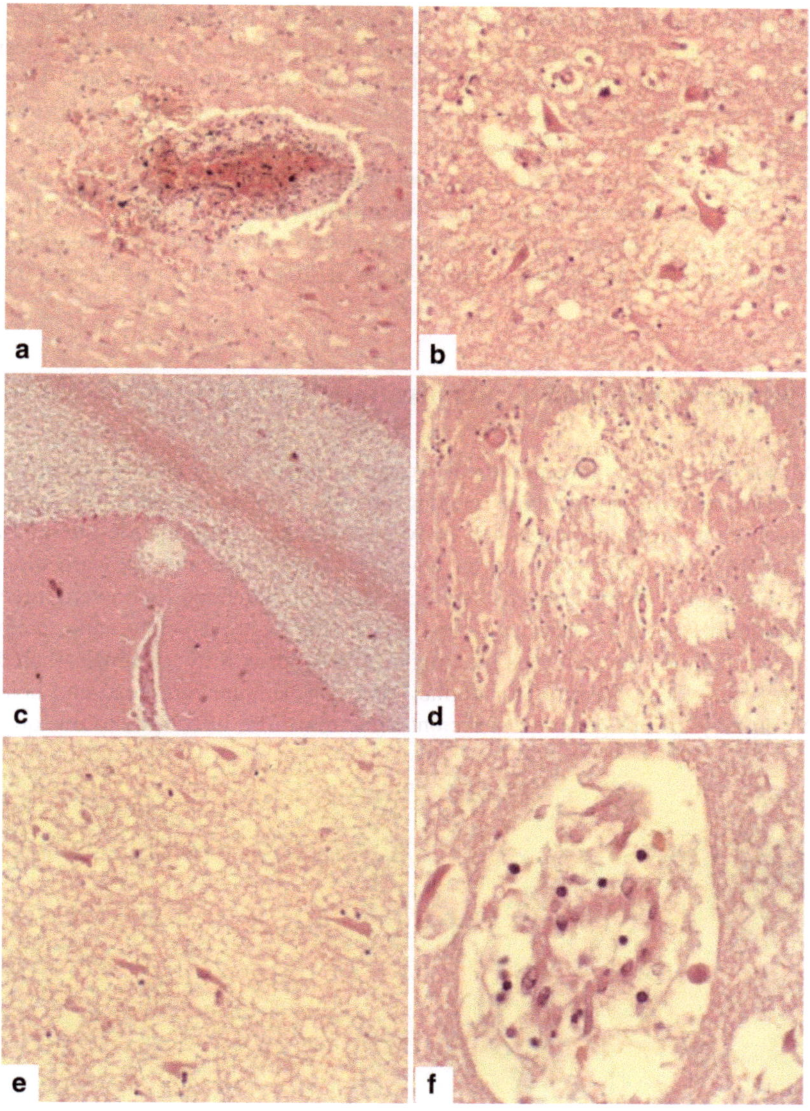

Fig. 3 Characteristic histopathology in a fatal case of Japanese encephalitis. (**a**) Perivascular inflammation and necrosis in the pons (×100). (**b**) Damaged, shrunken cells in the mid brain (×200). (**c**) An isolated, necrotic lesion related to a blood vessel in the cerebellum (×40). (**d**) Punched out, necrotic areas in the internal capsule. (**e**) Neuronal necrosis in the subiculum. (**f**) Perivascular lymphocytic infiltrate; the vessel is damaged (Adapted from German et al. 2006 with permission)

histological studies showed that there is a marked inflammatory response in the brain of patients who have died of JE (Fig. 3). Animal models of JE reflect the same inflammatory changes. There is typically glial cell upregulation and perivascular inflammation of lymphocytes and macrophages. In humans with JE there are punched out necrotic lesions, often close to blood vessels; however, these lesions are not seen in animal models (German et al. 2006; Miyake 1964). There is histological evidence of extensive apoptosis in post-mortem brain specimens from JE patients. How much brain damage is due to direct cytopathic virus effect, how much is due to (protective?) apoptosis and how much is due to immunopathology is not clear, but attenuating the CNS inflammatory response to JEV in mice with minocycline appears to result in improved survival suggesting that unregulated inflammation plays at least some part (Das et al. 2011; Mishra and Basu 2008).

5 Epidemiology and Transmission

JEV is transmitted by mosquitoes of the genus *Culex*. In its natural ecology, JEV is a virus of birds, particularly wading ardeids such as herons and egrets, and exists in an enzootic cycle passing between avian hosts through the bite of *Culex tritaeniorhynchus*. Pigs act as amplifying hosts for JEV because they are susceptible to infection with the virus, and generate sufficient viraemia for onward infection of mosquitoes (Fig. 4). Although most pig infections are asymptomatic, pregnant swines will abort. Because pigs live in much closer proximity to humans than birds do, they provide a reservoir for human infection. Humans are dead end hosts of JEV, as they do not generate sufficient viraemia to be infectious to mosquitoes.

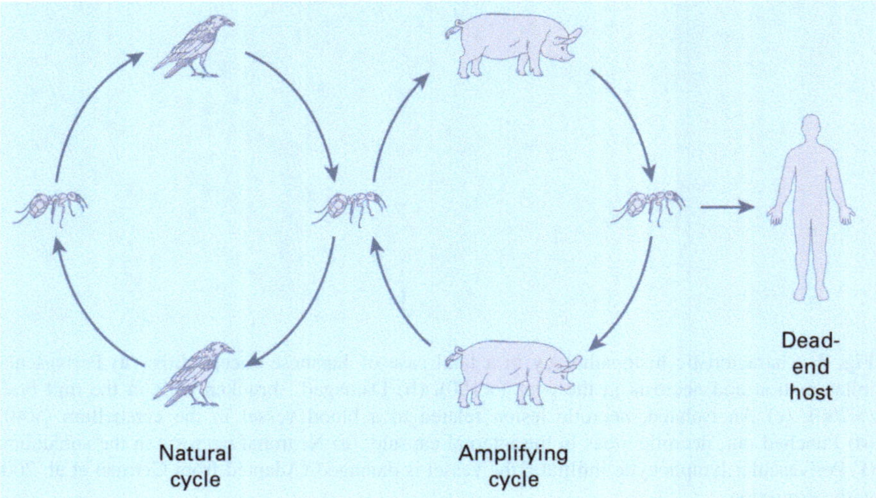

Fig. 4 The enzootic cycle and transmission of Japanese encephalitis virus. JEV is maintained in birds (particularly wading ardeids such as herons and egrets) transmitted by *Culex* mosquitoes. The virus can be transmitted to pigs, usually during the rainy season. In pigs there is sufficient viraemia to act as a reservoir for transmission to humans, from where it is no longer transmitted any further (Adapted from Solomon 2004a with permission)

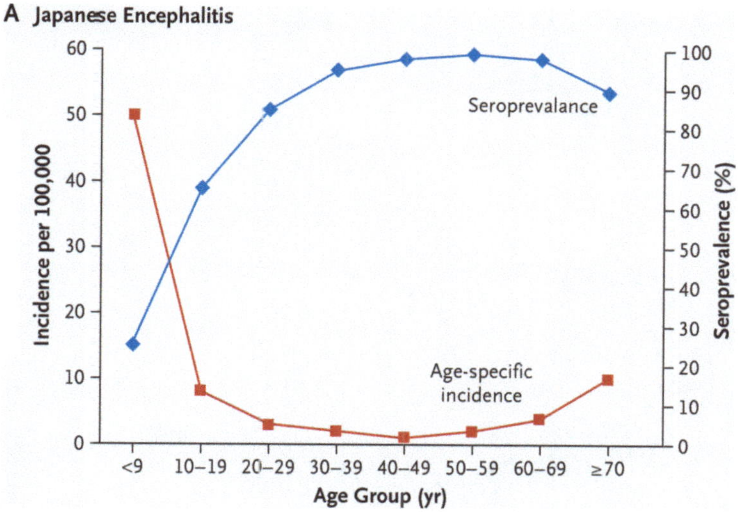

Fig. 5 Age specific incidence and seroprevalence of Japanese encephalitis virus. In endemic areas JE is principally a disease of children. The majority of the adult population are immune from silent exposure during childhood (From Solomon 2000b, with permission)

Human infection with JEV is, therefore, an accidental event in evolutionary terms. Because of the enzootic cycle involving *C. tritaeniorhynchus* (which breeds in rice paddies), birds and pigs, JE mainly affects rural and agricultural populations.

In endemic areas JE is predominantly a disease of children (Solomon 2004). It tends to occur in sporadic epidemics coinciding with periods of increased rainfall. The precise number of cases occurring every year is not known, but a recent estimate puts the figure at approximately 68,000 cases per year, with annual incidence 1.8 per 100 000 (Campbell et al. 2011). Only around 10% of these cases are reported to the World Health Organization.

Serological surveys show that in endemic areas the vast majority of adults are immune as a result of previous (clinically silent) exposure (Fig. 5). It is evident, therefore, that the majority of human infections with JEV do not result in disease. Estimates for the frequency of clinical JE in endemic areas vary from around one in 250 to one in 1000 infections (Solomon et al. 2000). There have been reports of disease resulting from migration to a JE endemic area in adult life indicating that it is probably past exposures resulting in immunity that protects adults. In adult individuals from non-endemic areas the chance of developing disease after JEV infection may also be higher.

The geographical region where JEV circulates and where JE is endemic stretches from Japan in the North to Pakistan in the West and Guinea in the Southeast (Fig. 6). JEV is still spreading and in recent years has been isolated from the mainland of Northern Australia (Van Den Hurk et al. 2006). Although JE has largely disappeared from Japan, this is as a consequence of large-scale vaccination, urbanisation and improvement in housing stock resulting in less exposure (Beasley et al. 2008). JEV still circulates in Japan and occasional cases of JE do still occur (Arai et al. 2008). The origin of JEV is not known for certain but based

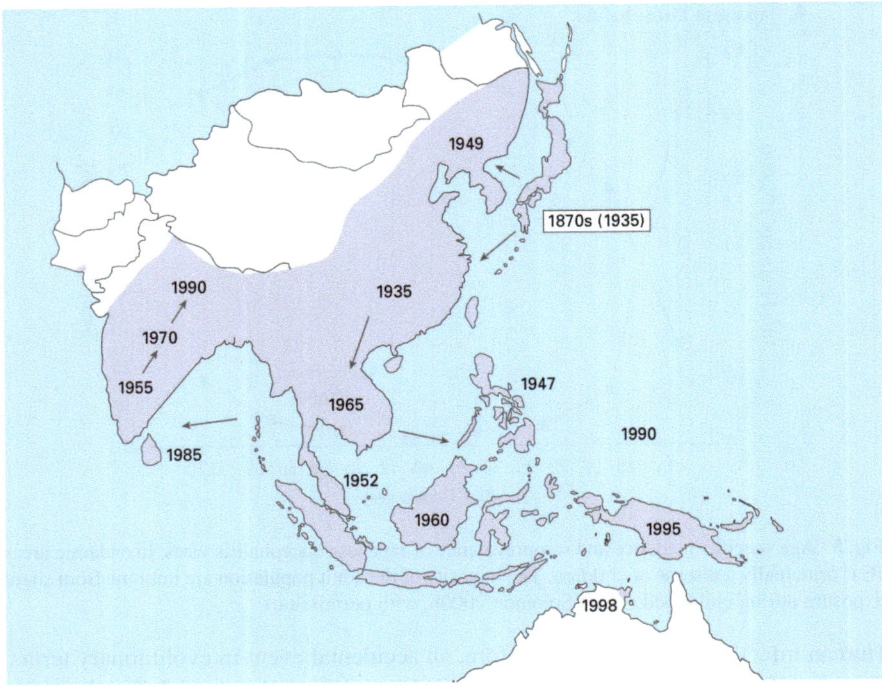

Fig. 6 Distribution of Japanese encephalitis virus. JEV is endemic to most of South and Southeast Asia, an area with a population of around 2.5 billion people. The historical spread of JE outbreaks from the first description in Japan in the 1870s to the appearance in Northern Australia in the late 1990s is shown (Adapted from Solomon and Winter 2004 with permission)

on sequence data it is thought that JEV arose from its common ancestor in the Malaysia–Indonesia region before spreading through Asia (Solomon et al. 2003b).

It is likely that climate plays a role in the epidemiology of JE, but with so many factors involved in the cycle this is likely to be a complex relationship. Both temperature and rainfall may drive JEV transmission through their influence on the mosquito vector life cycle, and effect on the extrinsic incubation period of the virus in mosquitoes (Murty et al. 2010). Early studies showed JEV transmission is associated with high temperatures and low rainfall (Mogi 1983). A recent study from Nepal showed an association of JE cases with low precipitation and the percentage of irrigated land (Impoinvil et al. 2011).

6 Clinical Aspects

After an incubation period of 5–15 days, JE starts as a non-specific febrile prodrome for a few days; other symptoms such as coryza and diarrhoea may also occur (Solomon et al. 2000). This is followed by headache, vomiting and clouding of consciousness, often accompanied by seizures. In older children and adults there

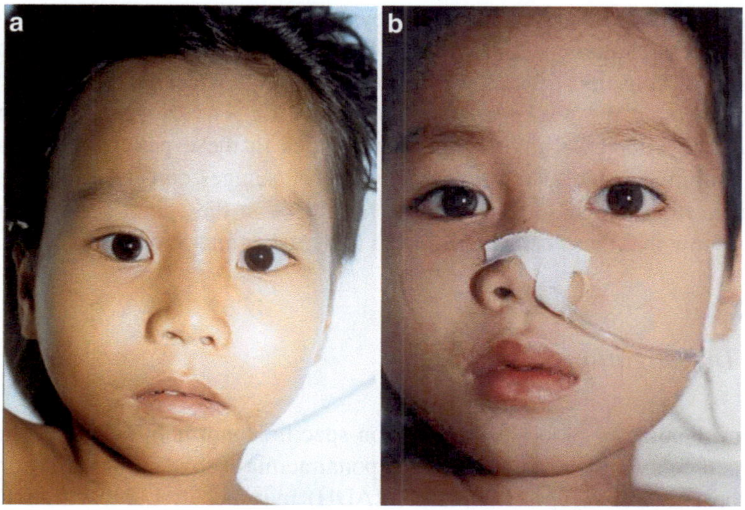

Fig. 7 Mask-like facies in a Vietnamese child with Japanese encephalitis. JE infection involves the extrapyramidal system leading to Parkinsonian features (Photograph: T Solomon, (adapted from Solomon et al. 2000) with permission)

may also be presenting psychiatric features. Some patients make a rapid recovery from this point, and a few show aseptic meningitis with full consciousness and no evidence of parenchymal involvement.

Seizures are frequent in JE, more so in children than in adults. Up to 85 % of children with JE may experience a seizure. Many seizure types can occur; generalised tonic-clonic seizures are more common than focal seizures. In some children there may be very subtle motor seizures such as twitching of a single digit or a facial muscle, eye deviation or nystagmus. Such seizures can be hard to diagnose but should be actively sought out and treated; multiple seizures and status epilepticus are associated with poorer prognosis (Solomon et al. 2002).

The typical appearance of a child with JE is of mask-like facies (Fig. 7) with an absent, unblinking stare, tremor, hypertonia and rigidity (often cogwheel rigidity). Other extrapyramidal signs may be seen, such as pill rolling, head nodding, lip smacking and facial grimacing. A Parkinsonian syndrome is also recognised to occur. Opisthotonos and rigidity spasms are sometimes seen and are associated with a worse outcome.

The brainstem can become involved in JE giving rise to clinical signs such as change in respiratory pattern, flexor or extensor posturing, which are poor prognostic signs. There are thought to be two mechanisms of brainstem involvement: direct viral infection of the brainstem and transtentorial herniation. In the latter rostrocaudal progression is seen in association with high CSF opening pressures, which can occasionally be reversible with aggressive measures to lower intracranial pressure (ICP) (Solomon et al. 2002).

Other focal neurological deficits occur in JE; for example 10 % of children have upper motor neurone facial palsy, which can be intermittent. JEV can cause a poliomyelitis-like illness characterised by flaccid paralysis with a normal level of consciousness (Solomon et al. 1998a); flaccid paralysis is seen in some patients with loss of consciousness as well. The lower limbs are affected more than the upper limbs and the weakness is often asymmetric. Occasionally the respiratory muscles are affected. Nerve conduction studies in these patients show reduced motor amplitudes and electromyography shows changes consistent with denervation suggesting damage to the anterior horn cells in the spinal cord; histological studies show that these cells are infected (Haymaker and Sabin 1947).

6.1 Investigations

Standard laboratory blood tests show non-specific abnormalities in JE. The white blood cell count may be elevated and hyponatraemia secondary to the syndrome of inappropriate anti-diuretic hormone (SIADH) may occur. Lumbar puncture can show elevated CSF pressure, pressures above 25 cm H_2O are associated with poor prognosis (Solomon et al. 2002). Typically the CSF cell count is 10–100 cells/mm^3, the protein is mildly elevated (50–200 mg%) and the glucose ratio is normal.

Imaging studies most frequently show bilateral lesions in the thalamus, though other structures can also be affected. Computed tomography (CT) scans show bilateral thalamic hypodensity in approximately 20% of patients (Fig. 8) (Dung et al. 2009). In one patient in this study the lesions disappeared 10 days after the acute phase (Fig. 9), an effect known as the "fogging effect". The fogging effect, first described in ischaemic cerebral infarcts, is thought to occur secondary to the influx of lipid laden macrophages which makes the lesions isodense with surrounding brain, so they are not seen on CT (Becker et al. 1979). MRI, which typically shows bilateral high signal on T2 weighted or T2 fluid attenuated inversion recovery (FLAIR) images, is more sensitive than CT and often reveals more lesions in other areas of the brain (Kalita and Misra 2000; Kalita et al. 2003). These imaging findings are insensitive but highly specific for JE in an endemic area. However, it should be noted that identical changes have been reported in other forms of arbovirus encephalitis caused by members of the *Flavivirus* genus, Murray Valley encephalitis virus, West Nile encephalitis, dengue encephalitis and tick borne encephalitis, so the specificity of these findings depends very much upon the local pattern of flavivirus circulation. The temporal lobes may also be affected on MRI in JE. Although the pattern of MRI changes is not identical to herpes simplex encephalitis, there can occasionally be diagnostic difficulty (Handique et al. 2006). MRI of the spinal cord in cases of flaccid paralysis due to JEV shows abnormal signal intensity on T2 weighted images (Solomon et al. 1998a). Electroencephalographic studies may show various abnormalities in JE, including theta and delta coma, burst suppression, epileptiform activity, and occasionally alpha coma. Diffuse slowing may help to distinguish JE from herpes simplex virus encephalitis, in which changes are characteristically frontotemporal. EEG can also be useful to identify subtle motor status epilepticus, as outlined above.

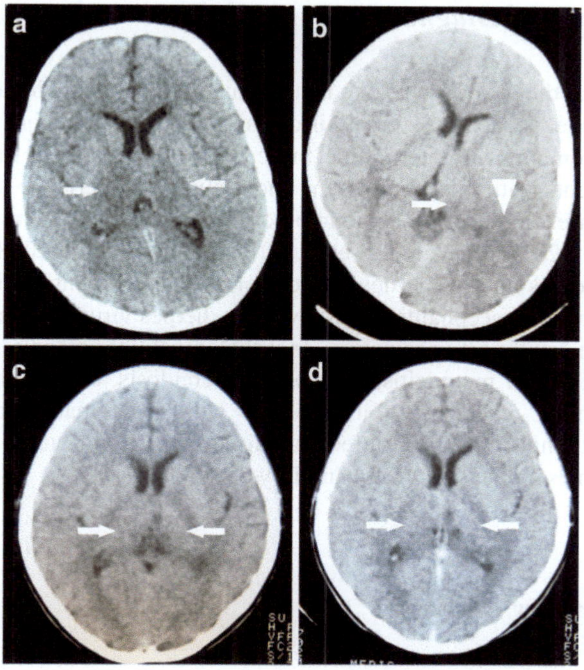

Fig. 8 Computed tomography (CT) scan abnormalities in Japanese encephalitis. Thalamic hypodensities are shown (*arrows*) and a cortical lesion (*arrowhead*, **b**). Lesions are bilateral, visible on an unenhanced scan (**a**, **c**) and do not enhance with contrast (**b**, **d**) (Adapted from Dung et al. 2009 with permission)

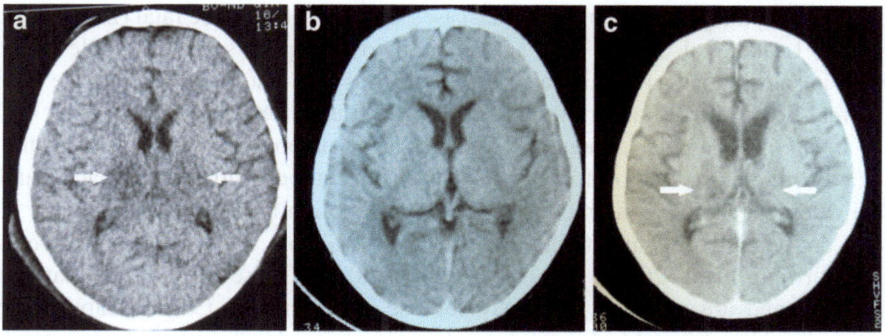

Fig. 9 CT scan showing the "fogging effect" in Japanese encephalitis. Bilateral thalamic hypodense lesions are seen during acute illness (**a**). Ten days later, the lesions have apparently disappeared (**b**) but they are visible again at 3 months (**c**) (Adapted from Dung et al. 2009 with permission)

6.2 Differential Diagnosis

The differential diagnosis of JE, like that of any encephalitis, is broad. Geographical location, age, occupation, travel history and vaccination history are all crucial pieces of information in determining the cause of encephalitis. In reality, most series of patients with encephalitis and most clinical experience shows that the majority of cases do not have an aetiological agent identified (Glaser et al. 2006; Granerod et al. 2010; Le Van Tan et al. 2010; Solomon et al. 2008). This is true both in the tropics and in the developed world. The differential diagnosis of JE includes other viral encephalitides (arboviruses, herpes viruses, enteroviruses, and post-infectious and post-vaccination encephalomyelitis), other CNS infections (bacterial and fungal meningitis, abscesses, tuberculosis, cerebral malaria, leptospirosis, tetanus), other infectious diseases with CNS manifestations (febrile convulsions, typhoid), and non-infectious diseases (tumours, toxic and alcoholic encephalopathies, epilepsy, strokes and Reye's syndrome).

6.3 Laboratory Diagnosis

None of the routine laboratory parameters in JE is specific, though they may be suggestive. A definitive diagnosis is made by finding JEV specific IgM in the CSF of a patient in the appropriate clinical context of acute encephalitis syndrome (Burke et al. 1985b). JEV is very rarely isolated from clinical specimens. The viraemia that occurs in humans is low and short lived (Sapkal et al. 2007); by the time encephalitis has developed the viraemic period is usually over. Virus can occasionally be grown from CSF; such cases are more likely to be fatal (Burke et al. 1985a; Leake et al. 1986). Viral antigen/antibody complexes are detectable in the CSF in some patients (Desai et al. 1995). The virus may also be detected in the CSF by real-time reverse transcriptase PCR (Swami et al. 2008).

If CSF is not available then the diagnosis can be supported by the finding of anti-JEV IgM in serum or a fourfold rise in neutralising antibody titre between acute and convalescent samples; in practice this last investigation is seldom carried out.

Commercially manufactured kits are now available for JEV sero-diagnosis (Lewthwaite et al. 2010b; Ravi et al. 2006, 2009). These are all IgM capture ELISAs and are validated variously for use in serum and CSF. A rapid filter paper based test suitable for field use has also been developed (Solomon et al. 1998b). The IgM ELISAs are fairly JEV specific though there are problems differentiating JEV from other flaviviruses in areas where different flaviviruses co-circulate, i.e., most of Asia. If a JE case is suspected in a non-endemic area the diagnostic laboratory should ideally be consulted prior to testing to ensure the appropriate tests are done and are interpreted correctly. In travellers who may have been vaccinated, a complete flavivirus vaccination history (JE, tick-borne encephalitis, yellow fever) is essential to allow the serology results to be correctly interpreted.

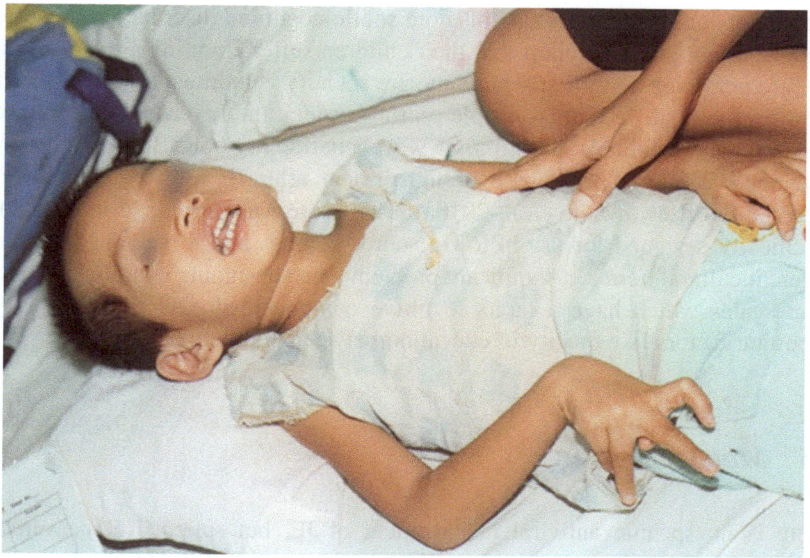

Fig. 10 Fixed flexion deformity of the upper limb following Japanese encephalitis (Photograph: T Solomon (in Solomon 2000), republished with permission)

In cases where samples are obtained after the acute phase, plaque reduction neutralisation titres (PRNT) represent the most sensitive and specific tool for determining prior infection. However, such assays require the culture of live virus and take around a week to perform each assay so their use remains restricted to research. Anti-JEV neutralising antibody is the standard endpoint for vaccine trials. IgG ELISA kits exist for some flaviviruses but their specificity has not been fully evaluated; they remain a research tool.

6.4 *Outcome*

JE is a serious disease and the outcome is generally poor. A wide variety of neurological sequelae are described. The commonest are focal motor deficits; these may involve both upper and lower motor neurons. Fixed flexion deformities of the upper limbs (Fig. 10) and hyperextension of the lower limbs with "equine feet" are common. Extrapyramidal movement disorders such as Parkinsonism can be seen as a consequence of damage to the substantia nigra, as can cerebellar signs. Twenty percent of survivors have severe learning difficulties and 20 % have seizures.

Classical teaching suggests around one-third of patients with JE die and another third are left with neurological sequelae (Huy et al. 1994; Kumar et al. 1993; Richter and Shimojyo 1961). If, however, very detailed assessments are undertaken, an even higher rate of more subtle cognitive disorders is found to be present. Many families report that their children suffer from impairment of memory, are not doing as well at school and are socially withdrawn to varying degrees (Lewthwaite et al. 2010a). A simple scoring system, the Liverpool outcome score (Table 1 and available at http://www.webcitation.org/65R5BbmkF), has been developed to assess the degree of impairment after JE (or any other form of encephalitis) (Lewthwaite et al. 2010a). This score was developed in Asia and is specifically designed for use in low resource settings and to be adaptable across different cultural norms, a significant problem when designing neurological assessment scales which have a large cognitive component due to the variability in important factors like quality of education between populations.

6.5 Management

There is no specific antiviral management of JE, but supportive measures are important and can have a significant impact on outcome (Solomon et al. 2000). Complications such as raised ICP and seizures should be actively sought out and treated. Nursing care and physiotherapy should aim to prevent the development of contractures (that lead to fixed flexion deformities) and bedsores. Careful hydration and nutrition are very important, though over-hydration should be avoided especially in the setting of raised ICP. Aspiration pneumonia is another potential complication in patients who are unconscious or with dysphagia.

Corticosteroids have been used for many years in the management of JE. However, a randomised controlled trial showed that they were of no benefit; though they were also safe (Hoke et al. 1992). This trial was small and some in the field regard the question as to whether or not corticosteroids are useful in JE as remaining unanswered; they are still widely used in many endemic areas and there have also been isolated reports of steroids being used in West Nile encephalitis though robust data are lacking (Pyrgos and Younus 2004). Although interferon-α (IFNα) inhibits JEV replication in vitro, levels are elevated in CSF of patients with JE and IFNα has some efficacy in animal models, a randomised placebo controlled trial in humans failed to show any benefit (Solomon et al. 2003a). Similarly the antiviral drug ribavirin, though theoretically promising, did not turn out to be effective in a randomised controlled trial (Kumar et al. 2009a).

7 Prevention

The potential methods of prevention of any vector borne zoonotic viral infection are reservoir reduction, vector control/avoidance and vaccination. Having any impact on the wild bird or rice paddy breeding mosquito populations of Asia is clearly unrealistic. Vector avoidance is wise if it is possible—it is likely that changes in

Table 1 A simple 15-point scoring system for assessing disability after encephalitis, based on Lewthwaite et al. (2010a)

History from child/parents/careers	
Speech	The same as other children of this age (5)/Reduced (3)/Not speaking or communicating (2)
Feeding	The same as other children (5)/Occasionally needs help (3)/Always needs more help (2)
Can the child be left alone?	Too young or Yes (5)/Yes briefly in familiar environment (3)/No (2)
Behaviour	Normal (5)/Gets angry easily (4)/Other behavioural problems (4)/Severely abnormal (2)
Recognition (other than the main carer)	Too young or yes (5)/Some (3)/None (2)
School/work	Now back to normal at school or work (5)/Not doing as well (4)/Dropped a school grade or no longer attending school or work (3)
	If too young: still able to do the same tasks at home and follow the same routine (5)/Not able to do as well as before (4)/not able to do at all (3)
Seizures in the last 2 months	No seizures and not on anti-epileptic drugs (5)/No seizures and on anti-epileptic drugs (4)/Yes has had seizures (3)/Yes, seizures most days (2)
Ability to dress	The same as other children the same age (5)/Occasionally needs extra help (3)/Always needs more help than other children of the same age (2)
Urinary and faecal continence	The same as other children the same age (5)/Needs more help or is incontinent of bowel or bladder (2)
Hearing	Normal (5)/reduced in one or both ears (4)/cannot hear at all (3)
Observation of the child	
Sitting	Too young or yes independently (5)/needs help (3)/Not at all (2)
Sitting to standing	Too young or yes independently (5)/needs help (3)/Not at all (2)
Observe the child walking for 5 m	Too young or Normal (5)/Abnormal, but independently +/− crutches/stick (3)/Not able to walk (2)
Put both hands on your head ask child to copy	Too young (5)/Normal both hands (5)/Abnormal one or both hands (4)/Unable one or both hands (3)
Pick up small object	Too young (5)/Normal pincer grasp both hands (5)/Unable one hand (3)/Abnormal one hand or both hands (3)/Unable both hands (2)

Specific scoring for each domain is shown. The overall score is the lowest score obtained for any question and therefore is scored from 5 (normal) to 1 (death). A score of 4 indicates mild disability such as behavioural problems, 3 indicates disability significant enough to affect function but leaves capacity for independent living and 2 indicates disability severe enough to prevent independent living. The score is archived (at the time of writing) at http://www.webcitation.org/65R5BbmkF. Check http://www.liv.ac.uk/infection-and-global-health/research/brain-infections-group/education/ for updated information

housing conditions have contributed to reducing JE cases in Japan for example, and mosquito nets and repellant can be advised for travellers to JE endemic areas. However, for the major public health burden of JE, these measures alone cannot be expected to be enough and vaccination remains the only viable option for large-scale prevention of JE.

JE vaccines have available for many decades. Historically most JE vaccines have used formalin inactivated JEV grown in mouse brain. Vaccines based on this strategy began to be developed in Japan in the 1930s shortly after the virus was discovered. Albert Sabin, the developer of the live poliomyelitis vaccine, made a similar vaccine for US troops during the Second World War (Solomon 2008). The Biken JE vaccine, manufactured in Japan, and the Korean Green Cross vaccine (both mouse brain derived) were made widely available after clinical trials in Thailand and Korea (Hoke et al. 1988; Solomon 2008). The trial in Thailand (Hoke et al. 1988) remains the only randomised controlled trial evidence of vaccine efficacy in the JE field (such a study would now be unethical). These vaccines were relatively poorly immunogenic and associated with frequent allergic and injection site reactions due to the presence of mouse brain derived tissue. There was also a small incidence of acute disseminated encephalomyelitis (ADEM) of approximately 1 in 1,000,000 associated with mouse brain derived JE vaccines (WHO 2005). Though this risk was outweighed by the benefits in endemic areas, for travellers this figure approaches equivalence to the risk of developing JE. In Japan the use and production of the Biken vaccine was suspended in 2005 after a case of ADEM in a child that was temporally associated with JE vaccination (Beasley et al. 2008).

In China in the 1960s a live attenuated strain of JEV, SA14-14-2, was developed by serial passage in animals and in cell culture. This live attenuated vaccine virus came into human use in China in the late 1980s and since then over 200 million doses have been administered in China. There has been little evidence of significant toxicity (Liu et al. 1997; WHO 2006), although by modern standards stricter monitoring would be required. Since the licensing and introduction of the vaccine in China several studies outside China have also demonstrated both safety and efficacy (Bista et al. 2001; Gatchalian et al. 2008; Kumar et al. 2009b; Ohrr et al. 2005), leading to the introduction of SA14-14-2 across much of the JE endemic region. This vaccine is very cheap to manufacture, another advantage in this setting. The vaccine is not yet licensed for use internationally.

In response to the need for a more suitable vaccine for travellers, where the risk of JE is low, a new killed vaccine based on the SA14-14-2 strain was developed and is now widely licensed under the name IXIARO® (JESPECT® in Australia and New Zealand). This vaccine is derived from vero cell culture (as is the live SA14-14-2 vaccine) and is formalin inactivated. This vaccine is effective (as shown by the development of neutralising antibody against JEV) and is also very safe, making it a better alternative than the old mouse brain derived vaccines (Tauber et al. 2007). In addition to this vaccine there are several other vaccines in development. Some of these are targeted at specific domestic markets and others are intended for international use (Beasley et al. 2008). Novel technologies have now been applied to the JE vaccine field, for example using genetically engineering viruses based on the yellow fever 17D vaccine backbone containing the envelope protein of JEV (Torresi et al. 2010). One such product is now available in Australia under the trade name IMOJEV®. All currently available JE vaccines are probably effective, and the choice between them will depend mostly on regional licensing arrangements and availability.

8 Conclusions

Since outbreaks of encephalitis were described in Japan nearly 150 years ago, Japanese encephalitis has become a major cause of illness and death across Asia. This highly virulent virus causes a greater loss of disability adjusted life years (DALYs) than any other arthropod-borne virus. Its complex zoonotic cycle involving a range of mosquito vectors and hosts means that control of the virus is difficult. Although vaccination offers hope for disease control, the virus will never be eradicated. The area endemic for JEV has grown steadily since its first recognition; the effects of global climate change make it unlikely that this spread will halt anytime soon. In the future we may expect an increasing JE problem. Good surveillance and rapid diagnostics are essential for identifying outbreaks, mapping the disease's spread and thus making maximal use of preventive vaccination. A better understanding of the host immune response and pathogenesis will point towards new treatments for better disease outcomes in the future.

References

Aleyas AG, Han YW, George JA et al (2010) Multifront assault on antigen presentation by Japanese encephalitis virus subverts CD8+ T cell responses. J Immunol 185:1429–1441. doi:10.4049/jimmunol.0902536

Arai S, Matsunaga Y, Takasaki T et al (2008) Japanese encephalitis: surveillance and elimination effort in Japan from 1982 to 2004. Jpn J Infect Dis 61:333–338

Beasley DWC, Lewthwaite P, Solomon T (2008) Current use and development of vaccines for Japanese encephalitis. Expert Opin Biol Ther 8:95–106. doi:10.1517/14712598.8.1.95

Becker H, Desch H, Hacker H, Pencz A (1979) CT fogging effect with ischemic cerebral infarcts. Neuroradiology 18:185–192

Bista MB, Banerjee MK, Shin SH et al (2001) Efficacy of single-dose SA 14-14-2 vaccine against Japanese encephalitis: a case control study. Lancet 358:791–795

Biswas SM, Ayachit VM, Sapkal GN et al (2009) Japanese encephalitis virus produces a CD4+ Th2 response and associated immunoprotection in an adoptive-transfer murine model. J Gen Virol 90:818–826. doi:10.1099/vir.0.008045-0

Burke DS, Lorsomrudee W, Leake CJ et al (1985a) Fatal outcome in Japanese encephalitis. Am J Trop Med Hyg 34:1203–1210

Burke DS, Nisalak A, Ussery MA et al (1985b) Kinetics of IgM and IgG responses to Japanese encephalitis virus in human serum and cerebrospinal fluid. J Infect Dis 151:1093–1099

Campbell G, Hills S, Fischer M et al (2011) Estimated global incidence of Japanese encephalitis. Bull World Health Organ 89:766–774. doi:10.2471/BLT.10.085233

Chambers TJ, Hahn CS, Galler R, Rice CM (1990) Flavivirus genome organization, expression, and replication. Annu Rev Microbiol 44:649–688. doi:10.1146/annurev.mi.44.100190.003245

Chen Y, Maguire T, Hileman RE et al (1997) Dengue virus infectivity depends on envelope protein binding to target cell heparan sulfate. Nat Med 3:866–871

Das S, Dutta K, Kumawat KL et al (2011) Abrogated inflammatory response promotes neurogenesis in a murine model of Japanese encephalitis. PLoS One 6:e17225. doi:10.1371/journal.pone.0017225.g008

Desai A, Shankar SK, Ravi V et al (1995) Japanese encephalitis virus antigen in the human brain and its topographic distribution. Acta Neuropathol 89:368–373

Desai A, Shankar SK, Jayakumar PN et al (1997) Co-existence of cerebral cysticercosis with Japanese encephalitis: a prognostic modulator. Epidemiol Infect 118:165–171

Dung NM, Turtle L, Chong WK et al (2009) An evaluation of the usefulness of neuroimaging for the diagnosis of Japanese encephalitis. J Neurol 256:2052–2060. doi:10.1007/s00415-009-5249-5

Fujii Y, Kitaura K, Nakamichi K et al (2008) Accumulation of T-cells with selected T-cell receptors in the brains of Japanese encephalitis virus-infected mice. Jpn J Infect Dis 61:40–48

Gatchalian S, Yao Y, Zhou B et al (2008) Comparison of the immunogenicity and safety of measles vaccine administered alone or with live, attenuated Japanese encephalitis SA 14-14-2 vaccine in Philippine infants. Vaccine 26:2234–2241. doi:10.1016/j.vaccine.2008.02.042

German AC, Myint KSA, Mai NTH et al (2006) A preliminary neuropathological study of Japanese encephalitis in humans and a mouse model. Trans R Soc Trop Med Hyg 100:1135–1145. doi:10.1016/j.trstmh.2006.02.008

Glaser CA, Honarmand S, Anderson LJ et al (2006) Beyond viruses: clinical profiles and etiologies associated with encephalitis. Clin Infect Dis 43:1565–1577. doi:10.1086/509330

Granerod J, Ambrose HE, Davies NW et al (2010) Causes of encephalitis and differences in their clinical presentations in England: a multicentre, population-based prospective study. Lancet Infect Dis 10:835–844. doi:10.1016/S1473-3099(10)70222-X

Handique SK, Das RR, Barman K et al (2006) Temporal lobe involvement in Japanese encephalitis: problems in differential diagnosis. AJNR Am J Neuroradiol 27:1027–1031

Haymaker W, Sabin AB (1947) Topographic distribution of lesions in central nervous system in Japanese B encephalitis; nature of the lesions, with report of a case on Okinawa. Arch Neurol Psychiatry 57:673–692

Hoke C, Nisalak A, Sangawhipa N et al (1988) Protection against Japanese encephalitis by inactivated vaccines. N Engl J Med 319:608–614

Hoke CH, Vaughn DW, Nisalak A et al (1992) Effect of high-dose dexamethasone on the outcome of acute encephalitis due to Japanese encephalitis virus. J Infect Dis 165:631–637

Huy BV, Tu HC, Luan TV, Lindqvist R (1994) Early mental and neurological sequelae after Japanese B encephalitis. Southeast Asian J Trop Med Public Health 25:549–553

Impoinvil DE, Solomon T, Schluter WW et al (2011) The spatial heterogeneity between Japanese encephalitis incidence distribution and environmental variables in Nepal. PLoS One 6:e22192. doi:10.1371/journal.pone.0022192

Kalita J, Misra UK (2000) Comparison of CT scan and MRI findings in the diagnosis of Japanese encephalitis. J Neurol Sci 174:3–8

Kalita J, Misra UK, Pandey S, Dhole TN (2003) A comparison of clinical and radiological findings in adults and children with Japanese encephalitis. Arch Neurol 60:1760–1764. doi:10.1001/archneur.60.12.1760

Kato H, Takeuchi O, Sato S et al (2006) Differential roles of MDA5 and RIG-I helicases in the recognition of RNA viruses. Nature 441:101–105. doi:10.1038/nature04734

Kimura-Kuroda J, Yasui K (1988) Protection of mice against Japanese encephalitis virus by passive administration with monoclonal antibodies. J Immunol 141:3606–3610

Konishi E, Kurane I, Mason PW et al (1995) Japanese encephalitis virus-specific proliferative responses of human peripheral blood T lymphocytes. Am J Trop Med Hyg 53:278–283

Kumar R, Mathur A, Singh KB et al (1993) Clinical sequelae of Japanese encephalitis in children. Indian J Med Res 97:9–13

Kumar P, Uchil PD, Sulochana P et al (2003) Screening for T cell-eliciting proteins of Japanese encephalitis virus in a healthy JE-endemic human cohort using recombinant baculovirus-infected insect cell preparations. Arch Virol 148:1569–1591. doi:10.1007/s00705-003-0118-5

Kumar P, Krishna VD, Sulochana P et al (2004a) Cell-mediated immune responses in healthy children with a history of subclinical infection with Japanese encephalitis virus: analysis of CD4+ and CD8+ T cell target specificities by intracellular delivery of viral proteins using the human immunodeficiency virus Tat protein transduction domain. J Gen Virol 85:471–482

Kumar P, Sulochana P, Nirmala G et al (2004b) Impaired T helper 1 function of nonstructural protein 3-specific T cells in Japanese patients with encephalitis with neurological sequelae. J Infect Dis 189:880–891. doi:10.1086/381768

Kumar P, Sulochana P, Nirmala G et al (2004c) Conserved amino acids 193-324 of non-structural protein 3 are a dominant source of peptide determinants for CD4+ and CD8+ T cells in a healthy Japanese encephalitis virus-endemic cohort. J Gen Virol 85:1131–1143

Kumar R, Tripathi P, Baranwal M et al (2009a) Randomized, controlled trial of oral ribavirin for Japanese encephalitis in children in Uttar Pradesh, India. Clin Infect Dis 48:400–406. doi:10.1086/596309

Kumar R, Tripathi P, Rizvi A (2009b) Effectiveness of one dose of SA 14-14-2 vaccine against Japanese encephalitis. N Engl J Med 360:1465–1466. doi:10.1056/NEJMc0808664

Larena M, Regner M, Lee E, Lobigs M (2011) Pivotal role of antibody and subsidiary contribution of CD8+ T Cells to recovery from infection in a murine model of Japanese encephalitis. J Virol 85:5446–5455. doi:10.1128/JVI.02611-10

Le Van Tan P, Do Quang Ha N, Lam Quoi Bao B et al (2010) Viral etiology of encephalitis in children in southern Vietnam: results of a one-year prospective descriptive study. PLoS Negl Trop Dis 4(10):e854

Leake CJ, Burke DS, Nisalak A, Hoke CH (1986) Isolation of Japanese encephalitis virus from clinical specimens using a continuous mosquito cell line. Am J Trop Med Hyg 35:1045–1050

Lewthwaite P, Begum A, Ooi MH et al (2010a) Disability after encephalitis: development and validation of a new outcome score. Bull World Health Organ 88:584–592. doi:10.2471/BLT.09.071357

Lewthwaite P, Shankar MV, Tio PH et al (2010b) Evaluation of two commercially available ELISAs for the diagnosis of Japanese encephalitis applied to field samples. Trop Med Int Health 15:811–818. doi:10.1111/j.1365-3156.2010.02537.x

Li J-K, Liang J-J, Liao C-L, Lin Y-L (2012) Autophagy is involved in the early step of Japanese encephalitis virus infection. Microbes Infect 14:159–168. doi:10.1016/j.micinf.2011.09.001

Libraty DH, Nisalak A, Endy TP et al (2002) Clinical and immunological risk factors for severe disease in Japanese encephalitis. Trans R Soc Trop Med Hyg 96:173–178

Lin R-J, Liao C-L, Lin E, Lin Y-L (2004) Blocking of the alpha interferon-induced Jak-Stat signaling pathway by Japanese encephalitis virus infection. J Virol 78:9285–9294. doi:10.1128/JVI.78.17.9285-9294.2004

Lin R-J, Chang B-L, Yu H-P et al (2006) Blocking of interferon-induced Jak-Stat signaling by Japanese encephalitis virus NS5 through a protein tyrosine phosphatase-mediated mechanism. J Virol 80:5908–5918. doi:10.1128/JVI.02714-05

Lindenbach BD, Thiel H-J, Rice CM (2007) *Flaviviridae*: the viruses and their replication. In: Knipe DM, Howley PM (eds) Fields virology. Lippincott Williams and Wilkins, Philadelphia, PA

Liu ZL, Hennessy S, Strom BL et al (1997) Short-term safety of live attenuated Japanese encephalitis vaccine (SA14-14-2): results of a randomized trial with 26,239 subjects. J Infect Dis 176:1366–1369

Lubiniecki AS, Cypess RH, Hammon WM (1973) Passive immunity for arbovirus infection. II. Quantitative aspects of naturally and artificially acquired protection in mice for Japanese (B) encephalitis virus. Am J Trop Med Hyg 22:535–542

Mathur A, Arora KL, Chaturvedi UC (1983) Host defence mechanisms against Japanese encephalitis virus infection in mice. J Gen Virol 64(Pt 4):805–811

Melian EB, Hinzman E, Nagasaki T et al (2010) NS1′ of flaviviruses in the Japanese encephalitis virus serogroup is a product of ribosomal frameshifting and plays a role in viral neuroinvasiveness. J Virol 84:1641–1647. doi:10.1128/JVI.01979-09

Mishra MK, Basu A (2008) Minocycline neuroprotects, reduces microglial activation, inhibits caspase 3 induction, and viral replication following Japanese encephalitis. J Neurochem 105:1582–1595. doi:10.1111/j.1471-4159.2008.05238.x

Miyake M (1964) The pathology of Japanese encephalitis. A review. Bull World Health Organ 30:153–160

Mogi M (1983) Relationship between number of human Japanese encephalitis cases and summer meteorological conditions in Nagasaki, Japan. Am J Trop Med Hyg 32:170–174

Mukhopadhyay S, Kuhn RJ, Rossmann MG (2005) A structural perspective of the flavivirus life cycle. Nat Rev Microbiol 3:13–22. doi:10.1038/nrmicro1067

Murali-Krishna K, Ravi V, Manjunath R (1996) Protection of adult but not newborn mice against lethal intracerebral challenge with Japanese encephalitis virus by adoptively transferred virus-specific cytotoxic T lymphocytes: requirement for L3T4+ T cells. J Gen Virol 77(Pt 4):705–714

Murty US, Rao MS, Arunachalam N (2010) The effects of climatic factors on the distribution and abundance of Japanese encephalitis vectors in Kurnool district of Andhra Pradesh, India. J Vector Borne Dis 47:26–32

Ohrr H, Tandan JB, Sohn YM et al (2005) Effect of single dose of SA 14-14-2 vaccine 1 year after immunisation in Nepalese children with Japanese encephalitis: a case-control study. Lancet 366:1375–1378. doi:10.1016/S0140-6736(05)67567-8

Pyrgos V, Younus F (2004) High-dose steroids in the management of acute flaccid paralysis due to West Nile virus infection. Scand J Infect Dis 36:509–512

Ravi V, Desai A, Balaji M et al (2006) Development and evaluation of a rapid IgM capture ELISA (JEV-Chex) for the diagnosis of Japanese encephalitis. J Clin Virol 35:429–434. doi:10.1016/j.jcv.2005.11.004

Ravi V, Robinson JS, Russell BJ et al (2009) Evaluation of IgM antibody capture enzyme-linked immunosorbent assay kits for detection of IgM against Japanese encephalitis virus in cerebrospinal fluid samples. Am J Trop Med Hyg 81:1144–1150. doi:10.4269/ajtmh.2009.09-0144

Richter RW, Shimojyo S (1961) Neurologic sequelae of Japanese B encephalitis. Neurology 11:553–559

Sapkal GN, Wairagkar NS, Ayachit VM et al (2007) Detection and isolation of Japanese encephalitis virus from blood clots collected during the acute phase of infection. Am J Trop Med Hyg 77:1139–1145

Solomon T (2000) In: Gilman S (ed) MedLink neurology. Medlink Corporation, Japanese Encephalitis. San Diego, CA. http://www.medlink.com

Solomon T (2004a) Encephalitis. In: Gill GV, Beeching N (eds) Lecture notes: tropical medicine, 5th edn. Wiley-Blackwell, Oxford

Solomon T (2004b) Flavivirus encephalitis. N Engl J Med 351:370–378. doi:10.1056/NEJMra030476

Solomon T (2006) Control of Japanese encephalitis – within our grasp? N Engl J Med 355:869–871. doi:10.1056/NEJMp058263

Solomon T (2008) New vaccines for Japanese encephalitis. Lancet Neurol 7:116–118. doi:10.1016/S1474-4422(08)70004-8

Solomon T, Winter PM (2004) Neurovirulence and host factors in flavivirus encephalitis – evidence from clinical epidemiology. Arch Virol Suppl (18):161–170

Solomon T, Kneen R, Dung NM et al (1998a) Poliomyelitis-like illness due to Japanese encephalitis virus. Lancet 351:1094–1097. doi:10.1016/S0140-6736(97)07509-0

Solomon T, Thao LT, Dung NM et al (1998b) Rapid diagnosis of Japanese encephalitis by using an immunoglobulin M dot enzyme immunoassay. J Clin Microbiol 36:2030–2034

Solomon T, Dung NM, Kneen R et al (2000) Japanese encephalitis. J Neurol Neurosurg Psychiatr 68:405–415

Solomon T, Dung NM, Kneen R et al (2002) Seizures and raised intracranial pressure in Vietnamese patients with Japanese encephalitis. Brain 125:1084–1093

Solomon T, Dung NM, Wills B et al (2003a) Interferon alfa-2a in Japanese encephalitis: a randomised double-blind placebo-controlled trial. Lancet 361:821–826

Solomon T, Ni H, Beasley DWC et al (2003b) Origin and evolution of Japanese encephalitis virus in southeast Asia. J Virol 77:3091–3098

Solomon T, Thao TT, Lewthwaite P et al (2008) A cohort study to assess the new WHO Japanese encephalitis surveillance standards. Bull World Health Organ 86:178–186

Sumiyoshi H, Mori C, Fuke I et al (1987) Complete nucleotide sequence of the Japanese encephalitis virus genome RNA. Virology 161:497–510

Swami R, Ratho RK, Mishra B, Singh MP (2008) Usefulness of RT-PCR for the diagnosis of Japanese encephalitis in clinical samples. Scand J Infect Dis 40:815–820. doi:10.1080/00365540802227102

Tauber E, Kollaritsch H, Korinek M et al (2007) Safety and immunogenicity of a Vero-cell-derived, inactivated Japanese encephalitis vaccine: a non-inferiority, phase III, randomised controlled trial. Lancet 370:1847–1853. doi:10.1016/S0140-6736(07)61780-2

Torresi J, McCarthy K, Feroldi E, Méric C (2010) Immunogenicity, safety and tolerability in adults of a new single-dose, live-attenuated vaccine against Japanese encephalitis: randomised controlled phase 3 trials. Vaccine 28:7993–8000. doi:10.1016/j.vaccine.2010.09.035

Turtle L, Griffiths MJ, Solomon T (2012) Encephalitis caused by flaviviruses. QJM 105:219–223. doi:10.1093/qjmed/hcs013

Uchil PD, Satchidanandam V (2003) Architecture of the flaviviral replication complex. Protease, nuclease, and detergents reveal encasement within double-layered membrane compartments. J Biol Chem 278:24388–24398. doi:10.1074/jbc.M301717200

Van Den Hurk AF, Montgomery BL, Northill JA et al (2006) Short report: the first isolation of Japanese encephalitis virus from mosquitoes collected from mainland Australia. Am J Trop Med Hyg 75:21–25

Wang K, Deubel V (2011) Mice with different susceptibility to Japanese encephalitis virus infection show selective neutralizing antibody response and myeloid cell infectivity. PLoS One 6:e24744. doi:10.1371/journal.pone.0024744

Wang T, Town T, Alexopoulou L et al (2004) Toll-like receptor 3 mediates West Nile virus entry into the brain causing lethal encephalitis. Nat Med 10:1366–1373. doi:10.1038/nm1140

Winter PM, Dung NM, Loan HT et al (2004) Proinflammatory cytokines and chemokines in humans with Japanese encephalitis. J Infect Dis 190:1618–1626. doi:10.1086/423328

World Health Organisation (2004) WHO world health report 2004, pp 1–96

World Health Organisation (2005) WER/GACVS June 2005 JE vaccine safety. Wkly Epidemiol Rec 80:241–248

World Health Organisation (2006) WER/2006 JE vaccine position statement. Wkly Epidemiol Rec 81:325–340

Chikungunya Virus Infection

Philippe Gasque

Abstract Chikungunya virus (CHIKV) belongs to the "old world" *Alphavirus* family and is an arbovirus transmitted by *Aedes* mosquitoes. It is a tropical disease originally described in central/east Africa in the 1950s, but the recent re-emergence from 2004 in Africa and rapid spread in and around the Indian Ocean (Reunion island, India, Malaysia) as well as Europe (Italy) led to several millions of cases with unprecedented neurological complications. Recent epidemics in 2005–2008 have revealed that there is a high titer viremia and that CHIKV can gain access to the brain in susceptible hosts such as neonates and elderly patients with comorbidities (chronic disease of liver, heart, and kidneys). New mutated forms were identified in the recent outbreaks, but their contribution to severe pathologies is only speculative at this stage. Classically, CHIKV causes an acute symptomatic illness over a period of 5–7 days with fever, skin rash, and incapacitating arthralgias, and can evolve into chronic rheumatoid arthritis-like diseases, particularly in elderly patients. CHIKV infection can also lead to severe meningoencephalitis, encephalitis with white matter lesions, peripheral neuropathies, optic neuritis, and death. In neonates central nervous system (CNS) infection can lead to long-term sequelae. The early and robust systemic innate and adaptive immune responses are able to protect the host. However, the virus has been shown to persist in tissue sanctuaries months after the initial infection in humans and animal models. In the brain, CHIKV preferentially targets astrocytes, ependymal cells, epithelial cells of the choroid plexus, and neurons. The route and the mechanisms involved in neuroinfection, the antiviral response mounted by resident cells and neuroinflammation are largely ill-characterized. CHIKV may critically polarize host-cell defense mechanisms such as IFN signaling pathway, apoptosis, and autophagy to its own advantage and studies to decipher the molecular mechanisms of immune escape are now highly warranted.

P. Gasque (✉)
GRI EA4517, CPER/FEDER and ICRES PCRD7 Immunopathology and Infectious Disease Research Grouping, University of la Reunion: CHU and CYROI, St. Denis, Reunion 97400, France
e-mail: gasque@univ-reunion.fr; gasquep@hotmail.com

A.C. Jackson (ed.), *Viral Infections of the Human Nervous System*,
Birkhäuser Advances in Infectious Diseases,
DOI 10.1007/978-3-0348-0425-7_12,

Keywords Chikungunya • Encephalitis • Glia • Inflammation • Innate immunity • Neuroinflammation • Neurons

1 Introduction: Arthritogenic as Well as Encephalitic Chikungunya Virus in Susceptible Hosts

The so-called "old-world" *Alphavirus* group can cause human joint disorders (arthralgias evolving to arthritis). This is the case for chikungunya virus (CHIKV), o'nyong-nyong virus (ONNV), Semliki forest virus (SFV), Ross River virus (RRV), Sindbis virus (SINV), and Mayaro virus (MAYV). The acute phase of the disease with old-world alphaviruses is highly symptomatic (over 90%) and is characterized mostly by fever, rash, generalized and excruciating myalgias, and arthralgias lasting for 1–2 weeks (Jaffar-Bandjee et al. 2009; Simon et al. 2011; Suhrbier and La Linn 2004). Arthralgias and crippling arthritis can persist for years, especially in elderly patients (Simon et al. 2011; Sissoko et al. 2009).

CHIKV has never been considered as a "true" neurotropic virus in contrast to "new-world" alphaviruses associated with encephalitis like eastern equine encephalitis virus (EEEV) (Zacks and Paessler 2010; Arpino et al. 2009). Remarkably, it is certainly the unprecedented incidence rate in the Indian Ocean in the epidemics of 2005–2006 with efficient clinical facilities that allowed a better description of severe cases with encephalitis, meningoencephalitis, peripheral neuropathies, and deaths among neonates (first ever reported cases of mother-to-child infection in la Reunion), and infants as well as in elderly patients (Economopoulou et al. 2009; Jaffar-Bandjee et al. 2010; Pialoux et al. 2007). In sharp contrast to *Flaviviruses* (e.g., West Nile Virus), the attack rate for CHIKV disease can be very high and ranging from 30 to 50% in several islands in the Indian Ocean (Reunion, Comoros, Mayotte, and Sri Lanka) (Jaffar-Bandjee et al. 2010).

Neonates are particularly at risk and developmental stages of infection and novel routes of infection need to be carefully considered in view of the experimental data supporting the capacity of alphaviruses to replicate into the olfactory epithelium (e.g., SINV, EEEV) (Das et al. 2010; Thach et al. 2000).

Elderly patients with severe comorbidities are likely to suffer from complex pathological mechanisms in addition to inflammation mediated by acute CHIKV infection (Economopoulou et al. 2009). Aging affects immunity and the elderly are generally at increased risk for pathogen infection (Linton and Dorshkind 2004). These individuals are less able to overcome viral infections and subsequently exhibit high morbidity and mortality after infection. Most studies have shown an impairment of adaptive T cell responses, but it is important to seek how aging modifies other host defenses such as innate immunity against viruses. The protective innate immune antiviral response may be compromised during aging and possibly facilitating virus evasion of immune privileged organs such as the brain through a damaged blood–brain barrier (BBB). Moreover, aberrancies in inflammatory cytokine responses result from impaired host–pathogen interactions and whether these aberrant responses affect the outcome of systemic viral infections in older individuals need to be clarified (Rosenstiel et al. 2008). Of note, it is known that elderly CHIKV patients (over 60 years of age) have much higher viral loads and higher levels of cytokines, which may contribute to more severe tissue injuries (Hoarau et al. 2010).

The main focus of ongoing CHIKV research is to understand virus emergence and to address the very diverse mechanisms allowing virus entry, replication, mobilization of the innate immune system through specific innate immune pattern recognitions receptors (PRRs), dissemination to preferred target cells, and the control of infection by subsequent adaptive immune responses (T cells and anti-virus antibodies) (Griffin 2003; Ryman and Klimstra 2008). To begin to address CHIKV neuroinfection, these aspects need to be carefully considered in light of previous observations on encephalitic viruses.

2 Epidemiology of CHIKV and Clinical Hallmarks

CHIKV was first isolated in 1952 in Tanganyika, which is now Tanzania (Robinson 1955). Recurrent epidemics have been reported primarily in Africa, Islands of the Indian Ocean, and Asia (Powers and Logue 2007). The largest epidemic of CHIKV disease ever recorded took place in 2004–2011 and was associated with the emergence of CHIKV efficiently transmitted by *Ae. Albopictus*, a vector that has seen a dramatic global expansion in its geographic distribution (Charrel et al. 2007). The epidemic began in Kenya, spread across the Indian Ocean Islands to India (with an estimated 1.4–6.5 million cases) and South East Asia (Ng and Hapuarachchi 2010). Remarkably, CHIKV has accumulated key mutations (such as E1-A226V and E2-I211T) that possibly contributed to the recently changed epidemiology and clinical impact, yet to be fully ascertained (Schuffenecker et al. 2006). The first autochthonous infections in Europe occurred in Italy in 2007 (less than 250 cases) and a few reported cases in France in 2010 (Jaffar-Bandjee et al. 2010; Schwartz and Albert 2010; Grandadam et al. 2011).

3 Severe Neurological Involvement Due to Chikungunya Virus Infection

Mother-to-child transmission of CHIKV with an estimated prevalence rate of 0.25% of all pregnancies was first reported during the 2005–2006 outbreak in La Réunion Island. This mode of transmission, occurring solely during the peripartum maternal infection, was responsible for a high rate of neurological morbidity (Ramful et al. 2007; Gerardin et al. 2008). Almost 50% of the neonates were infected from mothers with intrapartum viremia. Infected neonates presented with severe complications (30%), including encephalopathy with seizures in the acute phase of the disease. Pathological brain MRI was noted in 50% of infected neonates with brain swelling, scattered white matter lesions in the supratentorial regions, including the corpus callosum and the periventricular and subcortical areas, and parenchymal hemorrhages (Table 1) (Gerardin et al. 2008; Samperiz et al. 2007). CHIKV RNA was detected in the cerebrospinal fluid (CSF) even in apparently uncomplicated cases and in situations in which biochemical and cellular characteristics of the CSF were often unremarkable. Preliminary data concerning long-term clinical follow-up of the infected neonates confirm poor outcomes with a poor developmental quotient compared to the control group at 2 years of age (D. Ramful, personal communication). Reassuringly, no propensity to prematurity, growth restriction, fetal deaths, stillbirths, or congenital anomalies were reported in maternal infections (Fritel et al. 2010). It is clear that the placenta has an important protective role and this has been substantiated in several experimental settings (Chen et al. 2010; Couderc et al. 2008a). In infants, Robin et al. (2008) described a case series of pediatric patients with neurologic manifestations associated with CHIKV infection ranging from simple and complex febrile seizures to meningeal syndrome, acute encephalopathy, diplopia, acute disseminated encephalomyelitis, and encephalitis (almost 40%) with often unremarkable CSF findings and nonspecific electroencephalography. CSF pleocytosis was rare, although CSF reverse transcription polymerase chain reaction (RT-PCR) was positive in 61%. Similar findings were subsequently described in infants in India and Mayotte Island (Lewthwaite et al. 2009; Le Bomin et al. 2008). Overall, risk factors for residual neurologic deficits (20% of the patients) in the La Réunion cohort included young age (neonatal infection), severe initial clinical presentation with encephalitis and initial pathological MRI findings. Neurological clinical manifestations of CHIKV infection in the pediatric population added to growing evidence of the potential neurovirulence of this arboviral disease with an age-dependent morbidity (neonatal infection) and consistent laboratory (CHIKV RNA in CSF of patients) and imaging (pathological MRI) features. Histopathological investigations were not performed on these cases. However, biological and imaging findings are consistent with histopathological hallmarks recently reported in mouse and macaque models, in which young age is a risk factor for severe disease involving the CNS (see below).

Elderly patients with comorbidities (chronic disease of liver, heart, and kidneys) were also at risk of developing severe CHIKV disease with neurological complications. CHIKV infection becomes highly symptomatic (arthralgias and myalgias) over a period of days to weeks and recovery occurs in the large majority of cases (Borgherini et al. 2007, 2008). However, neurological manifestations described in adults requiring hospitalization involved cases of encephalopathy frequently associated with lymphopenia, thrombocytopenia, elevated C-reactive protein (CRP) and the presence of IgM anti-CHIKV and/or virus isolated from CSF (Lemant et al. 2008). These patients can be semiconscious with neck rigidity, absence of limb movements, generalized motor axonal neuropathy, and with abnormal encephalographic records and brain imaging. Similar neurological complications and fatalities were reported by several groups examining severe hospitalized CHIKV cases from India (Chandak et al. 2009; Rampal and Meena

Table 1 Cellular targets, immune response and long-term consequences following CNS alphavirus infection

Alphavirus	CNS target cell	Virus and immune responses	Pathological hallmarks
SINV (neuroadapted in mice)	Neurons (Johnson 1965; Griffin and Johnson 1977). Purkinje cells (Johnson 1965). Meningeal cells (Johnson 1965). Ependymal cells (Johnson 1965; Jackson et al. 1987; Jackson et al. 1988)	Production of IL-1β, IL-4, IL-6, IL-10, TNF-α, LIF, and TGF-β (Wesselingh et al. 1994) . Production of IFN-γ by $CD4^+$ and $CD8^+$ (Binder and Griffin 2001). Apoptosis of infected neurons (Levine et al. 1993; Nava et al. 1998). Bcl-2 reduces neuronal fatality (Levine et al. 1996)	Hindlimb paralysis and death (Griffin and Johnson 1977; Jackson et al. 1987). Acute encephalomyelitis (Jackson et al. 1987). Kyphoscoliosis (Jackson et al. 1988). Swelling of lumbar and thoracic neurons (Jackson et al. 1988). Death (Griffin and Johnson 1977; Jackson et al. 1987)
EEEV (human, North American strains and animal models)	Neurons (in mouse model and humans) (Griffin 2003; Deresiewicz et al. 1997)	Humans: Neuronophagia, demyelination CSF pleocytosis; microglial nodules (Deresiewicz et al. 1997). Periventricular white matter lesions (Deresiewicz et al. 1997)	Large spectrum of general neurologic complications (confusion, somnolence, focal weakness, epileptiform discharges, seizures, stupor, altered mental status, paresthesia accompanying paresis, hemiparesis) (Deresiewicz et al. 1997). Lesions of basal ganglia and cortex, encephalomalacia, focal intraparenchymal perivascular hemorrhage in the caudate nucleus and putamen, meningeal enhancement, leptomingeal vascular congestion, brain hemorrhage, cranial nerve palsies, focal necrosis; coma and death (Deresiewicz et al. 1997)
VEEV (human)	Neurons (Griffin 2003)	Meningeal infiltrates composed of lymphocytes, mononuclear cells, and neutrophils (Steele and Twenhafel 2010)	General neurological diseases (somnolence, confusion, disorientation, mental depression, convulsions, seizures, ataxia, paralysis) (Steele and Twenhafel 2010) ; predominant CNS pathologies

(continued)

Table 1 (continued)

Alphavirus	CNS target cell	Virus and immune responses	Pathological hallmarks
			(edema, congestion, hemorrhages, vasculitis, meningitis, encephalitis) (Zacks and Paessler 2010). Common neurological sequelae (Zacks and Paessler 2010) ; coma and death (Steele and Twenhafel 2010)
WEEV (macaque and human)	In the brain: microglia and Purkinje cells (Reed et al. 2005) In the spinal cord: motor neurons (Reed et al. 2005)	Monocytic inflammation in the CNS expanding perivascular spaces (Reed et al. 2005) Occasional infiltrates of lymphocytes, plasma cells, PMN leukocytes surrounding arterioles in the cerebral cortex (Anderson 1984)	Demyelination and myelitis in white matter areas (Reed et al. 2005) General neurological diseases (partial seizures, depression in level of consciousness, hyperreflexia, bilateral Babinski signs) (Delfraro et al. 2011) Severe CNS disorders (multifocal necrosis in the deep gray matter, dilation of the temporal horns of the lateral ventricles, and compression of the sylvian cisterns) (Anderson 1984; Delfraro et al. 2011) Coma, death (Delfraro et al. 2011)
CHIKV (Human and animal models)	Neurons Meningeal cells Ependymal cells Astrocytes (Das et al. 2010; Couderc et al. 2008a)	In mice: CHIKV RNA, multifocal inflammation, tissue necrosis, apoptotic neurons, gliosis In humans: CHIKV RNA and infectious virus and anti-CHIKV IgM in CSF, moderate or absent astrogliosis, sparse microglia response, but no neuronophagia in humans (Ganesan et al. 2008) Apoptosis of infected neurons (Wang et al. 2008)	In mice: Flaccid paralysis; death in 1–10 day old mice (Couderc et al. 2008a). In human neonates: Brain swelling, white matter lesions, parenchymal hemorrhages, long term developmental impairments, rare deaths (Gerardin et al. 2008; Ramful et al. 2007). In human infants: Disseminated encephalopathy, some encephalitis, rare deaths (Robin et al. 2008)

	In human adults: Optic neuritis (Mittal et al. 2007) Swollen human brain, subarachnoid cerebellar hemorrhages (Ganesan et al. 2008). Encephalomyeloradiculitis: Foci of demyelination in subcortical white matter and perivascular lymphocytic basal ganglia infiltrates (Ganesan et al. 2008) In adult macaques: Meningoencephalitis, death (if high titer viremia) (Labadie et al. 2010)

2007; Tandale et al. 2009). Other neurological complications associated with CHIKV infection included seizures, encephalitis, a Guillain–Barre syndrome-like clinical picture, and encephalomyeloradiculitis leading to rare deaths (Economopoulou et al. 2009; Lemant et al. 2008; Lebrun et al. 2009; Wielanek et al. 2007; Ganesan et al. 2008; Chandak et al. 2009; Tournebize et al. 2009). Some patients also had optic nerve inflammation (optic neuritis) during the acute phase of infection that was treated successfully with parenteral corticosteroids (Mittal et al. 2007). CHIKV infection in adults was also associated with bilateral frontoparietal white matter lesions with restricted diffusion, which are described as early features of viral encephalitis (Ganesan et al. 2008). Focal perivascular lymphocytic infiltrates were also present in areas of active demyelination and some degree of microglial activation was also noted in the gray matter, which may contribute to bystander neuronal loss. Surprisingly, viral inclusion bodies and reactive astrogliosis as ascertained by glial fibrillary acidic protein (GFAP) staining were absent in these CNS cases (Ganesan et al. 2008). Regrettably, this is one of the rare histopathological studies of a single case substantiating the contribution of CHIKV to neuroinfection and neuroinflammation, whereas most of the evidence comes from in vitro studies and animal models in mice and macaques.

4 Neurological Infection: Routes to CNS Infection and Tissue Injury

CNS infections by viruses are relatively uncommon, but are potentially devastating (Bruzzone et al. 2010). Viral invasion and successful infection of the CNS is an important step in the life cycle of many neurotropic viruses such as poliovirus, rabies virus, and measles virus (Griffin 2003). Viruses that invade the brain and peripheral nerves have been postulated to cause diseases through very diverse mechanisms with a unique mechanism of entry, replication, defense against the immune system, and dissemination to preferred target cells and developmental stages of infection (van den Pol 2006). Each of these aspects needs to be carefully addressed in order to understand the physiopathology of a given virus infection with neurological complications. Viruses can instigate neurological injuries not only by direct cytolytic actions on neurons and glia (oligodendrocytes and astrocytes), but also by inducing apoptosis, disrupting the protective BBB, polarizing resident innate immune cells (microglia) to produce pro-inflammatory and potentially cytotoxic cytokines, initiating an (auto)immune attack on specific cells, expressing viral genes and inhibiting cellular genes, altering neuronal migration, attenuating neural stem cells replication, and blocking CSF generation and flow (Fig. 1). These multiple mechanisms of viral induction of CNS neuroinfections and dysfunction further complicate our understanding of viral agents in brain disease.

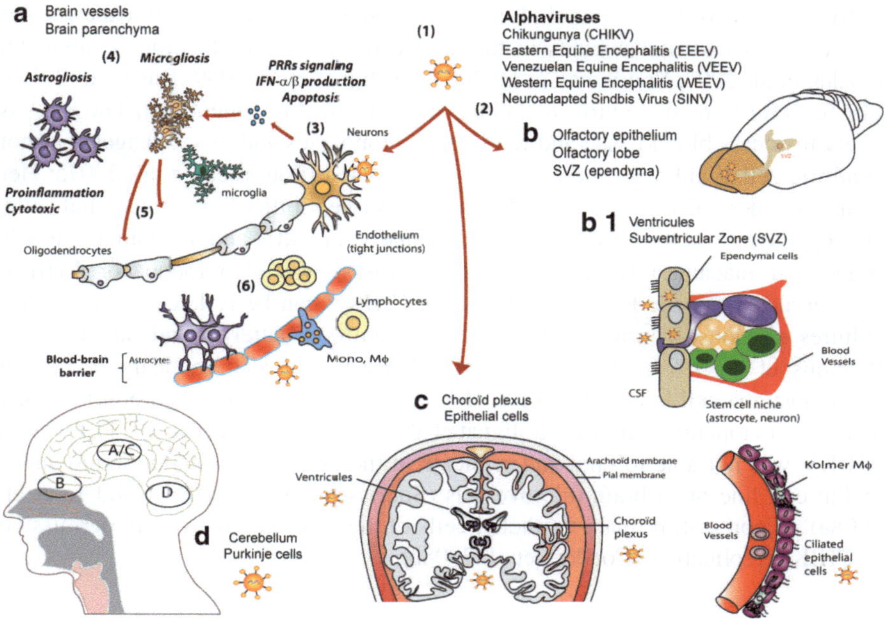

Fig. 1 Alphavirus infection and pathology of the CNS. The neurotropism of CHIKV has not been completely defined, but is substantiated by clinical hallmarks (with neurological involvement in neonates and in the elderly) together with data from experimental infections (mice and macaques) using different CHIKV strains. The pathological mechanisms can also be explored from observations using other encephalitic viruses [EEEV, VEEV, and WEEV (1)] and canonical neuroadapted Sindbis alphavirus. These viruses replicate at very high levels (10^{5-12} copies of viral RNA/ml of blood) and can gain access to the CNS through either the blood–brain barrier (BBB): (a6) possibly within infected monocytes (Mono), or through the choroid plexus: (c) and the meninges (2). In (**a**) regardless of the routes of entry, alphaviruses can infect neurons (3) and cause apoptosis together with mild gliosis [astrocytes and microglia, (4)]. Activated glial cells may contribute to the inflammatory response potentially cytotoxic to oligodendrocytes and, thus, contributing to demyelination (5). Cells of the adaptive immune response (T and B cells) recruited at the site of injury may have double-edged sword activity either to protect from infection (neutralizing noncytolytic antibodies) or to promote further neurotoxicity and demyelination (6). Of note (**b**), CHIKV and several other alphaviruses can infect ependymal cells: (b,b1) of the brain ventricles, which contribute to the stem cell niche in the postnatal brain. Whether this could have an effect on stem cell behavior and CNS developmental capacity is currently unknown. The capacity of CHIKV to infect purkinje cells of the cerebellum has not been reported (D,d)

4.1 Identification of CHIKV Target Cells In Vitro

In vitro studies are useful for the comprehension of CHIKV tissue tropism and can provide insights into the associated pathogenic mechanisms. CHIKV initially infects systemic tissues (skin, connective tissue, blood) before gaining access to the brain and nerve tissues. The main target cells for CHIKV infection and replication at the site of mosquito bites are skin fibroblasts (Hahon and Zimmerman 1970; Sourisseau et al. 2007a; Schilte et al. 2010; Krejbich-Trotot et al. 2011a).

CHIKV can also infect epithelial cells and skeletal muscle cells, but the role of keratinocytes and melanocytes in the epidermis in viral replication is unknown (Ozden et al. 2007; Sourisseau et al. 2007a; Solignat et al. 2009; Das et al. 2010). The receptor to mediate virus cell entry also remains to be identified. The virus is likely to access blood tissue and replicate in monocytes and macrophages (but not dendritic cells), although these findings are debated (Sourisseau et al. 2007a; Her et al. 2010; Krejbich-Trotot et al. 2011a). The susceptibility of neurons, glial cells, and ependymal cells, described experimentally in vitro and in animal models (mice and macaque) for the main alphaviruses, was also found for CHIKV (Das et al. 2010) (see Fig. 1 and Table 1). CHIKV can be isolated from primary cultures of mouse neurons and glial cells in vitro (Chatterjee and Sarkar 1965; Precious et al. 1974; Das et al. 2010). CHIKV can also replicate in human neuroblastoma cell lines (SH-SY5Y, IMR32/Kelly, our unpublished observations) and cause cytopathic activities (Solignat et al. 2009; Dhanwani et al. 2011). Of note, CHIKV was not able to bind and replicate in the hCMEC/D3 human brain endothelial cell line or in brain microvessels (Sourisseau et al. 2007b; Couderc et al. 2008a). In contrast, primary epithelial cells of the choroid plexus were permissive to CHIKV replication (Couderc et al. 2008a).

4.2 General Mechanisms of CNS Infection by Viruses

The BBB is composed of brain micro-vascular endothelial cells with intercellular tight junctions and supported by astrocytes, pericytes (smooth muscle cells), and a basement membrane (Fig. 1). The second interface is a highly vascularized and fenestrated barrier localized at the choroid plexus between blood and CSF, which also allows passage of some blood components, and, thirdly, an interface provided by avascular arachnoid epithelium, which underlies the dura and completely encloses the CNS (Fig. 1) (Abbott et al. 2010). There are at least four different mechanisms by which viruses can gain access into the CNS.

First, viruses may gain access by passing through a damaged BBB. In acute viral encephalitis, capillary and endothelial inflammation of cortical vessels is a striking pathological finding, occurring primarily in the gray matter or gray–white junction and this may facilitate virus entry. A number of viruses such as cytomegalovirus (CMV), human immunodeficiency virus (HIV), and several arboviruses are able to disrupt the BBB at least in vitro (van den Pol 2006). Recent investigations suggested that VEEV initially enters the CNS through the olfactory pathways and initiates viral replication in the brain, which subsequently induces the opening of the BBB, allowing a second wave of invading virus from the periphery to enter the brain (Schafer et al. 2011). Whether this is also the case for CHIKV infection is unknown.

Second, viruses may transmigrate across the BBB within virally infected leukocytes. For HIV, several studies have shown that virus shedding from infected $CD4^+$ T cells, macrophages, and monocytes during migration through the BBB can instigate CNS replication in the parenchyma (Nottet et al. 1996). As aforementioned, CHIKV can infect monocytes and macrophages, but their contribution to neuroinvasion remains to be addressed.

Third, viruses can also penetrate the CNS by taking advantage of incomplete closure of the BBB. Despite the intercellular tight junctions between the capillary endothelial cells in most regions of the BBB, certain areas of the CNS such as the choroid plexus, posterior pituitary, and circumventricular organs are not completely protected by the BBB due to a fenestrated endothelial cell layer and sparse basement membrane (Wolburg and Paulus 2010). A number of blood-borne viruses, including CHIKV, have been suggested to penetrate across the choroid plexus micro-vessels and infect the ciliated epithelium (Fig. 1) (Couderc et al. 2008a). In the CSF space, viruses can subsequently infect the ependymal cells and the surrounding brain tissue.

Finally, viruses can spread to the CNS through peripheral neuronal routes and providing a convenient route for neurotropic viruses (Mori et al. 2005; Tirabassi et al. 1998). This route of infection is unlikely to be involved in classical CHIKV disease.

4.3 *CHIKV Neuroinfection in Animal Models*

Although data are still scarce, the number of recent human cases with CNS involvement appears to support the neurotropic activity of CHIKV (Ganesan et al. 2008; Gerardin et al. 2008; Ramful et al. 2007; Chandak et al. 2009; Lewthwaite et al. 2009; Rampal and Meena 2007; Tandale et al. 2009). This unique CNS infection illustrated by subventricular white matter lesions and intraparenchymal hemorrhages has been described already experimentally and in clinical settings for other alphaviruses such as SFV, RRV, EEEV, and SINV (Jackson et al. 1987; Mims et al. 1973; Deresiewicz et al. 1997; Fazakerley et al. 2006).

Mice were the first animal models developed in order to mimic the CHIKV human disease and to gain a better understanding of the CHIKV-associated physiopathology. Using an inbred mouse model (C57B16) infected by the La Reunion CHIKV strain, Couderc and colleagues (2008a) found different pathophysiological features of CHIKV infection. The susceptibility of the virus was age-dependent as described for most of the alphaviruses and affecting neonates with high viral loads in the brain. Type-I IFN signaling was critical to control the infection as described for poliovirus and vesicular stomatitis virus (Detje et al. 2009; Kuss et al. 2008). CHIKV was able to disseminate in the CNS through the choroid plexus and the meninges, possibly with the contribution of epithelial cells and fibroblasts, respectively (Fig. 1) (Couderc et al. 2008a). Ependymal cells lining the subventricular zone (SVZ), also known as the neural stem cell niche, were also infected by CHIKV (Hauwel et al. 2005a, b). Of critical note, RRV was also shown to infect ependymal cells, leading to cortical thinning and hydrocephalus (Mims et al. 1973). To what extent CHIKV infection could affect the SVZ niche and, subsequently, the stem cells remains unknown.

Outbred young mice models (ICR and CD-1) infected by the La Reunion strain of CHIKV were also studied by Ziegler et al. (2008). Newborn and 14-day-old mice showed viremia and neurological signs of illness (lethargy, loss of balance and difficulty walking, and dragging of the hindlimbs) 7–10 days after CHIKV infection. In newborn mice, the virus replicated to high titers in the brain, sometimes causing death within 3–5 days. However, in older mice virus replication in the brain was more restricted, and mice often recovered. A third model was described by Weaver and colleagues, using CHIKV Ross strain (Wang et al. 2008). In this study, it was noted that infected mice showed severe multifocal inflammation and liquefactive necrosis in the cerebral cortex. Perivascular cuffs were also located diffusely throughout the cerebral cortex and CHIKV immunostaining was found in degenerating neurons located in the necrotic areas of the cerebral cortex. The limitations of these three studies are that they all referred to lethality as the measure of the disease process and, regrettably, the authors did not clearly identify the cell types involved in CHIKV replication and the process of neuroinflammation. In a more recent study, Dhanwani et al. (2011) confirmed CHIKV neuroinfection using 2–3 day old Swiss albino mice. The authors reported reactive gliosis, apoptosis and immunostaining of brain cells for CHIKV E2 proteins (10 days postinfection), and further explored the mechanisms of tissue injury through a comprehensive proteomics analysis. A total of 52 proteins showed significant changes in CHIKV-infected brains and particularly those involved in inflammation (kininogen, α2 HS glycoprotein, HMGB1/2), iron metabolism, cytoskeleton, oxidative stress (catalase), and apoptosis. Experiments are now highly warranted to specifically address the role of these proteins in the physiopathology of CHIKV.

Hence, several good mouse models have now been validated for the early acute CHIKV-induced disease, which result in some of the clinical signs in humans. Given that there are long-term developmental sequelae in neonates in human infections, the next step would be to refine these animal models to evaluate for chronic conditions post-CHIKV infection.

Non-human primates (NHP) have an immune system very close to that of the humans and can be used extensively to study pathophysiological mechanisms (Vierboom et al. 2007). We have known for a long time that NHP are susceptible to CHIKV infection and furthermore that they may represent a natural reservoir in Africa and possibly Asia (Peiris et al. 1993; Inoue et al. 2003). Old monkey models have been validated to study CHIKV-induced pathology and vaccine testing (Binn et al. 1967; Levitt et al. 1986). Although the viral replication pattern in macaques is much closer to that of humans, clinical signs and mechanisms of neuroinfection were not reported comprehensively. Infection of cynomolgus macaques (*Macaca fascicularis*) by CHIKV was reported by Labadie et al. (2010). The authors infected adult macaques with CHIKV strain LR-2006-OPY1 from the La Reunion outbreak and found that CHIKV infection recapitulated the viral, clinical, and pathological features observed in humans apart from chronic arthralgias and arthritis. During the acute phase of the disease, they showed that CHIKV targeted lymphoid tissues, liver, CNS, joints, and muscles and mainly infected macrophages, dendritic cells, and some endothelial cells. With regards to the later stages of infection, they found that CHIKV persisted in lymphoid organs, liver, joints, and muscles and was present in macrophages for up to 3 months after viral inoculation. The contribution of CHIKV persistence in these tissues and chronic inflammation was not investigated further. These results suggest that macrophages can be the main cellular reservoir during the chronic stages of CHIKV infection in vivo, potentially explaining long-lasting symptoms observed in humans. Pregnant female rhesus macaques (*Macaca mulatta*) were also tested for CHIKV infection to explore further the mechanisms of neonatal transmission and neurological complications (Chen et al. 2010). Maternal tissues (spleen, lymph nodes, joints, muscle, and liver) presented high viral loads, but not fetal tissues. Interestingly, the inflammatory cytokine/chemokine response was polarized with some being elevated (IL2, IL6, IL15, IL1ra, MCP-1, VEGF, and IL13) whereas others (IFN-γ, MIP1α, TNF-α, GM-CSF, IL8, and IL10) showed minimal increase as reported in human acute disease (Hoarau et al. 2010).

5 Protective Immune Responses and Inflammation

CHIKV was previously thought to cause only acute infections, but recent evidence suggests that host immunological, tissue-specific factors, and viral subversion strategies may act together in order to control viral persistence. Thus, it is essential to have a better understanding of both innate and adaptive immune responses that may be mobilized during CHIKV infection, but it should be stressed that most of the reported findings have been focused on non-neuronal tissues and cells.

Viral infection rapidly results in the induction of expression of the type I interferons (IFNs), i.e., IFNα and IFNβ, which constitute one of the first-line defenses against invading pathogens. The production of the protective type I IFNs responses is initiated by the specific recognition of conserved motifs such as pathogen-associated molecular patterns (PAMP) by a plethora of PRRs. Two types of PRRs are involved in innate immune sensing of RNA viruses (Takeuchi and Akira 2007): (1) The Toll-like receptors (TLRs), which are localized on membranes or endosomes and (2) the cytoplasmic retinoic acid-inducible gene I like receptors (RLRs). The main RLRs are the retinoic acid-inducible gene I (RIG-I) and the melanoma differentiation-associated protein (MDA-5). RIG-I and MDA-5 activate IPS-1 (IFN-β promoter stimulator-1, also known as MAVS, CARDIF, and VISA). Downstream this results in the translocation of the transcription factors IRF3/7 and NF-κB to promote innate immunity. CHIKV is an RNA virus of 12 kb, and may be detected by TLR7 and/or TLR8, but also TLR3. The helicase RIG-I and MDA-5 may be involved also in the recognition of CHIKV. Recently, White et al. (2011) showed that CHIKV infection of human fibroblasts leads to activation of the transcription factor IRF3 and subsequent transcription of IRF3-dependent antiviral genes, including IFNβ (White et al. 2011). IRF3 activation occurs via the engagement of IPS-1. Critically, it is the non-hematopoietic cells (e.g., fibroblasts and not professional innate immune cells) that contribute (in an IPS-1-dependent manner) to the control of CHIKV infection through production of type I IFNs immune cells (Schilte et al. 2010).

Release of type I IFN from infected cells results in autocrine and paracrine stimulation of the IFN-α/β receptor (IFNAR), which leads through associated tyrosine (TYK) and Janus (JAK) kinases to the phosphorylation of STAT (signal transducers and activators of transcription) 1 and 2. STAT1/2 heterodimers associate with IFN regulatory factor 9 (IRF9) and bind to IFN-stimulated response elements (ISREs) present in the promoter region of the IFN-stimulated genes (ISGs). ISG-encoded proteins represent the antiviral effector molecules that directly inhibit molecular and biochemical activities required for virus replication. Expectedly, type I IFNs were shown to be essential for the antiviral CHIKV response as demonstrated in vitro and in vivo (Briolant et al. 2004; Sourisseau et al. 2007a; Couderc et al. 2008b). The CHIKV-induced type I interferon response seems to be protective in vitro and possibly induces the expression of restriction factors such as oligoadenylate synthetase (OAS) (Brehin et al. 2009; Werneke et al. 2011). OAS has been shown to be active against CHIKV replication in HeLa cells, but its role in other cell types, including neuronal/glial cells, will need to be further established (Brehin et al. 2009). In patients and in the NHP model, the secreted IFNα response was very transient and dropped to basal levels within 24 h, whereas the peak of viremia persisted for several days (Werneke et al. 2011). These data suggest that CHIKV may be capable of somehow controlling the IFN response. Infection and replication of alphaviruses usually cause a general host translational shut off, leading to severe cytopathic activities in mammalian cells. White et al. (2011) confirmed this paradigm, demonstrating that expression of cellular (but not viral) genes is blocked during CHIKV infection. Fros et al. (2010) showed that CHIKV replication in Vero cells is resistant to inhibition by interferon once RNA replication has been initiated and that CHIKV actively suppresses the antiviral IFN response by preventing IFN-induced gene expression. CHIKV infection efficiently blocked STAT1 phosphorylation and/or nuclear translocation in mammalian cells induced by either type I or type II IFN. Nonstructural protein 2 (nsP2) was a potent inhibitor of IFN-induced JAK-STAT signaling. Hence, CHIKV, like many viruses, can evade or counteract the antiviral responses. Autophagy is first a fundamental cell surviving process during starvation conditions, but it also plays a role in both innate and adaptive immunity in response to pathogens (Schmid and Munz 2007). We have recently explored the role of the autophagy machinery in CHIKV infection and showed that cellular autophagy cells was promoted within CHIKV-HEK293 infected cells and, furthermore, contributed to enhance CHIKV replication (Krejbich-Trotot et al. 2011b). This is another good example of a subversion of an innate immune response by CHIKV.

There is little evidence indicating how CHIKV contributes to cellular and molecular mechanisms of brain tissue injury, which can be direct (neurotoxicity) or indirect through the mobilization of glial cell factors. To promote host survival, infected cells may undergo apoptosis, which can be qualified as an "altruistic" suicide in response to viral infections; however, neurons have limited capacity for regeneration. CHIKV can cause programmed-cell death through extrinsic apoptosis (death receptor/caspases 8 pathway) and intrinsic apoptosis (cytochrome C/caspases 9 pathway) of many cell types, including neuroblastoma cells (Krejbich-Trotot et al. 2011a, unpublished observations). This will need to be confirmed using primary cultures, but it should be stressed that this is likely salutary to the CNS tissue to limit virus spreading.

Astrogliosis and microgliosis have been reported in human and animal models of CHIKV neuroinfection and these responses may be essential to ward off the infectious challenge through the production of interferon and ISGs [for review see Ryman and Klimstra (2008)]. The immune response to the infection of the CNS has double-edged sword activity with, on the one hand, the capacity to protect from infection and, on the other, to promote further tissue injury if uncontrolled (Hauwel et al. 2005a). It is essential to have a better understanding of the plausible role of innate immune effectors such as cytokines and complement proteins produced at the site of injury. This is largely unknown in the case of CHIKV neuroinfection and important information should be obtained from other alphaviruses affecting brain cells and functions.

6 Conclusions

As highlighted throughout this review, our understanding of CNS pathology instigated by encephalitic alphaviruses is still in its infancy. CHIKV, in addition to its profound arthritogenic activity, has also encephalitic potential, particularly in newborns and elderly patients with severe comorbidities. Moreover, CHIKV is known to persist in tissue sanctuaries, although this remains to be ascertained in CNS tissues, which may contribute to chronic diseases. Although cell permissiveness and reactivity have been studied in great depth, the mechanisms of CHIKV persistence and associated tissue injuries remains largely ill-characterized. Therefore, with a proven potential to spread globally, it is now critical to devise strategies to circumvent infection of populations at risk and prevent new epidemics.

Acknowledgments This work was funded by CPER/FEDER and ICRES PCRD7.

References

Abbott NJ, Patabendige AA, Dolman DE, Yusof SR, Begley DJ (2010) Structure and function of the blood-brain barrier. Neurobiol Dis 37(1):13–25. doi:S0969-9961(09)00208-3 [pii], 10.1016/j.nbd.2009.07.030

Anderson BA (1984) Focal neurologic signs in western equine encephalitis. Can Med Assoc J 130 (8):1019–1021

Arpino C, Curatolo P, Rezza G (2009) Chikungunya and the nervous system: what we do and do not know. Rev Med Virol 19(3):121–129. doi:10.1002/rmv.606

Binder GK, Griffin DE (2001) Interferon-gamma-mediated site-specific clearance of alphavirus from CNS neurons. Science 293(5528):303–306. doi:10.1126/science.1059742, 293/5528/303 [pii]

Binn LN, Harrison VR, Randall R (1967) Patterns of viremia and antibody observed in rhesus monkeys inoculated with chikungunya and other serologically related group A arboviruses. Am J Trop Med Hyg 16(6):782–785

Borgherini G, Poubeau P, Staikowsky F, Lory M, Le Moullec N, Becquart JP, Wengling C, Michault A, Paganin F (2007) Outbreak of chikungunya on Reunion Island: early clinical and laboratory features in 157 adult patients. Clin Infect Dis 44(11):1401–1407

Borgherini G, Poubeau P, Jossaume A, Gouix A, Cotte L, Michault A, Arvin-Berod C, Paganin F (2008) Persistent arthralgia associated with chikungunya virus: a study of 88 adult patients on reunion island. Clin Infect Dis 47(4):469–475

Brehin AC, Casademont I, Frenkiel MP, Julier C, Sakuntabhai A, Despres P (2009) The large form of human 2′,5′-Oligoadenylate Synthetase (OAS3) exerts antiviral effect against Chikungunya virus. Virology 384(1):216–222

Briolant S, Garin D, Scaramozzino N, Jouan A, Crance JM (2004) In vitro inhibition of Chikungunya and Semliki Forest viruses replication by antiviral compounds: synergistic effect of interferon-alpha and ribavirin combination. Antiviral Res 61(2):111–117

Bruzzone R, Dubois-Dalcq M, Kristensson K (2010) Neurobiology of infectious diseases: bringing them out of neglect. Prog Neurobiol 91(2):91–94. doi:10.1016/j.pneurobio.2009.12.005

Chandak NH, Kashyap RS, Kabra D, Karandikar P, Saha SS, Morey SH, Purohit HJ, Taori GM, Daginawala HF (2009) Neurological complications of Chikungunya virus infection. Neurol India 57(2):177–180

Charrel RN, de Lamballerie X, Raoult D (2007) Chikungunya outbreaks—the globalization of vectorborne diseases. N Engl J Med 356(8):769–771

Chatterjee SN, Sarkar JK (1965) Electron microscopic studies of suckling mouse brain cells infected with Chikungunya virus. Indian J Exp Biol 3(4):227–234

Chen CI, Clark DC, Pesavento P, Lerche NW, Luciw PA, Reisen WK, Brault AC (2010) Comparative pathogenesis of epidemic and enzootic Chikungunya viruses in a pregnant Rhesus macaque model. Am J Trop Med Hyg 83(6):1249–1258. doi:10.4269/ajtmh.2010.10-0290

Couderc T, Chretien F, Schilte C, Disson O, Brigitte M, Guivel-Benhassine F, Touret Y, Barau G, Cayet N, Schuffenecker I, Despres P, Arenzana-Seisdedos F, Michault A, Albert ML, Lecuit M (2008a) A mouse model for Chikungunya: young age and inefficient type-I interferon signaling are risk factors for severe disease. PLoS Pathog 4(2):e29

Couderc T, Chretien F, Schilte C, Disson O, Brigitte M, Guivel-Benhassine F, Touret Y, Barau G, Cayet N, Schuffenecker I, Despres P, Arenzana-Seisdedos F, Michault A, Albert ML, Lecuit M (2008b) A mouse model for Chikungunya: young age and inefficient type-I interferon signaling are risk factors for severe disease. PLoS Pathog 4(2):e29. doi:10.1371/journal.ppat.0040029

Das T, Jaffar-Bandjee MC, Hoarau JJ, Krejbich Trotot P, Denizot M, Lee-Pat-Yuen G, Sahoo R, Guiraud P, Ramful D, Robin S, Alessandri JL, Gauzere BA, Gasque P (2010) Chikungunya fever: CNS infection and pathologies of a re-emerging arbovirus. Prog Neurobiol 91 (2):121–129. doi:10.1016/j.pneurobio.2009.12.006

Delfraro A, Burgueno A, Morel N, Gonzalez G, Garcia A, Morelli J, Perez W, Chiparelli H, Arbiza J (2011) Fatal human case of Western equine encephalitis, Uruguay. Emerg Infect Dis 17 (5):952–954. doi:10.3201/eid1705.101068

Deresiewicz RL, Thaler SJ, Hsu L, Zamani AA (1997) Clinical and neuroradiographic manifestations of eastern equine encephalitis. N Engl J Med 336(26):1867–1874. doi:10.1056/NEJM199706263362604

Detje CN, Meyer T, Schmidt H, Kreuz D, Rose JK, Bechmann I, Prinz M, Kalinke U (2009) Local type I IFN receptor signaling protects against virus spread within the central nervous system. J Immunol 182(4):2297–2304. doi:10.4049/jimmunol.0800596

Dhanwani R, Khan M, Alam SI, Rao PV, Parida M (2011) Differential proteome analysis of Chikungunya virus-infected new-born mice tissues reveal implication of stress, inflammatory and apoptotic pathways in disease pathogenesis. Proteomics 11(10):1936–1951. doi:10.1002/pmic.201000500

Economopoulou A, Dominguez M, Helynck B, Sissoko D, Wichmann O, Quenel P, Germonneau P, Quatresous I (2009) Atypical Chikungunya virus infections: clinical manifestations, mortality and risk factors for severe disease during the 2005-2006 outbreak on Reunion. Epidemiol Infect 137(4):534–541

Fazakerley JK, Cotterill CL, Lee G, Graham A (2006) Virus tropism, distribution, persistence and pathology in the corpus callosum of the Semliki Forest virus-infected mouse brain: a novel system to study virus-oligodendrocyte interactions. Neuropathol Appl Neurobiol 32 (4):397–409

Fritel X, Rollot O, Gerardin P, Gauzere BA, Bideault J, Lagarde L, Dhuime B, Orvain E, Cuillier F, Ramful D, Samperiz S, Jaffar-Bandjee MC, Michault A, Cotte L, Kaminski M, Fourmaintraux A (2010) Chikungunya virus infection during pregnancy, Reunion, France, 2006. Emerg Infect Dis 16(3):418–425. doi:10.3201/eid1603.091403

Fros JJ, Liu WJ, Prow NA, Geertsema C, Ligtenberg M, Vanlandingham DL, Schnettler E, Vlak JM, Suhrbier A, Khromykh AA, Pijlman GP (2010) Chikungunya virus nonstructural protein 2 inhibits type I/II interferon-stimulated JAK-STAT signaling. J Virol 84 (20):10877–10887. doi:10.1128/JVI.00949-10

Ganesan K, Diwan A, Shankar SK, Desai SB, Sainani GS, Katrak SM (2008) Chikungunya encephalomyeloradiculitis: report of 2 cases with neuroimaging and 1 case with autopsy findings. AJNR Am J Neuroradiol 29(9):1636–1637

Gerardin P, Barau G, Michault A, Bintner M, Randrianaivo H, Choker G, Lenglet Y, Touret Y, Bouveret A, Grivard P, Le Roux K, Blanc S, Schuffenecker I, Couderc T, Arenzana-Seisdedos F, Lecuit M, Robillard PY (2008) Multidisciplinary prospective study of mother-to-child chikungunya virus infections on the island of La Reunion. PLoS Med 5(3):e60

Grandadam M, Caro V, Plumet S, Thiberge JM, Souares Y, Failloux AB, Tolou HJ, Budelot M, Cosserat D, Leparc-Goffart I, Despres P (2011) Chikungunya virus, southeastern France. Emerg Infect Dis 17(5):910–913. doi:10.3201/eid1705.101873

Griffin DE (2003) Immune responses to RNA-virus infections of the CNS. Nat Rev Immunol 3(6):493–502. doi:10.1038/nri1105, nri1105 [pii]

Griffin DE, Johnson RT (1977) Role of the immune response in recovery from Sindbis virus encephalitis in mice. J Immunol 118(3):1070–1075

Hahon N, Zimmerman WD (1970) Chikungunya virus infection of cell monolayers by cell-to-cell and extracellular transmission. Appl Microbiol 19(2):389–391

Hauwel M, Furon E, Canova C, Griffiths M, Neal J, Gasque P (2005a) Innate (inherent) control of brain infection, brain inflammation and brain repair: the role of microglia, astrocytes, "protective" glial stem cells and stromal ependymal cells. Brain Res Brain Res Rev 48(2):220–233

Hauwel M, Furon E, Gasque P (2005b) Molecular and cellular insights into the coxsackie-adenovirus receptor: role in cellular interactions in the stem cell niche. Brain Res Brain Res Rev 48(2):265–272

Her Z, Malleret B, Chan M, Ong EK, Wong SC, Kwek DJ, Tolou H, Lin RT, Tambyah PA, Renia L, Ng LF (2010) Active infection of human blood monocytes by Chikungunya virus triggers an innate immune response. J Immunol 184(10):5903–5913. doi:10.4049/jimmunol.0904181

Hoarau JJ, Jaffar Bandjee MC, Krejbich Trotot P, Das T, Li-Pat-Yuen G, Dassa B, Denizot M, Guichard E, Ribera A, Henni T, Tallet F, Moiton MP, Gauzere BA, Bruniquet S, Jaffar Bandjee Z, Morbidelli P, Martigny G, Jolivet M, Gay F, Grandadam M, Tolou H, Vieillard V, Debre P, Autran B, Gasque P (2010) Persistent chronic inflammation and infection by Chikungunya arthritogenic alphavirus in spite of a robust host immune response. J Immunol 184 (10):5914–5927. doi:10.4049/jimmunol.0900255

Inoue S, Morita K, Matias RR, Tuplano JV, Resuello RR, Candelario JR, Cruz DJ, Mapua CA, Hasebe F, Igarashi A, Natividad FF (2003) Distribution of three arbovirus antibodies among monkeys (Macaca fascicularis) in the Philippines. J Med Primatol 32(2):89–94

Jackson AC, Moench TR, Griffin DE, Johnson RT (1987) The pathogenesis of spinal cord involvement in the encephalomyelitis of mice caused by neuroadapted Sindbis virus infection. Lab Invest 56(4):418–423

Jackson AC, Moench TR, Trapp BD, Griffin DE (1988) Basis of neurovirulence in Sindbis virus encephalomyelitis of mice. Lab Invest 58(5):503–509

Jaffar-Bandjee MC, Das T, Hoarau JJ, Krejbich Trotot P, Denizot M, Ribera A, Roques P, Gasque P (2009) Chikungunya virus takes centre stage in virally induced arthritis: possible cellular and molecular mechanisms to pathogenesis. Microbes Infect 11(14–15):1206–1218. doi:10.1016/j.micinf.2009.10.001

Jaffar-Bandjee MC, Ramful D, Gauzere BA, Hoarau JJ, Krejbich-Trotot P, Robin S, Ribera A, Selambarom J, Gasque P (2010) Emergence and clinical insights into the pathology of Chikungunya virus infection. Expert Rev Anti Infect Ther 8(9):987–996. doi:10.1586/eri.10.92

Johnson RT (1965) Virus invasion of the central nervous system: a study of sindbis virus infection in the mouse using fluorescent antibody. Am J Pathol 46:929–943

Krejbich-Trotot P, Denizot M, Hoarau JJ, Jaffar-Bandjee MC, Das T, Gasque P (2011a) Chikungunya virus mobilizes the apoptotic machinery to invade host cell defenses. FASEB J 25(1):314–325. doi:10.1096/fj.10-164178

Krejbich-Trotot P, Gay B, Li-Pat-Yuen G, Hoarau JJ, Jaffar-Bandjee MC, Briant L, Gasque P, Denizot M (2011b) Chikungunya triggers an autophagic process which promotes viral replication. Virol J 8(1):432. doi:10.1186/1743-422X-8-432

Kuss SK, Etheredge CA, Pfeiffer JK (2008) Multiple host barriers restrict poliovirus trafficking in mice. PLoS Pathog 4(6):e1000082. doi:10.1371/journal.ppat.1000082

Labadie K, Larcher T, Joubert C, Mannioui A, Delache B, Brochard P, Guigand L, Dubreil L, Lebon P, Verrier B, de Lamballerie X, Suhrbier A, Cherel Y, Le Grand R, Roques P (2010) Chikungunya disease in nonhuman primates involves long-term viral persistence in macrophages. J Clin Invest 120(3):894–906. doi:10.1172/JCI40104

Le Bomin A, Hebert JC, Marty P, Delaunay P (2008) [Confirmed chikungunya in children in Mayotte. Description of 50 patients hospitalized from February to June 2006]. Med Trop (Mars) 68 (5):491–495

Lebrun G, Chadda K, Reboux AH, Martinet O, Gauzere BA (2009) Guillain-Barre syndrome after chikungunya infection. Emerg Infect Dis 15(3):495–496

Lemant J, Boisson V, Winer A, Thibault L, Andre H, Tixier F, Lemercier M, Antok E, Cresta MP, Grivard P, Besnard M, Rollot O, Favier F, Huerre M, Campinos JL, Michault A (2008) Serious acute chikungunya virus infection requiring intensive care during the Reunion Island outbreak in 2005-2006. Crit Care Med 36(9):2536–2541

Levine B, Huang Q, Isaacs JT, Reed JC, Griffin DE, Hardwick JM (1993) Conversion of lytic to persistent alphavirus infection by the bcl-2 cellular oncogene. Nature 361(6414):739–742. doi:10.1038/361739a0

Levine B, Goldman JE, Jiang HH, Griffin DE, Hardwick JM (1996) Bc1-2 protects mice against fatal alphavirus encephalitis. Proc Natl Acad Sci USA 93(10):4810–4815

Levitt NH, Ramsburg HH, Hasty SE, Repik PM, Cole FE Jr, Lupton HW (1986) Development of an attenuated strain of chikungunya virus for use in vaccine production. Vaccine 4(3):157–162

Lewthwaite P, Vasanthapuram R, Osborne JC, Begum A, Plank JL, Shankar MV, Hewson R, Desai A, Beeching NJ, Ravikumar R, Solomon T (2009) Chikungunya virus and central nervous system infections in children, India. Emerg Infect Dis 15(2):329–331

Linton PJ, Dorshkind K (2004) Age-related changes in lymphocyte development and function. Nat Immunol 5(2):133–139. doi:10.1038/ni1033

Mims CA, Murphy FA, Taylor WP, Marshall ID (1973) Pathogenesis of Ross River virus infection in mice. I. Ependymal infection, cortical thinning, and hydrocephalus. J Infect Dis 127 (2):121–128

Mittal A, Mittal S, Bharati MJ, Ramakrishnan R, Saravanan S, Sathe PS (2007) Optic neuritis associated with chikungunya virus infection in South India. Arch Ophthalmol 125 (10):1381–1386. doi:10.1001/archopht.125.10.1381

Mori I, Nishiyama Y, Yokochi T, Kimura Y (2005) Olfactory transmission of neurotropic viruses. J Neurovirol 11(2):129–137. doi:G427587604635V16 [pii], 10.1080/13550280590922793

Nava VE, Rosen A, Veliuona MA, Clem RJ, Levine B, Hardwick JM (1998) Sindbis virus induces apoptosis through a caspase-dependent, CrmA-sensitive pathway. J Virol 72(1):452–459

Ng LC, Hapuarachchi HC (2010) Tracing the path of Chikungunya virus—evolution and adaptation. Infect Genet Evol 10(7):876–885. doi:10.1016/j.meegid.2010.07.012

Nottet HS, Persidsky Y, Sasseville VG, Nukuna AN, Bock P, Zhai QH, Sharer LR, McComb RD, Swindells S, Soderland C, Gendelman HE (1996) Mechanisms for the transendothelial migration of HIV-1-infected monocytes into brain. J Immunol 156(3):1284–1295

Ozden S, Huerre M, Riviere JP, Coffey LL, Afonso PV, Mouly V, de Monredon J, Roger JC, El Amrani M, Yvin JL, Jaffar MC, Frenkiel MP, Sourisseau M, Schwartz O, Butler-Browne G, Despres P, Gessain A, Ceccaldi PE (2007) Human muscle satellite cells as targets of Chikungunya virus infection. PLoS One 2(6):e527

Peiris JS, Dittus WP, Ratnayake CB (1993) Seroepidemiology of dengue and other arboviruses in a natural population of toque macaques (Macaca sinica) at Polonnaruwa, Sri Lanka. J Med Primatol 22(4):240–245

Pialoux G, Gauzere BA, Jaureguiberry S, Strobel M (2007) Chikungunya, an epidemic arbovirosis. Lancet Infect Dis 7(5):319–327

Powers AM, Logue CH (2007) Changing patterns of chikungunya virus: re-emergence of a zoonotic arbovirus. J Gen Virol 88(Pt 9):2363–2377

Precious SW, Webb HE, Bowen ET (1974) Isolation and persistence of Chikungunya virus in cultures of mouse brain cells. J Gen Virol 23(3):271–279

Ramful D, Carbonnier M, Pasquet M, Bouhmani B, Ghazouani J, Noormahomed T, Beullier G, Attali T, Samperiz S, Fourmaintraux A, Alessandri JL (2007) Mother-to-child transmission of Chikungunya virus infection. Pediatr Infect Dis J 26(9):811–815

Rampal SM, Meena H (2007) Neurological complications in Chikungunya fever. J Assoc Physicians India 55:765–769

Reed DS, Larsen T, Sullivan LJ, Lind CM, Lackemeyer MG, Pratt WD, Parker MD (2005) Aerosol exposure to western equine encephalitis virus causes fever and encephalitis in cynomolgus macaques. J Infect Dis 192(7):1173–1182. doi:JID34552 [pii], 10.1086/444397

Robin S, Ramful D, Le Seach F, Jaffar-Bandjee MC, Rigou G, Alessandri JL (2008) Neurologic manifestations of pediatric chikungunya infection. J Child Neurol 23(9):1028–1035

Robinson MC (1955) An epidemic of virus disease in Southern Province, Tanganyika Territory, in 1952-53 I. Clinical features. Trans R Soc Trop Med Hyg 49(1):28–32

Rosenstiel P, Derer S, Till A, Hasler R, Eberstein H, Bewig B, Nikolaus S, Nebel A, Schreiber S (2008) Systematic expression profiling of innate immune genes defines a complex pattern of immunosenescence in peripheral and intestinal leukocytes. Genes Immun 9(2):103–114. doi:10.1038/sj.gene.6364454

Ryman KD, Klimstra WB (2008) Host responses to alphavirus infection. Immunol Rev 225:27–45. doi:10.1111/j.1600-065X.2008.00670.x

Samperiz E, Gerardin P, Noormahomed T, Beullir G, Boya I (2007) Transmission perinatale du virus chikungunya, à propos de 47 cas à l'île de la Réunion. Bull Soc Pathol Exot Filiales 100:355

Schafer A, Brooke CB, Whitmore AC, Johnston RE (2011) The role of the blood-brain barrier during Venezuelan equine encephalitis virus infection. J Virol 85(20):10682–10690. doi:10.1128/JVI.05032-11

Schilte C, Couderc T, Chretien F, Sourisseau M, Gangneux N, Guivel-Benhassine F, Kraxner A, Tschopp J, Higgs S, Michault A, Arenzana-Seisdedos F, Colonna M, Peduto L, Schwartz O, Lecuit M, Albert ML (2010) Type I IFN controls chikungunya virus via its action on nonhematopoietic cells. J Exp Med 207(2):429–442. doi:10.1084/jem.20090851

Schmid D, Munz C (2007) Innate and adaptive immunity through autophagy. Immunity 27 (1):11–21. doi:10.1016/j.immuni.2007.07.004

Schuffenecker I, Iteman I, Michault A, Murri S, Frangeul L, Vaney MC, Lavenir R, Pardigon N, Reynes JM, Pettinelli F, Biscornet L, Diancourt L, Michel S, Duquerroy S, Guigon G, Frenkiel MP, Brehin AC, Cubito N, Despres P, Kunst F, Rey FA, Zeller H, Brisse S (2006) Genome microevolution of chikungunya viruses causing the Indian Ocean outbreak. PLoS Med 3(7):e263

Schwartz O, Albert ML (2010) Biology and pathogenesis of chikungunya virus. Nat Rev Microbiol 8(7):491–500. doi:10.1038/nrmicro2368

Simon F, Javelle E, Oliver M, Leparc-Goffart I, Marimoutou C (2011) Chikungunya virus infection. Curr Infect Dis Rep 13(3):218–228. doi:10.1007/s11908-011-0180-1

Sissoko D, Malvy D, Ezzedine K, Renault P, Moscetti F, Ledrans M, Pierre V (2009) Post-epidemic Chikungunya disease on Reunion Island: course of rheumatic manifestations and associated factors over a 15-month period. PLoS Negl Trop Dis 3(3):e389. doi:10.1371/journal.pntd.0000389

Solignat M, Gay B, Higgs S, Briant L, Devaux C (2009) Replication cycle of chikungunya: a re-emerging arbovirus. Virology 393(2):183–197. doi:10.1016/j.virol.2009.07.024

Sourisseau M, Schilte C, Casartelli N, Trouillet C, Guivel-Benhassine F, Rudnicka D, Sol-Foulon N, Le Roux K, Prevost MC, Fsihi H, Frenkiel MP, Blanchet F, Afonso PV, Ceccaldi PE, Ozden S, Gessain A, Schuffenecker I, Verhasselt B, Zamborlini A, Saib A, Rey FA, Arenzana-Seisdedos F, Despres P, Michault A, Albert ML, Schwartz O (2007a) Characterization of reemerging chikungunya virus. PLoS Pathog 3(6):e89. doi:10.1371/journal.ppat.0030089

Sourisseau M, Schilte C, Casartelli N, Trouillet C, Guivel-Benhassine F, Rudnicka D, Sol-Foulon N, Le Roux K, Prevost MC, Fsihi H, Frenkiel MP, Blanchet F, Afonso PV, Ceccaldi PE, Ozden S, Gessain A, Schuffenecker I, Verhasselt B, Zamborlini A, Saib A, Rey FA, Arenzana-Seisdedos F, Despres P, Michault A, Albert ML, Schwartz O (2007b) Characterization of reemerging chikungunya virus. PLoS Pathog 3(6):e89

Steele KE, Twenhafel NA (2010) Review paper: pathology of animal models of alphavirus encephalitis. Vet Pathol 47(5):790–805. doi:0300985810372508 [pii], 10.1177/0300985810372508

Suhrbier A, La Linn M (2004) Clinical and pathologic aspects of arthritis due to Ross River virus and other alphaviruses. Curr Opin Rheumatol 16(4):374–379

Takeuchi O, Akira S (2007) Recognition of viruses by innate immunity. Immunol Rev 220:214–224. doi:10.1111/j.1600-065X.2007.00562.x

Tandale BV, Sathe PS, Arankalle VA, Wadia RS, Kulkarni R, Shah SV, Shah SK, Sheth JK, Sudeep AB, Tripathy AS, Mishra AC (2009) Systemic involvements and fatalities during Chikungunya epidemic in India, 2006. J Clin Virol 46:145–149

Thach DC, Kimura T, Griffin DE (2000) Differences between C57BL/6 and BALB/cBy mice in mortality and virus replication after intranasal infection with neuroadapted Sindbis virus. J Virol 74(13):6156–6161

Tirabassi RS, Townley RA, Eldridge MG, Enquist LW (1998) Molecular mechanisms of neurotropic herpesvirus invasion and spread in the CNS. Neurosci Biobehav Rev 22(6):709–720. doi:S0149763498000098 [pii]

Tournebize P, Charlin C, Lagrange M (2009) Neurological manifestations in Chikungunya: about 23 cases collected in Reunion Island. Rev Neurol (Paris) 165(1):48–51

van den Pol AN (2006) Viral infections in the developing and mature brain. Trends Neurosci 29(7):398–406. doi:10.1016/j.tins.2006.06.002

Vierboom MP, Jonker M, Tak PP, t Hart BA (2007) Preclinical models of arthritic disease in non-human primates. Drug Discov Today 12(7–8):327–335. doi:10.1016/j.drudis.2007.02.012

Wang E, Volkova E, Adams AP, Forrester N, Xiao SY, Frolov I, Weaver SC (2008) Chimeric alphavirus vaccine candidates for chikungunya. Vaccine 26(39):5030–5039

Werneke SW, Schilte C, Rohatgi A, Monte KJ, Michault A, Arenzana-Seisdedos F, Vanlandingham DL, Higgs S, Fontanet A, Albert ML, Lenschow DJ (2011) ISG15 is critical in the control of Chikungunya virus infection independent of UbE1L mediated conjugation. PLoS Pathog 7(10):e1002322. doi:10.1371/journal.ppat.1002322

Wesselingh SL, Levine B, Fox RJ, Choi S, Griffin DE (1994) Intracerebral cytokine mRNA expression during fatal and nonfatal alphavirus encephalitis suggests a predominant type 2T cell response. J Immunol 152(3):1289–1297
White LK, Sali T, Alvarado D, Gatti E, Pierre P, Streblow D, Defilippis VR (2011) Chikungunya virus induces IPS-1-dependent innate immune activation and protein kinase R-independent translational shutoff. J Virol 85(1):606–620. doi:10.1128/JVI.00767-10
Wielanek AC, Monredon JD, Amrani ME, Roger JC, Serveaux JP (2007) Guillain-Barre syndrome complicating a Chikungunya virus infection. Neurology 69(22):2105–2107
Wolburg H, Paulus W (2010) Choroid plexus: biology and pathology. Acta Neuropathol 119(1):75–88. doi:10.1007/s00401-009-0627-8
Zacks MA, Paessler S (2010) Encephalitic alphaviruses. Vet Microbiol 140(3–4):281–286. doi: S0378-1135(09)00395-2 [pii], 10.1016/j.vetmic.2009.08.023
Ziegler SA, Lu L, da Rosa AP, Xiao SY, Tesh RB (2008) An animal model for studying the pathogenesis of chikungunya virus infection. Am J Trop Med Hyg 79(1):133–139

Nipah Virus Infection

Li-Yen Chang and Chong-Tin Tan

Abstract In 1998, a new zoonotic paramyxovirus emerged and caused an outbreak of severe febrile encephalitis among pig farm workers in Malaysia. The causative agent was named Nipah virus (NiV), after the village of Sungai Nipah in the state of Negeri Sembilan, Malaysia, which was the residential location from which the first virus isolate was obtained from cerebrospinal fluid. The disease is spread from the natural reservoir host, *Pteropus* spp. bats to pigs, and then to humans following close contact with the infected pigs. The mortality rate of NiV infection in humans is about 40 %. Since then, recurrent outbreaks of NiV encephalitis have been seen once in India and almost annually in Bangladesh. In India and Bangladesh, transmission of the virus was probably spread directly from bats to humans, with human-to-human spread as an important mode of infection. Also, there was no evidence of an intermediate animal host. The main pathological features in patients with NiV encephalitis is disseminated microinfarction associated with vasculitis and direct neuronal infection. Relapse of encephalitis was seen months to years later in 10 % of those who survived the initial illness.

Keywords Bangladesh • Bats • Encephalitis • Human-to-human spread • Magnetic resonance imaging • Malaysia • Mortality • Nipah virus • Pigs • Pneumonia • Relapse and late-onset • Ribavirin • Zoonotic paramyxovirus

L.-Y. Chang
Department of Medical Microbiology, Faculty of Medicine, Tropical Infectious Diseases Research and Education Centre (TIDREC), University Malaya, 50603 Kuala Lumpur, Malaysia
e-mail: changliyen@um.edu.my

C.-T. Tan (✉)
Department of Medicine, Faculty of Medicine, University Malaya, 50603 Kuala Lumpur, Malaysia
e-mail: chongtin.tan@gmail.com

A.C. Jackson (ed.), *Viral Infections of the Human Nervous System*,
Birkhäuser Advances in Infectious Diseases,
DOI 10.1007/978-3-0348-0425-7_13, © Springer Basel 2013

1 Introduction

Nipah virus (NiV), a paramyxovirus, was first discovered in 1998 following a severe outbreak of viral encephalitis among pig farm workers in Malaysia (Centers for Disease Control and Prevention 1999a; Chua et al. 1999, 2000a). The virus was isolated from the cerebrospinal fluid of a patient with severe encephalitis. The pig farm workers contracted the virus from infected pigs through airborne droplets or direct contact with the contaminated pig secretions and excretions. Because the putative reservoir host for NiV is *Pteropus* spp. bats, the suggested route of transmission of the virus to pigs occurred as a result of eating fruits and/or feed contaminated with bat secretions and excretions (Yob et al. 2001; Olson et al. 2002; Chua et al. 2002; Reynes et al. 2005; Wacharapluesadee et al. 2005; Rahman et al. 2010; Halpin et al. 2011). NiV infection has not been reported since then in Malaysia; however, infection by the virus has been reported almost annually in Bangladesh (International Centre for Diarrheal Diseases Research Bangladesh 2003, 2005; Hsu et al. 2004; Hossain et al. 2008; Montgomery et al. 2008). The transmission of the virus in Bangladesh is from bats directly to humans via consumption of food contaminated with bat secretions and excretions. Besides the difference in the mode of transmission of NiV in Malaysia and Bangladesh, the mortality rate was reported to be 70 % in Bangladesh as compared to 40 % in Malaysia.

2 History of Nipah Virus Infection

In February 1998, three pig farm workers were diagnosed with acute viral encephalitis in Perak, Malaysia while pigs in farms at the same area developed respiratory and neurologic disease (Centers for Disease Control and Prevention 1999a, b; Chua et al. 1999). Several months later, in September 1998, an outbreak of severe encephalitis with high death rates (approximately 56 %) occurred in a pig-farming community in Tambun, Perak. The outbreak then spread southward to several other pig-farming communities near or around Seremban in Negeri Sembilan (Centers for Disease Control and Prevention 1999a; Chua et al. 1999). By April 1999, at least 15 deaths were recorded in Tambun and 85 deaths were reported in Seremban (Centers for Disease Control and Prevention 1999b). The causative agent responsible for the outbreak was then discovered and named NiV after Kampung Sungai Nipah in Negeri Sembilan because the virus was isolated from cerebrospinal fluid from a patient from this village (Chua et al. 1999, 2002a).

During the outbreak, pigs in Seremban and Tambun were noted to present respiratory and neurologic syndrome (Centers for Disease Control and Prevention 1999a, b; Mohd Nor et al. 2000). A large proportion of the pigs appeared asymptomatic and the mortality rate was relatively low, between 1 and 5 %. However, as the development of NiV infection in humans seemed to be closely related to the occurrence of the respiratory and neurological disease in pigs, as well as it primarily affected those involved in pig-farming activities, the viral outbreak was quickly associated with the pig industry, with the pigs as the source of infection for humans (Chua et al. 1999, 2000a; Mohd Nor et al. 2000). The outbreak was subsequently controlled with the culling of almost one million pigs and strict quarantine measures (Centers for Disease Control and Prevention 1999b). At the end of the outbreak, more than 265 human cases were reported nationwide, with at least 105 deaths (Chua et al. 2000a).

Since the first reported outbreak of NiV encephalitis in Malaysia in 1998, sporadic outbreaks of NiV or NiV-like infections were reported in West Bengal, India in 2001 (Chadha et al. 2006; Harit et al. 2006), and almost annually in Bangladesh between 2001 and 2012 (International Centre for Diarrheal Diseases Research Bangladesh 2003, 2005; Hsu et al. 2004; Hossain et al. 2008; Montgomery et al. 2008; Pro-MED mail 2012). These outbreaks have been characterized with a higher mortality rate of 60–90 % as compared to 40 % in Malaysia. Several distinct features were associated with the NiV infection in Bangladesh and India, including acute respiratory distress syndrome and human-to-human transmission, particularly among family members and in healthcare settings. In addition, the route of transmission of the virus to humans is suggested to be directly from its reservoir host, which is *Pteropus* spp. bats, without a need of an intermediate amplifying host such as pigs.

3 Nipah Virus

NiV is a paramyxovirus, classified in the genus *Henipavirus* in the *Paramyxoviridae* family (Chua et al. 2000a; Harcourt et al. 2000). It is an enveloped, nonsegmented, negative-stranded RNA virus. NiV was initially isolated in Vero cells (ATCC, CCL-81) from cerebrospinal fluid from a patient with severe encephalitis during the outbreak in Malaysia (Chua et al. 2000a). The virus grows readily in various types of mammalian cells with the characteristic cytopathic effect of cell fusion and the formation of giant syncytial cells, with the aggregated nuclei forming rings that delineate the edge of the large syncytia when observed at the late stage of infection (Chua et al. 1999). However, the rate of virus replication and the titer of virus produced vary with the type of mammalian cells used (Chang et al. 2006a, b). Ultrastructurally, NiV contains a lipid bilayer envelope, is pleomorphic in shape with spherical to filamentous structures, and a size range of between 40 and 600 nm (Fig. 1a) (Chua et al. 2000a; Hyatt et al. 2001; Goldsmith et al. 2003). Tubule-like structures are also observed (Fig. 1b). In the infected cells, the virus forms intra-cytoplasmic inclusions of viral nucleocapsids, which under negative staining has the typical "herringbone" appearance that is characteristic for paramyxoviruses.

The full-length genomic sequence of NiV was determined for several isolates from Malaysia and Bangladesh (Harcourt et al. 2000, 2001, 2005; Chan et al. 2001; AbuBakar et al. 2004; Arankalle et al. 2011). Like other paramyxoviruses, the genome of NiV contains six genes [nucleocapsid protein (N) gene, phosphoprotein

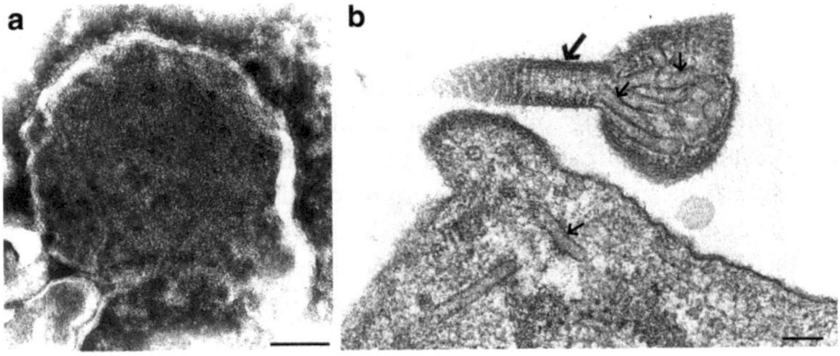

Fig. 1 Electron micrographs of (**a**) Nipah virus and (**b**) tubule-like structures (*small arrows*) associated with a maturing virus (*large arrow*) [from Hyatt et al. (2001) with permission]. *Bars* represent 100 nm

(P) gene, matrix protein (M) gene, fusion protein (F) gene, glycoprotein or attachment protein (G) gene, and the large protein (L) gene] flanked by a leader sequence at the 3′ end and a trailer sequence at the 5′ end (Harcourt et al. 2000, 2001). The complete genome of the virus is comprised of 18,246 nucleotides in Malaysian isolates (Harcourt et al. 2000, 2001; Chan et al. 2001; AbuBakar et al. 2004), whereas the NiVs isolated from India and Bangladesh outbreaks are 12 nucleotides longer (Harcourt et al. 2005; Arankalle et al. 2011), and this is similar in length with the Hendra virus (HeV), which is the other member of the genus *Henipaviruses* (Wang et al. 2000). The six additional nucleotides in the genome of the NiV isolate from Bangladesh maps to the 5′ nontranslated region of the F gene and thus, do not change the coding sequence (Harcourt et al. 2005).

Phylogenetic analysis of NiV and HeV demonstrates the close genetic relationship between these two viruses as they form a unique cluster, and are distinct from any of the established genera within the *Paramyxovirinae* (Fig. 2a) (Harcourt et al. 2000, 2001, 2005; Chan et al. 2001; AbuBakar et al. 2004). However, with more sequence data available from NiV isolates, a higher level of diversity among the NiV isolates is evident in the phylogenetic tree (Fig. 2b) (Rahman et al. 2010). Nonetheless, the NiV isolates from pigs and humans in Malaysia are closely related and nearly identical, forming a monophyletic clade together with the more recently characterized NiV isolated from *Pteropus vampyrus* in Malaysia. The NiV from humans in Bangladesh exhibits the most genetic variability and is more distantly related to the NiV isolates from Malaysia.

4 Epidemiology

The bat is the natural reservoir host for NiV, which was then transmitted to pigs in the Malaysian outbreak (Yob et al. 2001). Direct contact with the secretions and excretions from infected pigs or repeated use of the same needles without sterilization resulted in the spread of NiV among pigs in the farms (Mohd Nor et al. 2000). The spread of NiV among pig farms in Malaysia was attributed to the movement of

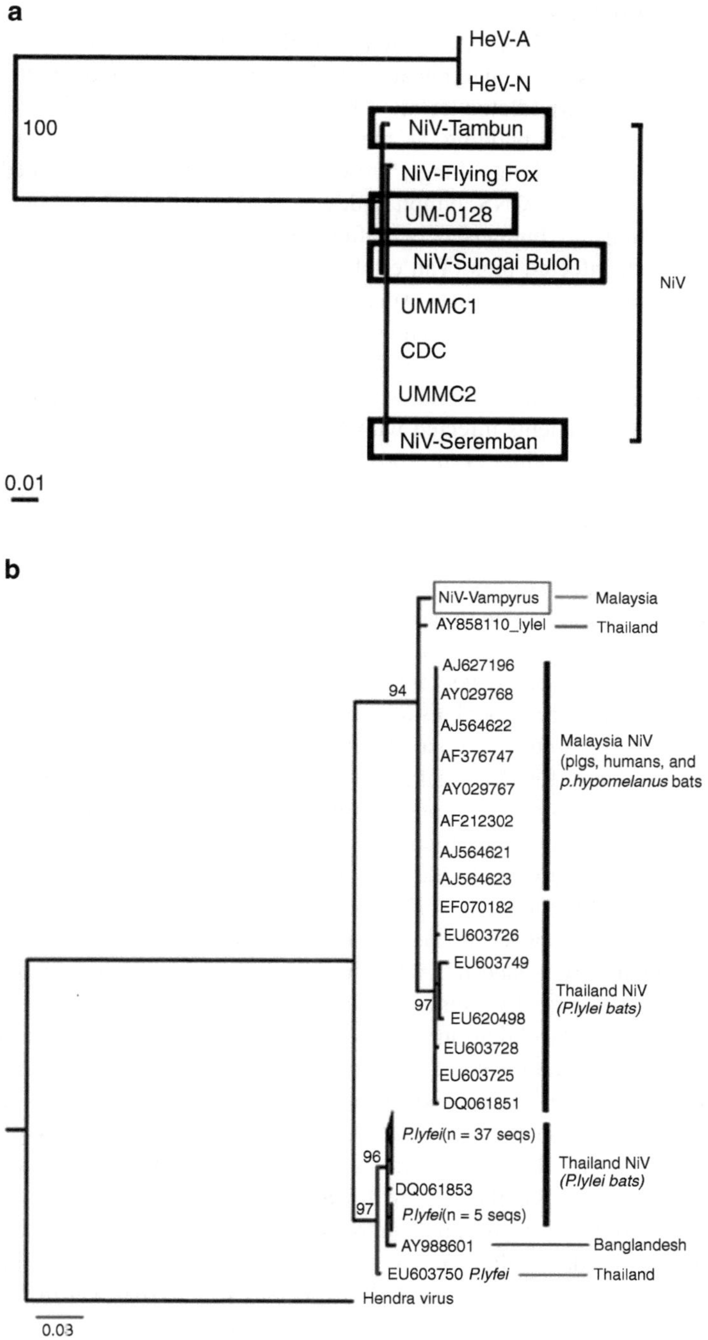

Fig. 2 (**a**) Phylogenetic tree illustrating the relationships of the Nipah virus isolates from Malaysia and Hendra virus and (**b**) phylogenetic tree illustrating a high level of sequence diversity among the Nipah viruses [from Rahman et al. (2010) with permission]. *Scale bar* indicates nucleotide substitutions per site

infected pigs from affected areas to other farms within and between the states. In infected pigs, NiV is primarily excreted through the oral–nasal routes in particular respiratory droplets, nasopharyngeal secretions, and urine (Hooper et al. 2001; Hyatt et al. 2001; Middleton et al. 2002; Weingartl et al. 2005). Viruses are occasionally observed budding at the plasma membrane within the alveolar epithelia in the infected pigs' lungs. Presence of the virus was also noted in other organs such as kidneys, CNS, lymph nodes, or spleen in the infected pigs.

NiV infections in humans occur after close contact with the secretions or excretions from infected pigs (Tan et al. 1999; Chew et al. 2000; Parashar et al. 2000). Infected pig farm workers were closely associated with pigs through pig-farming activities such as feeding and cleaning pigs, assisting in birth of piglets and pig breeding, as well as handling of dead pigs. Human-to-human transmission of NiV during the Malaysian outbreak was probably uncommon, although a study has demonstrated that patients infected with NiV excrete the virus in their respiratory secretions and urine (Chua et al. 2001). There was a report of a nurse who had previously cared for NiV encephalitis patients and subsequently seroconverted (Tan and Tan 2001). Brain magnetic resonance (MR) imaging showed multiple, discrete high signal lesions typical of infection with NiV. Nonetheless, the risk of transmission of NiV to healthcare workers was generally thought to be low (Mounts et al. 2001), and this could be due to early institution of strict barrier nursing and a low virus load in the respiratory secretions and urine from humans.

In the Bangladesh outbreaks, multiple mechanisms were probably involved in the transmission of virus from bats to humans. Contact with sick cows and herds of pigs was speculated to be the mode of transmission in the 2001 and 2003 outbreaks (International Centre for Diarrheal Diseases Research Bangladesh 2004; Hsu et al. 2004), whereas climbing trees, presumably resulting in direct contact with NiV-infected bat secretions, was found to be significant in the 2004 outbreak in Goalanda (Montgomery et al. 2008). Human-to-human transmission of NiV was apparent in the 2004 outbreak in Faridpur (Gurley et al. 2007a), as well as in the 2001 and 2003 outbreaks in Bangladesh (International Centre for Diarrheal Diseases Research Bangladesh 2004; Hsu et al. 2004), where index patients transmitted the disease to family members, relatives and friends who cared or visited the NiV-infected patients. Additionally, the transmission of NiV in the healthcare setting in the 2001 outbreak in Siliguri, India also suggested human-to-human spread (Chadha et al. 2006). However, NiV spread in hospitals to healthcare workers in Bangladesh is rarely seen (Gurley et al. 2007b). This may partly be explained by the healthcare workers having less direct physical contact with the infected patients in Bangladesh, with care being largely provided by family members and friends. In the 2005 outbreak in Tangail, Bangladesh, drinking of contaminated raw date palm sap was thought to be source of the development of NiV encephalitis (Luby et al. 2006, 2009; Khan et al. 2010). During the winter period, between the months of December and March, a pot or container is placed at the top of date palm tree to collect the juice overnight. These pots are frequently visited by bats. During collection of the juice in the morning, it is common to find a bat or its excrement floating in the juice. In addition, the juice is preferred by many to be consumed raw. Consumption of contaminated raw date palm juice may thus be an important explanation for the clustering of the outbreaks between January and April in Bangladesh. The other explanation is increased NiV shedding in bats during pregnancy at this time of the year.

On the other hand, HeV infection was first described in September 1994, before the discovery of NiV (Selvey et al. 1995). HeV was discovered and identified to be the causative agent responsible for the death of a horse trainer and 13 horses in Hendra, Brisbane, Queensland, Australia. A second outbreak that was retrospectively diagnosed occurred north of Brisbane in August 1994 and resulted in the death of a horse owner and two horses (Hooper et al. 1996). To date, more than 30 events of HeV infection have occurred sporadically, all involving infection of horses and with at least seven events reported to have spread to humans. The case fatality rate in humans is 60 %. Similar to NiV, the source of HeV is from *Pteropus* spp. bats and the transmission of the virus to horses may have occurred through exposure to bats secretions and excretions (Halpin et al. 2000). The spread of HeV among horses and from horses to humans is likely to be direct contact with respiratory secretions, and other contaminated tissue and fluids or from aerosol exposure (Playford et al. 2010).

5 Nipah Virus Reservoir

The natural reservoir host for NiV has been determined to be *Pteropus* spp. bats as supported by serological and virological evidence. Reactive and/or neutralizing antibodies to NiV were detected in samples collected from different *Pteropus* spp. bat populations found across a wide geographical area from Malaysia, to Cambodia and Thailand in the north, India and Madagascar in the west, and Papua New Guinea to the east (Yob et al. 2001; Olson et al. 2002; Reynes et al. 2005; Wacharapluesadee et al. 2005; Wacharapluesadee and Hemachudha 2007; Iehlé et al. 2007; Epstein et al. 2008; Breed et al. 2010). Besides *Pteropus* spp. bats, the presence of antibodies to henipaviruses was also found in nonpteropid bats in Madagascar, Ghana, and China (Iehlé et al. 2007; Hayman et al. 2008; Li et al. 2008). Isolation of NiV from specimens obtained from bats, particularly pooled urine samples collected in Tioman Island, Malaysia (Chua et al. 2002) and Cambodia (Reynes et al. 2005) further supported the likelihood of *Pteropus* spp. bats as the reservoir host for the virus. More recent evidence indicates that NiV can recrudesce in previously infected bats without causing disease in the reservoir host (Rahman et al. 2010; Halpin et al. 2011). The reactivation of the virus is suggested to be triggered by various ecological, physiological, and immunological factors. Shedding of the virus upon its reactivation is likely important and necessary for the viral maintenance in bat colonies. Although the mechanism of virus transmission from these bats to other animals still remains unclear, susceptible animals are likely to have contracted NiV as a result of coming into contact with contaminated secretions and excretions of bats that were shedding the infectious virus at high amounts.

6 Clinical Manifestations

The mean age of patients having NiV encephalitis during the Malaysian outbreak was 38 ± 14 years and the occupational activities of the patients were mainly related to pig-farming or the pig-farming industry, where they regularly had direct contact or were in close proximity to pigs (Chong et al. 2002). The incubation period for NiV infection appears to range from a few days to 2 weeks (Goh et al. 2000; Chong et al. 2002). The clinical manifestations are that of acute encephalitis with fever, headache, vomiting, and reduced level of consciousness (Chua et al. 1999; Lee et al. 1999; Paton et al. 1999; Goh et al. 2000). More distinctive clinical features are areflexia, hypotonia, and prominent autonomic changes such as tachycardia and hypertension. It was reported that at least 32 % of patients had segmental myoclonus, which is characterized by focal, rhythmic jerking of muscles, commonly involving the diaphragm and anterior muscles of the neck (Goh et al. 2000). In another report, respiratory tract involvement with cough was seen at presentation in 14 % of patients (Chong et al. 2002). In Singapore, 3 of 11 patients presented with atypical pneumonia with abnormal chest radiographs and one later developed encephalitis (Paton et al. 1999). However, none of the Malaysian patients had primary lung disease (Goh et al. 2000). On pathological examinations, other organs, including the lung, heart and kidney, were also affected as a result of the NiV infection (Wong et al. 2002). Viral antigens were detected in these organs, although to a lesser extent as compared to the viral antigen staining of the brain. It was clearly observed that the brain was the most severely affected organ, and strongly indicates that NiV infection is a predominantly neurological disease.

There were some patients who had nonencephalitic infection but presented with systemic symptoms and seroconverted. Of the 94 patients with NiV infection admitted to the University of Malaya Medical Center, Malaysia, 91 had acute encephalitis and three had nonencephalitic infection (Goh et al. 2000).

The overall mortality of acute NiV encephalitis was 40 % (Chua et al. 2000a). The interval between the onset of symptoms to death was 10 ± 6.8 days (Chong et al. 2002). Severe brain stem involvement appeared to be associated with poor prognosis, and included findings such as tachycardia, hypertension, high fever (Chong et al. 2002), and abnormal Doll's-eye reflexes (Goh et al. 2000). Concomitant diabetes mellitus was also related to mortality, probably due to immune dysfunction (Chong et al. 2001). Additionally, presence of infectious virus in the cerebrospinal fluid indicated that high viral replication was associated with high mortality (Chua et al. 2000b). Although the mortality was high, the outcome of patients who survived was good and at least 40 % of the patients recovered fully. Some of the surviving patients have mild residual neurological deficits such as cerebellar signs, monoparesis, and cranial nerve palsies, whereas a small percentage of the patients with higher mental function deficits have marked disabilities and are dependent on caregivers (Chong et al. 2002).

The demographic features of NiV-infected patients in India and Bangladesh were different from the patients in Malaysia. The age of patients ranged from 12 to

Table 1 Difference in epidemiologic and clinical features of Nipah virus encephalitis outbreaks between Malaysia–Singapore and Bangladesh–India (Chong et al. 2008)

	Malaysia–Singapore	Bangladesh–India
Epidemiology		
Outbreaks	One major outbreak in 1998/1999 involving more than 300 patients	Almost yearly outbreaks from 2001, in winter and spring, involving 4–66 patients
Age and occupation	Mainly adults pig farm workers	Adults and children; involve healthcare workers in Indian outbreak
Spread	Bats-to-pigs, pigs-to-human	Direct bats-to-human infection by consumption of date palm juice and fruits contaminated by bats; Bats-to-cows? Bats-to-pigs?
	Human-to-human occasional	Human-to-human spread important
Clinical features		
Respiratory involvement	14–29 %; 2 out of 11 patients in Singapore present with pneumonia without encephalitis	Cough (62 %), respiratory difficulty (69 %); chest radiographs with acute respiratory distress syndrome in some patients
Encephalitis	Segmental myoclonus seen in 32–54 %	Segmental myoclonus not reported
MR imaging	Disseminated small high-signal intensity lesion hall mark of MR imaging	Confluent high-signal brain lesion in limited MR imaging
Relapsed and late-onset encephalitis	About 10 %	Delayed onset neurological abnormalities in 4 out of 22 patients in a follow-up study
Mortality	32–41 %	73 %

70 years and there was no occupational linkage (Harit et al. 2006). On the other hand, the clinical features of NiV encephalitis outbreaks in India and Bangladesh were similar to the Malaysian outbreak, which is acute fatal encephalitis with a short incubation period, fever, coma, and seizures (Harit et al. 2006; Hossain et al. 2008). However, segmented myoclonus was not seen in the Indian and Bangladesh patients, although it was a prominent feature among Malaysian patients. Chest radiographic findings of five Bangladesh patients showed prominent lung changes consistent with acute respiratory distress syndrome, and this manifestation was again lacking in the Malaysian patients, and provides a possible explanation for the strong human-to-human spread of the disease in Bangladesh (Gurley et al. 2007a). Overall, respiratory difficulty was seen in at least 60 % of cases of NiV infection in Bangladesh (Hossain et al. 2008). Several other differences observed between the Malaysian outbreak and those that occurred in India and/or Bangladesh (Table 1) are the higher mortality in the Bangladesh and Indian outbreaks—70 % as compared to 40 % in the Malaysian outbreak (Chua et al. 2000a; Chadha et al. 2006; Chong et al. 2008; Hossain et al. 2008) and the occurrence among younger

patients with median age of 12 in Naogoan and Rajbari in Bangladesh (Hossain et al. 2008).

The clinical manifestations of HeV are that of influenza-like illness with fever, headache and myalgia, which can progress to severe respiratory infection or encephalitis with symptoms such as confusion and ataxia (O'Sullivan et al. 1997; Playford et al. 2010). Of the seven patients infected with HeV, three died from encephalitis within 2 weeks of the infection and the fourth died 13 months after the infection. The remaining three patients suffered mild illnesses and have recovered. HeV antigen and typical viral nucleocapsids were demonstrated in the brain tissues on postmortem examination (Hooper et al. 1996; Hyatt and Selleck 1996; O'Sullivan et al. 1997).

7 Pathology

Pathological studies of fatal acute NiV encephalitis on Malaysian patients were studied by light and electron microscopy, and immunohistochemistry (Chua et al. 1999, 2000a; Wong et al. 2002). Macroscopically, the features were, in general, rather nonspecific, although occasionally small lesions suggestive of necrosis were observed in the brain.

The blood vessels appeared to be one of the early major targets of NiV infection. Medium-sized to small blood vessels (arteries and veins) in major organs, including the brain, lung, heart and kidney, were susceptible to infection. The blood vessels in the brain were the most severely affected. The earliest lesion seemed to be the formation of multinucleated syncytium in the endothelium (Fig. 3a). More commonly, vascular damage took the form of endothelial ulceration with varying degrees of inflammation and fibrinoid necrosis. The inflammatory cell infiltration consisted of neutrophils, macrophages, and other cells within the vascular wall (Fig. 3b). Vasculitis was frequently associated with thrombosis and vascular occlusion. However, vasculitis was found in small blood vessels but was not found in medium-sized vessels or large arteries, and no vasculitis was seen in relapse NiV encephalitis case. Presence of viral antigen within the endothelium, endothelial syncytium, and tunica media was noted with immunohistochemical staining. These findings confirmed that NiV is able to directly infect blood vessels.

The brain is particularly susceptible to NiV infection and is the organ most severely affected. In the brain widespread vascular lesions were observed involving the meninges and the gray and white matter. Vasculitic vessels were seen with or without thrombosis, and were often adjacent to small areas of necrosis and ischemia (Fig. 4a, b). The inflammatory cells observed included neutrophils, macrophages, lymphocytes, and reactive microglia. Surviving neurons in these areas showed eosinophilic cytoplasmic or, less frequently, nuclear paramyxoviral-type inclusions. Although nuclear inclusions were less commonly found, they were found in 63 % of the cases examined and they were noted to occupy most of the nuclei except for a thin rim of chromatin that was pushed to the periphery. Immunohistochemical staining confirmed

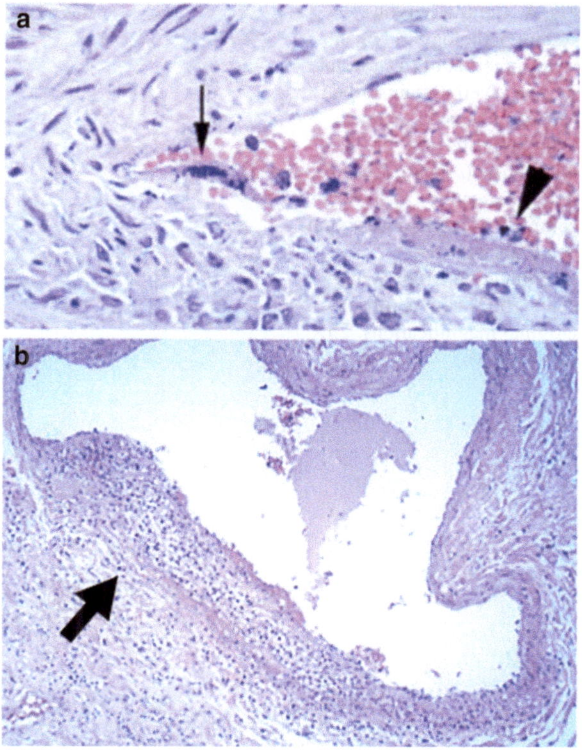

Fig. 3 Vascular pathology in Nipah virus infection of (**a**) pulmonary vessel showing endothelial syncytium (*arrow*) and ulceration (*arrow head*). (**b**) Lung artery showing vasculitis with inflammatory infiltrate (*large arrow*) [from Wong et al. (2002) with permission]. Original magnifications: 25× (**a**); 100× (**b**)

that these viral inclusions contained NiV antigens. However, no mature viral particles were observed. Parenchymal inflammation consisting of perivascular cuffing and neuronophagia could be seen, particularly in patients who died after more than a week of illness. In relapsing NiV encephalitis cases, viral inclusions were more prominent and abundant and the parenchymal lesions were also larger.

Outside of the central nervous system, except perhaps in the liver, vasculitis and parenchymal lesions are found in all the major organs, including lungs, kidneys and heart. After the central nervous system, the most severely affected organ is the lung. Focal fibrinoid necrosis involving several adjacent alveoli that is associated with vasculitis has been described. In addition, there is inflammation associated with multinucleated giant cells in the alveolar spaces. Pathologic changes in the kidney are characterized mainly by vasculitis and glomerular lesions consisting of microthrombi, scattered inflammatory cells, and occasionally multinucleated syncytia. In the heart, vasculitis is found in most cases. In the lymphoid organs, such as lymph nodes and spleen, multinucleated syncytia are observed. Viral antigen is detected in affected tissues from various organs.

Overall, symptomatic NiV patients usually present with an acute encephalitic syndrome because the central nervous system is severely affected as a result of the combination of ischemia and microinfarction, as well as direct neuronal infection

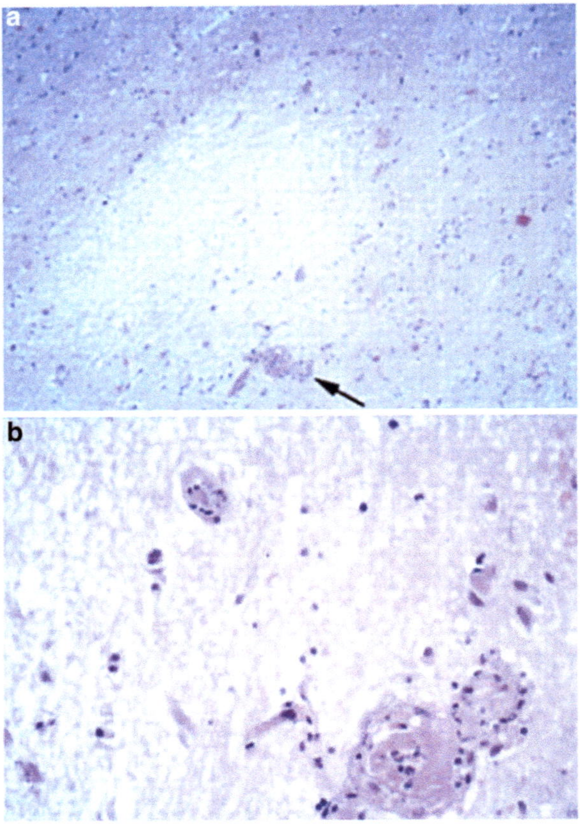

Fig. 4 Central nervous system pathology in Nipah virus infection of (**a**) arteriole showing well-circumscribed, necrotic plaque associated with vasculitis and thrombosis (*arrow*). (**b**) Arteriole seen in (**a**) showing complete arteriole obstruction at a higher-power magnification [from Wong et al. (2002) with permission]. Original magnifications: 25× (**a**); 100× (**b**)

by the virus. Direct neuronal infection leading to neurologic manifestations have been confirmed using immunohistochemical staining demonstrating the presence of viral antigens in neurons, the presentation of distinctive neurological signs such as segmental myoclonus (Goh et al. 2000), and the association between high mortality and the presence of virus in the cerebrospinal fluid (Chua et al. 2000b). No patients have been subjected to autopsy in the Bangladesh and India outbreaks.

8 Relapse and Late-Onset of Nipah Virus Encephalitis

Relapse of encephalitis has been reported to occur in a small number of patients with NiV encephalitis in the Malaysian outbreak (Tan et al. 2002; Chong and Tan 2003). These patients unexpectedly suffered a second or even a third neurologic episode following what appeared to be complete recovery. Asymptomatic or mild nonencephalitic patients also developed similar neurologic episodes for the first time several months later, indicating the occurrence of late-onset NiV encephalitis. The prevalence of relapsed encephalitis among the 160 patients who survived acute NiV encephalitis is approximately 9 %, while the prevalence is 5 % in 89 asymptomatic patients. Clinical, radiological and pathological findings suggest that relapse and late-onset NiV encephalitis are the same disease process, and are distinct from acute NiV encephalitis. The common clinical features are fever, headache, seizures, and focal neurological signs. There is an 18 % mortality rate. Confluent cortical lesions are seen in the brain MR imaging and there is a high degree of irreversible neuronal damage. Necropsy findings demonstrate the presence of viral antigen in the brain, thus suggesting that the relapse and late-onset of NiV encephalitis were the result of a recurrent infection rather than postinfectious demyelination described following measles and other viral infections. However, no virus was isolated from these cases.

Relapse of NiV encephalitis is similar to HeV infection, in which a single human case of HeV virus encephalitis developed fatal meningoencephalitis 1 year after recovery from the initial illness (O'Sullivan et al. 1997). HeV antigens were demonstrated in the brain tissue, but viral isolation was unsuccessful.

In NiV encephalitis in Bangladesh, a study reported at least 4 of 22 (18 %) survivors of NiV encephalitis developed new neurological symptoms months to years after the acute illness. Three of these patients presented with extraocular motor palsies, and one had cervical dystonia (Sejvar et al. 2007).

9 Laboratory Diagnosis

The cerebrospinal fluid examination of NiV encephalitis patients in Malaysia were abnormal in 75 % of patients, consisting of elevated protein concentrations or elevated white blood cell counts, whereas glucose concentrations were within normal limits (Goh et al. 2000; Chong et al. 2002). However, these features are nonspecific, and may be found in many patients with other etiologies of viral encephalitis.

IgM and IgG antibody detection using enzyme-linked immunosorbent assay (ELISA) testing, followed by confirmatory testing using serum neutralization tests (Daniels et al. 2001) were critical for the diagnosis of NiV infection. Initially, the antibody test utilized was an IgM-capture ELISA, while IgG antibodies were detected by an indirect IgG ELISA assay (Chua et al. 1999). In these tests, the rate of positive IgM was 60–71 % by day 4 and 100 % by day 12 of illness; for IgG, it was 7–29 % by day 1 and 100 % by day 25–26 of illness (Ramasundrum et al. 2000). The IgG was persistently positive whereas the IgM became negative after 10 years of illness (Siva et al. 2009). Subsequently, several other rapid, more sensitive and much safer tests were developed for diagnosis of NiV infection such as a rapid, immune plaque assay (Crameri et al. 2002), a solid-phase blocking ELISA (Kashiwazaki et al. 2004), protein-based ELISA (Yu et al. 2006), a comparative indirect ELISA (Chen et al. 2006), and molecular testing using polymerase chain reaction method (Guillaume et al. 2004b; Wacharapluesadee and

Hemachudha 2007). Apart from these serological and nucleic acid amplification-based methods, other methods including cell culture (Chua et al. 2000a), immunohistochemical staining (Tanimura et al. 2004) and electron microscopy (Chow et al. 2000) are also used for the diagnosis of NiV infection.

Brain MR imaging has proven to be a useful diagnostic aid for patients with acute NiV encephalitis (Sarji et al. 2000). Brain MR imaging of an acute NiV encephalitis patient shows multiple, disseminated, small discrete hyperintense lesions best seen in the fluid-attenuated inversion recovery (FLAIR) sequences, with abnormalities in the subcortical and deep white matter, and occasionally in the cerebral cortex. Lesions with a size of about 2–7 mm in diameter are attributed to the widespread microinfarctions noted in postmortem tissues. At least 16 % of asymptomatic patients with NiV infection are observed to have similar changes, demonstrating that subclinical cerebral involvement is not uncommon (Tan et al. 2000). The most common electroencephalographic abnormality is continuous diffuse, symmetrical slowing with or without focal discharges. The degree of slowing correlates with the severity of disease. Independent bitemporal periodic complexes are common among those patients who are deeply comatose, and are associated with 100 % mortality (Chew et al. 1999).

Brain MR imaging performed on three acute NiV encephalitis patients in Bangladesh showed confluent high signal lesions involving both gray and white matter (Quddus et al. 2004). This is different from the disseminated multiple discrete high-signal intensity lesions observed in the Malaysian patients, which was attributed to the widespread microinfarctions from underlying vasculitis of cerebral small vessels (Chew et al. 1999; Sarji et al. 2000; Tan et al. 2000). The difference in pathology and neuroimaging manifestations among Bangladesh patients as compared to that from Malaysia could be associated with the variation in the viral genomic sequence as the NiV isolate from Bangladesh shares 92 % homology to the Malaysian NiV isolates (Quddus et al. 2004; Harcourt et al. 2005).

10 Treatment

Ribavirin, an antiviral agent, which has a broad spectrum of activity against some DNA and RNA viruses, was utilized empirically in patients with NiV encephalitis (Chong et al. 2001). In an open-label trial of 140 patients, with 54 patients as controls (i.e., patients who refused treatment or were otherwise not given the drug), there were 45 deaths (32 %) in the ribavirin group versus 29 deaths (54 %) in the control group. This represented a 36 % reduction in mortality. Although this trial was disadvantaged by the use of historical controls, the study suggests that ribavirin may be useful in the treatment of acute NiV encephalitis. Moreover, there were no apparent serious adverse effects as a result of the administration of the drug to the patients. The survival benefit associated with the use of ribavirin given early in clinical illness was also suggested in a HeV-infected patient (Playford et al. 2010). More recently, a study reported that ribavirin showed strong antiviral activity against NiV replication in vitro, but not in vivo in hamster or guinea-pig models (Georges-Courbot et al. 2006). The difference in protective efficacy may have been due to several factors including differences in the amount of virus present that were higher in the animal study as compared to the infection in humans and the drug may be differently metabolized.

11 Prevention

Following mass culling of pigs infected and/or suspected to be infected with NiV during the outbreak in Malaysia, no cases of acute human infection have been further reported (Centers for Disease Control and Prevention 1999b). The last human patient that was infected and succumbed to acute NiV infection was reported on May 27, 1999. As the natural reservoir host for the virus is the pteropid bat, there is an ever-present risk for the NiV to reemerge in peridomestic or domestic animals, or even in humans as had been the case in Bangladesh. Thus, there was an early interest in developing vaccines as an anti-NiV strategy for livestock to prevent the reemergence of the virus. To date, several studies have been reported demonstrating the protective effect of the candidate vaccines employing the NiV glycoproteins to induce immune response in animal models (Guillaume et al. 2004a; Weingartl et al. 2006; Mungall et al. 2006). However, a good monitoring and surveillance program, particularly for the susceptible animal population including the natural reservoir host for NiV, may be more cost-effective in order to prevent the reemergence of the virus.

In the Bangladesh and Indian outbreaks, public education to avoid eating of fruits half-eaten by bats and drinking of raw date-palm juice as well as simple bamboo skirts to preclude access of bats to date palm sap can help to prevent spread from bats to humans (Khan et al. 2010). Most of the infection in Bangladesh occurs in small outbreaks, and sometimes associated with human-to-human spread. Education aiming at increased awareness of healthcare professionals in the region and close monitoring of the encephalitis outbreaks is thus crucial; barrier nursing should also be part of the protocol in the management plan of the affected patients.

References

AbuBakar S, Chang LY, MohdAli AR, Sharifah SH, Yusoff K, Zamrod Z (2004) Isolation and molecular identification of Nipah virus strains from pigs. Emerg Infect Dis 10:2228–2230

Arankalle VA, Bandyopadhyay BT, Ramdasi AY, Jadi R, Patil DR, Rahman M, Majumdar M, Banerjee PS, Hati AK, Goswami RP, Neogi DK, Mishra AC (2011) Genomic characterization of Nipah virus, West Bengal, India. Emerg Infect Dis 17:907–909

Breed AC, Yu M, Barr JA, Crameri G, Thalmann CM, Fa Wang L (2010) Prevalence of henipavirus and rubulavirus antibodies in pteropid bats, Papua New Guinea. Emerg Infect Dis 16:1997–1999

Centers for Disease Control and Prevention (1999a) Outbreak of Hendra-like virus—Malaysia and Singapore, 1998–1999. MMWR Morb Mortal Wkly Rep 48:265–269

Centers for Disease Control and Prevention (1999b) Update: outbreak of Nipah virus—Malaysia and Singapore. MMWR Morb Mortal Wkly Rep 48:335–337

Chadha MS, Comer JA, Lowe L, Rota PA, Rollin PE, Bellini WJ, Ksiazek TG, Mishra AC (2006) Nipah virus-associated encephalitis outbreak, Siliguri, India. Emerg Infect Dis 12:235–240

Chan YP, Chua KB, Koh CL, Lim ME, Lam SK (2001) Complete nucleotide sequences of Nipah virus isolates from Malaysia. J Gen Virol 82:2151–2155

Chang LY, MohdAli AR, Sharifah SH, AbuBakar S (2006a) Nipah virus RNA synthesis in cultured pig and human cells. J Med Virol 78:1105–1112

Chang LY, MohdAli AR, Sharifah SH, AbuBakar S (2006b) Quantitative estimation of Nipah virus replication kinetics in vitro. Virol J 3:47

Chen JM, Yu M, Morrissy C, Zhao YG, Meehan G, Sun YX, Wang QH, Zhang W, Wang LF, Wang ZL (2006) A comparative indirect ELISA for the detection of henipavirus antibodies based on a recombinant nucleocapsid protein expressed in *Escherichia coli*. J Virol Methods 136:273–276

Chew NK, Goh KJ, Tan CT, Ahmad Sarji S, Wong KT (1999) Electroencephalography in acute Nipah encephalitis. Neurol J Southeast Asia 4:45–51

Chew MH, Arguin PM, Shay DK, Goh KT, Rollin PE, Shieh WJ, Zaki SR, Rota PA, Ling AE, Ksiazek TG, Chew SK, Anderson LJ (2000) Risk factors for Nipah virus infection among abattoir workers in Singapore. Singapore J Infect Dis 181:1760–1763

Chong HT, Tan CT (2003) Relapsed and late-onset Nipah encephalitis, a report of three cases. Neurol Asia 8:109–112

Chong HT, Tan CT, Goh KJ, Chew NK, Kunjapan AR, Petharunam V, Thayaparan T (2001) Occupational exposure, age, diabetes mellitus and outcome of acute Nipah encephalitis. Neurol J Southeast Asia 6:7–11

Chong HT, Kunjapan SR, Thayaparan T, Tong J, Petharunam V, Jusoh MR, Tan CT (2002) Nipah encephalitis outbreak in Malaysia, clinical features in patients from Seremban. Can J Neurol Sci 5:61–67

Chong HT, Jahangir Hossain M, Tan CT (2008) Differences in epidemiologic and clinical features of Nipah virus encephalitis between the Malaysian and Bangladesh outbreaks. Neurol Asia 13:23–26

Chow VTK, Tambyah PA, Yeo WM, Phoon MC, Howe J (2000) Diagnosis of Nipah virus encephalitis by electron microscopy of cerebrospinal fluid. J Clin Virol 19:143–147

Chua KB, Goh KJ, Wong KT, Kamarulzaman A, Tan PSK, Ksiazek TG, Zaki SR, Paul G, Lam SK, Tan CT (1999) Fatal encephalitis due to Nipah virus among pig-farmers in Malaysia. Lancet 354:1257–1259

Chua KB, Bellini WJ, Rota PA, Harcourt BH, Tamin A, Lam SK, Ksiazek TG, Rollin PE, Zaki SR, Shieh W-J, Goldsmith CS, Gubler DJ, Roehrig JT, Eaton B, Gould AR, Olson J, Field H, Daniels P, Ling AE, Peters CJ, Anderson LJ, Mahy BWJ (2000a) Nipah virus: a recently emergent deadly paramyxovirus. Science 288:1432–1435

Chua KB, Lam SK, Tan CT, Hooi PS, Goh KJ, Chew NK, Tan KS, Kamarulzaman A, Wong KT (2000b) High mortality in Nipah encephalitis is associated with presence of virus in cerebrospinal fluid. Ann Neurol 48:802–805

Chua KB, Lam SK, Goh KJ, Hooi PS, Ksiazek TG, Kamarulzaman A, Olson J, Tan CT (2001) The presence of Nipah virus in respiratory secretions and urine of patients during an outbreak of Nipah virus encephalitis in Malaysia. J Infect 42:40–43

Chua KB, Koh CL, Hooi PS, Wee KF, Khong JH, Chua BH, Chan YP, Lim ME, Lam SK (2002) Isolation of Nipah virus from Malaysian Island flying-foxes. Microbes Infect 4:145–151

Crameri G, Wang LF, Morrissy C, White J, Eaton BT (2002) A rapid immune plaque assay for the detection of Hendra and Nipah viruses and anti-virus antibodies. J Virol Methods 99:41–51

Daniels P, Ksiazek T, Eaton BT (2001) Laboratory diagnosis of Nipah and Hendra virus infections. Microbes Infect 3:289–295

Epstein JH, Prakash V, Smith CS, Daszak P, McLaughlin AB, Meehan G, Field HE, Cunningham AA (2008) Henipavirus infection in fruit bats (*Pteropus giganteus*), India. Emerg Infect Dis 14:1309–1311

Georges-Courbot MC, Contamin H, Faure C, Loth P, Baize S, Leyssen P, Neyts J, Deubel V (2006) Poly(I)-poly ($C_{12}U$) but not ribavirin prevents death in a hamster model of Nipah virus infection. Antimicrob Agents Chemother 50:1768–1772

Goh KJ, Tan CT, Chew NK, Tan PSK, Kamarulzaman A, Sarji SA, Wong KT, Abdullah BJJ, Chua KB, Lam SK (2000) Clinical features of Nipah virus encephalitis among pig farmers in Malaysia. N Engl J Med 342:1229–1235

Goldsmith CS, Whistler T, Rollin PE, Ksiazek TG, Rota PA, Bellini WJ, Daszak P, Wong KT, Shieh WJ, Zaki SR (2003) Elucidation of Nipah virus morphogenesis and replication using ultrastructural and molecular approaches. Virus Res 92:89–98

Guillaume V, Contamin H, Loth P, Georges-Courbot MC, Lefeuvre A, Marianneau P, Chua KB, Lam SK, Buckland R, Deubel V, Wild TF (2004a) Nipah virus: vaccination and passive protection studies in a hamster model. J Virol 78:834–840

Guillaume V, Lefeuvre A, Faure C, Marianneau P, Buckland R, Lam SK, Wild F, Deubel V (2004b) Specific detection of Nipah virus using real-time RT-PCR (Taqman). J Virol Methods 120:229–237

Gurley ES, Montgomery JM, Hossain MJ, Bell M, Azad AK, Islam MR, Molla MA, Carroll DS, Ksiazek TG, Rota PA, Lowe L, Comer JA, Rollin P, Czub M, Grolla A, Feldmann H, Luby SP, Woodward JL, Breiman RF (2007a) Person-to-person transmission of Nipah virus in a Bangladeshi community. Emerg Infect Dis 13:1031–1037

Gurley ES, Montgomery JM, Hossain MJ, Islam MR, Molla MA, Shamsuzzaman SM, Akram K, Zaman K, Asgari N, Comer JA, Azad AK, Rollin PE, Ksiazek TG, Breiman RF (2007b) Risk of nosocomial transmission of Nipah virus in a Bangladesh hospital. Infect Control Hosp Epidemiol 28:740–742

Halpin K, Young PL, Field HE, Mackenzie JS (2000) Isolation of Hendra virus from pteropid bats: a natural reservoir of Hendra virus. J Gen Virol 81:1927–1932

Halpin K, Hyatt AD, Fogarty R, Middleton D, Bingham J, Epstein JH, Rahman SA, Hughes T, Smith C, Field HE, Daszak P, Henipavirus Ecology Research Group (2011) Pteropid bats are confirmed as the reservoir hosts of henipaviruses: a comprehensive experimental study of virus transmission. Am J Trop Med Hyg 85:946–951

Harcourt BH, Tamin A, Ksiazek TG, Rollin PE, Anderson LJ, Bellini WJ, Rota PA (2000) Molecular characterization of Nipah virus, a newly emergent paramyxovirus. Virology 271:334–349

Harcourt BH, Tamin A, Halpin K, Ksiazek TG, Rollin PE, Bellini WJ, Rota PA (2001) Molecular characterization of the polymerase gene and genomic termini of Nipah virus. Virology 287:192–201

Harcourt BH, Lowe L, Tamin A, Liu X, Bankamp B, Bowden N, Rollin PE, Comer JA, Ksiazek TG, Hossain MJ, Gurley ES, Breiman RF, Bellini WJ, Rota PA (2005) Genetic characterization of Nipah virus, Bangladesh, 2004. Emerg Infect Dis 11:1594–1597

Harit AK, Ichhpujani RL, Gupta S, Gill KS, Lal S, Ganguly NK, Agarwal SP (2006) Nipah/Hendra virus outbreak in Siliguri, West Bengal, India in 2001. Indian J Med Res 123:553–560

Hayman DT, Suu-Ire R, Breed AC, McEachern JA, Wang L, Wood JL, Cunningham AA (2008) Evidence of henipavirus infection in West African fruit bats. PLoS One 3:e2739

Hooper PT, Gould AR, Russell GM, Kattenbelt JA, Mitchell G (1996) The retrospective diagnosis of a second outbreak of equine morbillivirus infection. Aust Vet J 74:244–245

Hooper P, Zaki S, Daniels P, Middleton D (2001) Comparative pathology of the diseases caused by Hendra and Nipah viruses. Microbes Infect 3:315–322

Hossain MJ, Gurley ES, Montgomery JM, Bell M, Carroll DS, Hsu VP, Formenty P, Croisier A, Bertherat E, Faiz MA, Azad AK, Islam R, Molla MA, Ksiazek TG, Rota PA, Comer JA, Rollin PE, Luby SP, Breiman RF (2008) Clinical presentation of Nipah virus infection in Bangladesh. Clin Infect Dis 46:977–984

Hsu VP, Hossain MJ, Parashar UD, Ali MM, Ksiazek TG, Kuzmin I, Niezgoda M, Rupprecht C, Bresee J, Breiman RF (2004) Nipah virus encephalitis reemergence, Bangladesh. Emerg Infect Dis 10:2082–2087

Hyatt AD, Selleck PW (1996) Ultrastructure of equine morbillivirus. Virus Res 43:1–15
Hyatt AD, Zaki SR, Goldsmith CS, Wise TG, Hengstberger SG (2001) Ultrastructure of Hendra virus and Nipah virus within cultured cells and host animals. Microbes Infect 3:297–306
Iehlé C, Razafitrimo G, Razainirina J, Andriaholinirina N, Goodman SM, Faure C, Georges-Courbot MC, Rousset D, Reynes JM (2007) Henipavirus and Tioman virus antibodies in pteropid bats, Madagascar. Emerg Infect Dis 13:159–161
International Centre for Diarrheal Diseases Research Bangladesh (2003) Outbreaks of encephalitis due to Nipah/Hendra-like viruses, Western Bangladesh. Health Sci Bulletin 1:1–6
International Centre for Diarrheal Diseases Research Bangladesh (2004) Person-to-person transmission of Nipah virus during outbreak in Faridpur District, 2004. Health Sci Bulletin 2:5–9
International Centre for Diarrheal Diseases Research Bangladesh (2005) Nipah virus outbreak from date palm juice. Health Sci Bulletin 3:1–5
Kashiwazaki Y, Na YN, Tanimura N, Imada T (2004) A solid-phase blocking ELISA for detection of antibodies to Nipah virus. J Virol Methods 121:259–261
Khan MS, Hossain J, Gurley ES, Nahar N, Sultana R, Luby SP (2010) Use of infrared camera to understand bats' access to date palm sap: implications for preventing Nipah virus transmission. Ecohealth 7:517–525
Lee KE, Umapathi T, Tan CB, Tjia HT, Chua TS, Oh HM, Fock KM, Kurup A, Das A, Tan AK, Lee WL (1999) The neurological manifestations of Nipah virus encephalitis, a novel paramyxovirus. Ann Neurol 46:428–432
Li Y, Wang J, Hickey AC, Zhang Y, Li Y, Wu Y, Zhang H, Yuan J, Han Z, McEachern J, Broder CC, Wang LF, Shi Z (2008) Antibodies to Nipah or Nipah-like viruses in bats, China. Emerg Infect Dis 14:1974–1976
Luby SP, Rahman M, Hossain MJ, Blum LS, Husain MM, Gurley E, Khan R, Ahmed BN, Rahman S, Nahar N, Kenah E, Comer JA, Ksiazek TG (2006) Foodborne transmission of Nipah virus, Bangladesh. Emerg Infect Dis 12:1888–1894
Luby SP, Gurley ES, Hossain MJ (2009) Transmission of human infection with Nipah virus. Clin Infect Dis 49:1743–1748
Middleton DJ, Westbury HA, Morrissy CJ, van der Heide BM, Russell GM, Braun MA, Hyatt AD (2002) Experimental Nipah virus infection in pigs and cats. J Comp Path 126:124–136
Mohd Nor MN, Gan CH, Ong BL (2000) Nipah virus infection of pigs in peninsular Malaysia. Rev Sci Tech Off Int Epiz 19:160–165
Montgomery JM, Hossain MJ, Gurley E, Carroll DS, Croisier A, Bertherat E (2008) Risk factors for Nipah virus encephalitis in Bangladesh. Emerg Infect Dis 14:1526–1532
Mounts AW, Kaur H, Parashar UD, Ksiazek TG, Cannon D, Arokiasamy JT, Anderson LJ, Lye MS, Nipah Virus Nosocomial Study Group (2001) A cohort study of health care workers to assess nosocomial transmissibility of Nipah virus, Malaysia, 1999. J Infect Dis 183:810–813
Mungall BA, Middleton D, Crameri G, Bingham J, Halpin K, Russell G, Green D, McEachern J, Pritchard LI, Eaton BT, Wang LF, Bossart KN, Broder CC (2006) Feline model of acute Nipah virus infection and protection with a soluble glycoprotein-based subunit vaccine. J Virol 80:12293–12302
O'Sullivan JD, Allworth AM, Paterson DL, Snow TM, Boots R, Gleeson LJ, Gould AR, Hyatt AD, Bradfield J (1997) Fatal encephalitis due to novel paramyxovirus transmitted from horses. Lancet 349:93–95
Olson JG, Rupprecht C, Rollin PE, An US, Niezgoda M, Clemins T, Walston J, Ksiazek TG (2002) Antibodies to Nipah-like virus in bats (*Pteropus lylei*), Cambodia. Emerg Infect Dis 8:987–988
Parashar UD, Lye MS, Ong F, Mounts AW, Arif MT, Ksiazek TG, Kamaluddin MA, Mustafa AN, Kaur H, Othman G, Radzi HM, Kitsutani PT, Stockton PC, Arokiasamy J, Gary HE, Anderson LJ (2000) Case control study of risk factors for human infection with a new zoonotic Nipah virus during a 1998–1999 outbreak of Nipah virus encephalitis in Malaysia. J Infect Dis 181:1755–1759

Paton NI, Leo YS, Zaki SR, Auchus AP, Lee KE, Ling AE, Chew SK, Ang B, Rollin PE, Umapathi T, Sng I, Lee CC, Lim E, Ksiazek TG (1999) Outbreak of Nipah-virus infection among abattoir workers in Singapore. Lancet 354:1253–1256

Playford EG, McCall B, Smith G, Slinko V, Allen G, Smith I, Moore F, Taylor C, Kung Y-H, Field H (2010) Human Hendra virus encephalitis associated with equine outbreak, Australia, 2008. Emerg Infect Dis 16:219–223

Pro-MED mail (2012) Nipah encephalitis, human–Bangladesh: (Jipurhat). Pro-MED mail 20120125.1022056. http://www.promedmail.org Accessed 25 Jan 2012

Quddus R, Alam S, Majumdar MA, Anwar S, Khan MR, Zahid Mahmud Khan AKS, Arif SM, Alam R, Siddique FM, Tan CT, Faiz MA (2004) A report of 4 patients with Nipah encephalitis from Rajbari district, Bangladesh in the January 2004 outbreak. Neurol Asia 9:33–37

Rahman SA, Hassan SS, Olival KJ, Mohamed M, Chang LY, Hassan L, Saad NM, Shohaimi SA, Mamat ZC, Naim MS, Epstein JH, Suri AS, Field HE, Daszak P, Henipavirus Ecology Research Group (2010) Characterization of Nipah virus from naturally infected Pteropus vampyrus bats, Malaysia. Emerg Infect Dis 16:1990–1993

Ramasundrum V, Tan CT, Chua KB, Chong HT, Goh KJ, Chew NK, Tan KS, Thayaparan T, Kunjapan SR, Petharunam V, Loh YL, Ksiazek TG, Lam SK (2000) Kinetics of igM and IgG seroconversion in Nipah virus infection. Neurol J Southeast Asia 5:23–28

Reynes J-M, Counor D, Ong S, Faure C, Seng V, Molia S, Walston J, Georges-Courbot MC, Deubel V, Sarthou J-L (2005) Nipah virus in Lyle's flying foxes, Cambodia. Emerg Infect Dis 11:1042–1047

Sarji SA, Abdullah BJ, Goh KJ, Tan CT, Wong KT (2000) MR imaging features of Nipah encephalitis. AJR Am J Roentgenol 175:437–442

Sejvar JJ, Hossain J, Saha SK, Gurley ES, Banu S, Hamadani JD, Faiz MA, Siddiqui FM, Mohammad QD, Mollah AH, Uddin R, Alam R, Rahman R, Tan CT, Bellini W, Rota P, Breiman RF, Luby SP (2007) Long-term neurological and functional outcome in Nipah virus infection. Ann Neurol 62:235–242

Selvey LA, Wells RM, McCormack JG, Ansford AJ, Murray K, Rogers RJ, Lavercombe PS, Selleck P, Sheridan JW (1995) Infection of humans and horses by a newly described morbillivirus. Med J Aust 162:642–645

Siva SR, Chong HT, Tan CT (2009) Ten year clinical and serological outcome of Nipah virus infection. Neurol Asia 14:53–58

Tan CT, Tan KS (2001) Nosocomial transmissibility of Nipah virus. J Infect Dis 184:1367

Tan KS, Tan CT Goh KJ (1999) Epidemiological aspects of Nipah virus infection. Neurol J Southeast Asia 4:77–81

Tan KS, Ahmad Sarji S, Tan CT, Abdullah BJJ, Chong HT, Thayaparan T, Koh CN (2000) Patients with asymptomatic Nipah virus infection may have abnormal cerebral MR imaging. Neurol J Southeast Asia 5:69–73

Tan CT, Goh KJ, Wong KT, Sarji SA, Chua KB, Chew NK, Murugasu P, Loh YL, Chong HT, Tan KS, Thayaparan T, Kumar S, Jusoh MR (2002) Relapse and late-onset Nipah encephalitis. Ann Neurol 51:703–706

Tanimura N, Imada T, Kashiwazaki Y, Shahiruddin S, Sharifah SH, Aziz AJ (2004) Monoclonal antibody-based immunohistochemical diagnosis of Malaysian Nipah virus infection in pigs. J Comp Path 131:199–206

Wacharapluesadee S, Hemachudha T (2007) Duplex nested RT-PCR for detection of Nipah virus RNA from urine specimens of bats. J Virol Methods 141:97–101

Wacharapluesadee S, Lumlertdacha B, Boongird K, Wanghongsa S, Chanhome L, Rollin P, Stockton P, Rupprecht CE, Ksiazek TG, Hemachudha T (2005) Bat Nipah virus, Thailand. Emerg Infect Dis 11:1949–1951

Wang LF, Yu M, Hansson E, Pritchard LI, Shiell B, Michalski WP, Eaton BT (2000) The exceptionally large genome of Hendra virus: support for creation of a new genus within the family Paramyxoviridae. J Virol 74:9972–9979

Weingartl H, Czub S, Copps J, Berhane Y, Middleton D, Marszal P, Gren J, Smith G, Ganske S, Manning L, Czub M (2005) Invasion of the central nervous system in a porcine host by Nipah virus. J Virol 79:7528–7534

Weingartl H, Berhane Y, Caswell JL, Loosmore S, Audonnet J-C, Roth JA, Czub M (2006) Recombinant Nipah virus vaccines protect pigs against challenge. J Virol 80:7929–7938

Wong KT, Shieh WJ, Kumar S, Norain K, Abdullah W, Guarner J, Goldsmith CS, Chua KB, Lam SK, Tan CT, Goh KJ, Chong HT, Jusoh R, Rollin PE, Ksiazek TG, Zaki SR (2002) Nipah virus infection: pathology and pathogenesis of an emerging paramyxoviral zoonosis. Am J Pathol 161:2153–2167

Yob JM, Field H, Rashdi AM, Morrisy C, van der Heide B, Rota P, Adzhar A, White J, Daniels P, Jamaluddin A, Ksiazek TG (2001) Nipah virus infection in bats (order Chiroptera) in Peninsular Malaysia. Emerg Infect Dis 7:439–441

Yu F, Khairullah NS, Inoue S, Balasubramaniam V, Berendam SJ, The LK, Ibrahim NSW, Rahman SA, Hassan SS, Hasebe F, Sinniah M, Morita K (2006) Serodiagnosis using recombinant Nipah virus nucleocapsid protein expressed in *Escherichia coli*. J Clin Microbiol 44:3134–3138

Viral Hemorrhagic Fevers

Guey Chuen Perng and Marylou V. Solbrig

Abstract Viral hemorrhagic fever (VHF) agents are dominantly from four major RNA virus families: the Arenaviridae, Bunyaviridae, Filoviridae, and Flaviviridae. General and hemorrhagic signs and symptoms of these viral infections are well known, but their neurological complications and clinical variants with distinct neurologic syndromes are not so famous. Pathologic features of VHF disease: systemic rapid viral replication, abnormal immune and inflammatory responses culminating in hemorrhage, edema, coagulopathies, multiorgan failure, and molecular properties, such as those shared with encephalitic members of the same family, are now, being recognized to contribute to a wide spectrum of neurologic disorders. Efforts to characterize and understand the pathogenesis of these VHF-associated central nervous system disorders are underway.

Keywords Encephalitis • Hemorrhage • Inflammatory • Multiorgan failure • Neurotropic • Virus • Zoonosis

G.C. Perng
Department of Pathology and Laboratory Medicine, Emory Vaccine Center, Emory University, School of Medicine, Atlanta, GA, USA

Department of Microbiology and Immunology, National Cheng Kung University Medical College, Tainan, Taiwan
e-mail: gperng@emory.edu; pernggc@gmail.com

M.V. Solbrig (✉)
Department of Internal Medicine (Neurology), University of Manitoba, Winnipeg, MB, Canada

Department of Medical Microbiology, University of Manitoba, Winnipeg, MB, Canada
e-mail: msolbrig2@hsc.mb.ca

A.C. Jackson (ed.), *Viral Infections of the Human Nervous System*, Birkhäuser Advances in Infectious Diseases,
DOI 10.1007/978-3-0348-0425-7_14, © Springer Basel 2013

1 Introduction

Viral hemorrhagic fevers are febrile illnesses with abnormal vascular regulation and vascular damage. Although the combination of fever and hemorrhage can be caused by a number of human pathogens: viruses, rickettsiae, bacteria, protozoa, and fungi, the term hemorrhagic fever generally refers to a group of illnesses caused by several notorious RNA viruses. The group of VHF agents are dominantly from four major RNA virus families: the Arenavivridae, Bunyaviridae, Filoviridae, and Flaviviridae, with geographic distribution depending on their reservoir hosts (Table 1). Multiple mechanisms account for hemorrhage, and consistently involve leakage of the capillary walls due to damage to the endothelium lining of the vessels and blood clotting abnormalities (Table 1). Sequential events upon infection with VHF viruses are shown in Fig. 1. Often the neurologic aspects of these diseases are overshadowed by the hemorrhagic or shock syndrome, and, as in other critical illnesses, neurologic symptoms are most often attributed to cerebral edema, anoxia, hemorrhage, hyponatremia, hepatic or renal failure, microcapillary hemorrhage, and release of toxic substances. However, encephalitis can occur without hemorrhagic disease, as in Rift Valley fever (Alrajhi et al. 2004) or dengue (Borawake et al. 2011; Soares et al. 2011), and virus (infectious virus by culture), viral nucleic acids, viral antigen, and/or intrathecal antibody to members of each family of virus can be detected in the central nervous system (CNS) or eye (Table 2). Therefore, syndromic surveillance and pathogenesis studies are important for neurological medicine.

Now, as nearly half (49 %) of all emerging viruses are predicted to cause encephalitis or serious neurological clinical disease (Olival and Daszak 2005), we will look at some of the most dynamic groups of emerging viruses, the VHF, for the properties of virus and host that predispose to or drive CNS disease. Some agents share molecular or pathogenic features with another encephalitic family member. Others cause informative transitional illnesses recognizable as both encephalitic and hemorrhagic disease. Moreover for others, based on characteristic pantropism, associated blood–brain barrier breakdown, immunosuppression, and high viremia, it seems they should be causing CNS disease. Establishing the characteristics that make some agents both hemorrhagic and encephalitic viruses is challenging, given the rapid clinical course of the illnesses, remoteness of many epidemics, and lack of animal models that faithfully reproduce human disease. Nevertheless, it is worthwhile to try to learn from these agents: how to survive the systemic disease without permanent disability from the CNS disease.

2 Arenaviruses

Their name is from the Latin arenaceous "sandy," which refers to the EM appearance of virus with a granular or sandy surface. The granular appearance may represent presence of ribosomes in virus. Arenaviruses have a bisegmented negative-stranded RNA with a unique ambisense genomic organization. Viruses

Table 1 Summary of viral hemorrhagic fevers of humans

Family	Virus	Diseases	Localities	Natural hosts	Transmission	Key pathology and pathophysiology
Arenaviridae	Lassa	Lassa fever	Africa	Wild rodents	Direct contact; aerosolization of rodent excreta/body fluid; nosocomial	Bleeding diathesis—resulting from dysfunction of endothelial cells and platelets and hepatic injury. DIC almost never occurs. Thrombocytopenia is unremarkable. Leukopenia and hemoconcentration are common. Incubation: 5–21 days
	Junin	Argentine HF	Argentina	Wild rodents	Direct contact; aerosolization of rodent excreta/body fluid; nosocomial	Little inflammatory responses in affected organs. Small focal hemorrhages may occur on mucosal surfaces. Councilman-type bodies in the liver. DIC is not common. Leukopenia and hemoconcentration are common. Incubation: 7–16 days
	Machupo	Bolivian HF	Bolivia	Wild rodents	Direct contact; aerosolization of rodent excreta/body fluid; nosocomial	Same as Argentine HF
	Guanarito	Venezuelan HF	Venezuela	Cotton and cane rats	Direct contact; aerosolization of rodent excreta/body fluid; nosocomial	Same as Argentine HF
	Sabia	Brazilian HF	Brazil	Wild rodents	Direct contact; aerosolization of rodent excreta/body fluid; nosocomial	Same as Argentine HF
Bunyaviridae	Phlebovirus	Rift Valley fever	Africa	Wild and domestic animals	Mosquito; direct contact animal carcasses	Tropism in several cell types including hepatocytes. High viral found in lymphoid tissues. DIC is important in hemorrhagic phenomena. Interferon is critical to recovery. Incubation 2–12 days

(continued)

Table 1 (continued)

Family	Virus	Diseases	Localities	Natural hosts	Transmission	Key pathology and pathophysiology
	Nairovirus	Crimean-Congo HF	Africa, Asia, Europe	Hares, birds, ticks, domestic animals	Tick; direct contact with animals; nosocomial	Liver and spleen necrosis. Lymphocyte depletion in spleen. DIC is prominent Lymphopenia and marked thrombocytopenia are common. Incubation: 3–12 days
	Hantavirus	HF with renal syndrome	Asia, Europe, Eurasia	Wild rodents	Direct contact; aerosolization of rodent excreta/body fluids	Virus replicates in vascular endothelial cells. Renal dysfunction may result from Immune complex deposition. DIC occurs in varying degree. Shock and pulmonary edema may be due to capillary leak and extravasation of fluid Thrombocytopenia. Incubation: days to 4 weeks
	Puumala	HF with renal syndrome	Scandinavia and Finland, North Europe and Russia	Rodent	Aerosolization of body fluids	Immunological responses to viral antigens resulting in endothelial damage. Pulmonary leakage syndrome. Incubation: 2–4 weeks
Filoviridae	Ebola	Ebola HF	Africa	Unknown	Nosocomial	Focal necrosis with little inflammatory in affected organs. Councilman-like bodies in liver. Cytotoxic mediators from activated macrophages induce dysfunction of endothelial cells. Vasculopathic results in capillary leakage. Incubation: 2–21 days
	Marburg	Marburg HF	Africa	Unknown	Nosocomial	Same as Ebola. Viral antigen may impair host immune response. Virus may replicate in macrophages. Incubation 3–9 days

Flaviviridae	Yellow Fever	Yellow fever	Tropical Africa, South America	Monkeys	Mosquito	Virus can replicate in wide range of cells. Hepatocyte necrosis accompanied by significant inflammation in liver. In hemorrhagic cases, dysfunction of clotting factor synthesis and DIC may have a role. Leukopenia and thrombocytopenia are common. Incubation 3–6 days
	Dengue	Dengue fever	Asia, Africa, Pacific, Americas	Monkeys, human	Mosquito	Circulatory failure. Immune-complex deposition may account for enhancing viral entry to permissive cells. Lymphokines released from activated immune cells may contribute to vascular permeability and complement activation and abnormal coagulation cascades. Young children are more susceptible to severe illness. Thrombocytopenia and leucopenia are common. Incubation 2–15 days
	Kyasanur forest	Kyasanur forest disease	India	Monkeys, rodents, shrews	Tick	Leukopenia and thrombocytopenia are common. Distinct biphasic features compared to dengue. Incubation 3–8 days
	Omsk	Omsk HF	Russia	Rodents, muskrats	Tick	Similar to KFD

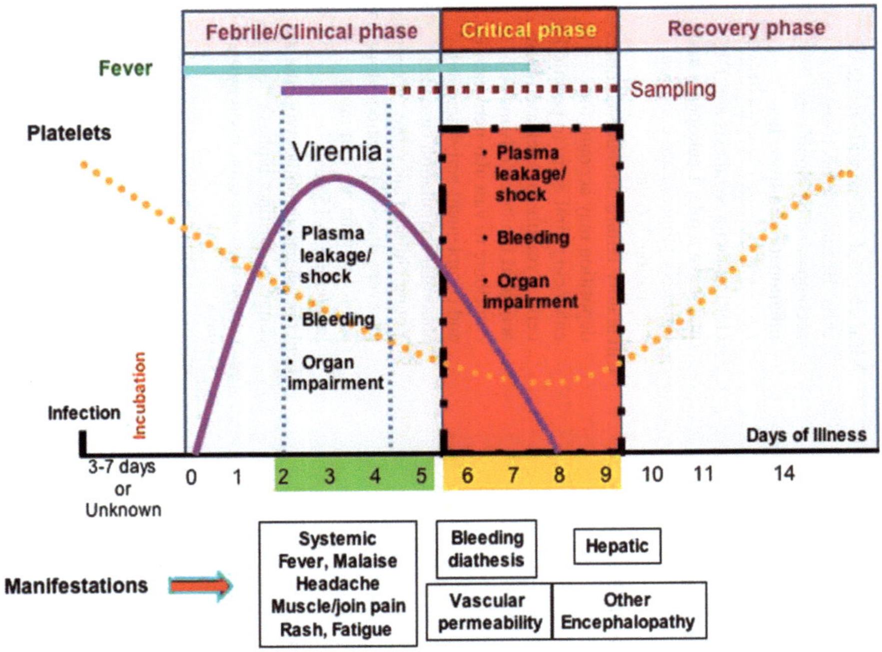

Fig. 1 Sequential events and clinical progression of a VHF. An incubation period, varying from 3 to 7 days (or unknown), dependent upon the route and the virus, is followed by fever and symptomatic viremia. In the case of dengue, illustrated here, fever lasts for about 5–7 days, and has probably decreased by the time the patient seeks medical attention. In dengue, severe clinical disease: DHF associated with DSS and low platelets, coincides with the defervescent stage. Recovery begins 10 days from start of illness, and is completed roughly within 2 weeks of illness. For other viruses, severe clinical presentations, bleeding, or plasma leakage associated with shock and organ failure can occur early in disease in the high viremic period. The timing of prodromal, clinical and critical phases are accelerated, expanded or contracted, depending on the virus. Laboratory evidence of DIC can accompany Hantaan, Rift Valley, DHF, but not arenaviruses. Jaundice is seen in yellow fever, CCHF, RVF, and filovirus hemorrhagic fever. Renal insufficiency, usually proportional to the degree of shock, occurs in HFRS

in the family contain two nucleic acid segments, one segment being slightly larger in the overall 5,000–7,400 base pair genome. Virions are enveloped, pleomorphic spherical structures with distinct, club-shaped surface projections (Buchmeier et al. 2007).

Arenaviruses are transmitted between rodents via urine, and humans are typically infected by contact with excreted virus. Humans are accidental, nonreservoir hosts. Highly evolved natural host–pathogen relationship with persistent infection has resulted in Arenaviruses causing chronic asymptomatic infections of rodents. Because of their rodent hosts, they are dispersed widely and found on all continents except Antarctica. In nature, there is extraordinary species specificity, usually with a single major reservoir host. Other mammals are easily infected by arenaviruses,

Table 2 Viral hemorrhagic fevers and the nervous system

Family; virus	Neurology	CNS virus detection
Arenaviridae		
Lassa	Seizures, dystonia, tremor, ataxia, hearing loss, encephalopathy, amnesia, psychosis, fatigue, depression, dementia	CSF (culture, PCR, virus-specific antibodies)
Lymphocytic choriomeningitis virus and "lymphocytic choriomeningitis virus-like" organ transplant arenavirus	Encephalopathy, seizures	
Lujo	Cerebral edema, coma	
Argentine HF	Ataxia, CN palsies	
Filoviridae		
Marburg	Uveitis, neuropathy, restless legs, psychosis, amnesia, depression, fatigue	Anterior eye chamber (culture)
Bunyaviridae		
Hantavirus		
Hantaan group		
Hemorrhagic fever with renal syndrome (Hantaan, Seoul)	Headache, dizziness/vertigo, confusion, aseptic meningitis, hypertensive encephalopathy, parenchymal or subarachnoid hemorrhage, pituitary apoplexy, reversible corpus callosum splenium lesion by MRI	Brain (viral antigen in neurons, glia by IHC) CSF (viral antigen and virus-specific antibodies)
Dobrava (HFRS)	Headache, blurred vision, vertigo, seizures, hemiparesis	CSF (virus-specific antibodies)
Sin Nombre group	Transverse myelitis, ADEM	
HPS (North America)		
Andes HPS (South America)	Headache, confusion, excitement, seizure, cortical and subcortical MRI abnormalities	CSF (virus-specific antibodies)
Puumala group		Brain (pituitary) (viral Ag by IHC)
Puumala	Encephalitis, seizures, Guillain–Barre, transient visual loss, cerebral hemorrhage, hypophyseal hemorrhage, panhypopituitarism, MRI-demonstrated white matter lesions, ADEM	CSF (PCR, virus-specific antibodies)

(continued)

Table 2 (continued)

Family; virus	Neurology	CNS virus detection
Phlebovirus	Encephalitis, disorientation, drowsiness, headache, seizures coma, hemiparesis, paraparesis retinal vasculitis	
Sandfly fever group		
Rift Valley		
Nairovirus		
CCHF group		
Crimean-Congo hemorrhagic fever	Encephalopathy	
Flaviviridae		
Dengue virus	Meningoencephalitis, encephalopathy, subdural hematoma, thalamic lesions, paraplegia, ADEM, ischemic stroke, Guillain–Barre, mononeuritis multiplex, brachial plexitis, myositis	CCSF (culture, PCR, virus-specific antibodies) brain (immunohistochemistry)
Kyasanur Forest disease, Omsk hemorrhagic fever, Alkhurma virus	Late-stage encephalitis, hearing loss, neuropsychiatric sequelae, encephalitis	

Neurologic complications independent of (not directly linked to) metabolic or hemorrhagic complications have been recognized for members of every viral hemorrhagic fever family (adapted from Solbrig and Perng 2011)

References: Lassa (Johnson et al. 1987; Cummins et al. 1990; Solbrig and McCormick 1991; Gunther et al. 2001), LCMV (Fischer et al. 2006; Palacios et al. 2008), Lujo (Briese et al. 2009), AHF (Maiztegui et al. 1979), Marburg (Gear et al. 1975; Bechtelsheimer et al. 1969; Jacob and Spalke 1971; Solbrig and Naviaux 1997), HFRS (Lukes 1954; Powell 1954; Kim 1965, Solbrig 1997, Baek et al. 2010), Dobrava (Cerar et al. 2007) Sin Nombre HPS (Huisa et al. 2009) Andes HPS (Talamonti et al. 2011) Puumala (Forslund et al. 1992; Gerding et al. 1995; Hautala et al. 2002, 2010, 2011; Mahonen et al. 2007), Rift Valley (Al-Hazmi et al. 2003, 2005; Alrajhi et al. 2004), CCHF (Schmaljohn and Nichol 2007), dengue (Kankirawatana et al. 2000; Solomon et al. 2000; Yeo et al. 2005; Liou et al. 2008, Borawake et al. 2011), KFD, OHF (Adhikari Prabhu et al. 1993; Monath 2001), Alkhurma virus (Madani et al. 2005, 2011) and the result may be death, persistent infection, or immunizing acute infection (Buchmeier et al. 2007).

Arenavirus infection of humans begins as initial replication at the site of infection, most often following aerosol deposition in the lung. Hilar lymph nodes, lung, and later, other parenchymal organs, are sites of virus growth. Macrophages are infected early, and many epithelial structures contain viral antigen and nucleic acids. The immune response may be protective, deleterious, or both (Buchmeier et al. 2007).

Lymphocytic choriomeningitic virus (LCMV), the first arenavirus isolated, was discovered during the St Louis encephalitis epidemic in 1933 (Armstrong 1941). To date, 5 Old World arenavirus species and 17 New World arenaviruses have been recognized by the International Committee for Taxonomy of Viruses, but the presence of additional arenaviruses in Africa is suspected. Some cause severe hemorrhagic fever; others are not pathogenic for or rarely infect humans.

Arenaviruses causing hemorrhagic syndromes are Lassa fever virus (Lassa fever) in West Africa, the recently discovered Lujo virus in South Africa (Briese et al. 2009; Paweska et al. 2009), agents of the South American hemorrhagic fevers including: Junin virus (Argentine hemorrhagic fever), Machupo (Bolivian hemorrhagic fever), Guanarito (Venezuelan hemorrhagic fever), Sabia virus in Brazil, and the White Water Arroyo virus in New Mexico and California in the USA.

LCMV, when transmitted by solid-organ transplantation, caused hemorrhagic syndromes in the United States (Fischer et al. 2006) and a new lymphocytic choriomeningitis virus-related arenavirus, Dandenong virus transmitted by solid-organ transplantation in Australia, caused a similar hemorrhagic syndrome (Palacios et al. 2008).

2.1 LCMV and LCMV-Like Viruses

LCMV infection causes febrile illness with accompanying viremia, followed by CNS disease, usually "aseptic meningitis." During CNS disease there is no viremia, but virus is still detectable in CSF and the pathogenesis is believed to be immune-mediated. Five to 34 % of patients hospitalized with documented LCMV have CNS complications beyond meningitis (Meyer et al. 1960). Severe neurologic cases with myelitis, encephalitis, bulbar weakness, hearing loss, arachnoiditis, ependymitis, and hydrocephalus have been reported.

In mice, LCMV's natural host, tissue-specific viral variants with different biological properties have been characterized. Neurotropism depends on retention of parental (wild-type) genotype of the viral glycoprotein (phenylalanine at residue 260). A single phenylalanine-to-leucine point mutation at the site is associated with macrophage tropism, viral persistence, and immune suppression (Villarete et al. 1994). Mutation in the virus surface glycoprotein (GP1:F260L) is responsible for the strong binding affinity to alpha-dystroglycan, the cellular receptor for both LCMV and Lassa virus that is expressed on dendritic cells of the immune system (Cao et al. 1998; Sullivan et al. 2011).

Our knowledge of arenavirus infection in man has been expanded by the recent cases of fatal LCMV infection in transplant patients. The cases, examples of a CNS agents causing hemorrhagic disease, reveal attenuated host resistance to infection to be the potential mechanism for the transition from CNS to hemorrhagic disease.

Both the prototypic lymphocytic choriomeningitis virus and a possibly more virulent "lymphocytic choriomeningitis-like virus," transmitted through organ transplantation, produced fatal febrile illnesses with hemorrhage and encephalopathy. However, the illnesses looked more like a classic VHF. In 2003 and 2005 in the United States, transplant recipients became ill with fever, thrombocytopenia, elevated transaminases, coagulopathy, encephalopathy, and seizures after receiving organs infected with lymphocytic choriomeningitis virus (Fischer et al. 2006). CNS abnormalities were seen but were eclipsed by the severe systemic illness that resembled Lassa fever; and then, within 2 years, a new transplant-associated arenavirus was reported in Australia. This new "lymphocytic choriomeningitis virus-like" arenavirus was associated with intraabdominal hematomas (in one patient) and encephalopathy (in all three patients) (Palacios et al. 2008). Considering these transplant cases, together with three cases of fatal LCMV following an attempted treatment of lymphoma by viral inoculation (Horton et al. 1971), researchers concluded the lack of effective T-cell responses permitted uncontrolled virus replication and prevented the development of the typical LCMV immunopathological meningitis. The immunosuppressed state created by antirejection therapy had opened new niches for infections caused by LCMV.

2.2 Lassa Fever

Lassa fever, a hemorrhagic fever of West Africa with a high mortality rate, causes over 300,000 infections per year with several thousand deaths. Lassa fever is a systemic infection with Lassa virus, producing fever, myalgia, prostration, nonicteric hepatitis, edema, bleeding, and shock. Hemorrhagic fever derives from direct virus-induced cellular necrosis in liver, and rarely disseminated intravascular coagulation (DIC). Fatal cases do not show overt destruction of vascular tissue. Instead, there is functional alteration of vascular endothelium, expression of cell adhesion molecules, coagulation abnormalities, and vasoactive mediators (Kunz 2009). In addition, 15–30 % of recovered patients have sensorineural hearing loss, representing the highest hearing loss morbidity caused by a virus (Cummins et al. 1990, 1992).

Just as hearing loss is generally a convalescent stage phenomena, the CNS syndromes of confusion, seizures, amnesia, movement disorders (dyskinesias and dystonia), tremor, or ataxia, and inflammatory CSF, occur in fully developed, clinically significant disease or follow the hemorrhagic fever (Solbrig and McCormick 1991). Lassa virus has been isolated from three out of three spinal fluids from patients with signs of meningitis (Johnson et al. 1987) and cultured from CSF, but not serum, in one patient with encephalopathy characterized by fever and a 30-min generalized seizure, then 2 days of disorientation and stupor (Gunther et al. 2001). Real-time PCR indicated higher virus load in CSF compared to serum and Lassa virus-specific IgM and IgG antibodies in CSF of that case (Gunther et al. 2001). Mechanisms used by Lassa to penetrate blood–brain barrier and cause neurologic disease are incompletely understood. CNS toxicity may be due to direct effects of the viral infection, indirect effects on vascular regulation, cell function, soluble mediators and immune regulation, or host genetics. Recovery from Lassa fever may be incomplete, as shown by prolonged or permanent sequelae of psychosis, chronic fatigue, mania, depression, and dementia (Solbrig and McCormick 1991; Solbrig 1993).

The similarities between late/convalescent Lassa syndromes: hearing loss, ataxia, encephalitis, and severe LCMV cases with hearing loss (Hirsch 1976; Ormay and Kovacs 1989), encephalitis or myelitis, raise the possibility of shared mechanisms of disease among arenaviruses, perhaps as immune dysregulation or suppression. The increased affinity of some LCMV strains and other Old World arenaviruses for α-dystroglycan (αDG), which has been identified as the cellular receptor for both LCMV and Lassa fever virus (Cao et al. 1998), allows the virus to infect and compromise dendritic cells, an essential component of viral immune response. Host genetic changes in biosynthesis of αDG are found in West African populations in Lassa fever-endemic regions, and may relate to clinical outcomes (Frazer et al. 2007; Sabeti et al. 2007; Sullivan et al. 2011).

2.3 *Argentine Hemorrhagic Fever*

Approximately 10 % of Argentine hemorrhagic fever (AHF) patients develop a late neurologic syndrome with fever, headache, cerebellar ataxia, and cranial nerve palsies after immune plasma treatment (Maiztegui et al. 1979). As in Lassa fever, hearing loss and encephalitis, the AHF syndrome appears time-locked to recovery phase or convalescence. Its occurrence after immune plasma treatment may represent antibody or immune-mediated CNS disease, but also raises the possibility of prior invasion of CNS by the virus. Perhaps, as in Lassa fever, certain patients achieve a level of host competence at handling infection at the expense of the nervous system.

3 Bunyaviruses

The Bunyaviridae family is the largest viral family, with over 350 viral species, and consists of five genera: the *Hantavirus* genus (including Hantaan, Dobrava, Seoul, Puumala, Sin Nombre, and Andes viruses); the *Nairovirus* genus (including Crimean-Congo hemorrhagic fever virus); the *Phlebovirus* genus (including Rift Valley fever virus and Toscana virus); the *Orthobunyavirus* genus (including encephalitic viruses of the California serogroup and Bunyamwera group, but not VHF) (Schmaljohn and Nichol 2007); and the *Tospovirus* genus (including plant viruses and has no human pathogens). Viruses in this family are negative-stranded RNA viruses with tripartite genomes consisting of large (L), medium (M), and small (S) RNA segments. The L segment encodes the RNA-dependent RNA polymerase, necessary for viral RNA replication and mRNA synthesis. The M segment encodes the viral glycoproteins, which project from the viral surface and aid the virus in attaching to and entering the host cell. The S segment encodes the nucleocapsid protein (N) (Schmaljohn and Nichol 2007).

Bunyaviruses are worldwide in distribution and infect a wide range of invertebrate and vertebrate hosts. Except for hantaviruses, which are spread from rodents, all of the other family members are vector-borne.

3.1 Crimean-Congo Hemorrhagic Fever

Crimean-Congo hemorrhagic fever (CCHF) was first described by a physician in Tajikistan in 1100 AD in a patient with hemorrhages (Hoogstraal 1979). CCHF is an acute, highly-contagious, and life-threatening disease caused by CCHF virus (Heyman et al. 2010; Vorou et al. 2007). CCHF virus circulates in nature in a tick-vertebrate-tick cycle. CCHF virus has the widest geographic range among all tick-borne viruses, being endemic in more than 30 countries in Euroasia and Africa (Leblebicioglu 2010; Vorou et al. 2007). Humans contract the disease results from the bite of infected ticks, mainly the genus *Hyalomma*, and from direct contact with blood of tissues of viremic patients or animals (Hoogstraal 1979; Vorou et al. 2007).

The incubation period of CCHF is 2–7 days. Typical clinical CCHF progresses rapidly with high fever, malaise, severe headache, and gastrointestinal symptoms. Hemorrhages are prominent later in disease, and case fatality rates range from 5 to 50 % (Centers for Disease Control 1988). Central nervous system symptoms: delirium, convulsion, cerebellar signs, coma, loss of hearing and loss of memory, and polyneuritis have been described (Hoogstraal 1979). Disseminated intravascular coagulation develops early and probably plays an important role in disease progression and encephalopathy.

3.2 Rift Valley Fever Virus

Rift Valley fever virus (RVFV), first described during an outbreak of enzootic hepatitis in ewes in 1931 in Kenya, has caused recurrent epizootics and epidemics in Africa, with loss of livestock and human illness. In 2000, outbreaks in Yemen and Saudi Arabia, showed the presence of RVFV infection outside Africa for the first time. RVFV is spread by numerous mosquito species and sandflies. Human infection is transmitted by infected mosquito bite, aerosols of blood, or other close contact with sick animals (Soldan and Gonzalez-Scarano 2005).

Introduced into the skin by arthropod bites, RVFV uses DC-SIGN (a C-type lectin pathogen receptor on the surface of dermal dendritic cells and some types of macrophages) to infect DCs and other DC-SIGN-expressing cells (Lozach et al. 2011). After infection, viral replication occurs in lymph nodes, resulting in primary viremia, spread via bloodstream, and infection of target organs. Major site of viral replication include liver, spleen, and brain of animals dying from encephalitis (Schmaljohn and Nichol 2007).

Systemic infection by RFVF is regulated by interferon and terminated by neutralizing antibody. In contrast, Lassa fever is resistant to interferon and LF is terminated by cellular immune effector mechanisms. Experimental animal studies show hemorrhagic fever derives from lytic virus–cell interaction and major effects by direct virus-induced cellular necrosis, particularly in liver, and DIC (Cosgriff et al. 1989; Morrill et al. 1991; Schmaljohn and Nichol 2007).

Human infection ranges from asymptomatic infection to severe disease with retinitis, hepatitis, renal failure, hemorrhagic fever, necrotic encephalitis, and death. Rift Valley fever can cause encephalitis and retinal vasculitis without overlap with the hemorrhagic fever syndrome (Alrajhi et al. 2004). RVFV encephalitis occurs in 1 % or less of infected humans (Hollidge et al. 2010) with fatal meningoencephalitis cases showing perivascular cuffing and round-cell infiltration in the postmortem brain (McIntosh et al. 1980). In less than 0.5 % of human Rift Valley fever infections, encephalitis may develop 1–4 weeks after recovery from the acute illness (Nichol 2001). Both direct viral and immune-mediated cell injury are thought to contribute to this complication (Anderson et al. 1987).

The pathogenesis of RVFV encephalitis in man is incompletely understood. Brains of experimentally infected calves show diffuse perivascular infiltrates of lymphocytes and macrophages, meningeal reaction, focal areas of neuronal necrosis, with neurons and glial cells containing viral antigen throughout brain and cervical cord (Rippy et al. 1992). Rhesus monkeys develop clinical disease similar to human cases of RVFV. Fatally infected rhesus monkeys may show mild multifocal perivascular encephalitis. The development of interferon response limits the severity of disease and CNS infection (Morrill et al. 1990). However, RVFV itself inhibits the interferon response.

RVFV NSs protein inhibits IFN-beta gene expression in the mammalian host by targeting the RNA polymerase II complex (Billecocq et al. 2004; Thomas et al. 2004). Like the NSs protein of La Crosse Virus (LACV, an important cause of pediatric arboviral encephalitis in the Midwestern USA), RVFV NSs protein is also a viral interferon antagonist, supporting the possibility of shared mechanisms of disease between a hemorrhagic and a neurotropic family member (Hollidge et al. 2010).

3.3 *Hantaviruses*

Hantaviruses are temperate-zone viruses. Two major clinical disorders are associated with hantavirus infections in human: hemorrhagic fever with renal syndrome (HFRS) and hantavirus pulmonary syndrome (HPS) (Papadimitriou 1995; Schmaljohn and Hjelle 1997). Hantavirus infections, possibly the agents of epidemic hemorrhagic fever of ancient China, are known today as causes of HFRS seen in Asia and Europe. HFRS is associated with Old World hantaviruses including Puumala, Dobrava, Hantaan, and Seoul. Hantaan virus is found in parts of eastern China, Korea, and Far Eastern Russia and is present in *Apodemus agrarius mantchuricus*, a mouse common in agriculture fields. Seoul virus is associated with *Rattus norvegicus* and *R. rattus* and is worldwide in distribution. Dobrava virus is found in *Apodemus flavicollis* and *A. agrarius* in the Balkans. Puumala virus causes an HFRS also called nephropathia epidemica (NE) in Europe. Puumala virus is associated with the bank vole, *Clethrionomys glareolus*, in Scandinavia (Schmaljohn and Nichol 2007). Recent studies in Hungary show that Puumala, Dobrava and Saaremaa, all pathogenic hantaviruses, co-circulate in the same geographic area and can be maintained side-by-side in different rodent species (Plyusnina et al. 2009). Hantaviruses use cell surface integrins for cell entry, thus rendering many cell types vulnerable (Gavrilovskaya et al. 1999).

A 2- to 3-week incubation period is followed by a flu-like illness, and symptoms relating to increased capillary permeability. Hantaviruses primarily affect blood vessels with infection of microvascular endothelial cells leading to increased vascular permeability and fluid leakage. Hantavirus hemorrhagic fever with renal syndrome is a severe systemic infection with shock, vascular leakage, hypotension, and acute renal failure. Its clinical course consists of five phases: febrile, hypotensive, oliguric, diuretic, and convalescent (Schmaljohn and Nichol 2007). Severity of Puumala illness has been associated with host genetics. Severe disease (and potentially CNS disease) is seen in B8 DRB1*0301 patients and mild disease in HLA B27 allele patients (Mustonen et al. 1998), the same HLA associations found in rapid and slow progressors, respectively, in HIV patients (Kanerva et al. 1998; Mustonen et al. 1998). Class II HLA DR15(2)-DQ6 is less common in CNS disease cases (Hautala et al. 2010). Severity of HFRS has been associated with human platelet alloantigen (HPA) alleles. The five common HPA gene variants produce differential susceptibility to cardiovascular disease and stroke. One of these polymorphisms, the HPA-3b allele, is found in patients with more severe HFRS (Liu et al. 2009).

CNS hemorrhages during life and found postmortem are well-documented for HFRS (Kim 1965; Lukes 1954; Powell 1954). The incidence of encephalitis in HFRS is around 1 % (Bergmann et al. 2002). In the absence of angiographic vascular anomalies, coagulopathy, thrombocytopenia, DIC, and immune complex damage to blood vessels contribute to parenchymal and subarachnoid hemorrhage (Xu et al. 2011). Focal anoxia that accelerates renal disease may also have a role in cerebral disease, along with direct infection. Viral antigens are present not only in capillary endothelium but can be present in neurons, glia in cases of HFRS (Liu 1989) and pituitary tissue in PUUV (Hautala et al. 2002), accessing the brain where the blood–brain barrier is absent. Direct demonstration of viral antigen in CNS tissue (Liu 1993), the presence of CNS symptoms and complications in HFRS and NE, inflammatory CSF, intrathecal production of virus-specific antibodies (Hautala et al. 2010; Launes and Hautanen 1988), and PCR detection of virus in CSF (Mahonen et al. 2007; Talamonti et al. 2011) are all consistent with true CNS infection rather than simply increased tissue permeability.

Hantavirus rodent models, like in human infections, have varying degrees of neurologic involvement. Newborn or immunodeficient rodents show cachexia and pareses (McKee et al. 1985; Yamanouchi et al. 1984) and adult immunocompetent mice experimentally infected with Hantaan virus die of acute encephalitis (Wichmann et al. 2002). These observations on CNS disease potential are important because of the ease with which hantaviruses can produce new or changing clinical symptoms as a consequence of having a different virus target organ. A recent example of this change was hantavirus pulmonary syndrome (HPS) in 1993.

HPS is caused by New World hantaviruses including Sin Nombre virus (SNV) and Andes viruses. The main reservoir for the virus is rodents of the subfamily Sigmodontinae. The abrupt onset of fever, headache, cough, and myalgia, associated with vascular leakage, noncardiogenic pulmonary edema and shock, HPS fatality rate for SNV and Andes viruses range from 30 to 50 % (Bi et al. 2008; Schmaljohn and Hjelle 1997; Vincent et al. 2000). Transverse myelitis and acute disseminated encephalomyelitis (ADEM-like syndrome) in HPS patients in USA have been reported (Huisa et al. 2009). Andes viruses, which are agents of HPS circulating in Argentina, Bolivia, Chile and Uruguay, are also implicated in encephalitic syndromes. In 2010, an Argentine patient with a febrile syndrome, respiratory distress and shock developed, on day 13 of illness, an encephalitic syndrome characterized by headache, confusion, excitation, generalized seizure, bilateral frontal, and parieto-occipital T1 and T2 signal abnormalities on MRI, greater in cortex than underlying white matter. Viral RNA was not detected in CSF sampled at 15 days postonset of symptoms, but virus-specific IgM and IgG antibodies were found in CSF (Talamonti et al. 2011).

4 Filoviruses

Filoviruses cause the most severe form of VHF in man with mortality rates to 88 % for Ebola-Zaire. Viruses in this family are single, negative-stranded RNA viruses with genome size at approximately 19 kb in length. The virions are pleomorphic, enveloped and consist of a nucleocapsid, a polymerase complex, and a matrix protein (Sanchez et al. 2007).

Filovirus disease was first seen in Marburg, Germany in 1967 in individuals that had handled monkey blood, tissues, or cell cultures from the African green (vervet) monkeys (*Cercopithecus aethiops*) imported from Uganda. The first known human outbreaks of Ebola were in Democratic Republic of Congo and Sudan in 1976. The geographic range of known Ebola human cases has included most of the African rain forest. In Central Africa, bats were suspected as potential reservoir species because of the association of index cases with caves (Marburg, Kenya 1980), a bat-infested gold mine (Marburg, Democratic Republic of Congo 1999), and a bat-infested factory (Ebola, Sudan 1976). Evidence of asymptomatic, natural infection by Ebola virus in three species of fruit bat in Africa (Leroy et al. 2005) confirmed the importance of these as reservoir species in Central Africa. The natural maintenance cycle is not clearly identified.

Spread of filovirus infections is by close contact with another case through body fluid contact, although transmission by fomites is possible. Aerosol transmission, shown for nonhuman primate infections, may also apply for humans. The incubation period is 7–9 days. Sexual transmission has been reported for Marburg virus; precautions should be taken during 3 months of convalescence (Sanchez et al. 2007).

Virus enters through lesions of the skin or mucous membranes from where it gains access to the vascular and/or lymphatic systems. The primary targets of filoviruses are cells of the mononuclear phagocytic system, in which virus lytically replicates, and endothelium. Vascular endothelial cells are targeted directly by virus infection, replication, and destruction and indirectly by viral or virus-mediated host-derived soluble factors that cause endothelial activation and dysfunction. Fatal cases are associated with shock (increased vascular permeability, hypotension, coagulation disorders, and variable degrees of hemorrhages and widespread focal tissue destruction) and immunosuppression by various mechanisms (lymphocyte apoptosis, interferon inhibition, cytokine dysregulation) (Mehedi et al. 2011; Sanchez et al. 2007; Schnittler et al. 1993). Ebola contains a retrovirus-like glycoprotein with an immunosuppressive motif found in oncogenic retroviruses (Volchkov et al. 1992).

4.1 Marburg and Ebola

During the 1967 Marburg outbreak acute and convalescent phase neurological syndromes were reported, along with abnormal CSF findings (WBC range from 0 to 24 cells/μL, protein to 68 mg/dL) and a positive viral culture from the anterior eye chamber in one recovered patient (Gear et al. 1975). Postmortem brains from this epidemic had widely distributed microglial nodules, diffuse glial cell proliferation, perivascular inflammation, and scattered petechial hemorrhages (Bechtelsheimer et al. 1969; Jacob and Spalke 1971).

Key features of filovirus infection are pantropism, high viremia, endothelial infection/injury and immunosuppression, which are shared with other VHFs. Given the overlap of critical elements of filoviral infection with other VHFs that cause CNS injury (such as hantaviruses), more cases may be associated with recognition of CNS disease. Ebola and Marburg viruses have been the only members of the Filoviridae family. Now, a revised taxonomy of the family Filoviridae with species named Reston ebolavirus, Sudan ebolavirus, Zaire ebolavirus, Tai Forest ebolavirus, Bundibugyo ebolavirus, Marburg marburgvirus, and Ravn marburgvirus, has been proposed, along with a new family member Lloviu cuenavirus, found in Schreiber's long-fingered bats in Spain to better reflect high-resolution phylogeny now available (Kuhn et al. 2010). Less pathogenic members and longer survival times may ultimately unmask more CNS disease.

Although some endemic transmission of the filoviruses may occur, the annual cases are usually less than 500 people. With the abrupt and acute nature of the illness, a comprehensive clinical CNS feature has not been totally completed. With available information, clinically, both Marburg and Ebola viruses infection manifest in a similar manner, with an abrupt onset of prostrating fever, headache, and myalgias. Patients may present with tender adenopathy, pleuritic chest pain, photophobia, and a dry or sore throat. Subjects frequently appear restless and anxious, which may progress to be disoriented and stuporous and exhibit other encephalopathic signs. The features of CNS involvement may include psychosis, delirium, seizure, and coma (Bausch and Ksiazek 2002; Sureau 1989).

5 Flaviviruses

The Flaviviridae family consists of genera Flavivirus, Pestivirus, Hepacivirus, and the genus of Hepatitis G virus. The name is from the Latin flavus meaning yellow, as this is the family of yellow fever virus. The *Flavivirus* genus consists of 67 identified human and animal viruses, most of which are arthropod-borne and have been divided into antigenic complexes and sub-complexes on the basis of serology, vector transmission and phylogenic analysis. The *Flavivirus* genus includes important human pathogens such as West Nile virus, Japanese encephalitis virus, tick-borne encephalitis virus, yellow fever virus, and dengue virus. *Hepacivirus* genus contains hepatitis C virus. The genus of *Pestivirus* has no human pathogens. Viruses in this family are all single, positive-stranded RNA viruses with linear nonsegmented genomes. The genomes of these viruses are on average 11,000 kb in length, encoding around ten genes (Gubler and Kuno 1997; Lindenbach et al. 2007; Weaver and Barrett 2004).

5.1 Dengue

Dengue is one the most important arboviral infections of man, causing more than 50 million cases each year in tropical and subtropical areas (Gould and Solomon 2008). The principal vector is the mosquito *Aedes aegypti*. In addition, *Ae. albopictus, Ae. polynesiensis* and *Ae. scutellaris* have a role in transmission (Gubler and Kuno 1997).

After introduction, it is generally believed that viral replication occurs in extraneural tissues and in lymph nodes, resulting in primary viremia, spread via bloodstream, and infection of target organs. Dengue virus infects several types of human cells: dendritic cells, monocytes/macrophages, B and T cells, hepatocytes, endothelial cells, bone marrow progenitor cells, and neuronal cells (Gubler and Kuno 1997; Kurane et al. 1990). Infection presents as dengue fever, an influenza-like illness (fever-arthralgia-rash), dengue hemorrhagic fever (in 1–10 % of severe cases), dengue shock syndrome, and encephalopathy or encephalitis. Flavivirus receptors on neurons have not been definitively identified (Das et al. 2009, 2010; Imbert et al. 1994; Ramos-Castaneda et al. 1997).

"Immunization," virus type and human genetic factors can influence severity and outcome of disease. Whereas levels of viremia are important for disease, many of the life-threatening complications come immediately after clearance of virus. Antibody and T-cell responses may be associated with protective or pathological immunity during infection (Halstead 1988; WHO 2010).

There are four distinctive serotypes of dengue virus, designated as DENV1, DENV2, DENV3, and DENV4. In each of the dengue virus infections, there is a wide spectrum of clinical manifestations, ranging from asymptomatic, mild undifferentiated fever, dengue fever (DF), to dengue hemorrhagic fever (DHF) with or without dengue shock syndrome (DSS), and a potential life-threatening condition can be registered. Clinically, two significant types of illness are the most frequently discussed: classic DF and DHF/DSS. Classic DF, most likely to occur in older children and adults, occurs with sudden onset of fever, which lasts for 2–5 days, accompanied by severe headache, intense myalgias, arthralgias, retro-orbital pain, anorexia, and rash. A diffuse morbilliform rash with bleeding phenomena may appear during the defervescent stage. DHF/DSS predominantly recognizes in children in tropical Asia. Clinical symptoms begin abruptly with fever, malaise, headache, anorexia, nausea and vomiting, cough, and facial flushing. During the defervescent period, conditions worsen with profound weakness and prostration, diaphoresis, restlessness, cool and clammy extremities, rapid but thread pulse, and a narrow pulse pressure. In this setting, hemorrhagic phenomena are frequent with petechiae and bleeding from the nose, gums, venipuncture sites, and the gastrointestinal, genitourinary, and respiratory tract. In severe cases, patients may develop subcutaneous ecchymotic eruption (Fig. 2a) and suffer circulatory failure and profound shock leading to severe end-organ dysfunction. It is estimated at 50 million dengue virus infection worldwide every year, with which about 500,000 people with DHF require hospitalization. The fatality rate is about 1 %, but can exceed 20 % if without proper treatment.

It is the general belief that infection with any one of these serotype viruses could provide a lifelong immunity against the same dengue serotype, heterologous cross-protection is believed to be brief, and may serve as a filter for emergence of more virulent strains. Observations that reinfection by a different serotype virus enhances the risk of dengue hemorrhagic fever/dengue shock syndrome DHF/DSS with internal bleeding and shock lead to the hypothesis of preexisting heterotypic dengue immune response as a risk factor for DHF/DSS (Halstead 2007; WHO 2010). Dengue patients who develop DHF/DSS generally do so at the time just after clearance of virus and at the defervescent period, consistent with DHF/DSS being immunopathologic diseases (Green and Rothman 2006; Halstead 1980; Noisakran et al. 2010).

Mechanisms that are thought to be important in pathogenesis are antibody-dependent enhancement of viral replication and abnormal lymphocyte responses to heterologous virus as cytotoxic T-cell-mediated cytokine production (Green and Rothman 2006; Halstead 1988). The degree of thrombocytopenia is associated with disease severity, with low platelet counts resulting from increased consumption, decreased production, direct viral infection or lysis by immune complex activation (Mitrakul 1987; Noisakran et al. 2009a, b; Scott et al. 1978; Srichaikul and

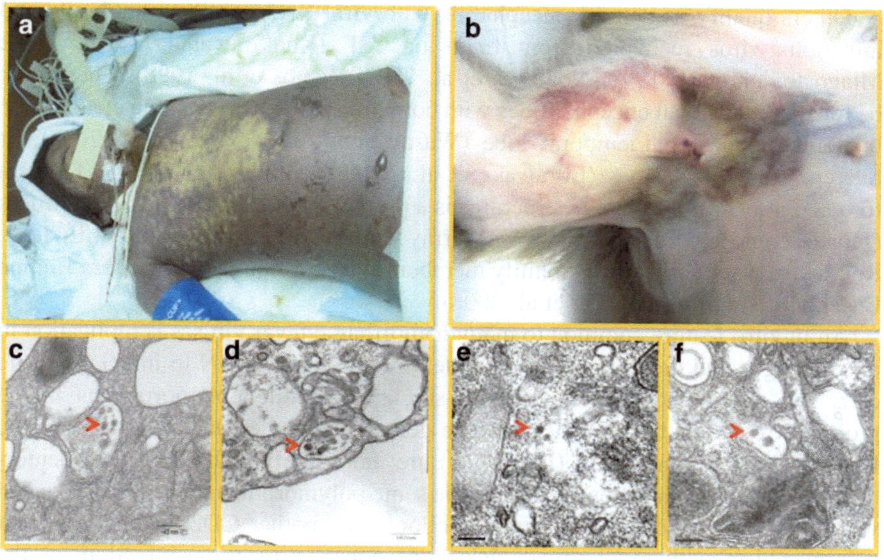

Fig. 2 Aspects of dengue virus infection. (**a**) Widespread ecchymosis in dengue hemorrhagic fever patient (courtesy of Kulkanya Chokephaibulkit, M.D.; Department of Pediatrics, Faculty of Medicine Siriraj Hospital, Mahidol University, Bangkok, Thailand). (**b**) Similar subcutaneous hemorrhage in a rhesus monkey after intravenous dengue virus infection. Hemorrhage correlates with infection of platelets by dengue virus. (**c, d**) Dengue viral particles in platelets isolated from dengue patient. (**e, f**) Dengue viral particles in platelets isolated from dengue virus-infected monkeys. *Arrowheads* (*red*) indicate dengue viral particles in platelet vesicles

Nimmannitya 2000; Tsai et al. 2011; Wang et al. 1995). The association of acute disseminated encephalomyelitis (ADEM), Guillain–Barre syndrome, brachial plexitis, and mononeuritis multiplex with dengue support the possibility of additional neurologic targets of immune-mediated disease (Chokephaibulkit et al. 2001; Kankirawatana et al. 2000; Nimmannitya et al. 1987; Solomon et al. 2000).

Most dengue virus infections cause DF or DHF, but a proportion of patients have a reduced level of consciousness due to either a complication of severe systemic disease or infection of the CNS. Encephalopathy in dengue virus infection is one of the most recognized and established scenarios in dengue patients. Encephalopathy is a clinical picture of reduced consciousness, which can be caused uncommonly by encephalitis but more commonly by other infections, metabolic derangements, alcohol, or drugs. Although acute encephalopathy is the most frequently reported in dengue patients, increasing evidence of dengue viral neurotropism suggests that a small proportion of cases are true encephalitis. However, affected DHF patients normally do not show any signs of sequelae after recovery. Serum AST elevations in dengue patients have been frequently registered but DIC development is a rare event, and CNS illnesses can occur without hemorrhagic disease since some of these patients showed similarity of clinical presentations to Reye syndrome (RS). Hepatic encephalopathy is a likely scenario in dengue patients (Solomon et al. 2000).

Others qualify as dengue encephalitis with viral RNA, antigen, antibody, and infectious virus recovered from brain or CSF (Kankirawatana et al. 2000; Miagostovich et al. 1997; Nogueira et al. 2002; Ramos et al. 1998). These may be focal or multifocal encephalitides with corpus callosum (Sreedharan et al. 2011), bilateral hippocampal/temporal lobe (Wasay et al. 2008; Yeo et al. 2005), or bilateral thalamic lesions (Borawake et al. 2011; Kamble et al. 2007). Tropism for deep gray or diencephalic structures is a property shared with other flaviviruses (Japanese encephalitis virus, WNV, Murray Valley encephalitis virus, SLE, TBE, Rocio virus) and *Togaviridae* family members Western Equine and Eastern Equine Encephalitis viruses (Solbrig et al. 2008).

An in vitro feature common to many hemorrhagic fever viruses is replication in macrophages and dendritic cells. Dengue virus uses DC-SIGN to infect dendritic cells and macrophages. Patients with single nucleotide polymorphisms in the promoter region of the CD209 gene, the gene that encodes dendritic cell-specific ICAM-grabbing nonintegrin (DC-SIGN) are more likely to have dengue hemorrhagic fever rather than dengue fever. The same polymorphism has been associated with more severe hepatitis C virus infection, and another CD209 gene promoter polymorphism with more severe tick-borne encephalitis virus infection, or the transition from meningitis to encephalitis (Barkhash et al. 2011). However, in vivo investigation suggested that dengue viral particles are likely associated with platelets isolated from dengue patients (Noisakran et al. 2009a, b and Fig. 2c, d). Importantly, the coagulopathy feature has been recapitulated in rhesus monkeys infected with dengue virus intravenously (Fig. 2b) and in platelets isolated from these animals (Noisakran et al. 2011 and Fig. 2e, f).

The incidence of neurological complications of dengue ranges from 0.5 to 18 % (in DHF cases) (Cam et al. 2001; Hendarto and Hadinegoro 1992; Kankirawatana et al. 2000; Varatharaj 2010) with DENV2 and 3 having the highest rate of neurologic manifestations (Cam et al. 2001). Mutations in DENV-2 envelope proteins are determinants of neurovirulence in mice (Bray et al. 1998). Organ and tissue tropism, host range, and virulence are expressions of multiple genes. Also, in a murine model, mouse-adapted dengue virus crosses the blood–brain barrier to directly infect brain cells possibly by using a cytotoxic factor and histamine released during infection (Chaturvedi et al. 1991). Mutations in noncoding regions affect virulence (Cahour et al. 1995) and NS genes influence activity of IFN host defenses and phenotype (Butrapet et al. 2000).

Dengue is an emerging disease. The global spread of dengue fever within and beyond tropical boundaries, combined with an increased severity of dengue-associated clinical outcomes, have made dengue virus of great public health importance. With human and environmental conditions for persistence and spread are present on all continents except Antarctica, dengue will continue to increase in geographic distribution and severity, threatening a significant portion of the world's population. In addition to our own activities, we have those of the virus to think about. Until recently, flavivirus evolution was thought to progress in a clonal manner, with diversity based on accumulation of mutational changes. However, intragenic recombination among positive-strand RNA viruses occurs. For example, heterologous recombination between two different alphaviruses, the New World Alphavirus eastern equine encephalitis virus and Sindbis virus, resulted in the appearance of western equine encephalitis virus (Hahn et al. 1988), and homologous recombination and complementation are now recognized in JE, SLE, and DEN (Aaskov et al. 2006; AbuBakar et al. 2002; Tolou et al. 2001; Twiddy and Holmes 2003; Uzcategui et al. 2001; Worobey et al. 1999). While recombination has not been thought a major factor in the evolution of dengue virus, so far, this phenomena and attendant risk of "hyperparasites" may pose additional risk to man and animals as more epidemics arise and more live vaccines are introduced (Gould and Solomon 2008).

5.2 *Yellow Fever*

Yellow fever virus is the prototype of the *Flavivirus* genus, transmitted by *Aedes aegypti* mosquitoes. Yellow fever is a zoonotic disease maintained by mosquito-monkey-mosquito transmission with occasional human infection when unvaccinated humans enter the forest (sylvatic or jungle yellow fever). Alternatively, in urban epidemics (urban yellow fever), there is interhuman transmission by the domestic mosquito, *A. aegypti*. It occurs throughout much of sub-Saharan Africa and tropical South America, where periodic epidemics have been occurring since 1647. There are an estimated 200,000 cases per year, causing 30,000 deaths each year (Monath 1987, 2001; WHO 2011).

After inoculation, the virus replicates in regional lymph nodes then spreads to liver, spleen, bone marrow, and cardiac and skeletal muscle. Bleeding is promoted by decreased synthesis of vitamin K-dependent coagulation factors by injured liver, disseminated intravascular coagulation, and altered platelet function (Monath 2001). Neurological manifestations of severe disease include delirium, stupor, convulsions, and coma. Neurotropic yellow fever infection of mice has been used as a model system for studies of flavivirus encephalitis (Gould et al. 1987). Azotemia, encephalopathy, progressive liver damage, and hemorrhage cause death in 30–50 % of patients who develop these signs of severe yellow fever (Monath 1987).

The 17D live attenuated yellow fever vaccine is considered one of the safest live virus vaccines, and yellow fever vaccine-associated neurotropic diseases are rare. The onset of illness is from 1 to 30 days after vaccination as a syndrome of meningoencephalitis, ADE, myelitis, Guillain–Barre, bulbar or Bell's palsy. Detection of YFV-specific IgM antibodies in CSF is required for diagnosis (Pires-Marczeski et al. 2011).

6 Treatment VHF

6.1 *Ribavirin and Plasma Therapy*

There are few licensed and effective broad-spectrum antivirals. Ribavirin has been used in the treatment of hemorrhagic fevers for more than 20 years. Alpha-interferon has significant side effects and is prohibitively expensive for wide usage. Ribavirin, the therapeutic choice for Lassa fever, is efficacious when administered to patients within the first 7 days of illness. Its use has been principally in Lassa fever patients with clinically significant illness and elevated serum aspartate transaminase levels (McCormick et al. 1986). Its penetration into the CNS is limited and its main side effect is a dose-related, reversible, hemolytic anemia (De Franceschi et al. 2000). The other potential risk is tetratogenicity and embryolethality during pregnancy (Rezvani and Koren 2006). Results with intravenous ribavirin have also been positive in hemorrhagic fever with renal syndrome (Huggins et al. 1991), Congo Crimean hemorrhagic fever (Centers for Disease Control 1988) and Argentine hemorrhagic fever (Enria and Maiztegui 1994). Therapy of Junin virus infections with ribavirin has been effective for the visceral phase of the infection. The administration of convalescent serum (usually 2–3 units with high neutralizing titer) within the first 8 days of illness, previously the treatment of choice (Enria et al. 1984), is also recommended supplemented by ribavirin (Geisbert and Jahrling 2004; Enria et al. 2008, reviewed in Solbrig and Perng 2011).

At present, ribavirin is approved for use in treating viral hemorrhagic fevers caused by arenaviruses and bunyaviruses but not filoviruses under the compassionate use provisions for investigational new drugs (Geisbert and Jahrling 2004). No drug has yet rescued or resuscitated a human or research primate with filovirus infection after onset of symptoms. While IV ribavirin was initially used in Congo Crimean hemorrhagic fever, subsequent studies have also shown clinical response to oral ribavirin. Patients with nausea and vomiting receive the drug via nasogastric tube or with antiemetics (World Health Organization 2001; Mardani et al. 2003; Ozkurt et al. 2006). Survival is improved with treatment started within 4 days of illness onset (Tasdelen Fisgin et al. 2009).

Overall, the efficacy of the ribavirin trials has been restricted to small observational studies and case series, and results of some larger trials generated disappointing results, thus, questions about methodological issues, such as no control for disease severity or day of initiation of treatment, have been raised (Bodur et al. 2011; Cevik et al. 2008; Chapman et al. 1999; Koksal et al. 2010; Maltezou et al. 2010; Mertz et al. 2004).

Expert opinions vary for the use of oral ribavirin for prophylaxis of primary contacts in case of exposure to a ribavirin-susceptible VHF (Crowcroft 2002). While some recommend immediate prophylaxis, others suggest observing contacts and treating if an individual develops fever (Kitching et al. 2009). Close personal contacts should be monitored for fever for a period of 3 weeks.

Therapy with convalescent plasma also gets mixed reviews. Some reports suggest clinical utility in some arenaviruses (Emond et al. 1977; Enria et al. 1984; Frame et al. 1984; Leifer et al. 1970; Maiztegui et al. 1979; Monath et al. 1974; Mupapa et al. 1999) and other studies show no benefit (Clayton 1977; McCormick et al. 1986; White 1972). At present, a well-designed and controlled clinical trial for the efficacy of immunotherapy is urgently needed to prepare in the event of unexpected mass outbreaks resulting from the natural cause or from deliberately spread of the VHF agents.

The rising number of emerging viral pathogens highlights our limited resources to develop treatments on a single-pathogen basis, and points to the need to develop broad-spectrum antiviral that target common components of large classes of viruses.

6.2 Prevention and Control Measures

The best prevention is knowledge and minimization of exposure to the virus. Prevention of rodent-borne arenaviruses and hantaviruses is by eliminating contact with rodents and their excreta, by rodent proofing homes, careful storage of food, disinfection, and removal of trapped rodents and excreta. For filoviruses, it is avoidance of contact with sick or dead monkeys. For those likely to contact Congo Crimean hemorrhagic fever-infected ticks, such as individuals with occupational exposure to livestock, treatment of clothing with pyrethroid compounds (organic compounds similar to natural pyrethrins) that repel or kill ticks, and mechanical barriers such as tucked-in pant legs are used. Immunization of livestock is the most effective way to prevent Rift Valley fever epizootics and human cases. Yellow fever is prevented by yellow fever 17D, a live viral vaccine that confers immunity within 10 days. Protective immunity may last for 30 years. Areas containing the domestic form of *A. aegypti* risk introduction and urbanization of yellow fever, and should be water-treated and sprayed. Prevention of dengue epidemics requires eradication of *A. aegypti* by breeding site elimination, use of larvicides, and insecticide spraying. Prevention of nosocomial outbreaks requires barrier nursing precautions, sterile instruments, and equipment. For high-risk exposure to a ribavirin-sensitive virus, the drug is used prophylactically or as expectant treatment at onset of fever. Several different inactivated virus vaccines for Hantaan and Seoul viruses have been licensed in Korea and China. Other classically-designed and molecular vaccines are in preclinical stages (Kruger et al. 2011). A Congo Crimean hemorrhagic fever formalin-inactivated vaccine has been developed in Eastern Europe (Papa et al. 2011). Formalin-inactivated Rift Valley fever vaccines have been used for immunization of at risk laboratory and field workers (Rusnak et al. 2011). Just as in CNS disease, the dual role of cellular immunity, protection versus immunopathogenicity, are important considerations for vaccines. At present, quarantine remains a means of controlling outbreaks that depend on person-to-person spread (Solbrig and Perng 2011).

7 VHF Agents as Biological Weapons

Biological weapons are unique in their invisibility and delayed effects, with sickness, death, fear, panic, insecurity, and paralyzing confusion. Potential threat agents have been divided into categories by CDC and NIAID panels of experts, with Category A containing those that pose the highest risk to national security and public health. They can be easily disseminated or transmitted from person to person; result in high mortality and have the potential for major public health impact; might cause public panic and social disruption; and require special action for public health preparedness. Category A agents are Arenaviruses (LCMV, Junin virus, Machupo virus, Guanarito virus, Lassa fever virus), Bunyaviruses (Hantaviruses and Rift Valley Fever), Flaviviruses (Dengue), and Filoviruses (Ebola/Marburg); a Category B agent is another Flavivirus (Kyasanur Forest virus); and Category C pathogens are tickborne hemorrhagic fever viruses (CCHF) and yellow fever (NIAID fact sheet Biodefense and Emerging Infectious Diseases 2011).

The presence of significant numbers of VHF agents on the list signifies the conviction that these viruses could be weaponized, that either wild-type or engineered viruses would lend themselves well for the purposes listed above. Ideal BW are lethal, associated with a recognizable sign or symptom such as hemorrhage, easy to produce (isolate and cultivate), easy to spread by natural (person to person) or technical means (aerosols), environmentally stable, and inducers of infections hard to treat and prevent. The availability of a vaccine and antiviral drugs that are safe would significantly remove these viruses as threats.

8 Conclusions

The viruses causing VHFs are not classical neurotropic viruses. Yet CNS complications have been reported for each family. CNS syndromes are rare with filoviruses, but are more common with arenaviruses. Traditionally hemorrhagic viruses, such as RVFV and hantaviruses, are frequently encephalitic in rodent species or laboratory animals. Dengue virus with or without hemorrhage causes encephalitis as well as the full range of central and peripheral neurological complications of infection. When coupled to modern investigational and clinical pathological techniques, or fully rendered by using in vitro or in vivo models, these infections are providing important lessons in CNS viral pathogenesis. Knowing these CNS complications occur and learning how they occur will better prepare us for patients and epidemics in the future.

References

Aaskov J, Buzacott K, Thu HM, Lowry K, Holmes EC (2006) Long-term transmission of defective RNA viruses in humans and Aedes mosquitoes. Science 311(5758):236–238

AbuBakar S, Wong PF, Chan YF (2002) Emergence of dengue virus type 4 genotype IIA in Malaysia. J Gen Virol 83(Pt 10):2437–2442

Adhikari Prabha MR, Prabhu MG, Raghuveer CV, Bai M, Mala MA (1993) Indian J Med Sci 47(5):124–130

Al-Hazmi M, Ayoola EA, Abdurahman M, Banzal S, Ashraf J, El-Bushra A et al (2003) Epidemic Rift Valley fever in Saudi Arabia: a clinical study of severe illness in humans. Clin Infect Dis 36(3):245–252

Al-Hazmi A, Al-Rajhi AA, Abboud EB, Ayoola EA, Al-Hazmi M, Saadi R et al (2005) Ocular complications of Rift Valley fever outbreak in Saudi Arabia. Ophthalmology 112(2):313–8
Alrajhi AA, Al-Semari A, Al-Watban J (2004) Rift Valley fever encephalitis. Emerg Infect Dis 10 (3):554–555
Anderson GW Jr, Slone TW Jr, Peters CJ (1987) Pathogenesis of Rift Valley fever virus (RVFV) in inbred rats. Microb Pathog 2(4):283–293
Armstrong C (1941) Studies on Choriomeningitis and Poliomyelitis: Harvey Lecture, October 31, 1940. Bull N Y Acad Med 17(4):295–318
Barkhash AV, Perelygin AA, Babenko VN, Brinton MA, Voevoda MI (2011) Single nucleotide polymorphism in the promoter region of the CD209 gene is associated with human predisposition to severe forms of tick-borne encephalitis. Antiviral Res 93(1):64–68
Baek SH, Shin DI, Lee HS, Lee SH, Kim HY, Shin KS, Lee SY, Han HS, Han HJ, Lee SS (2010) Reversible splenium lesion of the corpus callosum in hemorrhagic fever with renal failure syndrome. J Korean Med Sci 25(8):1244–1246
Bausch DG, Ksiazek TG (2002) Viral hemorrhagic fevers including hantavirus pulmonary syndrome in the Americas. Clin Lab Med 22(4):981–1020, viii
Bechtelsheimer H, Jacob H, Solcher H (1969) The neuropathology of an infectious disease transmitted by African green monkeys (*Cercopithecus aethiops*). Ger Med Mon 14(1):10–12
Bergmann F, Krone B, Bleich S, Prange H, Paulus W (2002) Encephalitis due to a hantavirus infection. J Infect 45(1):58–59
Bi Z, Formenty PB, Roth CE (2008) Hantavirus infection: a review and global update. J Infect Dev Ctries 2(1):3–23
Billecocq A, Spiegel M, Vialat P, Kohl A, Weber F, Bouloy M et al (2004) NSs protein of Rift Valley fever virus blocks interferon production by inhibiting host gene transcription. J Virol 78(18):9798–9806
Bodur H, Erbay A, Akinci E, Onguru P, Bayazit N, Eren SS et al (2011) Effect of oral ribavirin treatment on the viral load and disease progression in Crimean-Congo hemorrhagic fever. Int J Infect Dis 15(1):e44–e47
Borawake K, Prayag P, Wagh A, Dole S (2011) Dengue encephalitis. Indian J Crit Care Med 15(3):190–193
Bray M, Men R, Tokimatsu I, Lai CJ (1998) Genetic determinants responsible for acquisition of dengue type 2 virus mouse neurovirulence. J Virol 72(2):1647–1651
Briese T, Paweska JT, McMullan LK, Hutchison SK, Street C, Palacios G et al (2009) Genetic detection and characterization of Lujo virus, a new hemorrhagic fever-associated arenavirus from southern Africa. PLoS Pathog 5(5):e1000455
Buchmeier MJ, de la Torre JC, Peters CJ (2007) Arenviridae: the viruses and their replication. In: Knipe DM, Howley PM (eds) Fields virology, vol 2, 5th edn. Lippincott Williams and Wilkins, Philadelphia, PA, pp 1791–1827
Butrapet S, Huang CY, Pierro DJ, Bhamarapravati N, Gubler DJ, Kinney RM (2000) Attenuation markers of a candidate dengue type 2 vaccine virus, strain 16681 (PDK-53), are defined by mutations in the 5′ noncoding region and nonstructural proteins 1 and 3. J Virol 74(7):3011–3019
Cahour A, Pletnev A, Vazielle-Falcoz M, Rosen L, Lai CJ (1995) Growth-restricted dengue virus mutants containing deletions in the 5′ noncoding region of the RNA genome. Virology 207 (1):68–76
Cam BV, Fonsmark L, Hue NB, Phuong NT, Poulsen A, Heegaard ED (2001) Prospective case-control study of encephalopathy in children with dengue hemorrhagic fever. Am J Trop Med Hyg 65(6):848–851
Cao W, Henry MD, Borrow P, Yamada H, Elder JH, Ravkov EV et al (1998) Identification of alpha-dystroglycan as a receptor for lymphocytic choriomeningitis virus and Lassa fever virus. Science 282(5396):2079–2081
Cerar D, Avsic-Zupanc T, Jereb M, Strle F (2007) Case report: severe neurological manifestation of Dobrava hantavirus infection. J Med Virol 79(12):1841–1843

Centers for Disease Control (1988) Management of patients with suspected viral hemorrhagic fever. MMWR Morb Mortal Wkly Rep 37(Suppl 3):S1–S16
Cevik MA, Elaldi N, Akinci E, Onguru P, Erbay A, Buzgan T et al (2008) A preliminary study to evaluate the effect of intravenous ribavirin treatment on survival rates in Crimean-Congo hemorrhagic fever. J Infect 57(4):350–351
Chapman LE, Mertz GJ, Peters CJ, Jolson HM, Khan AS, Ksiazek TG et al (1999) Intravenous ribavirin for hantavirus pulmonary syndrome: safety and tolerance during 1 year of open-label experience. Ribavirin Study Group. Antivir Ther 4(4):211–219
Chaturvedi UC, Dhawan R, Khanna M, Mathur A (1991) Breakdown of the blood-brain barrier during dengue virus infection of mice. J Gen Virol 72(Pt 4):859–866
Chokephaibulkit K, Kankirawatana P, Apintanapong S, Pongthapisit V, Yoksan S, Kositanont U et al (2001) Viral etiologies of encephalitis in Thai children. Pediatr Infect Dis J 20(2):216–218
Clayton AJ (1977) Lassa immune serum. Bull World Health Organ 55(4):435–439
Cosgriff TM, Morrill JC, Jennings GB, Hodgson LA, Slayter MV, Gibbs PH et al (1989) Hemostatic derangement produced by Rift Valley fever virus in rhesus monkeys. Rev Infect Dis 11(Suppl 4):S807–S814
Crowcroft NS (2002) Management of Lassa fever in European countries. Euro Surveill 7(3):50–52
Cummins D, McCormick JB, Bennett D, Samba JA, Farrar B, Machin SJ et al (1990) Acute sensorineural deafness in Lassa fever. JAMA 264(16):2093–2096
Cummins D, Bennett D, Fisher-Hoch SP, Farrar B, Machin SJ, McCormick JB (1992) Lassa fever encephalopathy: clinical and laboratory findings. J Trop Med Hyg 95(3):197–201
Das S, Laxminarayana SV, Chandra N, Ravi V, Desai A (2009) Heat shock protein 70 on Neuro2a cells is a putative receptor for Japanese encephalitis virus. Virology 385(1):47–57
Das S, Chakraborty S, Basu A (2010) Critical role of lipid rafts in virus entry and activation of phosphoinositide 3′ kinase/Akt signaling during early stages of Japanese encephalitis virus infection in neural stem/progenitor cells. J Neurochem 115(2):537–549
De Franceschi L, Fattovich G, Turrini F, Ayi K, Brugnara C, Manzato F et al (2000) Hemolytic anemia induced by ribavirin therapy in patients with chronic hepatitis C virus infection: role of membrane oxidative damage. Hepatology 31(4):997–1004
Emond RT, Evans B, Bowen ET, Lloyd G (1977) A case of Ebola virus infection. Br Med J 2(6086):541–544
Enria DA, Maiztegui JI (1994) Antiviral treatment of Argentine hemorrhagic fever. Antiviral Res 23(1):23–31
Enria DA, Briggiler AM, Fernandez NJ, Levis SC, Maiztegui JI (1984) Importance of dose of neutralising antibodies in treatment of Argentine haemorrhagic fever with immune plasma. Lancet 2(8397):255–256
Enria DA, Briggiler AM, Sanchez Z (2008) Treatment of Argentine hemorrhagic fever. Antiviral Res 78(1):132–139
Fischer SA, Graham MB, Kuehnert MJ, Kotton CN, Srinivasan A, Marty FM et al (2006) Transmission of lymphocytic choriomeningitis virus by organ transplantation. N Engl J Med 354(21):2235–2249
Forslund T, Saltevo J, Anttinen J, Auvinen S, Brummer-Korvenkontio M, Korhonen A, Poutiainen M (1992) Complications of nephropathia epidemica: three cases. J Intern Med 232(1):87–90
Frame JD, Verbrugge GP, Gill RG, Pinneo L (1984) The use of Lassa fever convalescent plasma in Nigeria. Trans R Soc Trop Med Hyg 78(3):319–324
Frazer KA, Ballinger DG, Cox DR, Hinds DA, Stuve LL, Gibbs RA et al (2007) A second generation human haplotype map of over 3.1 million SNPs. Nature 449(7164):851–861
Gavrilovskaya IN, Brown EJ, Ginsberg MH, Mackow ER (1999) Cellular entry of hantaviruses which cause hemorrhagic fever with renal syndrome is mediated by beta3 integrins. J Virol 73(5):3951–3959
Gear JSS, Cassel GA, Gear AJ, Trappler B, Clausen L, Meyers AM et al (1975) Outbreak of Marburg virus disease in Johannesburg. Br Med J 4(5995):489–493

Geisbert TW, Jahrling PB (2004) Exotic emerging viral diseases: progress and challenges. Nat Med 10(12 Suppl):S110–S121

Gerding MN, Groen J, Jordans JG, Osterhaus AD (1995) Hantavirus nephropathy in The Netherlands: clinical, histopathological and epidemiological findings. Neth J Med 47(3):106–112

Gould EA, Solomon T (2008) Pathogenic flaviviruses. Lancet 371(9611):500–509

Gould EA, Buckley A, Groeger BK, Cane PA, Doenhoff M (1987) Immune enhancement of yellow fever virus neurovirulence for mice: studies of mechanisms involved. J Gen Virol 68(Pt 12):3105–3112

Green S, Rothman A (2006) Immunopathological mechanisms in dengue and dengue hemorrhagic fever. Curr Opin Infect Dis 19(5):429–436

Gubler DJ, Kuno G (eds) (1997) Dengue and dengue hemorrhagic fever, 1st edn. CAB International, Wallingford, UK

Gunther S, Weisner B, Roth A, Grewing T, Asper M, Drosten C et al (2001) Lassa fever encephalopathy: Lassa virus in cerebrospinal fluid but not in serum. J Infect Dis 184 (3):345–349

Hahn CS, Lustig S, Strauss EG, Strauss JH (1988) Western equine encephalitis virus is a recombinant virus. Proc Natl Acad Sci USA 85(16):5997–6001

Halstead SB (1980) Pathology and pathogenesis of DHF. Paper presented at the first ICMR seminar on dengue and dengue heomorrhagic fever, Kobe University School of Medicine, Kobe, Japan

Halstead SB (1988) Pathogenesis of dengue: challenges to molecular biology. Science 239(4839):476–481

Halstead SB (2007) Dengue. Lancet 370(9599):1644–1652

Hautala T, Sironen T, Vapalahti O, Paakko E, Sarkioja T, Salmela PI et al (2002) Hypophyseal hemorrhage and panhypopituitarism during Puumala Virus Infection: magnetic Resonance Imaging and detection of viral antigen in the hypophysis. Clin Infect Dis 35(1):96–101

Hautala T, Mahonen SM, Sironen T, Hautala N, Paakko E, Karttunen A et al (2010) Central nervous system-related symptoms and findings are common in acute Puumala hantavirus infection. Ann Med 42(5):344–351

Hautala T, Hautala N, Mähönen SM, Sironen T, Pääkkö E, Karttunen A et al (2011) Young male patients are at elevated risk of developing serious central nervous system complications during acute Puumala hantavirus infection. BMC Infect Dis 11:217

Hendarto SK, Hadinegoro SR (1992) Dengue encephalopathy. Acta Paediatr Jpn 34(3):350–357

Heyman P, Cochez C, Hofhuis A, van der Giessen J, Sprong H, Porter SR et al (2010) A clear and present danger: tick-borne diseases in Europe. Expert Rev Anti Infect Ther 8(1):33–50

Hirsch E (1976) Sensorineural deafness and labyrinth damage due to lymphocytic choriomeningitis. Report of a case. Arch Otolaryngol 102(8):499–500

Hollidge BS, Gonzalez-Scarano F, Soldan SS (2010) Arboviral encephalitides: transmission, emergence, and pathogenesis. J Neuroimmune Pharmacol 5(3):428–442

Hoogstraal H (1979) The epidemiology of tick-borne Crimean-Congo hemorrhagic fever in Asia, Europe, and Africa. J Med Entomol 15(4):307–417

Horton J, Hotchin JE, Olson KB, Davies JN (1971) The effects of MP virus infection in lymphoma. Cancer research." Cancer Res 31(8):1066–1068

Huggins JW, Hsiang CM, Cosgriff TM, Guang MY, Smith JI, Wu ZO et al (1991) Prospective, double-blind, concurrent, placebo-controlled clinical trial of intravenous ribavirin therapy of hemorrhagic fever with renal syndrome. J Infect Dis 164(6):1119–1127

Huisa BN, Chapin JE, Adair JC (2009) Central nervous system complications following Hanta virus cardiopulmonary syndrome. J Neurovirol 15(2):202–205

Imbert JL, Guevara P, Ramos-Castaneda J, Ramos C, Sotelo J (1994) Dengue virus infects mouse cultured neurons but not astrocytes. J Med Virol 42(3):228–233

Jacob H, Spalke G (1971) [Clinical aspects and neuropathology of central nervous system complications after acute water and electrolyte imbalance with special reference to central pontine myelinolysis]. Fortschr Neurol Psychiatr Grenzgeb 39(4):169–191

Johnson KM, McCormick JB, Webb PA, Smith ES, Elliott LH, King IJ (1987) Clinical virology of Lassa fever in hospitalized patients. J Infect Dis 155(3):456–464

Kamble R, Peruvamba JN, Kovoor J, Ravishankar S, Kolar BS (2007) Bilateral thalamic involvement in dengue infection. Neurol India 55(4):418–419

Kanerva M, Mustonen J, Vaheri A (1998) Pathogenesis of puumala and other hantavirus infections. Rev Med Virol 8(2):67–86

Kankirawatana P, Chokephaibulkit K, Puthavathana P, Yoksan S, Apintanapong S, Pongthapisit V (2000) Dengue infection presenting with central nervous system manifestation. J Child Neurol 15(8):544–547

Kim D (1965) Clinical analysis of 111 fatal cases of epidemic hemorrhagic fever. Am J Med 39:218–220

Kitching A, Addiman S, Cathcart S, Bischop L, Krahe D et al (2009) A fatal case of Lassa fever in London, January 2009. Euro Surveill 14(6):1–3

Koksal I, Yilmaz G, Aksoy F, Aydin H, Yavuz I, Iskender S et al (2010) The efficacy of ribavirin in the treatment of Crimean-Congo hemorrhagic fever in Eastern Black Sea region in Turkey. J Clin Virol 47(1):65–68

Kruger DH, Schonrich G, Klempa B (2011) Human pathogenic hantaviruses and prevention of infection. Hum Vaccin 7(6):685–693

Kuhn JH, Becker S, Ebihara H, Geisbert TW, Johnson KM, Kawaoka Y et al (2010) Proposal for a revised taxonomy of the family Filoviridae: classification, names of taxa and viruses, and virus abbreviations. Arch Virol 155(12):2083–2103

Kunz S (2009) The role of the vascular endothelium in arenavirus haemorrhagic fevers. Thromb Haemost 102(6):1024–1029

Kurane I, Kontny U, Janus J, Ennis FA (1990) Dengue-2 virus infection of human mononuclear cell lines and establishment of persistent infections. Arch Virol 110(1–2):91–101

Launes J, Hautanen A (1988) Nephropathia epidemica encephalitis. Acta Neurol Scand 78(3):234–235

Leblebicioglu H (2010) Crimean-Congo haemorrhagic fever in Eurasia. Int J Antimicrob Agents 36(Suppl 1):S43–S46

Leifer E, Gocke DJ, Bourne H (1970) Lassa fever, a new virus disease of man from West Africa. II. Report of a laboratory-acquired infection treated with plasma from a person recently recovered from the disease. Am J Trop Med Hyg 19(4):677–679

Leroy EM, Kumulungui B, Pourrut X, Rouquet P, Hassanin A, Yaba P et al (2005) Fruit bats as reservoirs of Ebola virus. Nature 438(7068):575–576

Lindenbach BD, Pragai BM, Montserret R, Beran RK, Pyle AM, Penin F et al (2007) The C terminus of hepatitis C virus NS4A encodes an electrostatic switch that regulates NS5A hyperphosphorylation and viral replication. J Virol 81(17):8905–8918

Liou LM, Lan SH, Lai CL (2008) Electroencephalography burst suppression in a patient with dengue encephalopathy: A case report. Clin Neurophysiol 119(10):2205–2208

Liu YF (1989) [The primary and secondary pathological changes in the brain in patients with epidemic hemorrhagic fever]. Zhonghua Bing Li Xue Za Zhi 18(4):244–246

Liu YF (1993) [Distribution of extracellular viral antigen of epidemic hemorrhagic fever in human autopsy cases]. Zhonghua Yi Xue Za Zhi 73(6):341–342, 380

Liu Z, Gao M, Han Q, Lou S, Fang J (2009) Platelet glycoprotein IIb/IIIa (HPA-1 and HPA-3) polymorphisms in patients with hemorrhagic fever with renal syndrome. Hum Immunol 70 (6):452–456

Lozach PY, Kuhbacher A, Meier R, Mancini R, Bitto D, Bouloy M et al (2011) DC-SIGN as a receptor for phleboviruses. Cell Host Microbe 10(1):75–88

Lukes RJ (1954) The pathology of thirty-nine fatal cases of epidemic hemorrhagic fever. Am J Med 16(5):639–650

Madani TA (2005) Alkhumra virus infection, a new viral hemorrhagic fever in Saudi Arabia. J Infect 51(2):91–97

Mahonen SM, Sironen T, Vapalahti O, Paakko E, Hautala N, Ilonen J et al (2007) Puumala virus RNA in cerebrospinal fluid in a patient with uncomplicated nephropathia epidemica. J Clin Virol 40(3):248–251

Maiztegui JI, Fernandez NJ, de Damilano AJ (1979) Efficacy of immune plasma in treatment of Argentine haemorrhagic fever and association between treatment and a late neurological syndrome. Lancet 2(8154):1216–1217

Maltezou HC, Andonova L, Andraghetti R, Bouloy M, Ergonul O, Jongejan F et al (2010) Crimean-Congo hemorrhagic fever in Europe: current situation calls for preparedness. Euro Surveill 15(10):19504

Mardani M, Jahromi MK, Naieni KH, Zeinali M (2003) The efficacy of oral ribavirin in the treatment of crimean-congo hemorrhagic fever in Iran. Clin Infect Dis 36(12):1613–1618

McCormick JB, King IJ, Webb PA, Scribner CL, Craven RB, Johnson KM et al (1986) Lassa fever. Effective therapy with ribavirin. N Engl J Med 314(1):20–26

McIntosh BM, Russell D, dos Santos I, Gear JH (1980) Rift Valley fever in humans in South Africa. S Afr Med J 58:803–806

McKee KT Jr, Kim GR, Green DE, Peters CJ (1985) Hantaan virus infection in suckling mice: virologic and pathologic correlates. J Med Virol 17(2):107–117

Mehedi M, Groseth A, Feldmann H, Ebihara H (2011) Clinical aspects of Marburg hemorrhagic fever. Future Virol 6(9):1091–1106

Mertz GJ, Miedzinski L, Goade D, Pavia AT, Hjelle B, Hansbarger CO et al (2004) Placebo-controlled, double-blind trial of intravenous ribavirin for the treatment of hantavirus cardiopulmonary syndrome in North America. Clin Infect Dis 39(9):1307–1313

Meyer HM Jr, Johnson RT, Crawford IP, Dascomb HE, Rogers NG (1960) Central nervous system syndromes of "vital" etiology. A study of 713 cases. Am J Med 29:334–347

Miagostovich MP, Ramos RG, Nicol AF, Nogueira RM, Cuzzi-Maya T, Oliveira AV et al (1997) Retrospective study on dengue fatal cases. Clin Neuropathol 16(4):204–208

Mitrakul C (1987) Bleeding problem in dengue haemorrhagic fever: platelets and coagulation changes. Southeast Asian J Trop Med Public Health 18(3):407–412

Monath TP (1987) Yellow fever: a medically neglected disease. Report on a seminar. Rev Infect Dis 9(1):165–175

Monath TP (2001) Yellow fever: an update. Lancet Infect Dis 1(1):11–20

Monath TP, Maher M, Casals J, Kissling RE, Cacciapuoti A (1974) Lassa fever in the Eastern Province of Sierra Leone, 1970–1972. II. Clinical observations and virological studies on selected hospital cases. Am J Trop Med Hyg 23(6):1140–1149

Morrill JC, Jennings GB, Johnson AJ, Cosgriff TM, Gibbs PH, Peters CJ (1990) Pathogenesis of Rift Valley fever in rhesus monkeys: role of interferon response. Arch Virol 110(3–4):195–212

Morrill JC, Czarniecki CW, Peters CJ (1991) Recombinant human interferon-gamma modulates Rift Valley fever virus infection in the rhesus monkey. J Interferon Res 11(5):297–304

Mupapa K, Massamba M, Kibadi K, Kuvula K, Bwaka A, Kipasa M et al (1999) Treatment of Ebola hemorrhagic fever with blood transfusions from convalescent patients. International Scientific and Technical Committee. J Infect Dis 179(Suppl 1):S18–S23

Mustonen J, Partanen J, Kanerva M, Pietila K, Vapalahti O, Pasternack A et al (1998) Association of HLA B27 with benign clinical course of nephropathia epidemica caused by Puumala hantavirus. Scand J Immunol 47(3):277–279

NIAID (2011) Biodefense category A,B,C pathogens. NIAID. NIH. www.niaid.nih.gov/topics/biodefenserelated/biodefense

Nichol ST (2001) Bunyaviruses. In: Knipe DM, Howley PM (eds) Fields virology, vol 2, 4th edn. Lippincott Williams and Wilkins, Philadelphia, pp 1603–1634

Nimmannitya S, Thisyakorn U, Hemsrichart V (1987) Dengue haemorrhagic fever with unusual manifestations. Southeast Asian J Trop Med Public Health 18(3):398–406

Nogueira RM, Filippis AM, Coelho JM, Sequeira PC, Schatzmayr HG, Paiva FG et al (2002) Dengue virus infection of the central nervous system (CNS): a case report from Brazil. Southeast Asian J Trop Med Public Health 33(1):68–71

Noisakran S, Chokephaibulkit K, Songprakhon P, Onlamoon N, Hsiao HM, Villinger F et al (2009a) A re-evaluation of the mechanisms leading to dengue hemorrhagic fever. Ann NY Acad Sci 1171(Suppl 1):E24–E35

Noisakran S, Gibbons RV, Songprakhon P, Jairungsri A, Ajariyakhajorn C, Nisalak A et al (2009b) Detection of dengue virus in platelets isolated from dengue patients. Southeast Asian J Trop Med Public Health 40(2):253–262

Noisakran S, Onlamoon N, Songprakhon P, Hsiao H-M, Chokephaibulkit K, Perng GC (2010) Cells in dengue virus infection in vivo. Adv Virol 2010:164878

Noisakran S, Onlamoon N, Hsiao HM, Clark KB, Villinger F, Ansari AA et al (2011) Infection of bone marrow cells by dengue virus in vivo. Exp Hematol 40(3):250.e4–259.e4

Olival KJ, Daszak P (2005) The ecology of emerging neurotropic viruses. J Neurovirol 11 (5):441–446

Ormay I, Kovacs P (1989) [Lymphocytic choriomeningitis causing unilateral deafness]. Orv Hetil 130(15):789–791

Ozkurt Z, Kiki I, Erol S, Erdem F, Yilmaz N et al (2006) Crimean-Congo hemorrhagic fever in Eastern Turkey: clinical features, risk factors and efficacy of ribavirin therapy. J Infect 52(3):207–215

Palacios G, Druce J, Du L, Tran T, Birch C, Briese T et al (2008) A new arenavirus in a cluster of fatal transplant-associated diseases. N Engl J Med 358(10):991–998

Papa A, Papadimitriou E, Christova I (2011) The Bulgarian vaccine Crimean-Congo haemorrhagic fever virus strain. Scand J Infect Dis 43(3):225–229

Papadimitriou M (1995) Hantavirus nephropathy. Kidney Int 48(3):887–902

Paweska JT, Sewlall NH, Ksiazek TG, Blumberg LH, Hale MJ, Lipkin WI et al (2009) Nosocomial outbreak of novel arenavirus infection, southern Africa. Emerg Infect Dis 15(10):1598–1602

Pires-Marczeski FC, Martinez VP, Nemirovsky C, Padula PJ (2011) Intrathecal antibody production in two cases of Yellow fever vaccine associated neurotropic disease in Argentina. J Med Virol 83:2208–2212

Plyusnina A, Ferenczi E, Racz GR, Nemirov K, Lundkvist A, Vaheri A et al (2009) Co-circulation of three pathogenic hantaviruses: Puumala, Dobrava, and Saaremaa in Hungary. J Med Virl 81(12):2045–2052

Powell GM (1954) Hemorrhagic fever: a study of 300 cases. Medicine (Baltimore) 33(2):97–153

Ramos C, Sanchez G, Pando RH, Baquera J, Hernandez D, Mota J et al (1998) Dengue virus in the brain of a fatal case of hemorrhagic dengue fever. J Neurovirol 4(4):465–468

Ramos-Castaneda J, Imbert JL, Barron BL, Ramos C (1997) A 65-kDa trypsin-sensible membrane cell protein as a possible receptor for dengue virus in cultured neuroblastoma cells. J Neurovirol 3(6):435–440

Rezvani M, Koren G (2006) Pregnancy outcome after exposure to injectable ribavirin during embryogenesis. Reprod Toxicol 21(1):113–115

Rippy MK, Topper MJ, Mebus CA, Morrill JC (1992) Rift Valley fever virus-induced encephalomyelitis and hepatitis in calves. Vet Pathol 29(6):495–502

Rusnak JM, Gibbs P, Boudreau E, Clizbe DP, Pittman P (2011) Immunogenicity and safety of an inactivated Rift Valley fever vaccine ina 19 year study. Vaccine 29(17):3222–3229

Sabeti PC, Varilly P, Fry B, Lohmueller J, Hostetter E, Cotsapas C et al (2007) Genome-wide detection and characterization of positive selection in human populations. Nature 449(7164):913–918

Sanchez A, Geisbert TW, Feldmann H (2007) Filoviridae: Marburg and Ebola viruses. In: Knipe DM, Howley PM (eds) Fields virology, vol 2. Lippincott Williams and Wilkins, Philadelphia, pp 1409–1448

Schmaljohn C, Hjelle B (1997) Hantaviruses: a global disease problem. Emerg Infect Dis 3(2):95–104

Schmaljohn CS, Nichol ST (2007) Bunyaviridae. In: Knipe DM, Howley PM (eds) Fields virology, vol 2, 5th edn. Lippincott Williams and Wilkins, Philadelphia, pp 1741–1788

Schnittler HJ, Mahner F, Drenckhahn D, Klenk HD, Feldmann H (1993) Replication of Marburg virus in human endothelial cells. A possible mechanism for the development of viral hemorrhagic disease. J Clin Invest 91(4):1301–1309

Scott RM, Nisalak A, Cheam-U-Dom U, Seridhoranakul S, Nimmannitya S (1978) A preliminary report on the isolation of viruses from the platelets and leukocytes of dengue patients. Asian J Infect Dis 2:95–97

Soares CN, Cabral-Castro MJ, Peralta JM, de Freitas MR, Zalis M, Puccioni-Sohler M (2011) Review of the etiologies of viral meningitis and encephalitis in a dengue endemic region. J Neurol Sci 303(1–2):75–79

Solbrig MV (1993) Lassa virus and central nervous system diseases. In: Salvato MS (ed) The Arenaviridae. Plenum, New York, pp 325–330

Solbrig MV, McCormick JB (1991) Lassa fever: central nervous system manifestations. J Trop Geogr Neurol 1:23–30

Solbrig MV (1997) The history and neurology of hantaviral infections. Neurol Infect Epidemiol 2:129–134

Solbrig MV, Naviaux RK (1997) Review of the neurology and biology of Ebola and Marburg virus infections. Neurol Infect Epidemiol 2:13–18

Solbrig MV, Perng G-C (2011) Neurologic complications of hemorrhagic fevers. In: Gilman S, Johnson RT (eds) Medlink neurology. Medlink, San Diego

Solbrig MV, Hasso AN, Jay CA (2008) CNS viruses – diagnostic approach. Neuroimaging Clin N Am 18(1):1–18, vii

Soldan SS, Gonzalez-Scarano F (2005) Emerging infectious diseases: the Bunyaviridae. J Neurovirol 11(5):412–423

Solomon T, Dung NM, Vaughn DW, Kneen R, Thao LT, Raengsakulrach B et al (2000) Neurological manifestations of dengue infection. Lancet 355(9209):1053–1059

Sreedharan SE, Chellenton J, Kate MP, Kesavadas C (2011) Reversible pancallosal signal changes in febrile encephalopathy: report of 2 cases. AJNR Am J Neuroradiol 32(9):E172–E174

Srichaikul T, Nimmannitya S (2000) Haematology in dengue and dengue haemorrhagic fever. Baillieres Best Pract Res Clin Haematol 13(2):261–276

Sullivan BM, Emonet SF, Welch MJ, Lee AM, Campbell KP, de la Torre JC et al (2011) Point mutation in the glycoprotein of lymphocytic choriomeningitis virus is necessary for receptor binding, dendritic cell infection, and long-term persistence. Proc Natl Acad Sci USA 108(7):2969–2974

Sureau PH (1989) Firsthand clinical observations of hemorrhagic manifestations in Ebola hemorrhagic fever in Zaire. Rev Infect Dis 11(Suppl 4):S790–S793

Talamonti L, Padula PJ, Canteli MS, Posner F, Marczeski FP, Weller C (2011) Hantavirus pulmonary syndrome: encephalitis caused by virus Andes. J Neurovirol 17(2):189–192

Tasdelen Fisgin N, Ergonul O, Doganci L, Tulek N (2009) The role of ribavirin in the therapy of Crimean-Congo hemorrhagic fever: early use is promising. Eur J Clin Microbiol Infect Dis 28 (8):929–933

Thomas D, Blakqori G, Wagner V, Banholzer M, Kessler N, Elliott RM et al (2004) Inhibition of RNA polymerase II phosphorylation by a viral interferon antagonist. J Biol Chem 279 (30):31471–31477

Tolou HJ, Couissinier-Paris P, Durand JP, Mercier V, de Pina JJ, de Micco P et al (2001) Evidence for recombination in natural populations of dengue virus type 1 based on the analysis of complete genome sequences. J Gen Virol 82(Pt 6):1283–1290

Tsai JJ, Jen YH, Chang JS, Hsiao HM, Noisakran S, Perng GC (2011) Frequency alterations in key innate immune cell components in the peripheral blood of dengue patients detected by FACS analysis. J Innate Immun 3(5):530–540

Twiddy SS, Holmes EC (2003) The extent of homologous recombination in members of the genus *Flavivirus*. J Gen Virol 84(Pt 2):429–440

Uzcategui NY, Camacho D, Comach G, Cuello de Uzcategui R, Holmes EC, Gould EA (2001) Molecular epidemiology of dengue type 2 virus in Venezuela: evidence for in situ virus evolution and recombination. J Gen Virol 82(Pt 12):2945–2953

Varatharaj A (2010) Encephalitis in the clinical spectrum of dengue infection. Neurol India 58(4):585–591

Villarete L, Somasundaram T, Ahmed R (1994) Tissue-mediated selection of viral variants: correlation between glycoprotein mutation and growth in neuronal cells. J Virol 68(11):7490–7496

Vincent MJ, Quiroz E, Gracia F, Sanchez AJ, Ksiazek TG, Kitsutani PT et al (2000) Hantavirus pulmonary syndrome in Panama: identification of novel hantaviruses and their likely reservoirs. Virology 277(1):14–19

Volchkov VE, Blinov VM, Netesov SV (1992) The envelope glycoprotein of Ebola virus contains an immunosuppressive-like domain similar to oncogenic retroviruses. FEBS Lett 305(3):181–184

Vorou R, Pierroutsakos IN, Maltezou HC (2007) Crimean-Congo hemorrhagic fever. Curr Opin Infect Dis 20(5):495–500

Wang S, He R, Patarapotikul J, Innis BL, Anderson R (1995) Antibody-enhanced binding of dengue-2 virus to human platelets. Virology 213(1):254–257

Wasay M, Channa R, Jumani M, Shabbir G, Azeemuddin M, Zafar A (2008) Encephalitis and myelitis associated with dengue viral infection clinical and neuroimaging features. Clin Neurol Neurosurg 110(6):635–640

Weaver SC, Barrett AD (2004) Transmission cycles, host range, evolution and emergence of arboviral disease. Nat Rev Microbiol 2(10):789–801

White HA (1972) Lassa fever. A study of 23 hospital cases. Trans R Soc Trop Med Hyg 66(3):390–401

WHO (2010) Dengue vaccine development: the role of the WHO South-East Asia Regional Office

WHO (2011) Yellow fever. Media Centre

Wichmann D, Grone HJ, Frese M, Pavlovic J, Anheier B, Haller O et al (2002) Hantaan virus infection causes an acute neurological disease that is fatal in adult laboratory mice. J Virol 76(17):8890–8899

World Health Organization (WHO) (2001) Crimean-Congo hemorrhagic fever. World Health Organization fact sheet no. 208. WHO, Geneva

Worobey M, Rambaut A, Holmes EC (1999) Widespread intra-serotype recombination in natural populations of dengue virus. Proc Natl Acad Sci USA 96(13):7352–7357

Xu Z, Xu P, Lei X, Wu Q, Zhang J (2011) Subarachnoid hemorrhage associated with epidemic hemorrhagic Fever: a rare case report. Int J Med Sci 8(8):640–642

Yamanouchi T, Domae K, Tanishita O, Takahashi Y, Yamanishi K, Takahashi M et al (1984) Experimental infection in newborn mice and rats by hemorrhagic fever with renal syndrome (HFRS) virus. Microbiol Immunol 28(12):1345–1353

Yeo PS, Pinheiro L, Tong P, Lim PL, Sitoh YY (2005) Hippocampal involvement in dengue fever. Singapore Med J 46(11):647–650

Part V
Unconventional Agents

Prion Diseases

Valerie L. Sim

Abstract Prion diseases are infectious neurodegenerative diseases that have fascinated scientists for decades. They can occur spontaneously, iatrogenically, or be inherited through mutation of the prion protein gene *Prnp*. Instead of a virus, the infectious agent is a corrupted form of the naturally occurring prion protein. Disease results when this corrupted form (PrP^{Sc}) converts the normally expressed prion protein (PrP^{C}) into the abnormal conformation, leading to protein aggregation and ultimately neuronal death. Different structures of PrP^{Sc} exist as different strains of prions, some of which can cross species barriers, and each strain can produce a unique phenotype. Clinical signs generally include dementia and ataxia, but because these diseases are relatively rare and can mimic other syndromes, diagnosis is often missed or delayed. While no one diagnostic test is 100 % accurate, new tests are on the horizon which may lead to earlier diagnosis. This, in turn, may facilitate therapy efforts, because as of now, there are no effective treatments for prion disease.

Keywords Clinical-pathological correlation • Conversion • Decontamination • Diagnostics • Pathogenesis • Prion disease • Protein misfolding • Prion strains • Therapeutics

1 Background

Prion diseases, otherwise known as transmissible spongiform encephalopathies (TSEs), are fatal neurodegenerative diseases. The most common human form is Creutzfeldt–Jakob disease (CJD), but prion diseases also exist in other mammals, including bovine spongiform encephalopathy (BSE or mad cow disease) in cattle,

V.L. Sim (✉)
Centre for Prions and Protein Folding Diseases, University of Alberta, 204 EEB, Edmonton, AB, Canada T6G 2M8
e-mail: valerie.sim@ualberta.ca

A.C. Jackson (ed.), *Viral Infections of the Human Nervous System*,
Birkhäuser Advances in Infectious Diseases,
DOI 10.1007/978-3-0348-0425-7_15,

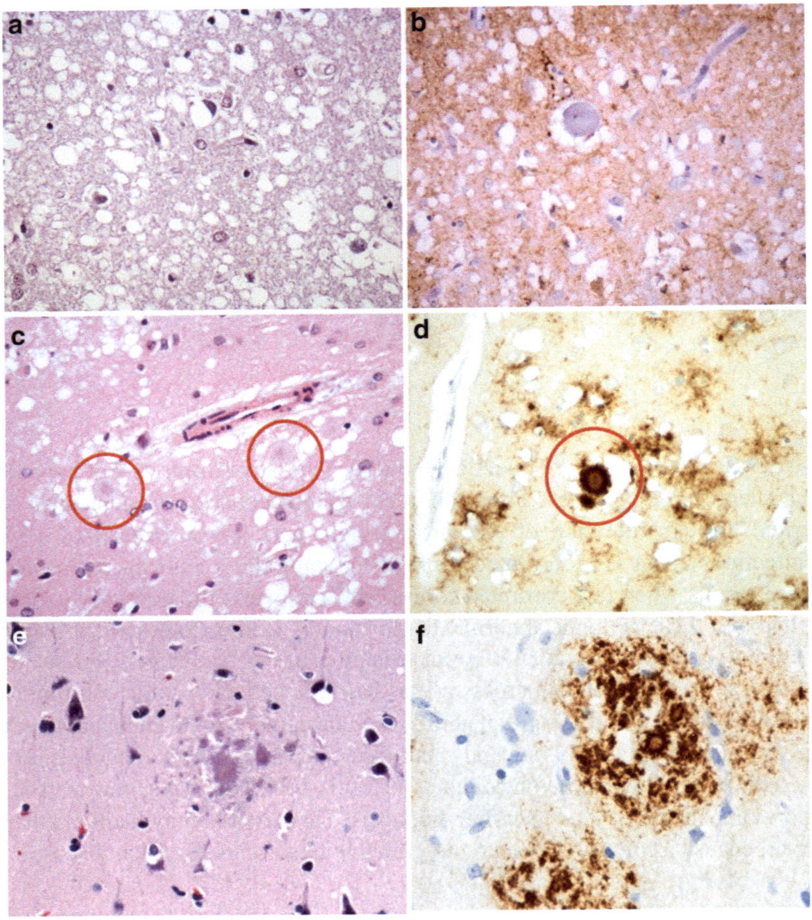

Fig. 1 Pathological findings in prion disease. (**a**) Sporadic Creutzfeldt–Jakob disease (sCJD) brain: Haematoxylin and eosin stain showing typical spongiform change; (**b**) sCJD brain: Immunostaining for prion protein (12F10 antibody) showing diffuse synaptic pattern (*brown*); (**c**) Variant Creutzfeldt–Jakob disease (vCJD): Periodic acid Schiff stain showing typical florid plaque surrounded by vacuoles (*red circles*); (**d**) vCJD: Immunostaining for prion protein (12F10 antibody) showing a florid plaque (*red circle*); (**e**) Gerstmann–Sträussler–Scheinker (GSS): Haematoxylin and eosin stain showing a typical multicentric plaque; (**f**) GSS: Immunostaining for prion protein (12F10 antibody) showing multicentric plaques. Courtesy of Dr. G. Jansen, University of Ottawa

chronic wasting disease (CWD) in cervids, and scrapie in sheep. Clinical presentations can vary, but patients generally present with a combination of progressive dementia and ataxia. The pathology is characterized by neuronal loss, spongiform change in the neuropil of the brain, and the accumulation of a misfolded form of the naturally occurring prion protein (PrP) (see Fig. 1). Unlike other neurodegenerative diseases associated with misfolded proteins (such as Alzheimer's and Parkinson's diseases), prion diseases are transmissible and can be acquired within and across species. Unlike other infectious diseases, prion diseases can be inherited and are caused by an infectious protein, not a virus.

1.1 A Historical Perspective

The oldest known prion disease is scrapie, a fatal neurodegenerative disease of sheep which was first reported in 1730 and which is infectious by nature. The human prion disease now known as sporadic Creutzfeldt–Jakob disease (sCJD) was first described in 1920 and 1921 by Hans Gerhard Creutzfeldt and Alfons Maria Jakob, respectively, but the concept of its being an infectious disease was not proposed until decades later after a chance occurrence. In 1957, the fatal neurodegenerative disease kuru was identified in the Foré people of Papua New Guinea by Vincent Zigas and Carleton Gajdusek. When a travelling exhibit of kuru neuropathology appeared in London in 1959, William Hadlow, a person very familiar with sheep scrapie neuropathology, noted that the kuru pathology was strikingly similar to that of scrapie: vacuolated neurons in the context of neuronal degeneration and intense astrocytosis with little, if any, inflammation. This led him to propose that kuru, like scrapie, may be an infectious disease (Hadlow 1959). This was proven when kuru (Gajdusek et al. 1966) and later CJD (Gibbs et al. 1968) were transmitted to chimpanzees. What followed were decades of research focussed on identifying the mysterious infectious agent.

1.2 Not a Virus...

From the early studies of scrapie, it was clear that this infectious agent was unusual, given its slow pathogenesis, lack of inflammatory response, and extreme resistance to disinfectants. Its resistance to ultraviolet irradiation suggested the absence of any nucleic acid (Alper et al. 1967) and raised the possibility that a protein alone could be infectious (Griffith 1967). Evidence for a protein agent started to build. First, a gene which influenced disease incubation time in mice (*Sinc*) was discovered (Dickinson et al. 1968). Then in 1981, amyloid protein fibrils called "scrapie-associated fibrils" were isolated from diseased brains and were found to be specifically associated with the disease (Merz et al. 1981, 1984). Stanley Prusiner coined the term "prion" to denote a small purely proteinaceous infectious particle (Prusiner 1982) and he and colleagues subsequently discovered a 27–30 kDa protein which was resistant to proteinase K digestion and which was associated with infectivity (Bolton et al. 1982). This led to the designation "PrP" for protease-resistant protein (McKinley et al. 1983). Pieces of the puzzle started to fit together; antibodies to PrP recognized the previously identified scrapie-associated fibrils (also called prion rods) (Barry et al. 1985; Merz et al. 1987) and the gene encoding PrP (*Prnp*) was found to be the same gene as *Sinc* (Carlson et al. 1986). Thereafter, genetic forms of human prion disease were traced to mutations in *Prnp* (Collinge et al. 1989; Owen et al. 1989) and it was found that knockout mice devoid of PrP were resistant

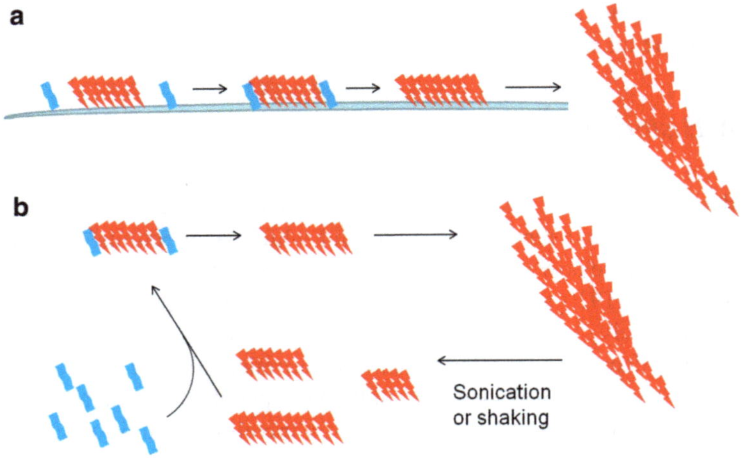

Fig. 2 Seeded polymerization. PrP^{C} is represented as *blue rectangles*, PrP^{Sc} as *red zigzags*. (**a**) Conversion in vivo likely occurs on the membrane surface or in endosomal compartments, where PrP^{C} interacts with PrP^{Sc} and is converted into the PrP^{Sc} structure. Large aggregates can accumulate intra- or extracellularly. (**b**) This process can be recreated in vitro with PMCA or QuIC, where sonication or shaking breaks the larger aggregates into smaller oligomers, which are more efficient at converting newly added PrP^{C}

to disease (Bueler et al. 1993; Sailer et al. 1994). However, despite the building evidence, the concept of a prion, as outlined in Prusiner's "prion hypothesis," still led to much scepticism. It also led to Prusiner's Nobel Prize in 1997.

1.3 Proving the Prion Hypothesis

The prion hypothesis remained difficult to prove for many years, but recent work has more definitively established the principle of a protein-based transmissible disease. In 2004, bacterially derived prion protein was turned into amyloid fibrils and caused a prion-like disease in transgenic mice, albeit at a much lower level of infectivity than *bona fide* PrP^{Sc} (Legname et al. 2004). In 2005, infectious prions were successfully propagated in vitro by an amplification procedure called protein misfolding cyclic amplification (PMCA) (Castilla et al. 2005). In this technique, new infectious agent is created by seeding normal brain homogenate with infectious agent, sonicating, and serial diluting into fresh normal brain homogenate (see Fig. 2). Later, brain-purified prion protein was successfully substituted for whole brain homogenate (Deleault et al. 2007), and then recombinant prion protein plus various RNA (Geoghegan et al. 2009) or lipid cofactors (Wang et al. 2010) was proven sufficient for propagation. Most recently, recombinant prion protein alone has been converted to infectious agent (Atarashi et al. 2007; Kim et al. 2010). Under specific conditions, PMCA can also generate infectious prions de novo, that is, without any seed, presumably by inducing a misfolding of normal prion protein by sonication (Deleault et al. 2007; Barria et al. 2009). As such, the prion hypothesis is well supported, and has actually generated new questions about the seeding and spread of other protein-misfolding neurodegenerative diseases such as Alzheimer's disease (Morales et al. 2011) and amyotrophic lateral sclerosis (ALS) (Polymenidou and Cleveland 2011).

2 The World of Prions

2.1 *Prion Terminology*

Linked to the prion hypothesis that a protein may act as an infectious agent is the question *how* can a protein act as an infectious agent? To answer this, one must turn to the biochemistry of the prion protein (PrP), and with it, to the confusing terminology and the use of superscripts to denote different forms of PrP. The normal cellular form of prion protein is generally referred to as PrP^{C}; in prion disease, PrP^{C} becomes misfolded into a disease-associated form, which is commonly referred to as PrP^{Sc}, where "Sc" denotes scrapie. Some researchers choose to use different superscripts for different diseases: PrP^{CJD}, PrP^{CWD}, PrP^{BSE}, but often the default PrP^{Sc} is used for all. Distinguishing PrP^{Sc} pathologically or biochemically has traditionally relied on the fact that PrP^{Sc} is partially resistant to proteinase K digestion, whereas PrP^{C} is sensitive. Because of this difference in resistance and sensitivity, some researchers refer to the resistant form as PrP^{res}, and the sensitive form as PrP^{sen}, both of which can be easily distinguished biochemically. However, in in vitro experiments, the only way to demonstrate true PrP^{Sc}, the form which will cause disease, is to infect animals and demonstrate pathology. While it is tempting to equate all PrP^{res} with PrP^{Sc}, protease-resistant forms of PrP can be generated which do not cause disease and some disease-causing PrP^{Sc} may be protease-sensitive (Tzaban et al. 2002; Silveira et al. 2005; Berardi et al. 2006; Pastrana et al. 2006; Gambetti et al. 2008, 2011).

2.2 *The Normal Prion Protein*

The cellular form of prion protein (PrP^{C}) is highly expressed in the central nervous system, but is also expressed on the cell membranes of most cell types, including dendritic cells, platelets, lymphocytes, and myoblasts. The protein itself is 23 kDa in size and is attached to the exterior cell surface by means of a glycophosphatidyl-inositol (GPI) anchor. PrP is variably N-glycosylated at two sites, which gives it a characteristic three-band pattern on Western blotting (see Fig. 3). The GPI anchor attaches to a globular C-terminal domain which contains three alpha helices, two beta strands, and a disulfide bond, and there is a flexible N-terminus that contains four or five octapeptide repeats which bind Cu^{2+}. The function of PrP^{C} remains unknown, but it can bind a wide variety of ligands and has been implicated in many roles, from memory and behavioural responses, to immunomodulatory functions, to

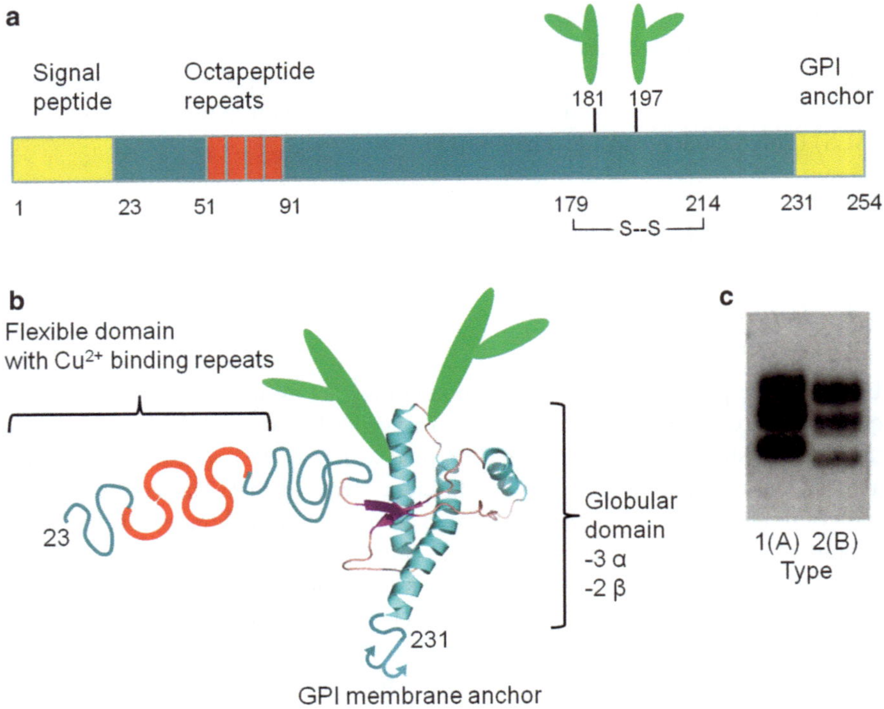

Fig. 3 PrP structure. (**a**) PrP^C residues 1–23 and 231–254 are cleaved during processing. The hamster sequence is shown as standard; human and mouse sequences are very similar but shorter by 1 amino acid. The *green ovals* represent glycans and the disulfide bond is represented as S–S; (**b**) PrP^C globular structure and flexible domain showing alpha helices, beta strands and copper binding regions; (**c**) Western blot profile of PrP^{Sc} after PK digestion showing the typical three-band glycosylation profile. Human strains 1A and 2B are shown. The molecular weight of the lowest band defines type 1 (21 kDa) or 2 (19 kDa); the different glycosylation band intensity distinguishes subtype A from B

oxidative stress resistance, metal binding, and signal transduction (reviewed in Linden et al. 2008). PrP^C may mediate the assembly of signalling modules composed of numerous molecular partners, thereby facilitating interactions and transmembrane signalling and leading to a number of different physiological and behavioural outcomes (Linden et al. 2008). From the perspective of prion disease, however, PrP^C is critical, as only animals which express PrP^C can develop disease (Bueler et al. 1993; Sailer et al. 1994).

2.3 The Abnormal Prion Protein

Unlike the structure of the PrP^C, the structure of PrP^{Sc} has yet to be solved. The insoluble and aggregated nature of PrP^{Sc} makes it difficult to study by traditional means such as NMR and X-ray crystallography. What is known is that PrP^{Sc} has a higher beta-sheet content (43–54%, Caughey et al. 1991b; Pan et al. 1993) when compared with PrP^{C} (which has only 3% beta sheet). Also, PrP^{Sc} is more resistant to proteinase K, with residues 90–231 remaining after digestion. Current structural models have been generated based on electron micrographs (Wille et al. 2002; Govaerts et al. 2004), computer modelling (DeMarco and Daggett 2004; DeMarco et al. 2006), EPR spectroscopy (Cobb et al. 2007) and even X-ray diffraction of infectious amyloid fibrils (Wille et al. 2009), but no one model is yet consistent with all experimental data regarding exposed epitopes and fibril sizes. The true structure(s) of PrP^{Sc} is one of the great unknowns in the prion field.

2.4 Conversion and Pathogenesis

How does PrP^{C} become PrP^{Sc}? Current evidence supports a seeded polymerization model (see Fig. 2). PrP^{Sc} is a structurally altered form of PrP^{C}, and it can bind and convert PrP^{C} to this altered form, thereby propagating or templating itself. This seeding process is also seen in other amyloid diseases, and the mechanism of prion infection may closely reflect the mechanisms of other protein-misfolding diseases. How the conversion or resulting PrP^{Sc} leads to disease is unclear, and a number of investigations have been done to examine signalling cascades that may be triggered, leading to apoptosis. Importantly, the PrP^{Sc} itself does not appear to be toxic on its own (Brandner et al. 1996). In fact, transgenic animals that shut off PrP^{C} expression at 8 weeks of age, after infection has already begun, have a halt of infection and a partial reversal of pathology despite the continued presence of PrP^{Sc} (Mallucci et al. 2003). In addition, mice that express PrP^{C} without its GPI anchor can accumulate large amounts of PrP^{Sc} plaque, but do not develop the clinical disease (Chesebro et al. 2005). Thus, the actual conversion process must occur for disease to follow (Bueler et al. 1993; Sailer et al. 1994) and this conversion must be mediated by GPI-anchored PrP^{C} on the cell surface or within endosomes (Caughey and Raymond 1991; Caughey et al. 1991a; Borchelt et al. 1992; Baron et al. 2002). The size of conversion product also is key, as there is evidence that soluble, protease-sensitive, oligomeric forms of PrP^{Sc} are more pathogenic (Silveira et al. 2005; Berardi et al. 2006; Pastrana et al. 2006) than the large insoluble resistant aggregates. This is reminiscent of other neurodegenerative diseases, such as Alzheimer's disease, where small oligomers are also thought to be the main pathogenic players.

2.5 From the Periphery to the Brain

It is important to recognize that in some orally acquired prion diseases (CWD, BSE, scrapie, and the variant form of CJD), prions appear to propagate in the peripheral tissues before invading the central nervous system and causing symptoms.

Thus asymptomatic carriers may exist. The gut dendritic cells play a major role in oral prion uptake (Montrasio et al. 2000) as does the gut-associated lymphoid mucosa. The spleen and other lymphoid tissues are also major sites of prion replication (Kimberlin and Walker 1989a; Brandner et al. 2000). From the gut, prions can follow splanchnic innervation (Kimberlin and Walker 1989b; Mabbott and Bruce 2001) or the autonomic nervous system (Glatzel et al. 2001; Prinz et al. 2003) to the spinal cord and brainstem. PrP^{C} expression on neurons is required for this neuroinvasion (Crozet et al. 2007). Different routes of exposure, such as inoculation into the tongue, oral or nasal mucosa can allow prions to follow cranial nerve pathways directly to the brain and bypass the lymphoreticular system completely (Bartz et al. 2005; Bessen et al. 2009). In cases of blood transmission, PrP^{Sc} may be able to enter the brain directly from the bloodstream, crossing the blood–brain barrier without even the requirement for host PrP^{C} expression (Urayama et al. 2011), although PrP^{C} is always required to develop disease.

2.6 From the Brain to the Periphery?

To state that peripheral involvement is only a result of route of exposure is an oversimplification. It appears that the distribution of PrP^{Sc} has more to do with the strain of prion, than with the route of infection. For instance, kuru is orally acquired but has no evidence of lymphoreticular involvement (Brandner et al. 2008). This is in contrast to the orally acquired variant CJD (vCJD, see below), where PrP^{Sc} is detectable in brain, tonsils, spleen, lymph node, retina, and proximal optic nerve with low levels in rectum, adrenal gland, and thymus (Bruce et al. 2001; Wadsworth et al. 2001). Also, careful analysis of the sporadic form of CJD has found evidence for low levels of PrP^{Sc} in spleen and skeletal muscle of long duration cases (Glatzel et al. 2003), and in the adrenal glands in primate transmission studies (Herzog et al. 2005). Together, these findings suggest that certain strains preferentially infect peripheral systems, and that there may be a secondary centrifugal spread of prions from the central nervous system back to the peripheral organs. This clearly has implications for iatrogenic transmission (see Sect. 3.2).

3 Human Prion Disease

3.1 The Rise of BSE

Prion disease has long fascinated clinicians and scientists by virtue of its association with an infectious protein. However, prion disease really rose in notoriety with the outbreak of bovine spongiform encephalopathy (BSE) in the UK in the 1980s. It had been presumed that scrapie in sheep did not transmit to other species, since it had been around for centuries. However, the BSE epidemic is thought to have occurred through the contamination of meat and bone meal, used to feed cattle, with scrapie. It is also possible that a sporadic case of BSE occurred in cattle and this was the source of contamination, as both sheep and cattle remnants were used to supplement feeds. Whatever the animal source, the incidence of BSE began dropping once meat and bone meal bans were put into place (Taylor 1996). Yet, while the BSE epidemic started winding down, a new epidemic soon began to emerge (see Fig. 4), that of a new human prion disease, variant Creutzfeldt–Jakob disease (vCJD).

3.2 *Transmission to Humans?*

Variant CJD (vCJD) is perhaps the most well-known type of acquired human prion disease, after kuru, and it has been strongly linked to eating BSE-contaminated food (Collinge et al. 1996; Bruce et al. 1997; Hill et al. 1997). vCJD can also transfer between humans, as demonstrated by five cases of transmission through blood transfusion (Hewitt et al. 2006; Peden et al. 2010). Despite its notoriety, vCJD has only affected a total of 222 patients worldwide as of April 2012 (http://www.cjd.ed.ac.uk). This is actually a very low number given the probability that over a million infected cattle entered the human food chain (Anderson et al. 1996). This inefficiency of transmission is actually not surprising, because in the field of prion transmission there is often a strong barrier to cross-species infection. However, once adapted to a species, transmission is highly efficient, with only a small infectious dose required to cause disease. Because of this, iatrogenic transmission remains a serious concern; there have been more than 400 iatrogenic cases of CJD resulting from corneal transplants, stereotactic electroencephalogram (EEG) electrodes, neurosurgery, cadaveric dura mater grafts, exposure to cadaveric human growth hormone and gonadotropin, and most recently transfusion of packed red blood cells (vCJD only) (Brown et al. 2006). No transmission has been seen with dentistry procedures and the risk is considered to be low (Head et al. 2003; Everington et al. 2007). However, there is realistic concern of transmission from endoscopy, although no cases have been documented, because prions are found in the gut lymphoid follicles and tonsils of vCJD patients. Sporadic CJD prions have not been found in the gut, but do occur in the spleen, so any endoscopy done on any CJD patient should be followed by incineration of the endoscope. The same approach is recommended for neurosurgical instruments used on a CJD brain. This has made many medical centres reluctant to perform brain biopsies on patients in whom CJD is in the differential diagnosis. Fortunately, person-to-person contact does not transmit this disease, sexual partners are not at higher risk, and there is no increased incidence in health care workers, pathologists or those who work in abattoirs. Also, vertical transmission has never been observed in cases of CJD; prions could not be detected in the uterus and gestational tissues of a pregnant sCJD patient (Xiao et al. 2009).

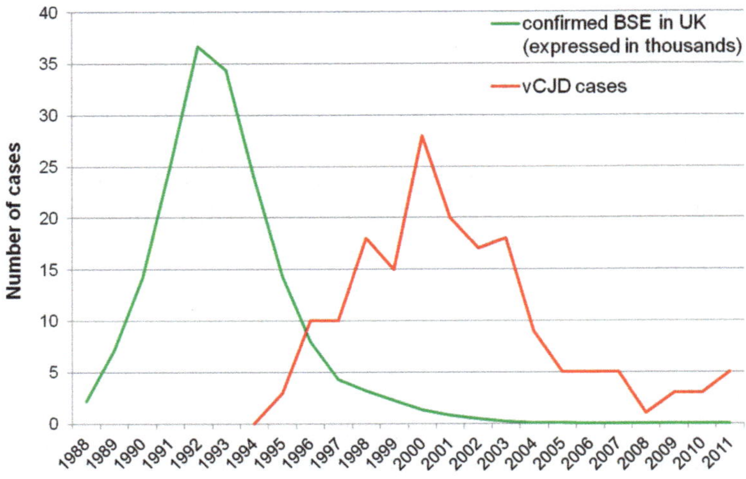

Fig. 4 Incidence of BSE in the UK and vCJD worldwide

3.3 *Crossing the Species Barrier*

For the most part, prions do not easily transmit across species barriers, as illustrated by the low incidence of vCJD after exposure to high levels of BSE. However, a process of adaptation can occur during the passage of a prion into a new host species (Pattison and Jones 1968; Kimberlin and Walker 1978; Kimberlin et al. 1987). This can generate a new prion "strain" which may now be able to infect a previously unsusceptible host. For example, scrapie is not pathogenic to humans, but if scrapie crossed into cattle to cause BSE, this newly adapted form or strain became pathogenic to humans, albeit it at low levels. This host adaptation and strain evolution have implications today, since one of the increasingly prevalent prion diseases is chronic wasting disease (CWD), which affects deer, elk and moose. CWD is shed in the environment and can be ingested by a number of other animals. So while CWD itself does not appear to transmit to humans, CWD which passes through other animals, such as the bank vole, which is susceptible to most known prion diseases, is a real concern.

3.4 *A Comment on Prion Strains*

Unlike in virology, the term "strain" is used more loosely in the prion field and is defined simply as any prion with a distinct incubation time, phenotype, and/or pathology (Fraser and Dickinson 1973; Bruce and Fraser 1991; Bruce 1993). Different species have slight differences in *Prnp* sequence, and polymorphisms do occur as well, all of which can affect the species barrier and produce some strain-like characteristics. However, sequence-identical strains can also be found within one host species, and it is suspected that the underlying cause for prion strains is a specific conformation of the protein itself which then induces the same structural change in the host PrP. Structural differences in strains have been detected by altered proteinase digestion and electrophoretic profiles (Bessen and Marsh 1992; Collinge et al. 1996; Telling et al. 1996), beta-sheet content (Caughey et al. 1998; Thomzig et al. 2004), denaturation curves (Peretz et al. 2001), antibody binding properties (Safar et al. 1998), and electron and atomic force microscopy (Sim and Caughey 2009a). Interestingly, there is evidence that more than one strain may be present within a single diseased sCJD brain (Parchi et al. 1999, 2009, Puoti et al. 1999; Polymenidou et al. 2005; Yull et al. 2006). One possible explanation for strain variation and the species barrier is that PrP^{C} can adopt multiple different conformations of PrP^{Sc} in any given host or cell type (Collinge and Clarke 2007). Some of these permissible conformations may overlap with those of different hosts. Such hosts may then be able to transfer PrP^{Sc} across their species barrier, where the conformation of PrP^{Sc} converts into the form preferred by the new species. This new form may then overlap with possible conformations within other species, and so the species barrier could be crossed again.

3.5 Classifying Human Prion Diseases: Sporadic CJD

The vast majority of human prion diseases, comprising 85–90 % of all cases, are sporadic, affecting 1–2 per million people per year. As in animal prion diseases, human prion diseases also have different strains with distinct PrP profiles. Attempts have been made to find clinical and pathological correlations with the molecular differences. This is complicated somewhat by the fact that two strains can coexist in up to 35 % of sCJD cases (Parchi et al. 2009). Nevertheless, the main strain categorization centres on two properties (1) the *Prnp* codon 129 polymorphism of the patient and (2) the electrophoretic pattern of PrP^{Sc} after proteinase K digestion (type 1 has a 21 kDa band, type 2 has a 19 kDa band; see Fig. 3). Codon 129 may either be methionine (M) or valine (V). While the relative frequencies in the Caucasian population are MM 37 %, MV 51 % and VV 12 % (Collinge et al. 1991), MM homozygosity is overrepresented in human prion diseases, implying a susceptibility influence. [M allele frequency does vary with ethnicity, and is 0.55 in Africans and >0.9 in Asians (Soldevila et al. 2003).] Using codon 129 and the electrophoretic profile, the following strains can be defined: MM1, MM2, MV1, MV2, VV1, and VV2. However, each of these does not act as a unique strain with respect to clinical and pathological findings. In animal transmission studies, MM1 and MV1 act similarly, as do MV2 and MM2 (Bishop et al. 2010). This confusing scenario has been well summarized in a recent review by Parchi et al. where strains are identified by molecular type, histopathological type, neuropathology, clinical features, prevalence, age of onset, and disease duration (Parchi et al. 2011).

A synopsis of the most common forms and the features relevant to clinical diagnosis (without biopsy information) are summarized in Table 1.

Important points to note include the different clinical features and diagnostic sensitivities. Myoclonus occurs in the most common forms, but is often a late symptom. When visual impairment occurs it can manifest as cortical blindness or Anton's syndrome with visual hallucinations. Dementia may be the first symptom, with or without ataxia, but 18 % have pure ataxia as the prominent initial feature. Often these ataxic patients are investigated for other forms of ataxia before prion disease is considered. Also of note is the fact that 14-3-3 can be falsely negative, especially in longer duration cases. Patients may also have a negative EEG, making Magnetic Resonance Imaging (MRI) the most helpful diagnostic tool. As will be discussed in the diagnostics section, there must be an index of suspicion and no test should be interpreted in isolation.

3.6 Truncated Fragments?

To complicate things further, there have been a number of truncated fragments of PrP^{Sc} identified on Western blot which correlate with specific subtypes of prion disease. These can be as small as 7–8 kDa up to 18.5 kDa and can be from the N or C terminus, or portions of the central region (reviewed by Parchi et al. 2011). While the presence of these fragments is not routinely used in strain typing, this is a definite possibility for the future.

3.7 Genetic Prion Disease

The first attribution of a genetic cause to a prion disease was made in 1950 (Jacob et al. 1950). After the sporadic CJD strains, genetic forms are the second most common, representing 10–15 % of cases. To date, all mutations have been found in the prion protein gene (*Prnp*) on chromosome 20, and are either point mutations or insertions into the octapeptide repeat region; E200K is the most common mutation. All are considered to be autosomal dominant and genetic screening is available. Historically, there are three main phenotypes of genetic prion disease: Gerstmann–Sträussler–Scheinker (GSS), fatal familial insomnia (FFI), and familial CJD (fCJD). Each has been shown to be transmissible (Roos et al. 1973; Rosenthal et al. 1976; Masters et al. 1981; Tateishi et al. 1995). Several different mutations may cause each phenotype, and it appears that molecular typing and codon 129 also influence clinical presentation. The classic example is the D178N mutation, which will produce a type 1 PrP molecular profile and a phenotype of fCJD if associated with 129 V on the same allele; if it is associated with 129 M it produces a type 2 PrP molecular profile and the FFI phenotype (reviewed in Goldfarb et al. 1992). As more of these rare cases are analyzed by molecular profiling, it is becoming apparent that clinical phenotypes of genetic CJD resemble their counterpart

Table 1 Characterization of the most common strains of sCJD by prevalence, molecular type, clinical, and diagnostic sensitivities

Prevalence (%)	Molecular type	Onset (years)	Duration (months)	Clinical presentation	Diagnostics
68	MM or MV Type 1 (40 %) or Type 1 AND 2 (28 %)	69 (42–89)	4 (1–26)	Rapid dementia Myoclonus 50 % ataxia at onset 30 % visual impairment	EEG positive 14-3-3 positive MRI: basal ganglia and cortex
18	VV Type 2 (15 %) or Type 1 AND 2 (3 %)	65 (45–85)	6 (3–18)	Ataxia at onset Dementia later	EEG negative 14-3-3 positive MRI: basal ganglia and thalamus
8	MV Type 2 (with plaques)	65 (48–81)	16 (5–48)	Dementia and ataxia	EEG negative 14-3-3 negative in 40 % MRI: basal ganglia and thalamus
3	MV Type 2 (with plaques and vacuoles)	n/a	n/a	n/a	n/a
1	MM or MV Type 2 (with vacuoles)	68 (61–75)	20 (12–36)	Dementia Myoclonus Pyramidal signs No ataxia	EEG negative 14-3-3 positive MRI: cortex
1	MM Type 2 (with thalamus and olive atrophy)	52 (36–71)	16 (8–24)	Insomnia Psychomotor hyper-activity Ataxia and motor signs	EEG negative 14-3-3 negative MRI: thalamus gliosis
1	VV Type 1	39 (24–49)	15 (14–16)	Dementia Myoclonus Pyramidal signs	EEG negative 14-3-3 positive MRI: cortex
<1	Variant CJD MM Type 2B[a]	27 (12–74)	14 (6–39)	Dementia and ataxia Myoclonus/chorea/dystonia Early psych changes Painful sensory symptoms (63 %)	EEG always negative 14-3-3 often negative MRI: pulvinar

vCJD is also listed for comparison. Codon 129 status (MM, MV, or VV) is available by genetic testing pre-mortem; PrP^{Sc} type (1 or 2) and plaque or vacuolar pathology is only available post-mortem or with biopsy but is included for completeness

[a]Type 2B in variant CJD refers to an altered glycoform ratio of PrP^{Sc}; all other types listed are type A. See text (vCJD) for details

sporadic CJD cases which share the same strain type (codon 129 and electrophoretic profile). For example, the fCJD phenotype resembles that of the most common form of sCJD (MM1), presenting with rapid dementia and myoclonus, and is generally seen when mutations are in *cis* with M at codon 129. The FFI phenotype is characterized by insomnia, dysautonomia, and thalamic atrophy (Medori et al. 1992) and has a sporadic counterpart in MM type 2 with thalamic and olive atrophy (see Table 1). GSS is more complicated in that the phenotypes can vary, as can mutations, there is no known sporadic counterpart, and the diagnosis relies on the pathology which demonstrates large multicentric amyloid plaques (see Fig. 1). The originally described GSS families develop a slowly progressive ataxia and late dementia, have a P102L mutation, and have codon 129 MM; however, a rapid phenotype resembling CJD also exists in this family, occurring with the same P102L mutation and codon 129 MM (Hainfellner et al. 1995). Interestingly, a case of P102L with 129 V has been reported, in which the presentation was largely psychiatric without ataxia (Bianca et al. 2003) and there have been two GSS cases resulting from an octapeptide repeat insertion with 129MM or 129MV and PrP^{Sc} type 1. These presented with a slow dementia, psychiatric symptoms, apraxia, and hypokinesis (Jansen et al. 2011).

3.8 Variant CJD

Variant CJD was first reported in 1996 (Will et al. 1996) and has several characteristics which distinguish it from other prion diseases (see Table 1). The diagnostic criteria are available through the World Health Organization; a recent assessment of these criteria has validated their ongoing use (Heath et al. 2010). Patients have a younger age of onset (average 27) and a prolonged course (average 14 months). Clinically, almost all cases present with early psychiatric symptoms such as depression, withdrawal or anxiety, and progress to dementia and ataxia. Movement disorders such as myoclonus, chorea, or dystonia are also prominent, and painful sensory symptoms are very specific when present (in 63 % of cases). Diagnostically, the periodic EEG changes are never seen, and 14-3-3 is much less reliable (Green et al. 2001). The MRI appearance classically involves the posterior pulvinar, which is positive in 91 % of cases, and unlike the other forms of CJD, vCJD can be diagnosed by tonsil biopsy with 93 % sensitivity (Heath et al. 2010). The brain pathology has a characteristic appearance also, one with amyloid plaques surrounded by vacuoles, called "florid plaques" (see Fig. 1). The molecular profile of vCJD is primarily MM type 2, but unlike other type 2 profiles, the three-band appearance on Western blotting shows a different glycoform ratio, with the upper band (diglycosylated band) being predominant (Collinge et al. 1996); in contrast, other forms of CJD have more dominant lower bands (see Fig. 3). As such, this appears to be very specific for vCJD, and is subclassified as type 2B. As for the codon 129 polymorphism, all proven symptomatic cases to date have been homozygous MM, but it appears that MV and VV may also be susceptible to vCJD, or at least are able to become infected. Two asymptomatic cases of MV have been detected after multiple blood transfusions (Peden et al. 2004, 2010), and a third MV case had symptoms of prion disease but did not go to autopsy so a confirmed diagnosis of prion disease could not be made (Kaski et al. 2009). Screening of appendices post-surgery also demonstrated PrP^{Sc} in two VV patients who remain asymptomatic (Ward 2006). Whether these patients will one day develop vCJD or remain carriers is not known. This has implications for surgical and endoscopy procedures with respect to decontamination or disposal of equipment (see Sect. 3.10).

3.9 Variably Protease-Sensitive Prionopathy

The traditional definition of prion disease has included the presence of protease-resistant PrP. However, a new entity "variably protease-sensitive prionopathy" (VPSPr) has been described, which may represent more than 3 % of all sCJD cases (Gambetti et al. 2008, 2011). The true prevalence is difficult to estimate because many cases have likely been diagnosed as non-Alzheimer's dementia. Clinical presentation includes behavioural and psychiatric features, aphasia, dementia, parkinsonism and ataxia, average onset at age 62, and a disease duration of 20 months. As in other prion diseases, there is spongiform change on pathology, but the pattern is unique. PrP still forms amyloid and is found in deposits, but in a distinct pattern. Most of the abnormal PrP is protease-sensitive, and the small amount which is resistant has a different electrophoretic profile from any other prion disease. No *Prnp* gene mutations have been found. It also preferentially occurs in the presence of codon 129 valine, in contrast with every other known prion disease. It remains to be seen whether this condition is infectious, as a true prion disease, or whether it is a protein-folding disease more akin to Alzheimer's or Parkinson's disease.

3.10 Prion Decontamination

The high efficiency of prion disease transmission, coupled with the resistance of PrP^{Sc} to standard decontamination techniques, makes iatrogenic transmission a real concern. Prions have an affinity for binding stainless steel (Flechsig et al. 2001; Edgeworth et al. 2009) and once dried are even harder to remove (Secker et al. 2011). Fortunately, rates of iatrogenic CJD are declining, likely because of increased awareness and screening of patients for CJD risk factors (Brown et al. 2006). Infectivity can be reduced to levels below detection in animal studies, and World Health Organization guidelines allow researchers to decontaminate instruments by soaking in 2 % chlorine for 1 h, or 1–2 M NaOH for 1 h followed by autoclaving at 134°C. However, none of these approaches is approved for human use; all surgical equipment that comes into contact with CJD tissue must be incinerated at 850°C. As such, most hospitals will not perform brain biopsies on patients suspected of having CJD.

3.11 *Diagnosis*

Pathologically, prion disease is characterized by spongiform change, astrocytosis, and the accumulation of PrP^{Sc}, diffusely or in plaques, depending on the strain and type of prion disease (see Fig. 1). However, without biopsy, the diagnosis requires clinical suspicion and supportive tests. No one test is 100 % accurate, and must be interpreted in context. The mainstay investigations are neuroimaging, electroencephalography, and cerebrospinal fluid (CSF) analysis. However, a new test, the "RT-QuIC" is on the horizon, and may provide the highest sensitivities and specificities of any test yet.

3.11.1 Magnetic Resonance Imaging (MRI)

Hyperintensities can be seen on MRI, but restricted diffusion appears to be the most useful parameter as it is detectable in 93–100 % of cases, with sensitivities and specificities of 91–96 % and 93–95 %, respectively (Young et al. 2005; Vitali et al. 2011). The reason for this restricted diffusion is not entirely clear, but the location of signal change correlates with areas of vacuolation and PrP^{Sc} deposition (Geschwind et al. 2009). The areas involved may also relate to the type of prion strain involved (see Table 1). Fifty-eight to 68 % of cases involve cortex (cortical ribboning) plus deep gray matter (striatum and/or thalamus), 24–33 % of cases involve only cortex, and 5 % have only deep gray matter involvement (Young et al. 2005; Meissner et al. 2008; Vitali et al. 2011). Examples of these patterns are shown in Fig. 5. The specificity is highest when corresponding hyperintensity is seen on both FLAIR and DWI images. Compared with patients having non-prion causes of rapidly progressive dementia, prion disease cases never have hyperintensities in only the limbic regions and rarely have changes in the precentral gyrus (Vitali et al. 2011).

3.11.2 Other Imaging Modalities

Positron emission tomography (PET) scans reportedly show extensive cortical hypometabolism in cases of sporadic CJD (Kim et al. 2011), but its utility in diagnosis has not been fully assessed.

3.11.3 Electroencephalogram

Characteristic changes can be seen on EEG in cases of sCJD (never in vCJD) although the findings often gradually evolve and then disappear. As shown in Fig. 6, the pattern includes sustained periodic sharp wave complexes (PSWC) with a variability of <500 ms, with generalized or lateralized periodic complexes (lasting 100–600 ms) demonstrating a bi- or triphasic morphology (Steinhoff et al. 1996). Age at disease onset and overall disease duration influence this test's sensitivity

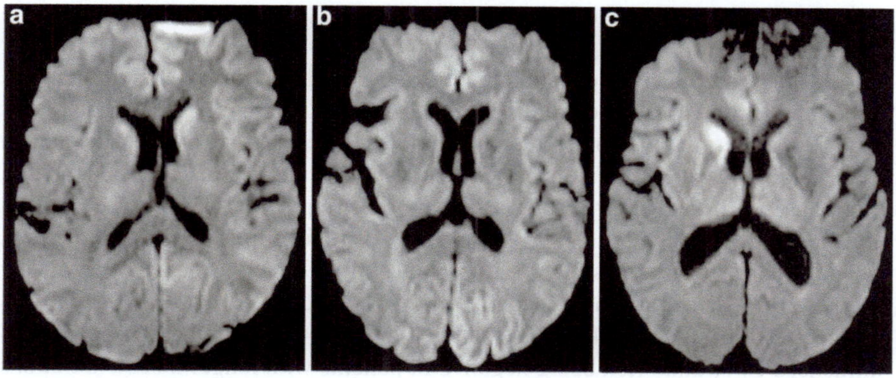

Fig. 5 Diffusion weighted MRI images from three CJD patients demonstrating different patterns of hyperintensity. (**a**) Cortical ribboning in left medial occipital lobe and insula plus hyperintense caudate bilaterally (the hyperintensity seen at the left frontal pole is artefact); (**b**) cortical ribboning most marked in left occipital lobe without deep gray involvement; (**c**) right caudate hyperintensity without significant cortical involvement

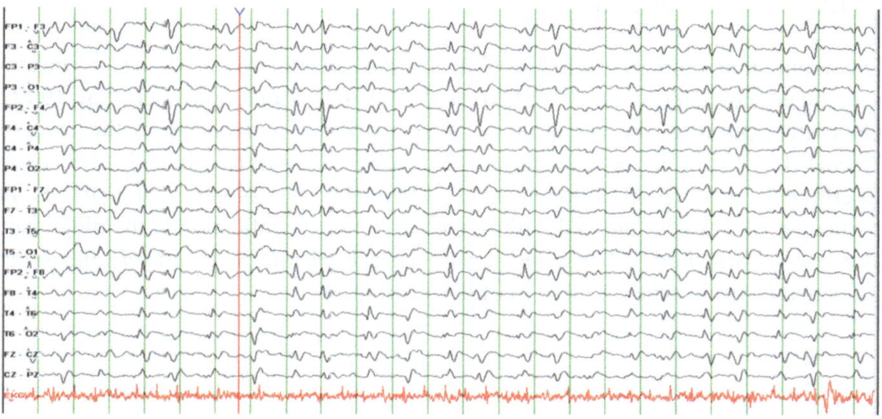

Fig. 6 EEG from a patient with sCJD showing generalized periodic positive sharp wave complexes

(Collins et al. 2006). Only one-third of patients under the age of 50, or whose course lasts longer than 12 months, have the classic EEG changes. Patients older than 70 or whose course lasts less than 6 months have changes 65 % of the time. Despite this low sensitivity, the test remains useful with a specificity of 86–91 % (Steinhoff et al. 1996, 2004). Again, context is always important (diagnosis is not made by EEG alone), and one must be sure that metabolic parameters are normal (hepatic encephalopathy may also produce triphasic complexes). Also, the generalized EEG complexes of sCJD can been misinterpreted as status epilepticus (Lapergue et al. 2010).

3.11.4 Cerebrospinal Fluid Analysis

In CJD there are no signs of inflammation and usually all parameters are normal, although the protein may be mildly elevated. Another commonly ordered marker is 14-3-3, which is highly expressed in the synapses of neurons and is often elevated in the CSF of CJD patients, but the protein is not specific to prion disease and anything that causes neuronal injury, such as stroke or encephalitis, can also lead to an elevated 14-3-3 (Hsich et al. 1996). There are several isoforms of 14-3-3, with β, γ, ε being more specific to sCJD (Wiltfang et al. 1999). A recent ten-year review measured the sensitivity and specificity of 14-3-3 for sCJD as 86 % and 74 % respectively (Chohan et al. 2010), but as shown in Table 1, these appear to be influenced by strain, often being negative in patients with longer disease durations (Parchi et al. 2011). 14-3-3 is much less sensitive for vCJD (Green et al. 2001). For the genetic forms of disease, sensitivity is 81 % for familial CJD but only 13 % and 10 % for FFI and GSS, respectively (Ladogana et al. 2009).

3.11.5 Other Biomarkers

Other potential biomarkers in the CSF are tau, S100b, Abeta1-42, and neuron-specific enolase (Zerr et al. 1995; Otto et al. 1997a, b; Van Everbroeck et al. 1999). Generally speaking, the elevation of these markers, or decrease in the case of Abeta1-42, has a higher specificity but lower sensitivity than 14-3-3 (Van Everbroeck et al. 2005; Chohan et al. 2010). The combination of markers, in particular 14-3-3 with tau and S100b, may give more confidence in the diagnosis. However, a standardized diagnostic ratio is lacking and there may again be an influence of strain type on these marker sensitivities.

3.11.6 Quake-Induced Conversion (QuIC)

An exciting new development in diagnosis occurred with the creation of the QuIC assay (Atarashi et al. 2008). This assay takes advantage of the seeded polymerization model of prion propagation (see Fig. 2). First, a small amount of infectious prion (within a CSF sample for example) is incubated with normal recombinant prion protein (rPrP) under shaking conditions in a tube. Over time, the prion converts some of the rPrP into a higher beta-sheet form, analogous to the infectious form. Next, an aliquot of this sample is diluted into a tube containing fresh rPrP and the process is continued. The process amplifies the amount of high beta-sheet rPrP with each step, until enough protease-resistant rPrP is generated to be detectable by Western blotting.

Real-Time *(RT-QuIC)*

The QuIC assay readout has been made easier with Thioflavin T (ThT), a compound which fluoresces when it binds the cross-beta-sheet structure of the converted form (Wilham et al. 2010; Atarashi et al. 2011b). Using this "real time" version, 200 CSF samples from subjects with sCJD or Alzheimer's disease were tested; no false positives were detected and the sensitivity was greater than 80 % (Atarashi et al. 2011a). However, the sensitivity for vCJD may be somewhat lower (Peden et al. 2011).

Enhanced *(eQuIC)*

The latest adaptation incorporates an antibody binding step to concentrate the prions in the sample. With this "enhanced" approach there is a remarkable increase in sensitivity for detecting vCJD (down to 2 attograms per mL of protein) including detection in blood samples spiked with vCJD prions (Orru et al. 2011). Applying this testing to clinical samples is now underway.

3.12 Prion Disease Mimics: The Differential Diagnosis

The challenge to any clinician is both to consider prion disease as a possible diagnosis and to exclude other possibilities which may mimic prion disease, especially those that may be treatable. While following the above diagnostic testing parameters is helpful, it is also reasonable to give a trial of high-dose steroids (1 g IV methylprednisolone daily for 5 days). This is because the main prion disease mimics are paraneoplastic or autoimmune syndromes, which are steroid-responsive. To date, none of these mimics manifest the diffusion changes on MRI (Vitali et al. 2011) but can have EEG changes and/or 14-3-3 positivity. Hashimoto's encephalitis, anti-voltage-gated potassium channel encephalopathy, anti-glutamic acid decarboxylase cerebellar ataxia, enteroviral meningoencephalitis, and autoimmune inflammatory meningoencephalitis are some of the recently identified mimics which generally respond to steroids (Hoffman Snyder et al. 2006; Chang et al. 2007; Geschwind et al. 2008b; Valcour et al. 2008; Santoro et al. 2011). Gluten sensitivity, associated with anti-gliadin, anti-transglutaminase, or anti-endomysial antibodies, may produce a dementia syndrome with ataxia and myoclonus (reviewed in Hu et al. 2006). Sjogren's syndrome or lupus can also present as a rapidly progressive dementia (Chin and Latov 2005). Prion diseases may mimic other diseases too; GSS, with its variable presentation and longer disease duration, has been mistaken for multiple sclerosis (Karmon et al. 2011). The characteristic EEG changes of CJD have been misinterpreted as non-convulsive status (Espinosa et al. 2010); conversely, CJD patients can actually present in non-convulsive status, although this is uncommon

Table 2 Workup for rapidly progressive dementias

Bloodwork	Complete blood count and differential, electrolytes, liver function, thyroid function
	Erythrocyte sedimentation rate, C-reactive protein
	Anti-nuclear antigens, extractable nuclear antigens
	Anti-thyroglobulin, anti-thyroperoxidase
	Vitamin B12, homocysteine
	Rapid plasmin reagin, HIV, Lyme serology
	Paraneoplastic panel
	Autoimmune panel: include anti-NMDA, VGKC, GAD, CV2
Urine	Urinalysis
	Optional: heavy metals, copper (for Wilson's disease)
CSF	Cells, protein, glucose
	IgG index, oligoclonal bands
	Cytology
	14-3-3, tau, S100b, neuron specific enolase
	VDRL
	Optional: Cryptococcal antigen, viral PCR, bacterial/ fungal cultures, acid fast bacteria, Whipple's PCR
MRI brain	Include FLAIR and DWI, +/− contrast
EEG	May do serially to detect evolution of periodic discharges
Other imaging	CT chest/abdomen/pelvis +/− contrast or body PET scan for occult neoplasm
Other	Testicular exam, breast exam/mammography

(Cohen et al. 2004; Aiguabella et al. 2010). [The general workup for rapidly progressive dementia is well reviewed in Geschwind et al. (2008a) and is summarized in Table 2.]

3.13 Treatment

At present there are no effective therapies, let alone cures, for any of the prion diseases, but many strategies have been explored to reduce disease: (1) decontamination of prion sources, (2) prophylaxis after exposure, (3) inhibition of prion propagation outside the central nervous system (applicable in orally acquired prion diseases), (4) prevention of neuroinvasion, (5) inhibition of conversion to and/or accumulation of PrP^{Sc}, (6) destabilization or enhanced clearance of PrP^{Sc}, (7) blockade of neurotoxic effects of PrP^{Sc}, and (8) compensation strategies for damaged or lost neurons. (For reviews of the numerous anti-prion compounds tried to date, please see Sim and Caughey 2009b; Sim 2012) Suffice it to say that most compounds have failed to demonstrate effect in humans, despite initial promise in animal studies. A few worth mentioning include pentosan polysulphate (PPS), quinacrine, and doxycycline.

3.13.1 Pentosan Polysulphate

This compound is a very effective inhibitor in cell culture (Caughey et al. 1994). It interferes with the binding of PrP^{Sc}, PrP^{C} and endogenous polyanions thus limiting conversion (Caughey and Raymond 1993), and it also promotes PrP^{C} internalization, thus reducing the amount of PrP^{C} available for conversion (Shyng et al. 1995). Because PPS does not cross the blood–brain barrier, intraventricular delivery into the CNS has been performed with success in animals, and has also been done in a number of human cases with mixed results (Todd et al. 2005; Rainov et al. 2007; Bone et al. 2008; Tsuboi et al. 2009). As these have not been prospective controlled studies and results are conflicting, it is impossible to conclude the extent of benefit in humans. A recent analysis of brains from patients treated with intraventricular PPS shows that there may be some effect on the accumulation of PrP^{Sc} oligomers but overall the pathology is similar in patients with and without treatment (Honda et al. 2011).

3.13.2 Quinacrine

This compound is also effective in cell culture (Doh-Ura et al. 2000; Korth et al. 2001), but lacks efficacy in animal models (Collins et al. 2002; Barret et al. 2003). Unlike PPS, however, 300 mg daily quinacrine treatment has been studied in a large human prospective study called PRION-1. Unfortunately, there was no evidence for an effect on lifespan or other clinical measures (Collinge et al. 2009; Mead et al. 2011) and preliminary data from a U.S. quinacrine trial have also been disappointing.

3.13.3 Doxycycline

The latest treatment to garner excitement has been doxycycline, which has also been studied in humans. Preliminary data demonstrate that 100 mg daily can double survival time in CJD patients, depending on the codon 129 polymorphism. However, before rushing to use this medication as a standard treatment, one has to recognize that prolonging the lifespan may not have a clinically significant effect on quality of life parameters. Final publications of the results are expected soon.

3.13.4 Symptomatic Treatments

Without curative agents, clinicians are left to treat symptoms. Impulsivity and hallucinations can be controlled with atypical antipsychotics such as quetiapine. Myoclonus can be treated with clonazepam to some extent.

3.14 *Conclusion*

Prions are unconventional infectious agents which, unlike viruses, are protein based. The fatal transmissible neurodegenerative diseases they cause are a challenge to physicians in terms of diagnosis and management, and are a challenge to researchers who strive to understand how the misfolding of a protein can have such dire consequences.

References

Aiguabella M, Falip M, Veciana M, Bruna J, Palasi A, Corral L, Herrero JI, Boluda S, Mora J, Iranzo A, Serrano C (2010) Refractory nonconvulsive status epilepticus in Creutzfeldt-Jakob disease. Epileptic Disord 12:239–242. doi:10.1684/epd.2010.0318

Alper T, Cramp WA, Haig DA, Clarke MC (1967) Does the agent of scrapie replicate without nucleic acid? Nature 214:764–766

Anderson RM, Donnelly CA, Ferguson NM, Woolhouse ME, Watt CJ, Udy HJ, MaWhinney S, Dunstan SP, Southwood TR, Wilesmith JW, Ryan JB, Hoinville LJ, Hillerton JE, Austin AR, Wells GA (1996) Transmission dynamics and epidemiology of BSE in British cattle. Nature 382:779–788. doi:10.1038/382779a0

Atarashi R, Sano K, Satoh K, Nishida N (2011a) Real-time quaking-induced conversion: a highly sensitive assay for prion detection. Prion 5:150–153. doi:10.4161/pri.5.3.16893

Atarashi R, Moore RA, Sim VL, Hughson AG, Dorward DW, Onwubiko HA, Priola SA, Caughey B (2007) Ultrasensitive detection of scrapie prion protein using seeded conversion of recombinant prion protein. Nat Methods 4:645–650. doi:10.1038/nmeth1066

Atarashi R, Wilham JM, Christensen L, Hughson AG, Moore RA, Johnson LM, Onwubiko HA, Priola SA, Caughey B (2008) Simplified ultrasensitive prion detection by recombinant PrP conversion with shaking. Nat Methods 5:211–212. doi:10.1038/nmeth0308-211

Atarashi R, Satoh K, Sano K, Fuse T, Yamaguchi N, Ishibashi D, Matsubara T, Nakagaki T, Yamanaka H, Shirabe S, Yamada M, Mizusawa H, Kitamoto T, Klug G, McGlade A, Collins SJ, Nishida N (2011b) Ultrasensitive human prion detection in cerebrospinal fluid by real-time quaking-induced conversion. Nat Med 17:175–178. doi:10.1038/nm.2294

Baron D, Assaraf YG, Cohen N, Aronheim A (2002) Lack of plasma membrane targeting of a G172D mutant thiamine transporter derived from Rogers syndrome family. Mol Med 8:462–474

Barret A, Tagliavini F, Forloni G, Bate C, Salmona M, Colombo L, De Luigi A, Limido L, Suardi S, Rossi G, Auvre F, Adjou KT, Sales N, Williams A, Lasmezas C, Deslys JP (2003) Evaluation of quinacrine treatment for prion diseases. J Virol 77:8462–8469

Barria MA, Mukherjee A, Gonzalez-Romero D, Morales R, Soto C (2009) De novo generation of infectious prions in vitro produces a new disease phenotype. PLoS Pathog 5:e1000421. doi:10.1371/journal.ppat.1000421

Barry RA, McKinley MP, Bendheim PE, Lewis GK, DeArmond SJ, Prusiner SB (1985) Antibodies to the scrapie protein decorate prion rods. J Immunol 135:603–613

Bartz JC, Dejoia C, Tucker T, Kincaid AE, Bessen RA (2005) Extraneural prion neuroinvasion without lymphoreticular system infection. J Virol 79:11858–11863. doi:10.1128/JVI.79.18.11858-11863.2005

Berardi VA, Cardone F, Valanzano A, Lu M, Pocchiari M (2006) Preparation of soluble infectious samples from scrapie-infected brain: a new tool to study the clearance of transmissible spongiform encephalopathy agents during plasma fractionation. Transfusion 46:652–658. doi:10.1111/j.1537-2995.2006.00763.x

Bessen RA, Marsh RF (1992) Biochemical and physical properties of the prion protein from two strains of the transmissible mink encephalopathy agent. J Virol 66:2096–2101
Bessen RA, Martinka S, Kelly J, Gonzalez D (2009) Role of the lymphoreticular system in prion neuroinvasion from the oral and nasal mucosa. J Virol 83:6435–6445. doi:10.1128/JVI.00018-09
Bianca M, Bianca S, Vecchio I, Raffaele R, Ingegnosi C, Nicoletti F (2003) Gerstmann-Straussler-Scheinker disease with P102L-V129 mutation: a case with psychiatric manifestations at onset. Ann Genet 46:467–469
Bishop MT, Will RG, Manson JC (2010) Defining sporadic Creutzfeldt-Jakob disease strains and their transmission properties. Proc Natl Acad Sci USA 107:12005–12010. doi:10.1073/pnas.1004688107
Bolton DC, McKinley MP, Prusiner SB (1982) Identification of a protein that purifies with the scrapie prion. Science 218:1309–1311
Bone I, Belton L, Walker AS, Darbyshire J (2008) Intraventricular pentosan polysulphate in human prion diseases: an observational study in the UK. Eur J Neurol 15:458–464. doi:10.1111/j.1468-1331.2008.02108.x
Borchelt DR, Taraboulos A, Prusiner SB (1992) Evidence for synthesis of scrapie prion proteins in the endocytic pathway. J Biol Chem 267:16188–16199
Brandner S, Isenmann S, Raeber A, Fischer M, Sailer A, Kobayashi Y, Marino S, Weissmann C, Aguzzi A (1996) Normal host prion protein necessary for scrapie-induced neurotoxicity. Nature 379:339–343. doi:10.1038/379339a0
Brandner S, Klein MA, Frigg R, Pekarik V, Parizek P, Raeber A, Glatzel M, Schwarz P, Rulicke T, Weissmann C, Aguzzi A (2000) Neuroinvasion of prions: insights from mouse models. Exp Physiol 85:705–712
Brandner S, Whitfield J, Boone K, Puwa A, O'Malley C, Linehan JM, Joiner S, Scaravilli F, Calder I, Alpers MP, Wadsworth JD, Collinge J (2008) Central and peripheral pathology of kuru: pathological analysis of a recent case and comparison with other forms of human prion disease. Philos Trans R Soc Lond B Biol Sci 363:3755–3763. doi:10.1098/rstb.2008.0091
Brown P, Brandel JP, Preece M, Sato T (2006) Iatrogenic Creutzfeldt-Jakob disease: the waning of an era. Neurology 67:389–393. doi:10.1212/01.wnl.0000231528.65069.3f
Bruce ME (1993) Scrapie strain variation and mutation. Br Med Bull 49:822–838
Bruce ME, Fraser H (1991) Scrapie strain variation and its implications. Curr Top Microbiol Immunol 172:125–138
Bruce ME, McConnell I, Will RG, Ironside JW (2001) Detection of variant Creutzfeldt-Jakob disease infectivity in extraneural tissues. Lancet 358:208–209
Bruce ME, Will RG, Ironside JW, McConnell I, Drummond D, Suttie A, McCardle L, Chree A, Hope J, Birkett C, Cousens S, Fraser H, Bostock CJ (1997) Transmissions to mice indicate that 'new variant' CJD is caused by the BSE agent. Nature 389:498–501. doi:10.1038/39057
Bueler H, Aguzzi A, Sailer A, Greiner RA, Autenried P, Aguet M, Weissmann C (1993) Mice devoid of PrP are resistant to scrapie. Cell 73:1339–1347
Carlson GA, Kingsbury DT, Goodman PA, Coleman S, Marshall ST, DeArmond S, Westaway D, Prusiner SB (1986) Linkage of prion protein and scrapie incubation time genes. Cell 46:503–511
Castilla J, Saa P, Hetz C, Soto C (2005) In vitro generation of infectious scrapie prions. Cell 121:195–206. doi:10.1016/j.cell.2005.02.011
Caughey B, Raymond GJ (1991) The scrapie-associated form of PrP is made from a cell surface precursor that is both protease- and phospholipase-sensitive. J Biol Chem 266:18217–18223
Caughey B, Raymond GJ (1993) Sulfated polyanion inhibition of scrapie-associated PrP accumulation in cultured cells. J Virol 67:643–650
Caughey B, Raymond GJ, Bessen RA (1998) Strain-dependent differences in beta-sheet conformations of abnormal prion protein. J Biol Chem 273:32230–32235
Caughey B, Raymond GJ, Ernst D, Race RE (1991a) N-terminal truncation of the scrapie-associated form of PrP by lysosomal protease(s): implications regarding the site of conversion of PrP to the protease-resistant state. J Virol 65:6597–6603

Caughey B, Brown K, Raymond GJ, Katzenstein GE, Thresher W (1994) Binding of the protease-sensitive form of PrP (prion protein) to sulfated glycosaminoglycan and congo red [corrected]. J Virol 68:2135–2141

Caughey BW, Dong A, Bhat KS, Ernst D, Hayes SF, Caughey WS (1991b) Secondary structure analysis of the scrapie-associated protein PrP 27-30 in water by infrared spectroscopy. Biochemistry 30:7672–7680

Chang CC, Eggers SD, Johnson JK, Haman A, Miller BL, Geschwind MD (2007) Anti-GAD antibody cerebellar ataxia mimicking Creutzfeldt-Jakob disease. Clin Neurol Neurosurg 109:54–57. doi:10.1016/j.clineuro.2006.01.009

Chesebro B, Trifilo M, Race R, Meade-White K, Teng C, LaCasse R, Raymond L, Favara C, Baron G, Priola S, Caughey B, Masliah E, Oldstone M (2005) Anchorless prion protein results in infectious amyloid disease without clinical scrapie. Science 308:1435–1439. doi:10.1126/science.1110837

Chin RL, Latov N (2005) Central nervous system manifestations of rheumatologic diseases. Curr Opin Rheumatol 17:91–99

Chohan G, Pennington C, Mackenzie JM, Andrews M, Everington D, Will RG, Knight RS, Green AJ (2010) The role of cerebrospinal fluid 14-3-3 and other proteins in the diagnosis of sporadic Creutzfeldt-Jakob disease in the UK: a 10-year review. J Neurol Neurosurg Psychiatry 81:1243–1248. doi:10.1136/jnnp. 2009.197962

Cobb NJ, Sonnichsen FD, McHaourab H, Surewicz WK (2007) Molecular architecture of human prion protein amyloid: a parallel, in-register beta-structure. Proc Natl Acad Sci USA 104:18946–18951. doi:10.1073/pnas.0706522104

Cohen D, Kutluay E, Edwards J, Peltier A, Beydoun A (2004) Sporadic Creutzfeldt-Jakob disease presenting with nonconvulsive status epilepticus. Epilepsy Behav 5:792–796. doi:10.1016/j.yebeh.2004.06.019

Collinge J, Clarke AR (2007) A general model of prion strains and their pathogenicity. Science 318:930–936. doi:10.1126/science.1138718

Collinge J, Palmer MS, Dryden AJ (1991) Genetic predisposition to iatrogenic Creutzfeldt-Jakob disease. Lancet 337:1441–1442

Collinge J, Sidle KC, Meads J, Ironside J, Hill AF (1996) Molecular analysis of prion strain variation and the aetiology of 'new variant' CJD. Nature 383:685–690. doi:10.1038/383685a0

Collinge J, Harding AE, Owen F, Poulter M, Lofthouse R, Boughey AM, Shah T, Crow TJ (1989) Diagnosis of Gerstmann-Straussler syndrome in familial dementia with prion protein gene analysis. Lancet 2:15–17

Collinge J, Gorham M, Hudson F, Kennedy A, Keogh G, Pal S, Rossor M, Rudge P, Siddique D, Spyer M, Thomas D, Walker S, Webb T, Wroe S, Darbyshire J (2009) Safety and efficacy of quinacrine in human prion disease (PRION-1 study): a patient-preference trial. Lancet Neurol 8:334–344. doi:10.1016/S1474-4422(09)70049-3

Collins SJ, Lewis V, Brazier M, Hill AF, Fletcher A, Masters CL (2002) Quinacrine does not prolong survival in a murine Creutzfeldt-Jakob disease model. Ann Neurol 52:503–506. doi:10.1002/ana.10336

Collins SJ, Sanchez-Juan P, Masters CL, Klug GM, van Duijn C, Poleggi A, Pocchiari M, Almonti S, Cuadrado-Corrales N, de Pedro-Cuesta J, Budka H, Gelpi E, Glatzel M, Tolnay M, Hewer E, Zerr I, Heinemann U, Kretszchmar HA, Jansen GH, Olsen E, Mitrova E, Alperovitch A, Brandel JP, Mackenzie J, Murray K, Will RG (2006) Determinants of diagnostic investigation sensitivities across the clinical spectrum of sporadic Creutzfeldt-Jakob disease. Brain 129:2278–2287. doi:10.1093/brain/awl159

Crozet C, Lezmi S, Flamant F, Samarut J, Baron T, Bencsik A (2007) Peripheral circulation of the prion infectious agent in transgenic mice expressing the ovine prion protein gene in neurons only. J Infect Dis 195:997–1006. doi:10.1086/512085

Deleault NR, Harris BT, Rees JR, Supattapone S (2007) Formation of native prions from minimal components in vitro. Proc Natl Acad Sci USA 104:9741–9746. doi:10.1073/pnas.0702662104

DeMarco ML, Daggett V (2004) From conversion to aggregation: protofibril formation of the prion protein. Proc Natl Acad Sci USA 101:2293–2298

DeMarco ML, Silveira J, Caughey B, Daggett V (2006) Structural properties of prion protein protofibrils and fibrils: an experimental assessment of atomic models. Biochemistry 45:15573–15582. doi:10.1021/bi0612723

Dickinson AG, Meikle VM, Fraser H (1968) Identification of a gene which controls the incubation period of some strains of scrapie agent in mice. J Comp Pathol 78:293–299

Doh-Ura K, Iwaki T, Caughey B (2000) Lysosomotropic agents and cysteine protease inhibitors inhibit scrapie-associated prion protein accumulation. J Virol 74:4894–4897

Edgeworth JA, Jackson GS, Clarke AR, Weissmann C, Collinge J (2009) Highly sensitive, quantitative cell-based assay for prions adsorbed to solid surfaces. Proc Natl Acad Sci USA 106:3479–3483. doi:10.1073/pnas.0813342106

Espinosa PS, Bensalem-Owen MK, Fee DB (2010) Sporadic Creutzfeldt-Jakob disease presenting as nonconvulsive status epilepticus case report and review of the literature. Clin Neurol Neurosurg 112:537–540. doi:10.1016/j.clineuro.2010.03.025

Everington D, Smith AJ, Ward HJ, Letters S, Will RG, Bagg J (2007) Dental treatment and risk of variant CJD–a case control study. Br Dent J 202:470–471. doi:10.1038/bdj.2007.126

Flechsig E, Hegyi I, Enari M, Schwarz P, Collinge J, Weissmann C (2001) Transmission of scrapie by steel-surface-bound prions. Mol Med 7:679–684

Fraser H, Dickinson AG (1973) Scrapie in mice. Agent-strain differences in the distribution and intensity of grey matter vacuolation. J Comp Pathol 83:29–40

Gajdusek DC, Gibbs CJ, Alpers M (1966) Experimental transmission of a Kuru-like syndrome to chimpanzees. Nature 209:794–796

Gambetti P, Puoti G, Zou WQ (2011) Variably protease-sensitive prionopathy: a novel disease of the prion protein. J Mol Neurosci 45:422–424. doi:10.1007/s12031-011-9543-1

Gambetti P, Dong Z, Yuan J, Xiao X, Zheng M, Alshekhlee A, Castellani R, Cohen M, Barria MA, Gonzalez-Romero D, Belay ED, Schonberger LB, Marder K, Harris C, Burke JR, Montine T, Wisniewski T, Dickson DW, Soto C, Hulette CM, Mastrianni JA, Kong Q, Zou WQ (2008) A novel human disease with abnormal prion protein sensitive to protease. Ann Neurol 63:697–708. doi:10.1002/ana.21420

Geoghegan JC, Miller MB, Kwak AH, Harris BT, Supattapone S (2009) Trans-dominant inhibition of prion propagation in vitro is not mediated by an accessory cofactor. PLoS Pathog 5: e1000535. doi:10.1371/journal.ppat.1000535

Geschwind MD, Shu H, Haman A, Sejvar JJ, Miller BL (2008a) Rapidly progressive dementia. Ann Neurol 64:97–108. doi:10.1002/ana.21430

Geschwind MD, Potter CA, Sattavat M, Garcia PA, Rosen HJ, Miller BL, DeArmond SJ (2009) Correlating DWI MRI with pathologic and other features of Jakob-Creutzfeldt disease. Alzheimer Dis Assoc Disord 23:82–87

Geschwind MD, Tan KM, Lennon VA, Barajas RF Jr, Haman A, Klein CJ, Josephson SA, Pittock SJ (2008b) Voltage-gated potassium channel autoimmunity mimicking Creutzfeldt-Jakob disease. Arch Neurol 65:1341–1346. doi:10.1001/archneur.65.10.1341

Gibbs CJ Jr, Gajdusek DC, Asher DM, Alpers MP, Beck E, Daniel PM, Matthews WB (1968) Creutzfeldt-Jakob disease (spongiform encephalopathy): transmission to the chimpanzee. Science 161:388–389

Glatzel M, Heppner FL, Albers KM, Aguzzi A (2001) Sympathetic innervation of lymphoreticular organs is rate limiting for prion neuroinvasion. Neuron 31:25–34

Glatzel M, Abela E, Maissen M, Aguzzi A (2003) Extraneural pathologic prion protein in sporadic Creutzfeldt-Jakob disease. N Engl J Med 349:1812–1820. doi:10.1056/NEJMoa030351

Goldfarb LG, Petersen RB, Tabaton M, Brown P, LeBlanc AC, Montagna P, Cortelli P, Julien J, Vital C, Pendelbury WW (1992) Fatal familial insomnia and familial Creutzfeldt-Jakob disease: disease phenotype determined by a DNA polymorphism. Science 258:806–808

Govaerts C, Wille H, Prusiner SB, Cohen FE (2004) Evidence for assembly of prions with left-handed beta-helices into trimers. Proc Natl Acad Sci USA 101:8342–8347. doi:10.1073/pnas.0402254101

Green AJ, Thompson EJ, Stewart GE, Zeidler M, McKenzie JM, MacLeod MA, Ironside JW, Will RG, Knight RS (2001) Use of 14-3-3 and other brain-specific proteins in CSF in the diagnosis of variant Creutzfeldt-Jakob disease. J Neurol Neurosurg Psychiatry 70:744–748

Griffith JS (1967) Self-replication and scrapie. Nature 215:1043–1044

Hadlow WJ (1959) Scrapie and kuru. Lancet 274:289–290. doi:10.1016/S0140-6736(59)92081-1

Hainfellner JA, Brantner-Inthaler S, Cervenakova L, Brown P, Kitamoto T, Tateishi J, Diringer H, Liberski PP, Regele H, Feucht M (1995) The original Gerstmann-Straussler-Scheinker family of Austria: divergent clinicopathological phenotypes but constant PrP genotype. Brain Pathol 5:201–211

Head MW, Ritchie D, McLoughlin V, Ironside JW (2003) Investigation of PrPres in dental tissues in variant CJD. Br Dent J 195:339–343. doi:10.1038/sj.bdj.4810536

Heath CA, Cooper SA, Murray K, Lowman A, Henry C, MacLeod MA, Stewart GE, Zeidler M, MacKenzie JM, Ironside JW, Summers DM, Knight RS, Will RG (2010) Validation of diagnostic criteria for variant Creutzfeldt-Jakob disease. Ann Neurol 67:761–770. doi:10.1002/ana.21987

Herzog C, Riviere J, Lescoutra-Etchegaray N, Charbonnier A, Leblanc V, Sales N, Deslys JP, Lasmezas CI (2005) PrPTSE distribution in a primate model of variant, sporadic, and iatrogenic Creutzfeldt-Jakob disease. J Virol 79:14339–14345. doi:10.1128/JVI.79.22.14339-14345.2005

Hewitt PE, Llewelyn CA, Mackenzie J, Will RG (2006) Creutzfeldt-Jakob disease and blood transfusion: results of the UK Transfusion Medicine Epidemiological Review study. Vox Sang 91:221–230. doi:10.1111/j.1423-0410.2006.00833.x

Hill AF, Desbruslais M, Joiner S, Sidle KC, Gowland I, Collinge J, Doey LJ, Lantos P (1997) The same prion strain causes vCJD and BSE. Nature 389:448–450, 526. doi:10.1038/38925

Hoffman Snyder C, Mishark KJ, Caviness JN, Drazkowski JF, Caselli RJ (2006) Nonvasculitic autoimmune inflammatory meningoencephalitis imitating Creutzfeldt-Jakob disease. Arch Neurol 63:766–768. doi:10.1001/archneur.63.5.766

Honda H, Sasaki K, Minaki H, Masui K, Suzuki SO, Doh-Ura K, Iwaki T (2011) Protease-resistant PrP and PrP oligomers in the brain in human prion diseases after intraventricular pentosan polysulfate infusion. Neuropathology. doi:10.1111/j.1440-1789.2011.01245.x

Hsich G, Kenney K, Gibbs CJ, Lee KH, Harrington MG (1996) The 14-3-3 brain protein in cerebrospinal fluid as a marker for transmissible spongiform encephalopathies. N Engl J Med 335:924–930. doi:10.1056/NEJM199609263351303

Hu WT, Murray JA, Greenaway MC, Parisi JE, Josephs KA (2006) Cognitive impairment and celiac disease. Arch Neurol 63:1440–1446. doi:10.1001/archneur.63.10.1440

Jacob H, Pyrkosch W, Strube H (1950) The hereditary form of Creutzfeldt-Jakob disease (the Backer family). Arch Psychiatr Nervenkr Z Gesamte Neurol Psychiatr 184:653–674

Jansen C, Voet W, Head MW, Parchi P, Yull H, Verrips A, Wesseling P, Meulstee J, Baas F, van Gool WA, Ironside JW, Rozemuller AJ (2011) A novel seven-octapeptide repeat insertion in the prion protein gene (PRNP) in a Dutch pedigree with Gerstmann-Straussler-Scheinker disease phenotype: comparison with similar cases from the literature. Acta Neuropathol 121:59–68. doi:10.1007/s00401-010-0656-3

Karmon Y, Kurzweil A, Lindzen E, Holmlund T, Weinstock-Guttman B (2011) Gerstmann-Straussler-Scheinker syndrome masquerading multiple sclerosis. J Neurol Sci 309:55–57. doi:10.1016/j.jns.2011.07.028

Kaski D, Mead S, Hyare H, Cooper S, Jampana R, Overell J, Knight R, Collinge J, Rudge P (2009) Variant CJD in an individual heterozygous for PRNP codon 129. Lancet 374:2128. doi:10.1016/S0140-6736(09)61568-3

Kim EJ, Cho SS, Jeong BH, Kim YS, Seo SW, Na DL, Geschwind MD, Jeong Y (2011) Glucose metabolism in sporadic Creutzfeldt-Jakob disease: a statistical parametric mapping analysis of (18) F-FDG PET. Eur J Neurol. doi:10.1111/j.1468-1331.2011.03570.x

Kim JI, Cali I, Surewicz K, Kong Q, Raymond GJ, Atarashi R, Race B, Qing L, Gambetti P, Caughey B, Surewicz WK (2010) Mammalian prions generated from bacterially expressed prion protein in the absence of any mammalian cofactors. J Biol Chem 285:14083–14087. doi:10.1074/jbc.C110.113464

Kimberlin RH, Walker CA (1978) Evidence that the transmission of one source of scrapie agent to hamsters involves separation of agent strains from a mixture. J Gen Virol 39:487–496

Kimberlin RH, Walker CA (1989a) The role of the spleen in the neuroinvasion of scrapie in mice. Virus Res 12:201–211

Kimberlin RH, Walker CA (1989b) Pathogenesis of scrapie in mice after intragastric infection. Virus Res 12:213–220

Kimberlin RH, Cole S, Walker CA (1987) Temporary and permanent modifications to a single strain of mouse scrapie on transmission to rats and hamsters. J Gen Virol 68:1875–1881

Korth C, May BC, Cohen FE, Prusiner SB (2001) Acridine and phenothiazine derivatives as pharmacotherapeutics for prion disease. Proc Natl Acad Sci USA 98:9836–9841. doi:10.1073/pnas.161274798

Ladogana A, Sanchez-Juan P, Mitrova E, Green A, Cuadrado-Corrales N, Sanchez-Valle R, Koscova S, Aguzzi A, Sklaviadis T, Kulczycki J, Gawinecka J, Saiz A, Calero M, van Duijn CM, Pocchiari M, Knight R, Zerr I (2009) Cerebrospinal fluid biomarkers in human genetic transmissible spongiform encephalopathies. J Neurol 256:1620–1628. doi:10.1007/s00415-009-5163-x

Lapergue B, Demeret S, Denys V, Laplanche JL, Galanaud D, Verny M, Sazdovitch V, Baulac M, Haik S, Hauw JJ, Bolgert F, Brandel JP, Navarro V (2010) Sporadic Creutzfeldt-Jakob disease mimicking nonconvulsive status epilepticus. Neurology 74:1995–1999. doi:10.1212/WNL.0b013e3181e39703

Legname G, Baskakov IV, Nguyen HO, Riesner D, Cohen FE, DeArmond SJ, Prusiner SB (2004) Synthetic mammalian prions. Science 305:673–676. doi:10.1126/science.1100195

Linden R, Martins VR, Prado MA, Cammarota M, Izquierdo I, Brentani RR (2008) Physiology of the prion protein. Physiol Rev 88:673–728. doi:10.1152/physrev.00007.2007

Mabbott NA, Bruce ME (2001) The immunobiology of TSE diseases. J Gen Virol 82:2307–2318

Mallucci G, Dickinson A, Linehan J, Klohn PC, Brandner S, Collinge J (2003) Depleting neuronal PrP in prion infection prevents disease and reverses spongiosis. Science 302:871–874. doi:10.1126/science.1090187

Masters CL, Gajdusek DC, Gibbs CJ Jr (1981) The familial occurrence of Creutzfeldt-Jakob disease and Alzheimer's disease. Brain 104:535–558

McKinley MP, Bolton DC, Prusiner SB (1983) A protease-resistant protein is a structural component of the scrapie prion. Cell 35:57–62

Mead S, Ranopa M, Gopalakrishnan GS, Thompson AG, Rudge P, Wroe S, Kennedy A, Hudson F, MacKay A, Darbyshire JH, Collinge J, Walker AS (2011) PRION-1 scales analysis supports use of functional outcome measures in prion disease. Neurology 77:1674–1683. doi:10.1212/WNL.0b013e3182364890

Medori R, Tritschler HJ, LeBlanc A, Villare F, Manetto V, Chen HY, Xue R, Leal S, Montagna P, Cortelli P (1992) Fatal familial insomnia, a prion disease with a mutation at codon 178 of the prion protein gene. N Engl J Med 326:444–449. doi:10.1056/NEJM199202133260704

Meissner B, Kallenberg K, Sanchez-Juan P, Krasnianski A, Heinemann U, Varges D, Knauth M, Zerr I (2008) Isolated cortical signal increase on MR imaging as a frequent lesion pattern in sporadic Creutzfeldt-Jakob disease. AJNR Am J Neuroradiol 29:1519–1524. doi:10.3174/ajnr.A1122

Merz PA, Somerville RA, Wisniewski HM, Iqbal K (1981) Abnormal fibrils from scrapie-infected brain. Acta Neuropathol 54:63–74

Merz PA, Kascsak RJ, Rubenstein R, Carp RI, Wisniewski HM (1987) Antisera to scrapie-associated fibril protein and prion protein decorate scrapie-associated fibrils. J Virol 61:42–49

Merz PA, Rohwer RG, Kascsak R, Wisniewski HM, Somerville RA, Gibbs CJ Jr, Gajdusek DC (1984) Infection-specific particle from the unconventional slow virus diseases. Science 225:437–440

Montrasio F, Frigg R, Glatzel M, Klein MA, Mackay F, Aguzzi A, Weissmann C (2000) Impaired prion replication in spleens of mice lacking functional follicular dendritic cells. Science 288:1257–1259

Morales R, Duran-Aniotz C, Castilla J, Estrada LD, Soto C (2011) De novo induction of amyloid-beta deposition in vivo. Mol Psychiatry. doi:10.1038/mp. 2011.120

Orru CD, Wilham JM, Raymond LD, Kuhn F, Schroeder B, Raeber AJ, Caughey B (2011) Prion disease blood test using immunoprecipitation and improved quaking-induced conversion. MBio 2:e00078–11. doi:10.1128/mBio.00078-11

Otto M, Wiltfang J, Tumani H, Zerr I, Lantsch M, Kornhuber J, Weber T, Kretzschmar HA, Poser S (1997a) Elevated levels of tau-protein in cerebrospinal fluid of patients with Creutzfeldt-Jakob disease. Neurosci Lett 225:210–212

Otto M, Stein H, Szudra A, Zerr I, Bodemer M, Gefeller O, Poser S, Kretzschmar HA, Mader M, Weber T (1997b) S-100 protein concentration in the cerebrospinal fluid of patients with Creutzfeldt-Jakob disease. J Neurol 244:566–570

Owen F, Poulter M, Lofthouse R, Collinge J, Crow TJ, Risby D, Baker HF, Ridley RM, Hsiao K, Prusiner SB (1989) Insertion in prion protein gene in familial Creutzfeldt-Jakob disease. Lancet 1:51–52

Pan KM, Baldwin M, Nguyen J, Gasset M, Serban A, Groth D, Mehlhorn I, Huang Z, Fletterick RJ, Cohen FE (1993) Conversion of alpha-helices into beta-sheets features in the formation of the scrapie prion proteins. Proc Natl Acad Sci USA 90:10962–10966

Parchi P, Strammiello R, Giese A, Kretzschmar H (2011) Phenotypic variability of sporadic human prion disease and its molecular basis: past, present, and future. Acta Neuropathol 121:91–112. doi:10.1007/s00401-010-0779-6

Parchi P, Strammiello R, Notari S, Giese A, Langeveld JP, Ladogana A, Zerr I, Roncaroli F, Cras P, Ghetti B, Pocchiari M, Kretzschmar H, Capellari S (2009) Incidence and spectrum of sporadic Creutzfeldt-Jakob disease variants with mixed phenotype and co-occurrence of PrPSc types: an updated classification. Acta Neuropathol 118:659–671. doi:10.1007/s00401-009-0585-1

Parchi P, Giese A, Capellari S, Brown P, Schulz-Schaeffer W, Windl O, Zerr I, Budka H, Kopp N, Piccardo P, Poser S, Rojiani A, Streichemberger N, Julien J, Vital C, Ghetti B, Gambetti P, Kretzschmar H (1999) Classification of sporadic Creutzfeldt-Jakob disease based on molecular and phenotypic analysis of 300 subjects. Ann Neurol 46:224–233

Pastrana MA, Sajnani G, Onisko B, Castilla J, Morales R, Soto C, Requena JR (2006) Isolation and characterization of a proteinase K-sensitive PrPSc fraction. Biochemistry 45:15710–15717. doi:10.1021/bi0615442

Pattison IH, Jones KM (1968) Modification of a strain of mouse-adapted scrapie by passage through rats. Res Vet Sci 9:408–410

Peden A, McCardle L, Head MW, Love S, Ward HJ, Cousens SN, Keeling DM, Millar CM, Hill FG, Ironside JW (2010) Variant CJD infection in the spleen of a neurologically asymptomatic UK adult patient with haemophilia. Haemophilia 16:296–304. doi:10.1111/j.1365-2516.2009.02181.x

Peden AH, Head MW, Ritchie DL, Bell JE, Ironside JW (2004) Preclinical vCJD after blood transfusion in a PRNP codon 129 heterozygous patient. Lancet 364:527–529. doi:10.1016/S0140-6736(04)16811-6

Peden AH, McGuire LI, Appleford NE, Mallinson G, Wilham JM, Orru CD, Caughey B, Ironside JW, Knight RS, Will RG, Green AJ, Head MW (2011) Sensitive and specific detection of sporadic Creutzfeldt-Jakob disease brain prion protein using real-time quaking induced conversion. J Gen Virol. doi:10.1099/vir.0.033365-0

Peretz D, Scott MR, Groth D, Williamson RA, Burton DR, Cohen FE, Prusiner SB (2001) Strain-specified relative conformational stability of the scrapie prion protein. Protein Sci 10:854–863. doi:10.1110/ps.39201

Polymenidou M, Cleveland DW (2011) The seeds of neurodegeneration: prion-like spreading in ALS. Cell 147:498–508. doi:10.1016/j.cell.2011.10.011

Polymenidou M, Stoeck K, Glatzel M, Vey M, Bellon A, Aguzzi A (2005) Coexistence of multiple PrPSc types in individuals with Creutzfeldt-Jakob disease. Lancet Neurol 4:805–814. doi:10.1016/S1474-4422(05)70225-8

Prinz M, Heikenwalder M, Junt T, Schwarz P, Glatzel M, Heppner FL, Fu YX, Lipp M, Aguzzi A (2003) Positioning of follicular dendritic cells within the spleen controls prion neuroinvasion. Nature 425:957–962. doi:10.1038/nature02072

Prusiner SB (1982) Novel proteinaceous infectious particles cause scrapie. Science 216:136–144

Puoti G, Giaccone G, Rossi G, Canciani B, Bugiani O, Tagliavini F (1999) Sporadic Creutzfeldt-Jakob disease: co-occurrence of different types of PrP(Sc) in the same brain. Neurology 53:2173–2176

Rainov NG, Tsuboi Y, Krolak-Salmon P, Vighetto A, Doh-Ura K (2007) Experimental treatments for human transmissible spongiform encephalopathies: is there a role for pentosan polysulfate? Expert Opin Biol Ther 7:713–726. doi:10.1517/14712598.7.5.713

Roos R, Gajdusek DC, Gibbs CJ Jr (1973) The clinical characteristics of transmissible Creutzfeldt-Jakob disease. Brain 96:1–20

Rosenthal NP, Keesey J, Crandall B, Brown WJ (1976) Familial neurological disease associated with spongiform encephalopathy. Arch Neurol 33:252–259

Safar J, Wille H, Itri V, Groth D, Serban H, Torchia M, Cohen FE, Prusiner SB (1998) Eight prion strains have PrP(Sc) molecules with different conformations. Nat Med 4:1157–1165. doi:10.1038/2654

Sailer A, Bueler H, Fischer M, Aguzzi A, Weissmann C (1994) No propagation of prions in mice devoid of PrP. Cell 77:967–968

Santoro D, Colombo I, Ghione I, Peverelli L, Bresolin N, Sciacco M, Prelle A (2011) Steroid-responsive Hashimoto encephalopathy mimicking Creutzfeldt-Jakob disease. Neurol Sci 32:719–722. doi:10.1007/s10072-011-0610-8

Secker TJ, Herve R, Keevil CW (2011) Adsorption of prion and tissue proteins to surgical stainless steel surfaces and the efficacy of decontamination following dry and wet storage conditions. J Hosp Infect 78:251–255. doi:10.1016/j.jhin.2011.03.021

Shyng SL, Lehmann S, Moulder KL, Harris DA (1995) Sulfated glycans stimulate endocytosis of the cellular isoform of the prion protein, PrPC, in cultured cells. J Biol Chem 270:30221–30229

Silveira JR, Raymond GJ, Hughson AG, Race RE, Sim VL, Hayes SF, Caughey B (2005) The most infectious prion protein particles. Nature 437:257–261. doi:10.1038/nature03989

Sim VL, Caughey B (2009a) Ultrastructures and strain comparison of under-glycosylated scrapie prion fibrils. Neurobiol Aging 30:2031–2042. doi:10.1016/j.neurobiolaging.2008.02.016

Sim VL, Caughey B (2009b) Recent advances in prion chemotherapeutics. Infect Disord Drug Targets 9:81–91

Sim VL (2012) Prion Disease: Chemotherapeutic Strategies. Infect Disord Drug Targets 12:144–160

Soldevila M, Calafell F, Andres AM, Yague J, Helgason A, Stefansson K, Bertranpetit J (2003) Prion susceptibility and protective alleles exhibit marked geographic differences. Hum Mutat 22:104–105. doi:10.1002/humu.9157

Steinhoff BJ, Zerr I, Glatting M, Schulz-Schaeffer W, Poser S, Kretzschmar HA (2004) Diagnostic value of periodic complexes in Creutzfeldt-Jakob disease. Ann Neurol 56:702–708. doi:10.1002/ana.20261

Steinhoff BJ, Racker S, Herrendorf G, Poser S, Grosche S, Zerr I, Kretzschmar H, Weber T (1996) Accuracy and reliability of periodic sharp wave complexes in Creutzfeldt-Jakob disease. Arch Neurol 53:162–166

Tateishi J, Brown P, Kitamoto T, Hoque ZM, Roos R, Wollman R, Cervenakova L, Gajdusek DC (1995) First experimental transmission of fatal familial insomnia. Nature 376:434–435. doi:10.1038/376434a0

Taylor DM (1995) Bovine spongiform encephalopathy–the beginning of the end? Br Vet J 152:501–518

Telling GC, Parchi P, DeArmond SJ, Cortelli P, Montagna P, Gabizon R, Mastrianni J, Lugaresi E, Gambetti P, Prusiner SB (1996) Evidence for the conformation of the pathologic isoform of the prion protein enciphering and propagating prion diversity. Science 274:2079–2082

Thomzig A, Spassov S, Friedrich M, Naumann D, Beekes M (2004) Discriminating scrapie and bovine spongiform encephalopathy isolates by infrared spectroscopy of pathological prion protein. J Biol Chem 279:33847–33854. doi:10.1074/jbc.M403730200

Todd NV, Morrow J, Doh-ura K, Dealler S, O'Hare S, Farling P, Duddy M, Rainov NG (2005) Cerebroventricular infusion of pentosan polysulphate in human variant Creutzfeldt-Jakob disease. J Infect 50:394–396. doi:10.1016/j.jinf.2004.07.015

Tsuboi Y, Doh-Ura K, Yamada T (2009) Continuous intraventricular infusion of pentosan polysulfate: clinical trial against prion diseases. Neuropathology 29:632–636. doi:10.1111/j.1440-1789.2009.01058.x

Tzaban S, Friedlander G, Schonberger O, Horonchik L, Yedidia Y, Shaked G, Gabizon R, Taraboulos A (2002) Protease-sensitive scrapie prion protein in aggregates of heterogeneous sizes. Biochemistry 41:12868–12875

Urayama A, Morales R, Niehoff ML, Banks WA, Soto C (2011) Initial fate of prions upon peripheral infection: half-life, distribution, clearance, and tissue uptake. FASEB J 25:2792–2803. doi:10.1096/fj.11-180729

Valcour V, Haman A, Cornes S, Lawall C, Parsa AT, Glaser C, Yagi S, Tihan T, Bhatnagar J, Geschwind M, Vanderbilt Continuing Medical Education online (2008) A case of enteroviral meningoencephalitis presenting as rapidly progressive dementia. Nat Clin Pract Neurol 4:399–403. doi:10.1038/ncpneuro0804

Van Everbroeck B, Boons J, Cras P (2005) Cerebrospinal fluid biomarkers in Creutzfeldt-Jakob disease. Clin Neurol Neurosurg 107:355–360. doi:10.1016/j.clineuro.2004.12.002

Van Everbroeck B, Green AJ, Pals P, Martin JJ, Cras P (1999) Decreased levels of amyloid-beta 1-42 in cerebrospinal fluid of Creutzfeldt-Jakob disease patients. J Alzheimers Dis 1:419–424

Vitali P, Maccagnano E, Caverzasi E, Henry RG, Haman A, Torres-Chae C, Johnson DY, Miller BL, Geschwind MD (2011) Diffusion-weighted MRI hyperintensity patterns differentiate CJD from other rapid dementias. Neurology 76:1711–1719. doi:10.1212/WNL.0b013e31821a4439

Wadsworth JD, Joiner S, Hill AF, Campbell TA, Desbruslais M, Luthert PJ, Collinge J (2001) Tissue distribution of protease resistant prion protein in variant Creutzfeldt-Jakob disease using a highly sensitive immunoblotting assay. Lancet 358:171–180

Wang F, Wang X, Yuan CG, Ma J (2010) Generating a prion with bacterially expressed recombinant prion protein. Science 327:1132–1135. doi:10.1126/science.1183748

Ward HJ (2006) Evidence of a new human genotype susceptible to variant CJD. Euro Surveill 11: E060601.3

Wilham JM, Orru CD, Bessen RA, Atarashi R, Sano K, Race B, Meade-White KD, Taubner LM, Timmes A, Caughey B (2010) Rapid end-point quantitation of prion seeding activity with sensitivity comparable to bioassays. PLoS Pathog 6:e1001217. doi:10.1371/journal.ppat.1001217

Will RG, Ironside JW, Zeidler M, Cousens SN, Estibeiro K, Alperovitch A, Poser S, Pocchiari M, Hofman A, Smith PG (1996) A new variant of Creutzfeldt-Jakob disease in the UK. Lancet 347:921–925

Wille H, Michelitsch MD, Guenebaut V, Supattapone S, Serban A, Cohen FE, Agard DA, Prusiner SB (2002) Structural studies of the scrapie prion protein by electron crystallography. Proc Natl Acad Sci USA 99:3563–3568. doi:10.1073/pnas.052703499

Wille H, Bian W, McDonald M, Kendall A, Colby DW, Bloch L, Ollesch J, Borovinskiy AL, Cohen FE, Prusiner SB, Stubbs G (2009) Natural and synthetic prion structure from X-ray fiber diffraction. Proc Natl Acad Sci USA 106:16990–16995. doi:10.1073/pnas.0909006106

Wiltfang J, Otto M, Baxter HC, Bodemer M, Steinacker P, Bahn E, Zerr I, Kornhuber J, Kretzschmar HA, Poser S, Ruther E, Aitken A (1999) Isoform pattern of 14-3-3 proteins in the cerebrospinal fluid of patients with Creutzfeldt-Jakob disease. J Neurochem 73:2485–2490

Xiao X, Miravalle L, Yuan J, McGeehan J, Dong Z, Wyza R, MacLennan GT, Golichowski AM, Kneale G, King N, Kong Q, Spina S, Vidal R, Ghetti B, Roos K, Gambetti P, Zou WQ (2009) Failure to detect the presence of prions in the uterine and gestational tissues from a Gravida with Creutzfeldt-Jakob disease. Am J Pathol 174:1602–1608. doi:10.2353/ajpath.2009.081045

Young GS, Geschwind MD, Fischbein NJ, Martindale JL, Henry RG, Liu S, Lu Y, Wong S, Liu H, Miller BL, Dillon WP (2005) Diffusion-weighted and fluid-attenuated inversion recovery imaging in Creutzfeldt-Jakob disease: high sensitivity and specificity for diagnosis. AJNR Am J Neuroradiol 26:1551–1562

Yull HM, Ritchie DL, Langeveld JP, van Zijderveld FG, Bruce ME, Ironside JW, Head MW (2006) Detection of type 1 prion protein in variant Creutzfeldt-Jakob disease. Am J Pathol 168:151–157. doi:10.2353/ajpath.2006.050766

Zerr I, Bodemer M, Racker S, Grosche S, Poser S, Kretzschmar HA, Weber T (1995) Cerebrospinal fluid concentration of neuron-specific enolase in diagnosis of Creutzfeldt-Jakob disease. Lancet 345:1609–1610

Index

A.C. Jackson (ed.), *Viral Infections of the Human Nervous System*, Birkhäuser Advances in Infectious Diseases, DOI 10.1007/978-3-0348-0425-7, © Springer Basel 2013

I

J

R

MIX
Papier aus verantwortungsvollen Quellen
Paper from responsible sources
FSC® C105338

If you have any concerns about our products,
you can contact us on
ProductSafety@springernature.com

In case Publisher is established outside the EU,
the EU authorized representative is:
Springer Nature Customer Service Center GmbH
Europaplatz 3, 69115 Heidelberg, Germany

Printed by Libri Plureos GmbH
in Hamburg, Germany